Y0-BTC-503

Lecture Notes in Computer Science 765

Edited by G. Goos and J. Hartmanis

Tor Helleseth (Ed.)

Advances in Cryptology – EUROCRYPT '93

Workshop on the Theory and Application
of Cryptographic Techniques
Lofthus, Norway, May 23-27, 1993
Proceedings

Springer-Verlag
Berlin Heidelberg New York
London Paris Tokyo
Hong Kong Barcelona
Budapest

Series Editors

Gerhard Goos
Universität Karlsruhe
Postfach 69 80
Vincenz-Priessnitz-Straße 1
D-76131 Karlsruhe, Germany

Juris Hartmanis
Cornell University
Department of Computer Science
4130 Upson Hall
Ithaca, NY 14853, USA

Volume Editor

Tor Helleseth
Department of Informatics, University of Bergen
Høyteknologisenteret, N-5020 Bergen, Norway

CR Subject Classification (1991): E.3-4, D.4.6, G.2.1

ISBN 3-540-57600-2 Springer-Verlag Berlin Heidelberg New York
ISBN 0-387-57600-2 Springer-Verlag New York Berlin Heidelberg

Printed in Germany

Typesetting: Camera-ready by author
45/3140-543210 - Printed on acid-free paper

Preface

Eurocrypt is a series of open workshops on the theory and application of cryptographic techniques. These meetings have taken place in Europe every year since 1982 and are sponsored by the International Association for Cryptologic Research (IACR).

Eurocrypt'93 was held on May 23–27 at Hotel Ullensvang, beautifully located in the village of Lofthus in the heart of Norway's fjord district. The conference attracted 266 participants from 29 countries. It is a pleasure to thank the local organizers of the conference and the general chair Kåre Presttun. A special acknowledgment to Leif Nilsen whose dedication and tremendous effort was crucial to make the conference a very successful one.

The call for papers resulted in 117 submissions with authors representing 27 different countries. The accepted papers were selected by the program committee after a blind refereeing process where the authors of the papers were unknown to the program committee members. Because of the large number of papers the members of the program committee were encouraged to ask reliable colleagues for assistance in the evaluation of the papers. The program committee had the difficult task selecting only 36 of these papers for presentation at the conference. In addition Professor Ernst Selmer was especially invited to present a talk at the conference.

The rump session this year was chaired by Ingemar Ingemarsson. Some of the presentations were, after a simplified review process, selected for publication in these proceedings and can be found at the end of this volume.

I would like to thank all the people who contributed to the work of putting together the program of Eurocrypt'93. I am indebted to the members of the program committee for their time and conscientious effort in the evaluation and selection of the papers for presentation at the conference. I am also grateful to the 31 additional reviewers who assisted the program committee members in their evaluation. A special thanks to my colleague Øyvind Ytrehus for his valuable assistance in handling the correspondence to the authors and preparing the proceedings. Finally, I would like to thank all the authors for submitting so many good papers and for their cooperation in preparing this volume.

Bergen, October 1993 Tor Helleseth

EUROCRYPT'93

General Chairman:
Kåre Presttun (Alcatel Telecom)

Organizing Committee:
Kenneth Iversen (KITH, Trondheim)
Torleiv Kløve (U. of Bergen)
Leif Nilsen (Alcatel Telecom)
Øystein Rødseth (U. of Bergen)
Øyvind Ytrehus (U. of Bergen)

Programme Chairman:
Tor Helleseth (U. of Bergen)

Programme Committee:
Ivan Damgård (U. of Aarhus)
Alfredo De Santis (U. of Salerno)
Yvo Desmedt (U. of Wisconsin)
Dieter Gollman (U. of London)
Ingemar Ingemarsson (U. of Linköping)
Kaoru Kurosawa (Tokyo Inst. of Techn.)
Jim Massey (ETH Zürich)
Bart Preneel (ESAT/COSIC)
Andrew Odlyzko (AT&T Bell Labs)
Claus Schnorr (U. of Frankfurt)
Jennifer Seberry (U. of Wollongong)

Contents

Stream ciphers

Digital signatures

Protocols I

Hash Functions

Payment Systems

Cryptanalysis

Protocols II

Rump Session

List of Authors

ON THE RELATION BETWEEN A-CODES AND CODES CORRECTING INDEPENDENT ERRORS

Thomas Johansson* Gregory Kabatianskii** Ben Smeets*

*Department of Information Theory
Lund University
Box 118
S-221 00 Lund, Sweden

**Inst for Problems of Information Transmission
Russian Academy of Sciences
Ermolovoy 19, Moscow, GSP-4
Russia

July 23, 1993

Abstract – In this paper we show an explicit relation between authentication codes and codes correcting independent errors. This relation gives rise to several upper bounds on A-codes. We also show how to construct A-codes starting from error correcting codes. The latter is used to show that if P_S exceeds P_I by an arbitrarily small positive amount, then the number of source states grows exponentially with the number of keys but if $P_S = P_I$ it will grow only linearly.

1 Introduction

The *authentication channel* was introduced by Simmons as a model for a communication situation with two trusting participants called the *transmitter* and the *receiver* who want to protect themselves against the actions of an active *opponent* who can insert its own messages or who can change messages already sent by the transmitter. To protect their communication, the transmitter and the receiver have agreed *secretly* upon a certain encoding rule e which is taken from a finite set $\mathcal{E}$ of n possible rules. This rule e enables them to transmit a piece of information s, hereafter called source state, which is taken from the finite set $\mathcal{S}$ of k possible source states. Each encoding rule maps s to a message m. In this paper we investigate only authentication codes (A-codes) for which the messages are pairs (s, z), $z \in \mathcal{Z}$. The coordinate z is called a tag or an authenticator. In authentication theory these codes are called appended authenticator schemes or Cartesian A-codes. People working in coding theory would use the term systematic codes. Furthermore, in this paper we will only deal with unconditionally secure A-codes.

Authentication theory deals with the analysis and design of A-codes and has since the publication of the paper by Gilbert, McWilliams and Sloane, [1], and the paper by Simmons, [2], developed into a discipline on its own. Various bounds on the probabilities of success for the various attacks by the opponent have been established and many constructions for obtaining A-codes are known. For an overview of the results between 1974 and 1991 we refer to [3] and [4].

At several occasions the view has been expressed that unconditionally secure authentication codes are in a strict mathematical sense dual to error detecting and correcting codes, see for example [3, page 397, 3rd par.]. However, the true nature of this duality seems never to be addressed. It is just this aspect that is one of the main subjects of this paper. Particularly we will show how to obtain a code for correcting independent errors (CIE-code) from an A-code and vice versa. By establishing this connection we obtain new results in authentication theory.

Another subject which we address is the question of how large we can make $\mathcal{S}$ for given n, P_I, and P_S, the latter two being the probability of a successful imitation attack, respectively, the probability of a successful substitution attack. In this paper we compute P_S as the *maximum probability* over all substitution messages. The above question has practical relevance if we want to authenticate, for example, long data files. It has previously been shown that if $P_S = P_I$, then the number of source states is linearly bounded by the number of keys, [5]. However, our results show that if $P_S = P_I + \varepsilon$, for arbitrary $\varepsilon > 0$, then the number of source states grows exponentially with the number of keys !

The organization of our paper will be as follows. First, in Section 2, we discuss how we can derive a CIE code from an A-code. We give some examples which illustrate the implications of the established relation. In this section we also give the bound on $|\mathcal{S}|$ when $P_S = P_I$. In Section 3 we show how we can construct an A-code starting with a CIE code.

2 From systematic A-codes to CIE-codes

In this section we will describe how we can derive a CIE-code from a systematic A-code and how we can state the probabilities P_I and P_S in terms of the properties of the CIE-code.

Consider a systematic A-code, i.e., a triple $(\mathcal{S}, \mathcal{E}, \mathcal{Z})$, where for any $s \in \mathcal{S}$ and $e \in \mathcal{E}$, the message $m = (s, z) \in \mathcal{S} \times \mathcal{Z}$ is defined by letting $z = e(s)$. We restrict ourselves to a uniform distribution of e. From definitions in [2] we derive that the probability of a successful impersonation attack, P_I, is given by the formula

$$P_I = \max_{\substack{s \in \mathcal{S} \\ z \in \mathcal{Z}}} \frac{|\{e \in \mathcal{E}\,;\, e(s) = z\}|}{|\mathcal{E}|}. \tag{1}$$

Let us consider A-codes with the property that

$$|\{e \in \mathcal{E}\,;\, e(s) = z\}| = \begin{cases} |\mathcal{E}|\, P_I, & \text{or,} \\ 0, & \text{otherwise,} \end{cases} \tag{2}$$

and let us call such codes *I-equitable A-codes*. These codes are optimal against the impersonation attack in the sence that only these codes meet the trivial bound $P_I \geq |\mathcal{S}|/|\mathcal{M}|$ with equality, [3]. In the sequel we assume that our A-codes are I-equitable and that $P_I = 1/q$, where q is a power of a prime.

Let $n = |\mathcal{E}|$ and let us enumerate the elements of $\mathcal{E}$ as $e_1, e_2, \ldots, e_n$. Consider the words (vectors)

$$\mathbf{v}^{(s)} = (e_1(s), e_2(s), \ldots, e_n(s)),$$

and the corresponding set of words

$$V = \left\{\mathbf{v}^{(s)}\,;\, s \in \mathcal{S}\right\}.$$

For given s let us also enumerate the values $e_j(s) \in \mathcal{Z}$ by the elements $b_1, b_2, \ldots, b_q$ from some q-ary alphabet B. It follows from property (2) that such an enumeration is possible. For each word $\mathbf{v}$ we define its composition as

$$\text{comp}(\mathbf{v}) = (c_1, c_2, \ldots, c_q), \quad \text{where } c_i = \frac{1}{n}\left|\{j\,;\, v_j = b_i\}\right|. \tag{3}$$

It follows from (2) that

$$\text{comp}\left(\mathbf{v}^{(s)}\right) = (P_I, \ldots, P_I) = \left(\frac{1}{q}, \ldots, \frac{1}{q}\right),$$

i.e., all words of V have a constant composition.

Now recall from the introduction that the probability P_S is given by

$$P_S = \max_{\substack{s\in\mathcal{S} \\ z\in\mathcal{Z}}} \max_{\substack{s\neq\widehat{s}\in\mathcal{S} \\ \widehat{z}\in\mathcal{Z}}} \frac{|\{e \in \mathcal{E}\,;\, e(s) = z, e(\widehat{s}) = \widehat{z}\}|}{|\{e \in \mathcal{E}\,;\, e(s) = z\}|}. \tag{4}$$

From (2) we get the inequality

$$\left|\left\{j\,;\, v_j^{(s)} = b, v_j^{(\widehat{s})} = \widehat{b}\right\}\right| \leq P_S P_I \cdot n. \tag{5}$$

By letting $b = \widehat{b}$ and letting b run through B, we have

$$\mathrm{d}\left(\mathbf{v}^{(s)}, \mathbf{v}^{(\widehat{s})}\right) \geq n - qP_I P_S \cdot n = n(1 - P_S), \tag{6}$$

where $\mathrm{d}(\mathbf{x}, \mathbf{y})$ denotes the usual Hamming distance between the vectors $\mathbf{x}$ and $\mathbf{y}$. But in general, we can let $b = c_1\widehat{b} + c_2$ for arbitrary $c_1 \neq 0, c_2 \in B$, if we consider B as a finite field. Then (5) gives

$$\mathrm{d}\left(\mathbf{v}^{(s)}, c_1\mathbf{v}^{(\widehat{s})} + c_2\mathbf{1}\right) \geq n - qP_I P_S \cdot n = n(1 - P_S). \tag{7}$$

This means that if we assume $P_S \neq 1$, we form from each codeword $\mathbf{v}^{(s)}$ in the code V new codewords by all the affine transformations $\phi : v \mapsto c_1 v + c_2$, where $c_1 \neq 0, c_2 \in B$. If $P_S \neq 1$ no two codewords from this transformation can be the same and the distance property of (6) still holds, *provided* we also assume that $P_S \geq P_I = 1/q$. Since all codewords have constant composition we also add multiples of the codeword $\mathbf{1}$ without changing the minimum distance. Thus we have a code V' given by

$$V' = \left\{c_1\mathbf{v}^{(s)} + c_2\mathbf{1}\,;\, \mathbf{v}^{(s)} \in V, c_1 \neq 0, c_2 \in B\right\} \cup \{c\mathbf{1}\,;\, c \in B\}$$

with the same distance property as V, i.e., the minimum distance d of the code V' is bounded by $d \geq n(1 - P_S)$. The number of codewords in V' is then $q(q-1)|\mathcal{S}| + q$.

Summary:
Given an I-equitable A-code with parameters $|\mathcal{E}|, |\mathcal{S}|$ and P_S, there exists a corresponding CIE-code with parameters $(n, M, d) = (|\mathcal{E}|, q(q-1)|\mathcal{S}| + q, |\mathcal{E}|(1 - P_S))$.

We can now apply upper bounds on the code V' to get upper bounds on the maximum number of source states in an I-equitable A-code with given P_S. For example by using the well-known Plotkin bound [7] we can prove

Theorem 1 ([5]):For an I-equitable A-code for which $P_I = P_S = 1/q$, the number of source states is upper bounded as

$$(q-1)|\mathcal{S}| \;\leq\; |\mathcal{E}| - 1. \tag{8}$$

■

Proof: The code obtained from the A-code has parameters $(n, M, d) = (|\mathcal{E}|, q(q-1)(|\mathcal{S}| + q, \theta \cdot |\mathcal{E}|)$, where $\theta = 1 - P_S = (q-1)/q$. Let $A_q(n,d)$ be the maximum number of codewords in an (n,d)-code. By the Plotkin bound, [7, pages 170-171], we have

$$A_q(n, \theta n) \leq qA_q(n-1, \theta n) \leq q\frac{\theta n}{\theta n - \theta(n-1)} = qn = q|\mathcal{E}|.$$

From this we have $q(q-1)|\mathcal{S}| + q \leq q|\mathcal{E}|$ and the result follows. □

Remark: This bound shows essentially what is happening for the case $P_S = \frac{1}{q}$, namely, the size of the source state space is, at the best, bounded by the size of the key space. Consequently we have to tolerate a large key when we want to authenticate many source states.

Let us return to equation (6). This is only a weak corollary to equation (5). One of the serious weaknesses of this result is that we calculate in (6) the total number of positions in which two words differ but the key property in the computation of P_S of an A-code is that the *ratio* of the number of positions in which two words differ must be almost the same for all given values of the coordinates, see equation (5).

Let us look at some examples which illustrate this.

Example 1: Let $q = 2$, then $P_I = 1/2$ and the vectors $\mathbf{v}^{(s)}$ of V have composition $(1/2, 1/2)$, i.e. they are binary vectors of constant weight $n/2$. We can rewrite (5) as

$$\left|\left\{j\,;\, v_j^{(s)} = b, v_j^{(\hat{s})} = \hat{b}\right\}\right| \leq P_S\frac{n}{2}.$$

Letting $b = \hat{b}$ we have $d\left(\mathbf{v}^{(s)}, \mathbf{v}^{(\hat{s})}\right) \geq n - qP_I P_S n = n(1-P_S)$. On the other hand by letting $b = \hat{b} \oplus 1$ (complement), we have $d\left(\mathbf{v}^{(s)}, \mathbf{v}^{(\hat{s})} \oplus \mathbf{1}\right) \geq n(1-P_S)$, where $\mathbf{1}$ is the word (vector) $(1, 1, \ldots, 1)$. The latter inequality can also be written as $d\left(\mathbf{v}^{(s)}, \mathbf{v}^{(\hat{s})}\right) \leq n - n(1 - P_S) = P_s n$. It can be shown that if for some binary code V, $\alpha n \leq d(\mathbf{v}, \mathbf{v}') \leq (1-\alpha)n$ and $\mathbf{v}$, $\mathbf{v}'$ have weight $n/2$, that this code gives us an A-code with $P_I = 1/2$ and $P_S = \alpha$. Hence in the case $P_I = 1/2$ we have a one-to-one correspondence between I-equitable A-codes and binary weight $n/2$ codes in the so-called antipodal Hamming space, (we define $d_{\text{antipodal}}(\mathbf{x}, \mathbf{y}) = \min(d(\mathbf{x}, \mathbf{y}), n - d(\mathbf{x}, \mathbf{y}))$ and $\mathbf{x} \equiv \mathbf{y}$ if and only if $\mathbf{x} = \mathbf{y} \oplus \mathbf{1}$). In particular, if

$$S(\mathcal{E}, P_I, P_S) = \max |S|, \tag{9}$$

for A-codes with given $\mathcal{E}$, P_I and P_S, we have

$$\lim_{n \to \infty} \frac{\log S(n, 1/2, P_S = \alpha)}{n} = f(\alpha) = \lim_{n \to \infty} \frac{1}{n} \log A_2(n, \alpha n),$$

where $f(\alpha)$ is a final answer in coding theory. If the Varshamov-Gilbert (V-G) bound, [6], is tight in the binary case, then $f(\alpha) = 1 - H(\alpha)$, where $H(\alpha) = -\alpha \log(\alpha) - (1 - \alpha) \log(1 - \alpha)$ is the binary entropy function.

Note that when $P_S \to 1/2$, then $f(P_S) \to 0$ as it should since when $P_S = P_I = 1/2$ we have by Theorem 1 that $\lim_{n \to \infty} \log |S| / n \leq \lim_{n \to \infty} \log(n-1)/n = 0$. ◻

Example 2: Let $q = 3$, and let the alphabet $B = \{0, -1, 1\}$. Consider two different A-codes and their corresponding 3-ary codes of which we only list two codewords.

$$V : \begin{cases} 1\ \ 1\ \ 1\ \ \text{-1 -1 -1}\ 0\ \ 0\ \ 0 \\ 1\ \ 0\ \ \text{-1}\ 1\ \ 0\ \ \text{-1}\ 1\ \ 0\ \ \text{-1} \end{cases} \qquad d = 6, \qquad P_I = 1/3, \quad P_S = 1/3,$$

$$\tilde{V} : \begin{cases} 1\ \ 1\ \ 1\ \ \text{-1 -1 -1}\ 0\ \ 0\ \ 0 \\ \text{-1 -1}\ 0\ \ 0\ \ 0\ \ 1\ \ 1\ \ 1\ \ \text{-1} \end{cases} \qquad d = 9, \qquad P_I = 1/3, \quad P_S = 2/3.$$

We can discover that $\tilde{V}$ is a bad A-code by permuting the assignment of the symbols of B. Thus we get 3! words from the first and and 3! words from the second word and by pairwise checking their distances we find that $d = 3$. This renumeration technique can be used to improve the bound of equation (7). It is important to note that we can only check the distance between words that stem from different codewords in $\tilde{V}$, since the minimum distance of two words that stem from the same codeword may be less than $|\mathcal{E}|(1 - P_S)$.

If we want to consider all the obtained codewords as a code with minimum distance unchanged, we must check that the minimum distance of any two words from the same codeword in $\tilde{V}$ is at least $|\mathcal{E}|(1-P_S)$. Thus it will depend on the value of P_S which permutations we can apply. For example, if $P_S \geq (q-2)/q$ we can apply all the $q!$ permutations without changing the minimum distance and thus get a code with parameters

$$(n, M, d) = (|\mathcal{E}|, q!\,|S| + q, |\mathcal{E}|(1 - P_S)), \quad \text{for } P_S \geq \frac{q-2}{q}.$$

However, the following example shows that even with this renumeration we do not get a tight bound.

$$V' : \begin{cases} 1\ \ 1\ \ 1\ \ \text{-1 -1 -1}\ 0\ \ 0\ \ 0 \\ 1\ \ 1\ \ 1\ \ 0\ \ 0\ \ \text{-1 -1 -1}\ 0 \end{cases} \qquad d = 4$$

After renumeration we get

$$\begin{cases} 1 \ 1 \ 1 \ \text{-1} \ \text{-1} \ \text{-1} \ 0 \ 0 \ 0 \\ 1 \ 1 \ 1 \ \text{-1} \ \text{-1} \ 0 \ 0 \ 0 \ \text{-1} \end{cases} \quad d = 2,$$

but we have $P_S = 1$!

Example 3: We now give a new A-code construction.

Consider the following construction of a systematic A-code.

Construction 1: Let $\mathcal{S} = \{s = (s_1, ..., s_k)\,;\ s_i \in \mathbb{F}_q\}$. Now define the source state polynomial to be

$$s(x) = s_1 x + s_2 x^2 + \ldots + s_k x^k.$$

Let $\mathcal{E} = \{e = (a, b)\,;\ a, b \in \mathbb{F}_q\}$. For the transmission of source state s we generate the message m which is obtained as

$$m = (s, a + s(b)) = (s_1, s_2, \ldots, s_k, a + s(b)).$$

Theorem 2: Construction 1 gives an A-code which has parameters:

$$P_I = \frac{1}{q}, \quad P_S = \frac{k}{q}.$$

Proof: Here $|\mathcal{E}| = q^2$ and from (1) we have

$$P_I = \max_{\substack{s \in \mathcal{S} \\ z \in \mathcal{Z}}} \frac{|\{e \in \mathcal{E}\,;\ e(s) = z\}|}{|\mathcal{E}|} = \max_{\substack{s \in \mathcal{S} \\ z \in \mathcal{Z}}} \frac{|\{a, b\,;\ s(b) + a = z\}|}{q^2}.$$

For a given value of b is a uniquely determined by $a = z - s(b)$ for any value of (s, z) and thus $P_I = q/q^2 = 1/q$. For the substitution attack, we have from (4) that

$$P_S = \max_{\substack{s \in \mathcal{S} \\ z \in \mathcal{Z}}} \max_{\substack{s \neq \widehat{s} \in \mathcal{S} \\ \widehat{z} \in \mathcal{Z}}} \frac{|\{a, b\,;\ s(b) + a = z, \widehat{s}(b) + a = \widehat{z}\}|}{|\{a, b\,;\ s(b) + a = z\}|} =$$

$$= \max_{\substack{s \in \mathcal{S} \\ z \in \mathcal{Z}}} \max_{\substack{s \neq \widehat{s} \in \mathcal{S} \\ \widehat{z} \in \mathcal{Z}}} \frac{|\{a, b\,;\ s(b) + a = z, (s - \widehat{s})(b) + (z - \widehat{z}) = 0\}|}{q}.$$

Now a is uniquely determined by $a = z - s(b)$ and since $(s - \widehat{s})(x) + (z - \widehat{z})$ is a non-zero polynomial of degree at most k it has at most k zeros. Thus for any $(s, z), (\widehat{s}, \widehat{z})$ we have $|\{a, b\,;\ s(b) + a = z, (s - \widehat{s})(b) + (z - \widehat{z}) = 0\}| \leq k$ and $P_S = k/q$. □

Consider again an I-equitable A-code. It is also possible to associate a binary code to our A-code. Let us assign to every source state $s \in \mathcal{S}$ q binary vectors of length n

$$\mathbf{b}^{(s,0)}, \mathbf{b}^{(s,1)}, \ldots, \mathbf{b}^{(s,q-1)}, \quad \text{where } b_i^{(s,j)} = \begin{cases} 1, & \text{if } e_i(s) = j, \\ 0, & \text{otherwise,} \end{cases}$$

i.e., the characteristic functions. It follows from the properties of our A-code that these vectors have weight n/q, $n = |\mathcal{E}|$, and any two distinct vectors have not more than $P_S(n/q)$ "common" 1-s. Hence the distance d of the code obtained from these vectors is at least $2(1 - P_S)(n/q)$ and thus

$$S(n = |\mathcal{E}|, 1/q, P_S) \leq \frac{1}{q} A_2 \left(n, \underbrace{2(1 - P_S)\frac{n}{q}}_{2\delta}, \underbrace{\frac{n}{q}}_{w} \right), \tag{10}$$

see (9) and $A_2(n, d, w)$ in [6]. For $w - \delta =$ constant, $w/n =$ constant and $n \to \infty$ we have, [6, page 527],

$$A_2(n, 2\delta, w) \leq \lfloor \frac{n}{w} \lfloor \frac{n-1}{w-1} \cdots \lfloor \frac{n - (w - \delta)}{\delta} \rfloor \cdots \rfloor \rfloor \approx \left(\frac{n}{w} \right)^{w - \delta + 1}.$$

We see that the A-codes in Construction 1 are asymptotically optimal because we have $|S| = q^k$, $n = |\mathcal{E}| = q^2$, $w = q$, $w - \delta + 1 = k + 1$ and

$$\frac{1}{q} A_2(q^2, 2(q - k), q) \sim q^{k+1-1} = |S|, \quad \text{for } k \text{ fixed and } q \to \infty.$$

Summarizing, we have shown how we can derive CIE codes from A-codes and how we can apply some of the known results of coding theory to get new results for the parameters of feasible A-codes.

3 Construction of an A-code from a q-ary linear code

Assume that we have a code C over $\mathbb{F}_q$ with the property

$$\forall \mathbf{c} \in C, \lambda \in \mathbb{F}_q \qquad \mathbf{c} + \lambda \mathbf{1} \in C. \tag{11}$$

Assume that the code C has parameters (n, M, d).

If $b - \hat{b}$ is constant we compute, see (5),

$$\begin{aligned} \left| \{ j \,;\, c_j = b, \hat{c}_j = \hat{b} \} \right| &= \left| \{ j \,;\, c_j - \hat{c}_j = b - \hat{b} \} \right| \\ &= n - \mathrm{d}\big(\mathbf{c} - \hat{\mathbf{c}}, (b - \hat{b})\mathbf{1}\big) \\ &\leq n - d. \end{aligned}$$

However we have

$$d\left(\mathbf{c} - \widehat{\mathbf{c}}, (b - \widehat{b})\mathbf{1}\right) = 0, \text{ if and only if } \mathbf{c} - \widehat{\mathbf{c}} = \lambda\mathbf{1}.$$

Thus the words of C whose difference is a multiple of $\mathbf{1}$, would result in an A-code for which $P_S = 1$. We have to factor these words out of C.

The code C can be partitioned into equivalence classes by introducing the equivalence relation R as

$$\mathbf{a}\, R\, \mathbf{b} \text{ if } \mathbf{a} - \mathbf{b} = \lambda\mathbf{1} \text{ for some } \lambda \in \mathbb{F}_q.$$

Clearly each equivalence class contains q elements. Now let $[u]$ denote the equivalence class containing the codeword u. We form the quotient set

$$\widehat{U} = C/\{\mathbf{1}\} = \{[u]\,;\, u \in C\}.$$

Now U is the code obtained by replacing each equivalence class by a specific representative. Clearly U has parameters $(n, M/q, d)$.

Now extend the code U to a new code V of length nq where

$$V = \{(\mathbf{u}, \mathbf{u} + \alpha_1\mathbf{1}, \mathbf{u} + \alpha_2\mathbf{1}, \ldots, \mathbf{u} + \alpha_{q-1}\mathbf{1})\,;\, u \in U, \mathbb{F}_q = \{0, \alpha_1, \ldots, \alpha_{q-1}\}\}. \tag{12}$$

By reversing the reasoning in the previous section we can use the code V to construct an A-code. Each codeword in V corresponds to a source state. Thus denote V as

$$V = \left\{\mathbf{v}^{(s)}\,;\, s \in \mathcal{S}\right\},$$

where $\mathbf{v}^{(s)} = (e_1(s), e_2(s), \ldots, e_{nq}(s))$. Now the property (12) gives that $\text{comp}(\mathbf{v}) = (1/q, \ldots, 1/q)$ for any $\mathbf{v} \in V$, so $P_I = 1/q$. To find the probability of successful substitution, let $\mathbf{v}^{(s)}$ and $\mathbf{v}^{(\widehat{s})}$ be the two codewords corresponding to two distinct source states that maximize P_S. Assume that we have observed s with the tag b from $\mathbf{v}^{(s)}$ and replace it with $\widehat{s}$ and tag $\widehat{b}$ from $\mathbf{v}^{(\widehat{s})}$. Then

$$\begin{aligned} P_S &= \frac{\left|\left\{j\,;\, v_j^{(s)} = b, v_j^{(\widehat{s})} = \widehat{b}\right\}\right|}{\left|\left\{j\,;\, v_j^{(s)} = b\right\}\right|} \\ &= \frac{\left|\left\{j\,;\, v_j^{(s)} = b, v_j^{(s)} - v_j^{(\widehat{s})} = b - \widehat{b}\right\}\right|}{\left|\left\{j\,;\, v_j^{(s)} = b\right\}\right|} \\ &= \frac{\left|\left\{j\,;\, u_j^{(s)} - u_j^{(\widehat{s})} = b - \widehat{b}\right\}\right|}{n}. \end{aligned}$$

Thus P_S is the maximal value of the composition values in $\text{comp}\left(\mathbf{u}^{(s)} - \mathbf{u}^{(\widehat{s})}\right)$. Also

$$\text{comp}\left(\mathbf{v}^{(s)} - \mathbf{v}^{(\widehat{s})}\right) = \text{comp}\left(\mathbf{u}^{(s)} - \mathbf{u}^{(\widehat{s})}\right).$$

Let α be the index element for the maximal composition value. Consider instead $\text{comp}\left(\mathbf{u}^{(s)} - (\mathbf{u}^{(\widehat{s})} - \alpha\mathbf{1})\right)$ in the code C. Here the maximum composition value is located at the index element 0 and then the maximum composition value is actually $1 - d/n$. Thus the maximum value of $\text{comp}\left(\mathbf{v}^{(s)} - \mathbf{v}^{(\widehat{s})}\right)$ over all pairs in the code V is $1 - d/n$ and we have $P_S = 1 - d/n$.

Summarizing we have:

Theorem 3: Given a code C with parameters (n, M, d) such that if $\mathbf{c} \in C$, then $\mathbf{c}+\lambda\mathbf{1} \in C$ for all $\lambda \in \mathbb{F}_q$. Then there exists a corresponding A-code with parameters

$$|\mathcal{S}| = Mq^{-1}, \quad |\mathcal{E}| = nq, \quad P_I = 1/q \text{ and } P_S = 1 - d/n.$$

■

We refer to this construction as the *q-twisted* construction.

Now suppose we have a *linear* $(n, k+1)$ q-ary code C' with the property

$$\mathbf{1} \in C'. \tag{13}$$

It follows from the linearity of the code that property (11) holds. Thus we again get in the end a code V with the parameters given in Theorem 3 by applying the q-twisted construction.

Example 4: Let C be a Reed-Solomon (R-S) code of length q. Let $L = \{$all polynomials of degree $< k+1$ in $\mathbb{F}_q[x]\}$. Then the R-S code C can be described as

$$C = \{(f(0), f(\alpha_1), f(\alpha_2), \ldots, f(\alpha_{q-1}))\,;\, f \in L, \mathbb{F}_q = \{0, \alpha_1, \ldots, \alpha_{q-1}\}\}\,.$$

If we now form the quotient code U, it will be as above but over all non-constant polynomials of degree $< k+1$ in $\mathbb{F}_q[x]$.

After the extension, the parameters are $|\mathcal{E}| = q^2$, $|\mathcal{S}| = q^k$, and since $d = q-(k+1)+1 = q-k$, we have $P_S = 1 - d/q = k/q$. We see that we obtained an A-code with the same parameters as in Construction 1. In fact, looking more closely, the two codes are the same up to renaming of source states and encoding rules. □

Let $A_q(n, d)$ be as usual, [7], the maximum number of codewords in a q-ary code of length n and with distance d. Now let $A_q^*(n, d)$ be the corresponding quantity if we add the property that $\mathbf{c} \in C$ implies $\mathbf{c} + \lambda\mathbf{1} \in C$ for all $\lambda \in \mathbb{F}_q$. Then

$$S(n, 1/q, P_S) \geq A_q^*\left(\frac{n}{q}, (1-P_S)\frac{n}{q}\right)/q,$$

where $S(n, 1/q, P_S)$ denotes the maximum number of source states as in (9).

For the special case of linear codes we have the following lemma:

Lemma 4: If there exists a linear code C with a codeword c such that every element in c is nonzero, then there also exists a linear code C' with the same parameters and such that $\mathbf{1} \in C'$. ■

Proof: Let the code C have generator matrix G. We can now do some elementary manipulations on G without changing the minimum distance. These operations include multiplication of columns by a nonzero scalar. Thus we multiply each column of G with the inverse of c's element in that column, getting a new code C'. Then $\mathbf{1} \in C'$. □

For nonlinear codes we can prove the Varshamov-Gilbert (V-G) bound even with the special restriction that $c \in C$ implies $c + \lambda\mathbf{1} \in C$.

Lemma 5: The maximum number of codewords in a q-ary code C of length n and minimum distance d such that $c \in C$ implies $c + \lambda\mathbf{1} \in C$, satisfies

$$A_q^*(n,d) \geq \frac{q^n}{V_q(n,d-1)},$$

where $V_q(n,d-1) = \sum_{i=0}^{d-1} \binom{n}{i}(q-1)^i$ is the size of a usual Hamming sphere around a codeword. ■

Proof: Consider the code with cardinality $A_q^*(n,d)$. If $A_q^*(n,d)V_q(n,d-1) < q^n$ there exists a point c which does not lie in any of the spheres. But this implies that $c + \lambda\mathbf{1}$ does not lie in any sphere either! This because if $c + \lambda\mathbf{1}$ is in the sphere around c' then $\mathrm{d}(c + \lambda\mathbf{1}, c') < d$ and thus $\mathrm{d}(c, c' - \lambda\mathbf{1}) < d$ and c is in the sphere around $c' - \lambda\mathbf{1}$ which is a contradiction. Thus we can add these q points as codewords and still have a code for which we have that if $c \in C$ then $c + \lambda\mathbf{1} \in C$, which contradicts the maximality of $A_q^*(n,d)$. □

Actually, this slightly strengthen the usual bound since $A_q^*(n,d)$ must be divisible by q.

Any known asymptotic bound[1] for q-ary codes is also valid for this extra condition, [8], since this is not a strong restriction when $n \to \infty$. Thus we have the same asymptotic behavior and the asymptotic V-G bound gives

$$S(|\mathcal{E}|, 1/q, P_S) \gtrapprox q^{g_q(P_S)|\mathcal{E}|/q}, \quad \text{for } P_S > 1/q, \tag{14}$$

where $g_q(x) = -x\log_q(x) - (1-x)\log_q(1-x) + x\log_q(q-1)$. Thus if we allow P_S to be larger than $P_I = 1/q$, then we can get A-codes that have an exponential number of source states.

Example 5: For $q = 2$, it follows from Example 1 that $|S|$ can be very close to $2^{\xi n}$, where $\xi \approx 1$ when $P_S \approx 1$. □

Example 6: Let us consider again Construction 1. Unfortunately, the upper bound (10) gives us not a coincidence of lower and upper bound for the case $|\mathcal{E}| \to \infty$, $P_S > 1/q$ fixed. □

For the binary case we give the upper bound (10). From the work of Wegman of Carter, [9, in the abstract], we have the following bound

$$\log|\mathcal{E}| \geq \log q + \log\log|S| - \log\log q.$$

From this we would obtain

$$|S| \leq q^{|\mathcal{E}|/q}. \tag{15}$$

But this bound is *not correct* since it does not depend on the P_S at all. Moreover, for $q = 2$. we get from (15) that $|S| \leq 2^{n/2}$, but we know from Example 5 that the size of S can be very close to 2^n.

[1] The V-G bound is known to be not optimal for $q = 49$

4 Conclusion

We have shown a relation between authentication codes and codes correcting independent errors and we have also shown how to construct A-codes from CIE codes. This is used to show that if P_S exceeds P_I by an arbitrarily small positive amount, then the number of source states in S grows exponentially with the number of keys and if $P_S = P_I$ it will grow only linearly. Furthermore we falsified a statement in Reference [9].

References

[1] E.N. Gilbert, F.J. MacWilliams and N.J.A. Sloane, "Codes which detect deception", Bell Syst. Tech. J., Vol. 53, no. 3, 1974, pp. 405-424.

[2] G.J. Simmons, "Authentication theory/coding theory", in *Advances in Cryptology, Proceedings of CRYPTO 84*, G.R. Blakley and D. Chaum, Eds. Lecture Notes in Computer Science, No. 196. New York, NY: Springer, 1985, pp. 411–431.

[3] G.J. Simmons, "A survey of Information Authentication", in *Contemporary Cryptology, The science of information integrety*, ed. G.J. Simmons, IEEE Press, New York, 1992.

[4] D.R. Stinson, "The combinatorics of authentication and secrecy codes", Journal of Cryptology, Vol. 2, no. 1, 1990, pp. 23-49.

[5] D.R. Stinson, "Universal hashing and authentication codes" *Proceedings of Crypto 91*, Santa Barbara, USA, 1991, pp. 74-85.

[6] F.J. McWilliams, N. Sloane, *The Theory of Error-Correcting Codes*, North-Holland, New-York, 1977.

[7] S. Roman, *Coding and Information Theory*, Springer-Verlag, New York, 1992.

[8] M.A. Tsfasman, S.G. Vlăduţ, *Algebraic-Geometric Codes*, Kluwer Academic Publ., Dortrecht, 1991.

[9] M.N. Wegman, J.L. Carter, "New hash functions and their use in authentication and set equality", J. Computer and System Sciences, Vol. 22, 1981, pp. 265-279.

Optimal Authentication Systems

R. Safavi-Naini *
L. Tombak **

Department of Computer Science University of Wollongong
Northfields Ave., Wollongong 2522, AUSTRALIA

Abstract. In this paper we define an optimal authentication systems as a system whose minimum probability of deception is k/M, k and M being the number of source states and cryptograms respectively, and satisfies information theoretic bounds on the value of impersonation and substitution games. We will characterize order-1 perfect systems and δ-perfect systems and prove their optimality when E, the number of encoding rules, satisfies certain bounds. We will show that both types of systems, in this case, also have best game theoretic performance. This will be used to prove that optimal systems exist only if $E \geq M^2/k^2$ and for less value of E probability of deception is always greater than k/M. We will prove that doubly perfect codes are optimal systems with minimum value of E and perfect systems are not optimal. Characterization of doubly perfect systems follows from characterization theorems mentioned earlier. We give constructions for each class.

1 Introduction

In this paper we will study authentication systems (A-systems) with optimum performance and characterize two classes of such systems. In an A-system the enemy has the option of playing impersonation game (I-game), substitution game (S-game) or combined game (C-game). Values of these games are P_I, P_S and P_C. Defining optimality for an A-system is not straight forward. It is important to note that minimizing value of the games does not ensure efficient use of redundancy added during coding process. For any A-system P_C is at least equal to probability of success in randomly selecting a cryptogram from cryptogram space. We define optimality of an A-system by requiring the system to satisfy an information theoretic bound on the value of I-game, a same kind of bound on the value of S-game and having $P_C = k/M$ where k is the number of source states and M is the number of cryptograms. For impersonation game the only information theoretic bound is due to Simmons [8]. Massey [3] and Sgarro [2] gave shorter proofs of the bound and necessary and sufficient conditions for achieving the

* Support for this project was partly provided by Australian Research Council grant A49030136.

** Support for this project was provided by Australian Research Council grant A49030136.

bound with equality. For substitution game we have two such bound; Simmons-Brickell bound is derived in [1] and later extended by Stinson [11] where he gives necessary conditions on A-systems that satisfy the bound with equality. The second bound for substitution is by Pei [5] (we give a sketch of the proof of this bound in appendix 7.3). Pei gives necessary and sufficient conditions for systems satisfying the bound with equality. We transform Pei's conditions into equivalent ones which are similar to necessary and sufficient conditions of Simmons' bound when substitution game is played. Hence the bound can be considered as the counterpart of Simmons' bound for substitution. We define order-1 perfect systems as systems for which P_I satisfies Simmons' bound and P_S satisfies Pei's bound. Similarly δ-perfect A-systems are those for which P_I and P_S satisfy Simmons' bound and Stinson's bound respectively. We will prove that for $E \geq E_0 = M(M-1)/(k(k-1))$ order-1 perfect systems are optimal and for $M^2/k^2 \leq E \leq E_0$, δ-perfect systems are optimal. Moreover in each case the value of I-game, S-game and C-game is minimal, that is, optimal systems have best game theoretic performance too. We give a complete characterization of the two classes and list properties of them. Next we define δ-doubly perfect systems as A-systems that are δ-perfect and have minimum value of E for a given δ, that is, $E = \delta M^2/k^2$. For $\delta = 1$ we have doubly perfect system of Brickell which is in fact the optimal system with minimum possible E. This implies that for $E < M^2/k^2$ the value of the combined game is always greater than k/M. Application of our characterization theorems, mentioned earlier, gives a characterization of doubly perfect A-systems, not known before. We also examine properties of perfect A-systems, as defined by Simmons, and show that they are not optimal as the value of substitution game does not satisfy any information theoretic bound. Finally we give some construction for each class and present some concluding remarks.

2 Preliminaries

We consider an authentication system in which a transmitter wants to send the states of a source to a distant receiver over a public channel. An encoding rule is a mapping from the set S, $|S| = k$, of source states into the set $\mathcal{M}$, $|\mathcal{M}| = M$, of codewords (cryptogram). An authentication code (A-code) is a collection $\mathcal{E}$ of mappings (encoding rules), indexed by key information, such that each mapping specifies one (or a number of) cryptogram for every $s \in S$. We assume the code is without splitting, that is, a source state and a key uniquely determines a cryptogram. We use $s(e,m)$ to denote the source state which is mapped into m by key e and $P_s(e,m)$ to denote its probability. We define the *incidence matrix* of an A-code to be a zero-one matrix A, the rows of which correspond to encoding rules and columns to cryptogram, and

$$a_{em} = \begin{cases} 1 & \text{m is authentic under key } e_i \\ 0 & \text{otherwise.} \end{cases}$$

Let $E(m)$ denotes the subset of keys that are incident with $m \in \mathcal{M}$ and $M(e)$ the subset of cryptograms incident with the key $e \in \mathcal{E}$. The communicants secretly choose the encoding rule e. Enemy can use an ***impersonation attack*** in which he/she attempts to find an $m \in M(e)$ or a ***substitution attack*** in which he/she intercepts a cryptogram $m \in \mathcal{M}$ and wants to substitute it with another cryptogram $m' \in \mathcal{M}_m$ where $\mathcal{M}_m = \mathcal{M} \setminus m$. We also refer to these attacks as order zero and one attack respectively. Let $E(m, m') = \{e : e \in E(m) \bigcap E(m')\}$. Simmons showed that A-systems can be modeled using game theory [7, 9]. Enemy has the choice of playing ***impersonation game*** (I-game), ***substitution game*** (S-game) or ***combined game*** (C-game) whose game matrix is the concatenation of the game matrices of I-game and S-game [9]. Let P_I,P_S and P_C denote the value of the game in each case. Communicant's strategy is always a probability distribution $\pi = (\pi_1, \pi_2, ..., \pi_E)$ on the encoding rules but enemy's strategy depends on the kind of game that he/she plays. In general we have $P_C \geq max(P_I, P_S)$ but if the best strategy of the communicants' is the same for I-game and S-game (which implies the same best strategy for C-game) then $P_C = max(P_I, P_S)$ and enemy's best strategy reduces to his/her best impersonation or substitution strategy. Game matrix of I-game is the incidence matrix of the A-code. For S-game payoff of replacing m by m' is,

$$payoff(m, m') = \frac{\sum_{i=1}^{E} \pi_i a_{im} a_{im'} P_s(e_i, m)}{P(m)},$$

where $P_s(e_i, m)$ is the probability of $s(e_i, m)$ and,

$$V_S(m) = max_{m'}(payoff(m, m')).$$

An authentication system provides ***perfect protection for impersonation*** if enemy's best strategy is random selection from $\mathcal{M}$. Probability of success in this case is equal to k/M. An A-code provides ***perfect protection for substitution*** if for all $m \in \mathcal{M}$ enemy's best strategy, when m is intercepted, is random selection from $\mathcal{M}_m$. His/her probability of success is equal to $(k-1)/(M-1)$. Stinson gave the characterization of A-systems that provide perfect protection for impersonation and the ones that provide perfect protection for substitution (Theorems 2.1 and 2.4 in [13]). Perfect protection for impersonation depends only on the incidence matrix of the A-code and is independent of the source. However perfect protection for substitution depends on the incidence matrix of the code and the probability distribution of the source.

3 Bounds on Probability of Deception

3.1 Simmons-Pei bound

The first (and the only) information theoretic bound on P_I is due to Simmons [8].

Theorem 1. *For an A-code without splitting,*

$$P_I \geq 2^{-(H(E)-H(E|\mathcal{M}))}. \tag{1}$$

Equality holds if and only if,

1. *the A-code provides perfect protection for impersonation;*
2. *$P_s(e_i, m)$ is independent of e_i and $P_s(e_i, m) = P_s(m)$, that is, all the source states that map to a cryptogram m have the same probability.*

Hence to obtain equality in (1), one must use an A-code whose incidence matrix accords with theorem 2.1 of [13] with a source whose first order statistics *matches* (in the sense of condition 2) the A-code. Pei's bound for substitution is the counterpart of Simmons bound for impersonation attack. We give the sketch of the proof of this bound, first presented in Asiacrypt '91, in appendix 7.3.

Theorem 2 Pei, [5].

$$P_S \geq 2^{-(H(E|\mathcal{M})-H(E|\mathcal{M}^2))}, \tag{2}$$

and equality holds if and only if

$$\frac{P(e|m)}{P(e|m, m')}, \tag{3}$$

is independent of m, m', e, for all $m, m' \in M(e)$.

In proposition 5 we show that condition (3) can be transformed into two conditions similar to theorem 1. We need the following generalization of perfect protection for substitution. Let C_m denotes the set of the cryptograms m' that, when substituted for m, have non-zero probability of success, i.e., $C_m = \{m' \in \mathcal{M} : E(m, m') \neq \emptyset\}$ and $|C_m| = N_m$.

Definition 3. An A-code provides near-perfect protection for substitution if the enemy's best strategy, when a cryptogram m is received, is random selection from C_m.

Near-perfect protection is weaker than perfect protection as enemy's strategy is random selection from C_m for which $N_m = |C_m| \leq M - 1 = |\mathcal{M}_m|$. Although probability of success depends on the intercepted cryptogram it is easy to see that for an A-code with near-perfect protection for substitution, for all $m \in \mathcal{M}$, we have

$$V_S(m) = \frac{k-1}{N_m}.$$

Corollary 4. *An A-code provides perfect protection for substitution if and only if it provides near-perfect protection for substitution and $N_m = M - 1$.*

Near-perfect protection can be defined for higher order attacks but for impersonation it reduces to perfect protection if we assume that every cryptogram is authentic under at least one key. An *A*-code provides *uniform near-perfect protection* if it provides a near-perfect protection and N_m does not depend from the actual intercepted cryptogram,i.e, $N_m = N_1$

Proposition 5. *The necessary and sufficient conditions for equality in (2) are*

1. *the A-code provides uniform near-perfect protection for substitution;*
2. $P_s(m'|e_i, m)$ *is independent of* e_i *for all* $m, m' \in \mathcal{M}$ *with* $E(m, m') \neq \emptyset$.

Proof: See appendix 7.1.

Proposition 5 shows that an authentication system that satisfies bound (2) will provide near-perfect protection for substitution. Equality in (2) requires an A-code with uniform near-perfect protection for substitution together with a source whose second order statistics satisfy condition 2 of proposition 5.

Definition 6. An authentication system is called 0-perfect if it satisfies Simmons' bound and 1-perfect if it satisfies Pei's bound.

Corollary 7 is an immediate result of proposition 5 and theorem 1.

Corollary 7. *An authentication system is i-perfect* ($i = 0, 1$) *if and only if it provides uniform near-perfect protection for order i attack and source statistics of order* $i+1$ *is 'matched' to the A-code.*

3.2 Simmons-Brickell-Stinson Bound

Simmons and Brickell derived the following bound on P_S:

Theorem 8 Simmons-Brickell, [1].

$$P_S \geq 2^{-H(E|M)}. \tag{4}$$

If equality holds then

1. $V_S(m) = \dfrac{\pi_i P_s(e_i, m)}{P(m)}$, *where* $m \in \mathcal{M}$ *and* $e_i \in \mathcal{E}$ *such that* $e_i \in E(m)$;
2. $P_S = V_S(m),\ m \in \mathcal{M}$;
3. $|E(m, m')| \leq 1$, *and* $m \in M$, $m' \in \mathcal{M}_m$.

Stinson gave a more general form of this bound,

$$P_S \geq \delta 2^{-H(E;M)}, \tag{5}$$

where δ is equal to,

$$\delta = min_{m,m',e_i}(\delta(e_i, m, m')) = min_{m,m',e_i}\left(\frac{\sum \pi_i a_{im} a_{im'} P_s(e_i, m)}{\pi_i P_s(e_i, m)}\right)$$

and proved the following.

Theorem 9 Stinson [11], Theorem 2.8. *In an A-system without splitting that satisfies bound (5) with equality we have:*

1. $|E(m)| = kE/M$,
2. $|E(m, m')| = 0$ *or* λ,
3. $\delta(e_i, m, m') = \delta = \lambda$ *for all* $e_i \in E(m, m')$,
4. $P_S = \delta M/(kE)$.

Simmons-Brickell bound is a special case of Stinson's bound when $\delta = 1$ and hence equality in Simmons-Brickell's bound implies $P_S = M/(kE)$.

4 Optimal A-Systems

As noted in previous section i-perfect A-systems satisfy the information theoretic bound and provide uniform near-perfect protection and so can be considered optimal when only one type (impersonation or substitution) of attack is considered. In this section we will consider both types of attacks. For an A-system $P_C \geq P_I \geq k/M$.

Definition 10. An A-system is optimal if it has $P_C = k/M$, satisfies Simmons' bound for P_I and an information theoretic bound (2) or (5) on P_S.

We define order-1 perfect A-systems and δ-perfect systems and show that they are optimal. We prove that they also achieve minimum value of the I-game, S-game and C-game. Our major results are complete characterization of both types of system (theorems 14, 18). Simmons' definition of perfect A-systems and Brickell's definition of doubly perfect systems are studied in this context. In particular we show that perfect A-systems are not optimal and doubly perfect systems are optimal with least possible E.

Let the communicants use their optimum strategy for combined game.

Definition 11. An authentication system is called order-1 perfect if P_I and P_S satisfy Simmons' bound and Pei's bound respectively and $|E(m, m')| \geq \lambda > 0$ for all $m, m' \in \mathcal{M}$.

Let $E_0 = \frac{M(M-1)}{k(k-1)}$.

Proposition 12. *For an order-1 perfect A-system we have,*

$$E \geq \lambda E_0.$$

Hence such systems can exist if $E \geq E_0$.

Proof: Follows from counting pairs of cryptograms and using the minimum value of $\lambda = 1$. □

Note that for order-1 perfect systems, in general, we do not have $P_I = P_S$. Enemy's best strategy for I-game and S-game are random strategies and the best C-game strategy is the same as the best I-game strategy. Proposition 13 shows that for order-1 perfect A-systems source must be uniform. It also specifies other properties of such systems.

Proposition 13. *If an A-system is order-1 perfect the followings hold,*

1. *best enemy's strategy in impersonation or substitution is random strategy and the overall best strategy is the same as the best impersonation strategy;*
2. $\sum_{j=1}^{E} \pi_j a_{jm} a_{jm'} = 1/E_0$;
3. *probability of a cryptogram m occurring in the channel is uniform ($P(m) = 1/M$);*

4. source is uniform;
5. $P_s(m'|m, e_j) = P_s(m'|m)$ *for all* $m, m' \in \mathcal{M}$ *and* $e_j \in E(m, m')$.

Proof: Let the A-system be order-1 perfect. Then property 1 follows from the definition of order-1 perfect systems and theorems 1 and 2. Enemy's overall optimal strategy is random selection from $\mathcal{M}$ (impersonation) as $P_I = k/M > (k-1)/(M-1) = P_S$ and P_I and P_S are achievable for the same communicants' strategy. Property 3 follows from Theorem 3.2 of [14] where it was proved that perfect protection for impersonation and substitution implies $P(m) = 1/M$. Also property 2 follows from the same theorem when the source is uniform. To prove property 4 we note that,

$$P(m) = \sum_{j=1}^{E} \pi_j a_{jm} P_s(e_j, m),$$

but equality in (1) implies $P_s(e_j, m) = P_s(m)$ and we have,

$$P(m) = P_s(m) \sum_{j=1}^{E} \pi_j a_{jm} = P_s(m)(k/M), \tag{6}$$

where the last equality holds because A-system provides perfect protection for impersonation. Using $P(m) = 1/M$ with (6) we have $P_s(m) = 1/k$. Finally property (5) is true because the A-system satisfies Pei bound (2).
□

Communicants' optimal strategy can be obtained by solving a system of linear equations [6] which depends on the incidence matrix of the A-code and is independent of the source. Conditions (2) to (5) of the proposition 13 are sufficient for an A-system to be order-1 perfect. Theorem 14 characterizes such systems.

Theorem 14. *An order-1 perfect A-system satisfies conditions 1 to 5 of proposition 13. Moreover conditions 2 to 5, or equivalently, 1 (or 2), 4 and 5 are sufficient conditions.*

Proof: See appendix 7.2.

Proposition 13 and theorem 14 show that A-systems that are order-1 perfect are obtained from A-codes whose incidence matrices satisfy certain conditions together with a *'matched'* source.

Corollary 15. *An order-1 perfect A-system achieves minimum values for I-game, S-game and C-game and hence has the best game theoretic performance, that is,*

1. $P_C = k/M$;
2. $P_I = k/M$;
3. $P_S = (k-1)/(M-1)$.

Moreover we have

- *enemy's best combined strategy is the same as his/her best strategy for impersonation and is actually a random selection from $\mathcal{M}$;*
- *enemy's best substitution strategy is random selection from all the remaining cryptograms;*
- *communicants best strategy can be calculated by solving a set of linear equations whose coefficients are derived from the incidence matrix of the A-code.*

Using proposition 12 we conclude that order-1 perfect systems are optimal if $E \geq E_0$. An A-system that satisfies Simmons' bound for impersonation has $P_I = k/M$ and it is shown in [14] if $E < E_0$ then

$$P_S \geq \max[\frac{k-1}{M-1}, \delta \frac{M}{kE}]. \tag{7}$$

δ-perfect A-systems, defined below, are optimal when $M^2/k^2 \leq E \leq E_0$.

Definition 16. An A-system is δ-perfect if it satisfies Simmons bound for P_I and
$P_S = \delta 2^{-H(E|M)}$.

Proposition 17. *A δ-perfect A-system satisfies the following,*

1. $|E(m)| = const = kE/M$;
2. $|E(m,m')| = \delta = \lambda$ *or* 0;
3. $P_s(e_j, m) = P_s(m)$;

Proof: Properties one and two follows from theorem 9 and property 3 follows from theorem 1. □

Theorem 18. *δ-perfect A-systems are optimal if $E_0 \geq E \geq M^2/k^2$. In this case communicants best strategy for combined game is uniform distribution on the key space. Moreover, this condition together with 1, 2 and 3 of proposition 17 are sufficient for an A-system to be δ-perfect.*

Proof: The system satisfies Simmons' bound and hence $P_I = k/M$. It satisfies Stinson's bound and so $P_S = \delta M/(kE)$. If $E \geq M^2/k^2$ we have $P_S \leq P_I$ and $P_C = P_I = k/M$ and the system is optimal. Communicants' best strategy for combined game will be uniform distribution on the key space.

To prove sufficiency, we note that if condition 1 of proposition 17 holds then communicants uniform strategy provides perfect protection for impersonation. Using the uniform strategy for substitution and taking into account conditions 1, 2 and 3 of proposition 17 we show that $payoff(m,m') = 0$ or $\delta M/(kE)$. This is true because,

$$payoff(m,m') = \frac{\sum_{j=1}^{E} \pi_j a_{jm} a_{jm'} P_s(e_j, m)}{\sum_{j=1}^{E} \pi_j a_{jm} P_s(e_j, m)} = \frac{\sum_j a_{jm} a_{jm'}}{\sum_j a_{jm}} = \frac{M \sum_j a_{jm} a_{jm'}}{kE}.$$

So $V_S(m) = \delta M/(kE)$ and $P_S = \delta M/(kE)$, which means that uniform strategy is the best communicants' substitution strategy. Using the same conditions it is

easy to see that $2^{-H(E|M)} = \delta M/(kE) = P_S$ and $2^{-(H(M)-H(M|E))} = k/M = P_I$, which proves the result.
□

Corollary 19. *A δ-perfect A-system achieves minimum possible value for P_S and satisfies*

1. $P_C = P_I = k/M$;
2. $P_S = \delta M/(kE)$.

Moreover enemy and communicants' best strategies are given by

- *enemy's best combined strategy is random selection from $\mathcal{M}$;*
- *enemy's best impersonation strategy is random selection from $\mathcal{M}$ and his/her best substitution strategy is random selection from keys that are incident with the received cryptogram and then randomly selecting a cryptogram which is authentic under the chosen key;*
- *communicants best strategy is uniform distribution on $\mathcal{E}$.*

Definition 20. An optimal δ-perfect A-system with minimum number of encoding rule is called δ-doubly perfect system.

Proposition 21. *For a δ-doubly perfect system $P_S = P_I = P_C$.*

Proof: We have $E = \delta M^2/k^2$ and $P_I = k/M = \delta M/(kE) = P_S$. □

4.1 Doubly Perfect A-systems

A *doubly perfect* A-system, as defined by Brickell [1], is a perfect A-system that satisfies

$$P_C = 2^{-H(E|M)}.$$

Doubly perfect A-systems are special case of δ-doubly perfect A-systems when $\delta = 1$. This is true because for perfect A-systems $P_C = P_I$ and hence we have $P_I = P_S = P_C$. Doubly perfect A-codes have all properties mentioned in corollary 19.

4.2 Perfect A-systems

Simmons defined a *perfect A-system* as an A-system that satisfies the following bound:

$$P_C = 2^{-I(M;E)}.$$

For an A-system using theorem 1, we have $P_I \geq 2^{-I(M;E)}$. So for a perfect A-system $2^{-I(M;E)} = P_C \geq P_I \geq 2^{-I(M;E)}$ and hence $P_I = 2^{-I(M;E)}$ and the A-system is 0-perfect. Moreover $P_I = P_C \geq P_S$. Hence enemy's best combined strategy is random selection from $\mathcal{M}$. However the enemy's best chance of success in substitution is not known. Communicants' optimal strategy for C-game can be obtained by solving a system of linear equations [6]. We note that for perfect A-systems we have $P_C = P_I = k/M$ and $P_S < P_I$.

Corollary 22. *Perfect A-systems are not optimal as P_S does not satisfy any bound.*

Corollary 23. *Optimal A-systems exist only if $E \geq M^2/k^2$.*

Proof: For optimal A-systems $P_C = P_I = k/M$. In this case P_S is lower bounded as in (7). If $E < M^2/k^2$ then $E < E_0$ and the best achievable value of P_S is $M/(kE)$ but in this case $M/(kE) > k/M$ and hence $P_C > k/M$. □

We summarize these results in the following corollary.

Corollary 24. *For a given E, M, k we have,*

1. *If $E < M^2/k^2$ then $P_C > k/M$.*
2. *If $E = M^2/k^2$ then doubly perfect A-systems are optimal and have $P_C = P_S = P_I = k/M = M/(kE)$.*
3. *If $M^2/k^2 < E < E_0$ then δ-perfect A-systems are optimal. We have $P_C = P_I = k/M > P_S = \lambda M/(kE)$.*
4. *If $E \geq E_0$ then order-1 perfect A-systems are optimal and we have $P_C = P_I > P_S = (k-1)/M - 1$.*

This corollary can be re-stated for A-systems for which $E(m, m') \geq \lambda$ when $E(m, m') \neq 0$.

Corollary 25. *For a given E, M, k λ we have,*

1. *If $E = \lambda M^2/k^2$ then δ-doubly perfect A-systems are optimal and $P_C = P_S = P_I = k/M = \lambda M/(kE)$.*
2. *If $\lambda M^2/k^2 < E < \lambda E_0$ then δ-perfect A-systems are optimal and have $P_C = P_I = k/M > P_S = \lambda M/(kE)$.*
3. *If $E \geq \lambda E_0$ then order-1 perfect A-systems are optimal.*

5 Construction of Optimal Codes

In this section we will give some constructions for optimal A-systems.

5.1 δ-doubly perfect A-systems and δ-perfect systems

Definition 26. $A(v, k, r, \lambda)$-PBIB is a pair (M, E),where $|M| = v$ is a set of elements called points and E is a set of blocks,where block is a k-element subset of M; such that each point occurs in exactly r-blocks, and each pair of points occurs in exactly λ blocks or does not occur at all.

Using proposition 17 we immediately get the following result.

Proposition 27. *If there exists a δ-perefct system for a uniform source then there exist a (v, k, r, λ)-PBIB. Conversely if there exist a (v, k, r, λ)-PBIB then there exist a δ-perefct A-system with k equiprobable source states, $v = M$ cryptograms and $E = rv/k$ keys.*

In order to construct a δ-doubly perfect system we need another construction called transversal design.

Definition 28. A transversal design $TD(k, \lambda, n)$ is a triple (X, G, A), which satisfies the following properties

1. X is a set of kn elements called points;
2. G is a partition of X into k subsets of n points, called groups;
3. A is a set of λn^2 subsets of X (called blocks) such that a group and a block contain at most one common point;
4. every pair of points from distinct groups occurs in exactly λ blocks.

Using this combinatorial design we can construct a Cartesian δ-doubly perfect A-system.

Proposition 29 [11], theorem 3.5. *If there is a $TD(k, \lambda, n)$ then there is a δ-doubly perfect Cartesian A-system with $M = kn$ cryptograms, k-source states, $E = \lambda n^2$ keys for which $P_S = P_I = 1/n$. Conversely if there exist a δ-doubly perfect Cartesian A-system with no splitting then there exist a transversal design $TD(k, \delta, n = M/k)$.*

5.2 Order-1 perfect authentication codes

We can construct order-1 perfect authentication codes using a well known combinatorial construction called balance incomplete block design-BIBD.

Definition 30. A (v, k, r, λ)-BIBD is a collection of k-subsets, called blocks, of a v-set, called points, such that each such that each point occurs in exactly r-blocks, and each pair of points occurs in exactly λ blocks.

Proposition 31. *If there exists a (v, k, r, λ)-BIBD then there exists an order-1 perfect A-system with k-equiprobable source states, $v = M$ cryptograms and $E = vr/k$ keys.*

For a fixed parameters M, k, E, λ this order-1 perfect system has minimum possible number of keys.

6 Concluding Remarks

We have defined optimal performance of an A-system using information theory and game theory measures and have characterized them when E is within different ranges. In particular we have proved that these systems can only exist for $E \geq M^2/k^2$ and for less number of encoding rules the chance of enemy's success is always greater than k/M. We noted that perfect A-systems of Simmons are not optimal but doubly perfect systems of Brickell are optimal with least number of encoding rules. We have given some construction for each case. Further research is needed to construct larger classes of optimal systems.

7 Appendix

7.1 A

Proof of proposition 5: Necessity: We show that the above conditions could be derived from (3). To obtain first condition we have

$$\sum_{e_j \in E(m,m')} P(e_j|m,m') = 1$$

and

$$\frac{P(e_j|m)}{P(e_j|m,m')} = \frac{P(e_j|m)}{P(e_j|m,m')} \sum_{e_j \in E(m,m')} P(e_j|m,m') \tag{8}$$

$$= \frac{\sum_{e_j \in E(m,m')} P(e_j|m,m')P(e_j|m)}{P(e_j|m,m')}$$

$$= \sum_{e_j \in E(m,m')} P(e_j|m) = const \tag{9}$$

Hence $payoff(m,m')$, given in (9), is independent of m and m' and the A-code provides uniform near-perfect protection for substitution, i.e.,

$$\frac{P(e_j|m)}{P(e_j|m,m')} = \frac{k-1}{N_1}, \quad E(m,m') = 0,$$
$$= 0, \quad E(m,m') = \emptyset.$$

To get second condition we have

$$\frac{P(e_j|m)}{P(e_j|m,m')} = \frac{P(e_j,m)/P(m)}{P(e_j,m,m')/P(m,m')},$$
$$= \frac{\pi_j a_{jm} P_s(e_j,m)}{P(m)} \times \frac{P(m,m')}{\pi_j a_{jm} a_{jm'} P_s(e_j,m,m')},$$
$$= \frac{P(m,m')}{P(m)P_s(e_j,m'|m)} = const,$$

where $P_s(m'|m,e_j)$ is the conditional source probability $P_s(s(e_j,m')|s(e_j,m))$. That is, the source states that are mapped into a cryptogram m', when m is received, are equiprobable.

Sufficiency: We show that conditions 1 and 2 result in equality in (2). Using condition 1 we have

$$\frac{\sum_{j=1}^{E} \pi_j a_{jm} a_{jm'} P_s(e_j,m)}{P(m)} = const.$$

Multiplying numerator and denominator by $P_s(e_j, m'|m)$ and using condition 2 we have

$$\frac{\sum_{j=1}^{E} \pi_j a_{jm} a_{jm'} P_s(e_j, m) P_s(e_j, m'|m)}{P(m) P_s(e_j, m'|m)} =$$
$$\frac{\sum_{j=1}^{E} \pi_j a_{jm} a_{jm'} P_s(e_j, m, m')}{P(m) P_s(e_j, m'|m)} =$$
$$\frac{P(m, m')}{P(m) P_s(e_j, m'|m)} = \frac{P(m, m') P_s(e_j, m)}{P(m) P_s(e_j, m'|m) P_s(e_j, m)} =$$
$$\frac{\pi_j a_{jm} P_s(e_j, m)}{P(m)} \times \frac{P(m, m')}{\pi_j a_{jm} a_{jm'} P_s(e_j, m)} = \frac{P(e_j|m)}{P(e_j|m, m')}.$$

□

7.2 B

Proof of theorem 14:

Necessity has been already given in the proof of proposition 13. Sufficiency: using conditions 2 and 4 we have,

$$\sum_{j=1}^{E} \pi_j a_{jm} P_s(e_j, m) = 1/M = 1/k \sum_{j=1}^{E} \pi_j a_{jm},$$
$$\sum \pi_j a_{jm} = k/M.$$

So the code provides perfect protection for impersonation and $P_I = k/M$. Moreover

$$2^{-I(E,M)} = 2^{H(S)-H(M)} = k/M,$$

and so $P_I = 2^{-I(E,M)}$ and the code is 0-perfect.

To show that the code is 1-perfect we note that,

$$\frac{P_s(e_j, m, m')}{P_s(e_j, m)} = P_s(m'|m, e_j) = P_s(m'|m),$$

where the last equality follows from condition 5. Using condition 4 we have

$$P_s(e, m, m') = P_s(m, m'),$$

which results in

$$\frac{P(e_j|m)}{P(e_j|m, m')} = \frac{P(e_j, m)/P(m)}{P(e_j, m, m')/P(m, m')}. \tag{10}$$

We make the following substitutions in (10),

$$P(e_j, m) = (1/k)\pi_j a_{jm},$$
$$P(m) = 1/M,$$
$$P(e_j, m, m') = \pi_j P_s(e_j, m, m') a_{jm} a_{jm'},$$
$$P(m, m') = \sum_{j=1}^{E} \pi_j a_{jm} a_{jm'} P_s(e_j, m, m'),$$

which gives,

$$\frac{P(e_j|m)}{P(e_j|m, m')} = (k/M) \sum_j \pi_j a_{jm} a_{jm'} = (k-1)/(M-1),$$

and hence using theorem 2 the A-system is 1-perfect. □

7.3 C

Pei proved the following bound on probability of the enemy's success for spoofing attack of order r.

$$P_r \geq 2^{H(E|M^{r+1}) - H(E|M^r)}$$

The following is a proof for $r = 1$ which can be generalized for spoofing of order r. We need the following propositions.

Proposition 32. *Suppose that P and Q are probability vectors of the same dimension with non zero coordinates. So $p_i > 0$ and $q_i > 0$ $(1 < i < n)$, and*

$$\sum_{i=1}^{n} p_i = \sum_{i=1}^{n} q_i \doteq 1.$$

Then

$$\sum_{i=1}^{n} p_i \log(p_i/q_i) \geq 0$$

Moreover equality holds if and only if $p_i = q_i$

Proposition 33. *Suppose that $|E(m', m)| > 0$ then,*

$$\log(payoff(m, m')) \geq \sum_{e_j \in E(m', m)} p(e_j|m, m') \log \frac{p(e_j|m)}{p(e_j|m, m')}$$

Moreover equality holds if and only if

$$payoff(m, m') = p(e_j|m)/p(e_j|m, m')$$

for any $e_j \in E(m, m')$

Proof.
Let

$$p_j = p(e_j|m, m'),$$
$$q_j = \frac{p(e_j|m)}{\sum_{e_i \in E(m,m')} p(e_i|m)},$$

for any $e_j \in E(m, m')$ and use proposition 32. □

Proposition 34.

$$P_S \geq \sum_{m,m' \in \mathcal{M}} p(m, m') 2^{\sum_{e_i \in E(m,m')} p(e_i|m) \log \frac{p(e_i|m)}{p(e_i|m,m')}},$$

and equality holds if and only if

$$\frac{p(e_i|m)}{p(e_i|m, m')} = const = C,$$

for any $e_i \in E(m, m')$. In the case of equality $P_S = C$.

Proof. From proposition 33 we have

$$payoff(m, m') \geq 2^{\sum_{e_i \in E(m,m')} p(e_i|m) \log \frac{p(e_i|m)}{p(e_i|m,m')}}.$$

Averaging over all m and m' we have the desired result. □

Now we can prove the main theorem.

Theorem 35.

$$P_S \geq 2^{H(E|M^2) - H(E|M)}.$$

And equality holds if and only if the following conditions is satisfied: for any $m, m' \in \mathcal{M}$ and $e_J \in E(m, m')$ $p(e_j|m)/p(e_j|m, m') = const = C$. In the case of equality $P_S = C$

Proof.
Using proposition 34 and Jensen's inequality we have

$$\log P_S \geq \log(\sum_{m,m'} p(m, m') 2^{\sum_{e_j \in E(m',m)} p(e_j|m,m') log \frac{p(e_j|m)}{p(e_j|m,m')}}$$

$$\geq \sum_{m,m'} p(m, m') \sum_{e_j \in E(m',m)} p(e_j|m, m') \log \frac{p(e_j|m)}{p(e_j|m, m')}$$

$$= \sum_{m,m'} \sum_{e_j \in E(m',m)} p(e_j, m, m') \log \frac{p(e_j|m)}{p(e_j|m, m')} = H(E|M^2) - H(E|M).$$

which completes the proof.

References

1. E. Brickell, *A Few Results in Message Authentication* , Congressus Numerantium, vol 43, 1984, pp 141-154.
2. R.Johansen, A. Sgarro *Strengthening Simmons' Bound in Impersonation* , IEEE Transactions on Information Theory, vol 37, No 4, July 1991, pp 1182-1185.
3. J.L. Massey, *Cryptography - A Selective Survey,* Proc. of 1985 Int. Tirrenia Workshop on Digital Communication, Tirrenia, 1985, Digital Communications, ed.E. Biglieri and G. Pratti, Elsevier Science Publ., 1986, North-Holland, pp 3-25
4. J.L. Massey, *Introduction to Contemporary Cryptography* , Proceedings of the IEEE, vol 76, No 5, May 1988, pp 533-549.
5. D. Pei *Information - Theoretic Bounds for Authentication Codes and PBIB* , Asiacrypt 1991, Ramp Session.
6. R. Safavi, L. Tombak, *Authentication Codes under Impersonation Attack,* Proc. of Auscrypt 1992, to appear.
7. G.J. Simmons, *Message Authentication: A Game on Hypergraphs,* Congressus Numerantium, Vol 45, 1984, pp 161-192.
8. G.J. Simmons, *Authentication Theory/Coding Theory,* Proc. of Crypto 84, Lecture Notes in Computer Science 196, Springer 1985, pp 411-432.
9. G.J. Simmons, *A Game Theory Model of Digital Message Authentication,* Congressus Numerantium, Vol. 34, 1982, pp 413-424.
10. G.J. Simmons, *A Survey of Information Authentication* , Proceedings of the IEEE vol 76, No 5, May 1988, pp 603-619.
11. D.R. Stinson, *Some Constructions and Bounds for Authentication Codes,* Journal of Cryptology, No 1, 1988, pp 37-51.
12. D.R. Stinson, *The Combinatorics of Authentication and Secrecy Codes* , Journal of Cryptology, No 2 (1990), pp 23-49.
13. D.R. Stinson, *Combinatorial Characterization of Authentication Codes* , Proceedings Crypto 91, Lecture Notes in Computer Science 576, Springer 1992, pp 62-72.
14. L. Tombak, R. Safavi, *Authentication Codes with Perfect Protection,* Proc. of Auscrypt 1992, to appear.
15. M. Walker *Information-Theoretic Bounds for Authentication Schemes,* Journal of Cryptology, No 2, 1990, pp 131-143.

Factoring Integers Using SIMD Sieves

Brandon Dixon[1] and Arjen K. Lenstra[2]

[1] Department of Computer Science, Princeton University, Princeton, NJ 08544, USA, *E-mail*: bdd@cs.princeton.edu

[2] Room MRE-2Q334, Bellcore, 445 South Street, Morristown, NJ 07960, USA, *E-mail*: lenstra@bellcore.com

Abstract. We describe our single-instruction multiple data (SIMD) implementation of the multiple polynomial quadratic sieve integer factoring algorithm. On a 16K MasPar massively parallel computer, our implementation can factor 100 digit integers in a few days. Its most notable success was the factorization of the 110-digit RSA-challenge number, which took about a month.

1 Introduction

Usually one distinguishes two types of integer factoring algorithms, the *general purpose algorithms* whose expected run time depends solely on the size of the number n being factored, and the *special purpose algorithms* whose expected run time also depends on properties of the (unknown) factors of n. To evaluate the security of factoring-based cryptosystems, it is important to study the practical behavior of general purpose factoring algorithms. In this paper we present an efficient SIMD-implementation of the multiple polynomial quadratic sieve (QS) factoring algorithm [12], still the most practical general purpose method for integers in the range from 80 to 120 digits.

The largest number factored by QS is a 116-digit number. This factorization was carried out in a few months on a widely distributed network of workstations, and took a total computation time of approximately 400 mips-years (1 mips-year is about $3.15 \cdot 10^{13}$ instructions) [10]. The previous QS record for a single-machine implementation had 101 digits, and was carried out on one processor of a four processor Cray Y-MP4/464 in 475 CPU-hours [14]. This record was broken by our factorization of the 110-digit RSA-challenge number.

As will be explained in Section 3, QS consists of two main steps: the sieving step, and the matrix elimination step. All successful parallel implementations of QS that we know of have followed the approach described in [2]: distribute the sieving step over any number of available processors, which work independently of each other, and collect their results and perform the matrix elimination at a central location. In [2] the sieving step was done on a local network of workstations using the Ethernet for the communication, in [9] the workstations are scattered around the world, and communicate with the central location using electronic mail. In both of these implementations the network of participating machines can be viewed as a loosely coupled multi-processor machine, where the processors work asynchronously in *multiple-instruction multiple data* (MIMD) mode; i.e., each processor carries out its own set of instructions on its own set of data. Very powerful and fairly expensive massively parallel MIMD machines, with more computational power than was ever achieved using the approach from [9], are currently available. It should not be hard to break the 116-digit QS record on such a machine, but getting enough computing time might be prohibitively expensive, and is certainly more expensive than it was to use the donated cycles from the internet network.

Another type of massively parallel machine is the *single-instruction multiple data* (SIMD) machine. These machines usually consist of some particular network (hyper-cube, mesh) of several thousand small processors, each with its own fairly limited amount of memory. Unlike the processors on MIMD machines which can work more or less on their own, the SIMD processors simultaneously carry out the same instructions, but each processor on its own data. Furthermore, arbitrary subsets of processors can be made inactive or reactivated at any time. Although the rough computational power of large SIMD machines is comparable to that of supercomputers or large MIMD machines, SIMD machines are only efficient for computations that can be cast as a SIMD process.

SIMD machines have proven to be very useful for the matrix elimination step of QS [6; 8]. For that application, as well as for various other factoring applications (cf. [6]), it suffices to have a fairly restrictive type of SIMD machines, namely machines that only allow direct addressing (where the memory address is part of the instruction). For efficient sieving applications, however, it is essential that indirect addressing is available as well, i.e., the memory address depends on a value local to the processor. Although this is not the case for all SIMD machines, we will assume throughout this paper that this requirement is met (as it is for the 16K MasPar SIMD machine[3] that we used for our work, cf. Section 2).

Nevertheless, at first sight the sieving step does not look like the type of operation that would run well on any SIMD machine (cf. Section 3). Indeed, a MasPar-implementation of the QS-sieving step which was attempted in [5] seems to support this supposition. In this paper we describe a different approach to the SIMD-implementation of the sieving step which works quite efficiently: on a 16K MasPar 100-digit integers can be factored within three CPU-days. A 110-digit number took one CPU-month, where we used only 5/8 of the total available memory; using all the memory this factorization would have taken about 20

[3] It is the policy of Bellcore to avoid any statements of comparative analysis or evaluation of products or vendors. Any mention of products or vendors in this presentation or accompanying printed materials is done where necessary for the sake of scientific accuracy and precision, or to provide an example of a technology for illustrative purposes, and should not be construed as either a positive or negative commentary on that product or vendor. Neither the inclusion of a product or a vendor in this presentation or accompanying printed materials, nor the omission of a product or a vendor, should be interpreted as indicating a position or opinion of that product or vendor on the part of the presenter or Bellcore.

days. With a later, faster version of the program we were able to do a 105-digit number in 5.3 CPU-days, using half of the available memory. This shows that relatively inexpensive SIMD machines are much better for general purpose factoring than was previously expected. For SIMD-implementations of special purpose factoring algorithms (like the elliptic curve method) we refer to [4].

The success of this implementation prompted work on a SIMD-implementation of the general number field sieve factoring method [1]. With this number field sieve implementation we broke the record set by the factorization of the ninth Fermat number, by factoring the 151-digit number $(2^{503}+1)/3$ and the 158-digit number $2^{523}-1$.

The remainder of this paper is organized as follows. In Section 2 a short description of the SIMD machine that we used is given. Section 3 contains a general description of QS. A simple algorithm for the redistribution of data on a SIMD machine is given in Section 4. An overview of our SIMD QS implementation is presented in Section 5, and Section 6 contains an example.

2 The Hardware

This section contains a short overview of the 16K MasPar, the massively parallel computer that we have used for the implementation to be described in this paper. Our description is incomplete, and only covers those aspects of the machine that are referred to in the following sections. For a complete description of the MasPar we refer to the manuals, such as [11].

The 16K MasPar is a SIMD machine, consisting of, roughly, a *front end*, an *array control unit* (ACU), and a 128×128 array of *processing elements* (PE array). Masks, or conditional statements, can be used to select and change a subset of active processors in the PE array, the so-called *active set*. The fact that it is a SIMD machine means that instructions are carried out sequentially, and that instructions involving parallel data are executed simultaneously by all processors in the active set, while the other processors in the PE array are idle. The instructions involving singular (i.e., non-parallel) data are executed either on the front end or on the ACU; for the purposes of our descriptions the front end and the ACU play the same role.

According to our rough measurements, each PE can carry out approximately $2 \cdot 10^5$ additions on 32 bit integers per second, and can be regarded as a 0.2 MIPS processor. Furthermore, each PE has 64KBytes of memory, which implies that the entire PE array has 1GByte of memory. PE's cannot address each other's memory, but as mentioned in the introduction PE's can do indirect addressing. Each processor can communicate efficiently with its north, northeast, east, southeast, south, southwest, west, and northwest neighbor, with toroidal wraparound. Actually, a processor can send data to a processor at any distance in one of these eight directions, with the possibility that all processors that lie in between also get a copy of the transmitted data. There is also a less efficient global router that allows any processor to communicate with any other processor, but we never needed it.

Each job has a size between 4K and 64K, reflecting the amount of PE-memory it uses. Only those jobs which together occupy at most 64K are scheduled in a round robin fashion, giving each job 10 seconds before it is preempted, while the others must wait. This means that jobs are never swapped out of PE-memory.

For our implementations we used the MasPar Parallel Application Language MPL, which is, from our perspective, a simple extension of C.

3 The Quadratic Sieve Factoring Method

Let $n > 1$ be an odd positive integer that is not a prime power. For each random integer x satisfying

$$x^2 \equiv 1 \bmod n \tag{3.1}$$

there is a probability of at least 1/2 that $\gcd(n, x-1)$ is a non-trivial factor of n. To factor n it therefore suffices to construct several such x's in a more-or-less random manner.

In many factoring algorithms, solutions to (3.1) are sought by collecting integers v such that

$$v^2 \equiv \prod_{p \in P} p^{e_p(v)} \bmod n, \tag{3.2}$$

where the *factor base* P is some finite set of integers that are coprime to n, and $e_p(v) \in \mathbf{Z}$ for $p \in P$. A pair $(v, e(v))$ satisfying (3.2), with $e(v) = (e_p(v))_{p \in P} \in \mathbf{Z}^{\#P}$, is called a *relation*, and will be denoted by v for short. If V is a set of relations with $\#V > \#P$, then there exist at least $2^{\#V - \#P}$ distinct subsets W of V with $\sum_{v \in W} e(v) = (2w_p)_{p \in P}$ and $w_p \in \mathbf{Z}$; these subsets can be found using Gaussian elimination modulo 2 on the set of vectors $e(v)$ mod 2. Each such W leads to an $x \equiv (\prod_{v \in W} v) \cdot (\prod_{p \in P} p^{-w_p}) \bmod n$ satisfying (3.1).

In the original quadratic sieve factoring algorithm [12] relations are collected as follows. Let P consist of -1 and the primes $\leq B$ with Legendre symbol $\left(\frac{n}{p}\right) = 1$, for some bound B. An integer is called B-smooth if it can be written as a product over P. Relations are collected by looking for small integers i such that $f(i) = (i + [\sqrt{n}])^2 - n$ is B-smooth; for such i we have that $v = i + [\sqrt{n}]$ satisfies (3.2). Because a prime p divides $f(i)$ if and only if it divides $f(i+kp)$ for any integer k, smooth values can be found efficiently using a sieve, if the roots of f modulo the primes in P are known. For this reason, the relation collecting step is called the *sieving step*. Notice that only primes p with $\left(\frac{n}{p}\right) = 1$ can divide $f(i)$; this explains the definition of P.

The second step, finding subsets W as above, is called the *matrix elimination step*. In this paper we will not pay any further attention to this step, as it is well known how it can be dealt with for $\#P$ up to, say, 200000 (cf. [6; 8]).

Let

$$L[\gamma] = \exp((\gamma + o(1))\sqrt{\log n \log\log n}), \tag{3.3}$$

for a real number γ, and $n \to \infty$. With $B = L[1/2]$ it can be shown on loose heuristic grounds that both steps of the quadratic sieve algorithm can be completed in expected time $L[1]$.

Because in the original quadratic sieve only one polynomial f is used to generate all $> \#P$ relations, the interval of i-values to be inspected is rather large. Since $f(i) \approx 2i[\sqrt{n}]$ grows linearly with i, the probability of $f(i)$ being B-smooth decreases with increasing i. Davis [3] suggested using more than one polynomial, thus allowing a smaller i-interval per polynomial which should increase the yield (per i) of the sieving step. In our implementation we used Montgomery's version of this same idea [16]. Let P be as above, and let

$$f(i) = a^2 i^2 + bi + (b^2 - n)/(4a^2), \tag{3.4}$$

for integers a, b with $b^2 \equiv n \bmod 4a^2$. This requires n to be $1 \bmod 4$, which can be achieved by replacing n by $3n$ if necessary; in practice it might even be advantageous to use some other multiplier, cf. [16]. If $f(i)$ is B-smooth, then $v = (ai + b/(2a)) \bmod n$ satisfies (3.2), and B-smooth $f(i)$'s can again be found using a sieve once the roots of f modulo the primes in P are known. Thus, polynomials satisfying (3.4) can be used to generate relations efficiently. The expected run-time of the resulting factoring algorithm, however, is still $L[1]$.

3.5 Constructing Polynomials

(cf. [7; 16]). We show how polynomials as in (3.4) can be constructed. Let M be such that $f(i)$'s with $i \in [-M, M)$ will be tested for smoothness in the sieve. Let $a \equiv 3 \bmod 4$ be a probable prime with $a^2 \approx \sqrt{n/2}/M$ and Jacobi symbol $\left(\frac{a}{n}\right) = 1$. Since a is free of primes in P, the polynomial f has two roots modulo all primes in the factor base P. To find b such that $b^2 \equiv n \bmod 4a^2$, we first set $b = n^{(a+1)/4} \bmod a$ so that $b^2 \equiv n \bmod a$. Next we replace b by

$$b + a\big((2b)^{-1}((n - b^2)/a) \bmod a\big),$$

and finally, if b turns out to be even we replace b by $b - a^2$. It follows that $b^2 \equiv n \bmod 4a^2$, and $|f(i)| = O(i\sqrt{n})$ for $i \in [-M, M)$. Notice that the roots of $f \bmod p$ are $(-b \pm \sqrt{n})/(2a^2) \bmod p$, so that computation of the roots requires one inversion modulo p for all primes p in P; the values of $r_p \equiv (\sqrt{n}) \bmod p$ should be precomputed and stored in a table.

Another method to generate polynomials is presented in [13], and has the advantage that the roots of a polynomial modulo the primes in the factor base can be derived easily from the roots of the previous polynomial. We have no practical experience with this method.

In implementations of QS one usually also collects values of i for which $f(i)$ factors almost completely using the elements of P, i.e., except for one (or two) larger primes. If the large primes in these so-called *partial relations* match, they can be combined to form relations as in (3.2). In practice this enhancement leads to a speed-up factor of 4 to 6. For a detailed description of how the number of useful combinations among the partial relations can be counted, and how the combinations can be actually formed, we refer to [10].

3.6 The Sieving Step

Summarizing, for some fixed choice of factor base P, sieving bound M and $n \equiv 1 \bmod 4$, the QS-sieving step can be carried out by performing steps (a) through (h).

(a) For all primes p in P compute $r_p \equiv (\sqrt{n}) \bmod p$;

(b) Set $a_{\text{low}} = [\sqrt[4]{n/2}/\sqrt{M}]$;

(c) Compute the smallest $a > a_{\text{low}}$ that satisfies the requirements in (3.5), compute the corresponding b as in (3.5), and let f be as in (3.4);

(d) For all primes p in P compute the roots r_{p1} and r_{p2} of $f \bmod p$ as $(-b \pm r_p)/(2a^2) \bmod p$;

(e) For all integers i with $i \in [-M, M)$ set $s(i)$ to zero;

(f) For all primes p in P and $v = 1, 2$ replace $s(i)$ by $s(i) + [\log p]$ for all $i \in [-M, M)$ which are equal to r_{pv} modulo p (this is the actual sieving step);

(g) For all $i \in [-M, M)$ for which $s(i)$ is sufficiently close to the report bound $\log|f(i)|$, try to factor $f(i)$ using the elements of P, and store the resulting relations and partial relations (an i for which $s(i)$ is close to the report bound is called a *report*);

(h) If more relations and partial relations are needed, replace a_{low} by a and go back to step (c); otherwise terminate.

3.7 Practical Remarks

In step (f) one often does not sieve with the small primes in P, or replaces them by sufficiently large powers, to increase the speed. This lowers the number of reports in step (g), so that the report bound has to be lowered accordingly. For a 100-digit n, the factor base P will have approximately 50000 elements, and the largest element of P will be about $1.3 \cdot 10^6$. Because a small multiple of this largest prime is a good choice for M, several million $s(i)$'s have to be stored. Although each $s(i)$ is usually represented by a single byte (8 bits), the $s(i)$'s together might not fit in memory. In that case, the interval $[-M, M)$ is broken into smaller subintervals, which are processed consecutively as above. In practice the report bound $\log|f(i)|$ is often replaced by some appropriately chosen fixed bound.

3.8 Parallelization

The sieving step can easily be parallelized on any number of independent processors, by restricting each of them to a unique interval of candidate a-values, disjoint from the intervals assigned to other processors. Notice that two different identical processors that run the same sieving program and that started at the same time, each on its own interval of candidate a-values, are most likely to be at entirely different points in the program, even after a very short run: one processor might find a 'good' a-value earlier than the other in step (c), and thus begin earlier with the next steps, or one processor might find more reports in step (g) and spend more time on the trial divisions of the corresponding $f(i)$'s. Also at other points the precise instruction stream that gets executed might differ (for instance, in the inversions modulo p in step (d)), but these differences are minor compared to the entirely different things that might happen in steps (c) and (g). In a situation where several copies of (3.6) are processed simultaneously in SIMD-mode, this might lead to major inefficiencies, because the process that happens to be the slowest for a particular step sets the pace for that step. In the next section it is shown how these inefficiencies can be avoided at the expense of some memory.

4 Redistributing Data

In our SIMD-implementations of (3.6)(c) and (3.6)(g) we find ourselves in the following situation. We have a toroidal mesh M of m SIMD processing elements (PE's) that allows fast communication between each PE and its eight nearest neighbors. For the 16K MasPar described in Section 2, for instance, M would be the array of PE's, and m would be 16K. Furthermore, there is an inexpensive SIMD-process G, such that each PE running G has a fairly small probability p, independent from the other PE's, to generate a useful packet of information. These packets have to be processed by a time-consuming SIMD-process B. The goal is to process as many packets as possible, by repeating G and B indefinitely.

Clearly, since p is small it is quite inefficient to perform B right after G because then only the few PE's that have found a packet would be processing B. Fortunately, in our situation we can take advantage of the following.

(i) G can be repeated an arbitrary number of times before B is executed (i.e., packets do not have to be used immediately after they have been generated);

(ii) It is irrelevant for B on which PE a particular packet was found (i.e., packets may be generated and processed on different PE's);

(iii) It is not crucial that *all* packets that have been generated are also actually processed by B, but of course generating packets that will not be used leads to inefficiencies.

Using (i) we could keep a stack of packets per PE, and apply G until each stack contains at least one packet. At that point all m top of stack elements could be popped and processed by B, after which G is again applied, and so on. For small p and large m this approach would require rather large stacks on the PE's unless many packets are discarded, using (iii).

A better solution that uses much smaller stacks and that avoids discarding too many packets redistributes the packets after every application of G, thus making use of (ii) as well. There are many ways to do this; for us the following worked entirely satisfactorily.

4.1 Random Redistribution

On all m PE's simultaneously, do the following in SIMD-mode. Fix some arbitrary ordering N_1, N_2, ..., N_8 of the eight nearest neighbors (the same ordering for all PE's). For $i = 1, 2, \ldots, 8$, set $N = N_i$ and perform steps (a) and (b).

(a) Get the number of packets n on N's stack;

(b) If $n + 1$ is smaller than the number of packets on the PE's own stack, then perform steps (b1) through (b3);

(b1) Set e equal to the top of stack packet, and pop this packet from the stack;

(b2) Push e on the top of the stack of N;

(b3) Go back to step (a).

This approach resulted in the following behavior. Starting from empty stacks on all PE's it took on average $2/p$ applications of G (each followed by (4.1)) until none of the PE's had an empty stack. From that point on it takes, after each execution of B on all m PE's, on average $1/p$ applications of G (plus (4.1)), with a very small variance, before B can be applied again to process m packets. Except for the start-up stage, this is the best one could hope for.

Although we tried several orderings of the neighbors, we never noticed a significant difference in the performance. With stacks of at most 5 packets we occasionally lost packets but this introduced only a minor inefficiency. A simpler variant of (4.1) would be to remove the jump back to step (a) in the case that at least one packet has been moved to a neighbor. Similarly, G can be repeated a few times, before (4.1) is applied (with the jump). We have no experience with these simplifications, but we suspect that they work equally well. Notice that (4.1) uses only communication with nearest neighbors, with toroidal wraparound, which keeps communication costs to a minimum.

5 A SIMD Implementation of the QS-Sieving Step

Given the redistribution algorithm from the previous section, there are various ways to implement the QS-sieving step efficiently on a SIMD-machine, as long as the machine provides reasonably fast communication between neighbors. The simplest approach would be to let each PE generate polynomials as in (3.6)(c) (making sure that they try different a-values), until each PE got at least one polynomial (using a stack of polynomials and (4.1)), after which each PE performs steps (3.6)(d)-(g) using the polynomial it ended up with (on the top of its stack of polynomials). This works efficiently in SIMD mode and without further inter-processor communication, except for the trial divisions in (3.6)(g); the

$f(i)$'s, however, can again be redistributed using (4.1), so that trial division too can be performed on all PE's simultaneously.

Although these applications of (4.1) solve the synchronization problems caused by SIMD execution of (3.6)(c) and (3.6)(g), this approach is inefficient on the 16K MasPar, because there is not enough memory per PE to store a sufficiently large chunk of the interval $[-M, M)$. Furthermore, the roots in (3.6)(d) would have to be recomputed for each subinterval of $[-M, M)$ to be processed, because there is not enough memory on a PE to store them.

The opposite approach would be to process one polynomial at a time, and to spread the interval $[-M, M)$ over the PE's. This is a feasible approach if there if an ordering PE_0, PE_1, ..., PE_{m-1} such that PE_i and PE_{i+1} can communicate quickly (with indices modulo m). On the 16K MasPar this would not be impossible, but it would lead to a fairly small subinterval per PE with a very small hit-probability during the sieving step, unless the combined interval $[-M, M)$ is chosen exceedingly long.

One row of 128 PE's on a 16K MasPar has a total amount of memory of $128 \times 64\text{K}=$ 8MBytes, which is just about the right amount of memory to store the sieving interval $[-M, M)$ for the factorization of a 100-digit n. This observation suggests that on the 16K MasPar it might be a good idea to process 128 polynomials at a time, with each of the 128 rows of 128 processors taking care of one polynomial. This is the approach that we opted for. Since there will be no communication between processors in different rows, except for the redistributions (cf. (4.1)) during the polynomial generation and trial division steps, we restrict our description to what happens in a single row of 128 processors.

Let PE_0, PE_1, ..., PE_{127} be a row of 128 processors, such that PE_j and PE_{j+1} can communicate quickly (with indices modulo 128, i.e., with toroidal wraparound). We remove -1 and the small primes from the factor base P, choose the remaining P such that $\#P = 128 \cdot k$, for some integer k, and we partition P over the row of processors in such a way that each processor stores k primes (but see Remark (5.2)(f)). Furthermore, each processor contains k square roots of n modulo its k primes (the r_p from (3.6)(a)). Finally, we choose M such that $M = 64 \cdot L$ for some integer L, and we divide the sieving interval $[-M, M)$ over the processors in such a way that PE_j stores the length L interval $I_j = [(j-64)L, (j-63)L)$.

Suppose that we have repeatedly applied (3.6)(c) combined with (4.1) on the entire 128×128 processor array simultaneously until each of the 16K processors has a non-empty stack of polynomials. In particular, we suppose that each PE_j contains a unique polynomial f_j, for $0 \leq j \leq 127$. These polynomials are processed one at a time in 128 iterations. To process f_j processor PE_j first broadcasts f_j to the other processors in the row, so that all 128 processors share the same polynomial, after which the pipelined sieving from (5.1) is performed.

5.1 Pipelined Sieving

Suppose that PE_0 through PE_{127} all have the same polynomial f, represented by a and b as in (3.4). The values for m and l below are the same on all processors and can thus be taken care of by the ACU (cf. Section 2). Perform steps (a) through (c) on PE_0 through PE_{127} simultaneously.

(a) For all $i \in I_j$ set $s(i)$ to zero (cf. (3.6)(e));

(b) For $m = 1, 2, \ldots, k$ in succession (where $\#P = 128 \cdot k$), perform steps (b1) through (b5);

(b1) Let p be the mth prime on PE_j and r_p the corresponding squareroot of n (with different p's for different j's);

(b2) Compute r_{p1} and r_{p2} as the smallest integers $\geq (j-64)L$ which are equal to $(-b+r_p)/(2a^2)$ and $(-b-r_p)/(2a^2)$ modulo p, respectively (cf. (3.6)(d)), and compute $\tilde{r}_{p1}$ and $\tilde{r}_{p2}$ as the smallest integers $\geq -64L$ which are equal to r_{p1} and r_{p2} modulo p, respectively;

(b3) Set $l = 0$;

(b4) For $v = 1, 2$, as long as $r_{pv} < (j-63)L$, replace $s(r_{pv})$ by $s(r_{pv}) + [\log p]$ and next r_{pv} by $r_{pv} + p$ (cf. (3.6)(f));

(b5) Replace l by $l+1$. If $l < 128$, replace the 5-tuple $(p, r_{p1}, r_{p2}, \tilde{r}_{p1}, \tilde{r}_{p2})$ by the corresponding 5-tuple from the left neighbor (with wraparound), on PE_0 only replace r_{p1} and r_{p2} by $\tilde{r}_{p1}$ and $\tilde{r}_{p2}$, respectively, and return to step (b4);

(c) For all $i \in I_j$ for which $s(i)$ is sufficiently close to the report bound $\log|f(i)|$, push i, a, and b on the top of a stack of reports (cf (3.6)(g)).

This finishes the description of (5.1). Notice that in (5.1) 128 polynomials are processed simultaneously on 128 rows of processors. After every execution of (5.1) the elements of the stack built in step (c) will be redistributed using (4.1). As soon as all 16K processors have at least one report, the $f(i)$'s are computed and the actual trial divisions (using the original factor base including the small primes) are carried out on 16K processors simultaneously. And after 128 applications of (5.1) new polynomials are generated using (3.6)(c) and (4.1).

5.2 Practical Remarks

(a) In our implementation we removed the first 50 primes from P, to speed up the sieving step; the threshold is best determined experimentally.

(b) We gained considerable additional efficiency by distributing P over the PE_i in such a way that its smallest 128 elements are taken care of by $m = 1$ in step (b), the next smallest 128 by $m = 2$, etc.: as soon as for some m all p's are $> L$, at most one sieve location per PE has to be updated in step (b4), which means that for larger m simpler code can be used. An additional advantage of this approach is that $[\log p]$ in (5.1)(b4) can be replaced by a sufficiently close approximation which is the same over the entire array of processing elements, and which can be changed depending on the value of m. As usual, instead of $\log|f(i)|$ we used some fixed value for the report bound in (5.1)(c).

(c) In our implementation we only kept track of r_{p1}, $\tilde{r}_{p1}$ and $r_{p1}-r_{p2}$. Although this leads to slightly more complicated computations, it also resulted in fewer communications (a 4-tuple instead of a 5-tuple in (5.1)(b5)) and thus greater speed.
(d) If the interval $[-M, M)$ is too large, (5.1) can be applied repeatedly to subintervals of $[-M, M)$ simply by changing the definition of I_j appropriately (but see Remark (f) below). In this way we could decrease the memory requirements of our implementation considerably, at the expense of an acceptable slow-down. So, instead of the full 64K per PE, we normally use only about 28K, thus allowing other people to share the machine with us.
(e) In the trial division, the PE's simultaneously process the primes in the complete factor base to build a list of primes occurring in their $f(i)$. After that, each PE works on this list of primes to determine the multiplicities and the remaining factor. All PE's for which the remaining factor is 1, or a sufficiently small prime $> B$, or a sufficiently small composite, dump their trial division results to disk. The remaining composites (which lead to partial relations with two large primes) are factored by a separate program that runs independently on the front end.
(f) In the above description each processor stores k elements of P and the corresponding roots, where $\#P = 128 \cdot k$, and processors in the same column store the same subsets. We found it more efficient to distribute these subsets over the columns, so that processors store at most $[(k-1)/128]+1$ elements of P plus the corresponding roots. This leads to some extra communication (in Step (5.1)(b1)), but it makes more memory available for the sieve. If the sieve is broken into smaller subintervals, as suggested in Remark (d), then we either need $O(k)$ memory per processor to remember various useful values for the next subinterval, or these values have to be recomputed for each subinterval. This leads to loss of memory (which could have been used for the sieve) or loss of time. In particular for large $\#P$ we found it more efficient *not* to use Remark (d) at all, even though the sieve gets so small compared to $\max P$ that many primes do not even hit it once.

6 Example

The largest number we have factored using our SIMD QS implementation is the 110-digit number from the RSA-challenge list [15]:

$$\begin{aligned}\mathrm{RSA}(110) =& 35794234179\ 72586877499\ 18078325684\ 55403003778\ 02422822619\\ & 35329081904\ 84670252364\ 67741151351\ 61112045040\ 60317568667\\ =& 58464182144\ 06154678836\ 55318297916\ 23841986105\ 05601062333\cdot\\ & 61224210904\ 93547576937\ 03731756141\ 88412257585\ 54253106999\end{aligned}$$

At the time of writing, this is the largest number factored on a single machine using a general purpose factoring method. This factorization took approximately a month of computing time on the 16K MasPar, where we used only 40K of the

64K bytes per PE. The factor base consisted of approximately 80,000 primes, and the sieving interval $[-M, M)$ was broken into two consecutive pieces (cf. (5.2)(d)). We did not use the suggestions from Remark (5.2)(f) for this factorization; we later found out that they would have saved us quite some time. They were used, however, for the factorization of a 105-digit number in 5.3 CPU-days using 30K per PE. Extrapolation to a 110-digit number would give at most 20 CPU-days, still with 30K per PE.

References

1. Bernstein, D. J., Lenstra, A. K.: A general number field sieve implementation (to appear)
2. Caron, T. R., Silverman, R. D.: Parallel implementation of the quadratic sieve. J. Supercomputing **1** (1988) 273–290
3. Davis, J. A., Holdridge, D. B.: Factorization using the quadratic sieve algorithm. Tech. Report SAND 83-1346, Sandia National Laboratories, Albuquerque, NM, 1983
4. Dixon, B., Lenstra, A. K.: Massively parallel elliptic curve factoring. Advances in Cryptology, Eurocrypt'92, Lecture Notes in Comput. Sci. **658** (1993) 183–193
5. Gjerken, A.: Faktorisering og parallel prosessering (in norwegian), Bergen, 1992
6. Lenstra, A. K.: Massively parallel computing and factoring. Proceedings Latin'92, Lecture Notes in Comput. Sci. **583** (1992) 344–355
7. Lenstra, A. K., Lenstra, H. W., Jr.: Algorithms in number theory. Chapter 12 in: van Leeuwen, J. (ed.): Handbook of theoretical computer science. Volume A, Algorithms and complexity. Elsevier, Amsterdam, 1990
8. Lenstra, A. K., Lenstra, H. W., Jr., Manasse, M. S., Pollard, J. M.: The factorization of the ninth Fermat number. Math. Comp. **61** (1993) (to appear)
9. Lenstra, A. K., Manasse, M. S.: Factoring by electronic mail. Advances in Cryptology, Eurocrypt '89, Lecture Notes in Comput. Sci. **434** (1990) 355–371
10. Lenstra, A. K., Manasse, M. S.: Factoring with two large primes. Math. Comp. (to appear)
11. MasPar MP-1 principles of operation. MasPar Computer Corporation, Sunnyvale, CA, 1989
12. Pomerance, C.: Analysis and comparison of some integer factoring algorithms. 89–139 in: Lenstra, H. W., Jr., Tijdeman, R. (eds): Computational methods in number theory. Math. Centre Tracts **154/155**, Mathematisch Centrum, Amsterdam, 1983
13. Pomerance, C., Smith, J. W., Tuler, R.: A pipeline architecture for factoring large integers with the quadratic sieve algorithm. SIAM J. Comput. **17** (1988) 387–403
14. te Riele, H., Lioen, W., Winter, D.: Factorization beyond the googol with mpqs on a single computer. CWI Quarterly **4** (1991) 69–72
15. RSA Data Security Corporation Inc., sci.crypt, May 18, 1991; information available by sending electronic mail to challenge-rsa-list@rsa.com
16. Silverman, R. D.: The multiple polynomial quadratic sieve. Math. Comp. **84** (1987) 327–339

A New Elliptic Curve Based Analogue of RSA

N. Demytko

Telecom Australia Research Laboratories
770 Blackburn Road, Clayton, Victoria, 3168
Australia

Abstract. A new public key cryptosystem based on elliptic curves over the ring Z_n is described. The scheme can be used for both digital signature and encryption applications, does not expand the amount of data that needs to be transmitted and appears to be immune from homomorphic attacks. The main advantage of this system over other similar elliptic curve based systems is that there is very little restriction on the types of elliptic curves and types of primes (comprising the arithmetic modulus, n) that can be used. In addition, the system works on fixed elliptic curves. Problems associated with imbedding plaintext onto a curve are avoided by working within a multiple group structure. This enables the encryption and decryption operations to be performed on only the first coordinate of points on the given curve. The security of the system relies on the difficulty of factorising large composite numbers.

1 Introduction

An analogue of the Diffie-Hellman key exchange protocol [1] based on the use of elliptic curves was first proposed by Miller [2] in 1985. Elliptic curve based analogues of the ElGamal scheme and the Massey-Omura scheme followed in 1987 and are described in [3]. The first elliptic curve based analogue to the RSA scheme was introduced in 1991 [4]. Three trapdoor one-way functions (TOFs), based on elliptic curves over the ring Z_n, were proposed. The first class of function, denoted a "type 0" TOF, can only be used in a digital signature scheme, and not in a public key cryptosystem. The second, denoted a "type 1" TOF, has the commutative property and can be used for the same applications as RSA, however, its use restricts the types of primes (forming the arithmetic modulus) and the types of elliptic curves that can be used. The third class, denoted "type 2", is the Rabin generalisation of the type 1 scheme.

In this paper a new public key cryptographic scheme (or TOF) based on elliptic curves over a ring Z_n is proposed that overcomes most of the limitations of the schemes proposed in [4]. In common with RSA, security is based on the difficulty of factorising composite numbers formed by the product of two large primes (and not the discrete logarithm problem on elliptic curves on which the schemes presented in [2] and [3] are based). In the new scheme the message or plaintext is represented by the first (or x) coordinate of a point P = (x, y) on an elliptic curve, $y^2 \equiv x^3 + ax + b$ modulo n, with fixed parameters (a and b). Ciphertext, x_e, is produced by computing the first coordinate only of the point P multiplied by e (the encryption multiplier). The plaintext is recovered by computing the first co-ordinate only of the point eP = (x_e, y_e) multiplied by one of four possible decryption multipliers, d_i, i = 1 to 4 (assuming n is the product of two large primes, p and q). The appropriate value of d_i to be used is determined by the values of the two Legendre symbols $\left(\frac{w}{p}\right)$ and $\left(\frac{w}{q}\right)$ where $w \equiv x_e^3 + ax_e + b$ modulo n. Digital signatures are produced in a similar fashion. The Chinese Remainder Algorithm may also be used to reduce the computation time involved in the decryption procedure and in the production of digital signatures.

A brief review of the basic definitions and facts about elliptic curves over a finite field is given in Section 2. Section 3 introduces the concept of a "complementary" group on an elliptic curve over a finite field. The proposed encryption scheme and the rules used to compute the first coordinate of a point on an elliptic curve are described in Sections 4 and 5, respectively. Sections 6 and 7 summarise the encryption and digital signature schemes in terms of the first coordinate of a point on an elliptic curve. Finally, the basis for scheme's immunity to homomorphic attack is given in Section 8.

2 Elliptic Curves (mod p)

Let p be a prime, greater than 3, and let a and b be integers chosen such that

$$4a^3 + 27b^2 \not\equiv 0 \pmod{p}. \tag{1}$$

Then $E_p(a,b)$ denotes the elliptic group modulo p whose elements, (x,y), are pairs of non-negative integers less than p satisfying

$$y^2 \equiv x^3 + ax + b \pmod{p}, \tag{2}$$

together with a special (identity) element denoted ∞ and called the point at infinity. The operation on two points, P and Q, to produce a third point, R, is termed "addition" and is written as

$$P + Q = R. \tag{3}$$

If $P = (x_1,y_1)$ and $Q = (x_2,y_2)$, then $R = (x_3,y_3)$ is determined by the following rules:

$$x_3 \equiv \lambda^2 - x_1 - x_2 \pmod{p} \tag{4}$$

$$y_3 \equiv \lambda(x_1 - x_3) - y_1 \pmod{p} \tag{5}$$

where

$$\lambda \equiv \begin{cases} \dfrac{y_1 - y_2}{x_1 - x_2} & \text{if } x_1 \not\equiv x_2 \pmod p \\ \dfrac{3x_1^2 + a}{2y_1} & \text{if } x_1 \equiv x_2 \text{ and } y_1 \not\equiv -y_2 \pmod p \end{cases} \tag{6}$$

If Q is the identity element, then P + Q = Q + P = P.

If $x_1 \equiv x_2$ and $y_1 \equiv -y_2 \pmod p$, then $P + Q = \infty$, i.e., $P = -Q$ or $(x_2,-y_2) \equiv -(x_2,y_2) \pmod p$.

The order of the group, denoted $|E_p(a,b)|$, is given by:

$$|E_p(a,b)| = 1 + \sum_{x=1}^{p}\left(\left(\frac{z}{p}\right) + 1\right) \tag{7}$$

where $\left(\frac{z}{p}\right)$ is the Legendre symbol and $z \equiv x^3 + ax + b \pmod p$.

This equation is easy to verify by noting that, in addition to the point at infinity, for a given value of x:

(1) there are two values of y that correspond to that value of x, if z is a quadratic residue modulo p;

(2) there is one value of y that corresponds to that value of x, if $z \equiv 0$ modulo p; and

(3) there are no values of y that correspond to that value of x, if z is a quadratic non-residue modulo p.

A polynomial-time algorithm, due to Schoof [5], for computing the order of an elliptic group over a finite field exists. However, even though it is far more efficient than computing (7) directly, it is not practical for large p. Practical techniques for computing the order of an elliptic group modulo p, for large p with stated properties, are discussed in [6]. Two particular cases using these techniques are described in [7] and are as follows.

In the first case, if p is an ordinary prime which is congruent to 1 modulo 4, r is a complex prime that divides p and is congruent to 1 modulo 2 + 2i, and D is any integer not divisible by p then the order of $E_p(-D,0)$ is

$$|E_p(-D,0)| = p + 1 - \overline{\left(\frac{D}{r}\right)_4}\, r - \left(\frac{D}{r}\right)_4 \overline{r} \tag{8}$$

where $\left(\frac{x}{r}\right)_4$ is the fourth power symbol and $\overline{r}$ is the conjugate of the complex integer r.

For example, if p = 13 and r = 3 + 2i, then

$|E_{13}(-1,0)| = 14 - (1)(3 + 2i) - (1)(3 - 2i) = 8$
$|E_{13}(1,0)| = 14 - (-1)(3 + 2i) - (-1)(3 - 2i) = 20$
$|E_{13}(-2,0)| = 14 - (i)(3 + 2i) - (-i)(3 - 2i) = 18$
$|E_{13}(2,0)| = 14 - (-i)(3 + 2i) - (i)(3 - 2i) = 10.$

In the second case, if p is an ordinary prime which is congruent to 1 modulo 3, r is a cubic prime that divides p and is congruent to 2 modulo 3, and D is any integer not divisible by p then the order of $E_p(0,D)$ is

$$|E_p(0,D)| = p + 1 + \overline{\left(\frac{4D}{r}\right)_6 r} + \left(\frac{4D}{r}\right)_6 \overline{r} \tag{9}$$

where $\left(\frac{x}{r}\right)_6$ is the sixth power symbol and $\overline{r}$ is the conjugate of the cubic integer r.

For example, if p =13 and r = -4 - 3ω, where $\omega = e^{2\pi i/3}$, then

$|E_{13}(0,1)| = 14 + (\omega^2)(-4 - 3\omega) + (\omega)(-1 + 3\omega) = 12$
$|E_{13}(0,2)| = 14 + (-1)(-4 - 3\omega) + (-1)(-1 + 3\omega) = 19$
$|E_{13}(0,3)| = 14 + (1)(-4 - 3\omega) + (1)(-1 + 3\omega) = 9$
$|E_{13}(0,4)| = 14 + (\omega)(-4 - 3\omega) + (\omega^2)(-1 + 3\omega) = 21$
$|E_{13}(0,5)| = 14 + (-\omega^2)(-4 - 3\omega) + (-\omega)(-1 + 3\omega) = 16$
$|E_{13}(0,6)| = 14 + (-\omega)(-4 - 3\omega) + (-\omega^2)(-1 + 3\omega) = 7$

Note: It is well known that

$$|E_p(a,b)| = p + 1 + \alpha, \quad \text{where } |\alpha| \le 2\sqrt{p} \tag{10}$$

for every elliptic curve over F_p.

3 Complementary Group on a Given Elliptic Curve (mod p)

Definition: Let p be a prime, greater than 3, and, again, let a and b be integers chosen such that (1) holds. In addition, let $E_p(a,b)$ denote the elliptic group modulo p whose elements, (x,y), satisfy equation (2), as before, but where y is an indeterminant in the field F_p for non-negative integer values of x. That is, y is of the form $y \equiv u\sqrt{v}$ (mod p), where u is a non-negative integer less than p and v is a fixed quadratic non-residue modulo p. The identity element, ∞, and the "addition" operation are identical to those defined in the previous section.

It is easy to show that all group axioms hold for the above definition. For example, if $P = (x_1,y_1) = (x_1,u_1\sqrt{v})$ and $Q = (x_2,y_2) = (x_2,u_2\sqrt{v})$ are two elements in the group, then $R = (x_3,y_3) = (x_3,u_3\sqrt{v})$ is also in the group (closure), i.e.,

$$(x_1,y_1) + (x_2,y_2) \equiv (x_3,y_3) \pmod{p}, \tag{11}$$

where, if $x_1 \not\equiv x_2 \pmod p$,

$$x_3 \equiv \left(\frac{u_1 - u_2}{x_1 - x_2}\right)^2 v - x_1 - x_2 \pmod p \tag{12}$$

$$y_3 \equiv \left(\left(\frac{u_1 - u_2}{x_1 - x_2}\right)(x_1 - x_3) - u_1\right)\sqrt{v} \pmod p, \tag{13}$$

or, if $x_1 \equiv x_2$ and $y_1 \not\equiv -y_2 \pmod p$,

$$x_3 \equiv \left(\frac{3x_1^2 + a}{2u_1 v}\right)^2 v - x_1 - x_2 \pmod p \tag{14}$$

$$y_3 \equiv \left(\left(\frac{3x_1^2 + a}{2u_1 v}\right)(x_1 - x_3) - u_1\right)\sqrt{v} \pmod p \tag{15}$$

The order of this "complementary" group is given by

$$|\overline{E_p(a,b)}| = 1 + \sum_{x=1}^{p}\left(1 - \left(\frac{z}{p}\right)\right) \tag{16}$$

where $\left(\frac{z}{p}\right)$ is the Legendre symbol and $z \equiv x^3 + ax + b \pmod p$.

In this case, in addition to the point at infinity, for a given value of x:

(1) there are two values of y that correspond to that value of x, if z is a quadratic non-residue modulo p;

(2) there is one value of y that corresponds to that value of x, if $z \equiv 0$ modulo p; and

(3) there are no values of y that correspond to that value of x, if z is a quadratic residue.

Suppose there are A values of x for which $\left(\frac{z}{p}\right) = 1$, B values of x for which $\left(\frac{z}{p}\right) = 0$ and C values of x for which $\left(\frac{z}{p}\right) = -1$. In addition, since x must be in one of p possible residue classes,

$$A+B+C = p. \tag{17}$$

From (7) and (10),

$|E_p(a,b)| = 1 + 2A + B = 1 + p + \alpha$,

$$\text{i.e., } 2A + B = p + \alpha \tag{18}$$

Consequently, from (16), (17) and (18),

$$|\overline{E_p(a,b)}| = 1 + 2C + B = 1 + 2p - (2A + B) = 1 + p - \alpha \qquad (19)$$

4 Encryption Scheme

Select two primes, p and q, and let n = pq denote the arithmetic modulus. Select an elliptic curve with the parameters a and b where $\gcd(4a^3 + 27b^2, n) = 1$. Let $|E_p(a,b)| = 1+p+\alpha$, $|\overline{E_p(a,b)}| = 1+p-\alpha$, $|E_q(a,b)| = 1+q+\beta$ and $|\overline{E_q(a,b)}| = 1+q-\beta$. It is assumed that the order of these groups can be determined using the techniques referred to in Section 2. In addition, let x represent the plaintext and s the ciphertext (where $0 \leq x, s \leq n-1$).

Encryption is then defined as

$$(s,t) \equiv (x,y)\#e \pmod{n}, \qquad (20)$$

where (x,y)#e (or eP) denotes the point P = (x,y) "multiplied" by e. Multiplication of a point P by i is defined as the addition of the point P to itself i times.

Decryption is defined as

$$(x,y) \equiv (s,t)\#d_i \pmod{n}, \qquad (21)$$

where

$$e.d_i \equiv 1 \pmod{N_i}, \quad i = 1 \text{ to } 4, \qquad (22)$$

$$\gcd(e, N_i) = 1, \quad i = 1 \text{ to } 4, \qquad (23)$$

$$N_1 = \operatorname{lcm}(p+1+\alpha, q+1+\beta) \quad \text{if } \left(\frac{w}{p}\right) = 1 \text{ and } \left(\frac{w}{q}\right) = 1, \qquad (24)$$

$$N_2 = \operatorname{lcm}(p+1+\alpha, q+1-\beta) \quad \text{if } \left(\frac{w}{p}\right) = 1 \text{ and } \left(\frac{w}{q}\right) \neq 1, \qquad (25)$$

$$N_3 = \operatorname{lcm}(p+1-\alpha, q+1+\beta) \quad \text{if } \left(\frac{w}{p}\right) \neq 1 \text{ and } \left(\frac{w}{q}\right) = 1, \qquad (26)$$

$$N_4 = \operatorname{lcm}(p+1-\alpha, q+1-\beta) \quad \text{if } \left(\frac{w}{p}\right) \neq 1 \text{ and } \left(\frac{w}{q}\right) \neq 1, \qquad (27)$$

$$z \equiv x^3 + ax + b \pmod{n}, \qquad (28)$$

$$y = \sqrt{z}, \qquad (29)$$

$$w \equiv s^3 + as + b \pmod{n}, \text{ and} \qquad (30)$$

$$t = \sqrt{w}. \qquad (31)$$

Alternatively, the decryption time may be reduced, by a factor approaching 4, by computing (21) modulo p and modulo q and then combining the results via the Chinese Remainder Theorem.

Note that only the first coordinates, x and s, have to be computed in this scheme. Computation of the second coordinates, y and t, can be avoided using the rules and algorithm described in [7]. The rules are summarised in the following section.

Note also that if p, q, a and b are chosen so that $\alpha = \beta = 0$ in equations (24) to (27), then $N_i = \mathrm{lcm}(p+1, q+1)$ remains fixed for all i. Consequently, d_i is fixed for all i, and decryption is independent of the Legendre symbols, $\left(\frac{w}{p}\right)$ and $\left(\frac{w}{q}\right)$.

5 Rules for Computing the First Coordinate of a Point on an Elliptic Curve

In the elliptic group $E_p(a,b)$ (or $\overline{E_p(a,b)}$), let $(x_i,y_i) \equiv (x,y)\#i \pmod p$. If $y_i \not\equiv 0 \pmod p$, then

$$x_{2i} \equiv \frac{(x_i^2 - a)^2 - 8bx_i}{4(x_i^3 + ax_i + b)} \pmod p. \tag{32}$$

In addition, if $x_i \not\equiv x_{i+1}$ and $x \not\equiv 0 \pmod p$, then

$$x_{2i+1} \equiv \frac{(a - x_i x_{i+1})^2 - 4b(x_i + x_{i+1})}{x(x_i - x_{i+1})^2} \pmod p \tag{33}$$

Unfortunately congruence (33) cannot be used if $x \equiv 0$ modulo p (or q). However, it can be shown that the congruence can be rearranged to give

$$x_{2i+1} \equiv \frac{4b + 2(a - x_i x_{i+1})(x_i + x_{i+1})}{(x_i - x_{i+1})^2} - x \pmod p \tag{34}$$

which is valid for all $0 \le x \le p-1$ (and consequently for all $0 \le x \le n-1$ when computations are performed modulo n). It can be shown that x_i is never congruent to x_{i+1} modulo p (or q) during the course of computing $s \equiv x_e$ modulo n, as given by (20). Similarly s_i is never congruent to s_{i+1} modulo p (or q) during the course of computing (21). However, it is possible (although extremely unlikely) that y_i may become congruent to 0 modulo p (or q) during the course of computations and therefore for (32) to become undefined. The way around this problem is to use homogeneous coordinates and therefore avoid division until the final stage of the encryption or decryption procedure.

Homogeneous coordinates are formed by setting $x \equiv \frac{X}{Z}$ (mod p) and $y \equiv \frac{Y}{Z}$ (mod p). If $(x_i, y_i) \equiv (X_i/Z_i, Y_i/Z_i) \equiv (X/Z, Y/Z)\#i$ (mod p), computational rules (32) and (34) can be restated in the following form.

$$X_{2i} \equiv (X_i^2 - aZ_i^2)^2 - 8bX_iZ_i^3 \pmod{n} \tag{35}$$

$$Z_{2i} \equiv 4Z_i(X_i^3 + aX_iZ_i^2 + bZ_i^3) \pmod{n} \tag{36}$$

$$X_{2i+1} \equiv Z[4bZ_i^2Z_{i+1}^2 + 2(aZ_iZ_{i+1} + X_iX_{i+1})(X_iZ_{i+1} + X_{i+1}Z_i)] - X(X_iZ_{i+1} - X_{i+1}Z_i)^2 \pmod{n} \tag{37}$$

$$Z_{2i+1} \equiv Z(X_iZ_{i+1} - X_{i+1}Z_i)^2 \pmod{n} \tag{38}$$

6 The Encryption Scheme in Terms of the First Coordinate of a Point on an Elliptic Curve

Encryption and Decryption as defined in (20) and (21) can be rewritten in terms of the notation of Section 5 as:

$$s \equiv x_e \equiv X_e/Z_e \pmod{n} \qquad \text{where } X = x \text{ and } Z = 1\text{, and} \tag{39}$$

$$x \equiv s_{d_i} \equiv S_{d_i}/Z_{d_i} \pmod{n} \tag{40}$$

where $S = s$, $Z = 1$ and d_i is as defined by (22) to (31).

7 Digital Signature Scheme in Terms of the First Coordinate of a Point on an Elliptic Curve

A digital signature, s, is formed by computing:

$$s \equiv X_{d_i}/Z_{d_i} \pmod{n} \tag{41}$$

where $X = x$ is the message or plaintext, $Z = 1$ and d_i is as defined by (22) to (31) with $z \equiv x^3 + ax + b$ (mod n) replacing w in (24) to (27).

Signature verification is performed by computing:

$$x \equiv S_e/Z_e \pmod{n} \qquad \text{where } S = s \text{ and } Z = 1. \tag{42}$$

8 Homomorphic Attack

Let s_1 and s_2 represent two signatures produced for the messages x_1 and x_2 respectively. If it is possible to determine the second coordinates, t_1, t_2, y_1 and y_2, corresponding to the the above first coordinates, then it is possible to create a new signature, s, on a new message, x, by using the addition rules given in Section 2 (using modulo n rather than modulo p arithmetic), i.e., a new signature point:

$$(s,t) = (s_1,t_1) + (s_2,t_2) \tag{43}$$

can be computed that corresponds to the new message point:

$$(x,t) = (x_1,y_1) + (x_2,y_2). \tag{44}$$

In fact, only t_1 and t_2 need to be determined since x can be found, once s is known, by using congruence (42).

Values for t_1 and t_2 can be treated as indeterminants and computed using (30) and (31), i.e.,

$$t_1 = \sqrt{w_1} \quad \text{where} \quad w_1 \equiv s_1^3 + as_1 + b \pmod{n}$$

$$t_2 = \sqrt{w_2} \quad \text{where} \quad w_2 \equiv s_2^3 + as_2 + b \pmod{n}$$

The next step is to eliminate one of the indeterminants, say t_2. If t_2 is written in the form, $t_2 = u\sqrt{w_1}$, then s is given by (see Section 3):

$$s \equiv \left(\frac{1 - u}{s_1 - s_2}\right)^2 w_1 - s_1 - s_2 \pmod{n} \tag{45}$$

The only remaining problem is to determine the value of $u \equiv \sqrt{w_2/w_1} \pmod{n}$. However, it is impossible to find a square root modulo n unless the prime factors of n are known. Consequently, u cannot be determined from w_2 and w_1, even if w_2/w_1 is a quadratic residue modulo n (in most cases, an integer value of u will not exist, since w_2/w_1 will not be a quadratic residue modulo n). Thus, whilst it is possible to add a point to itself any number of times in this scheme, it is impossible to add two arbitrary points together if only the first coordinates of the points are known (unless the primes (p and q) comprising the arithmetic modulus, n, are also known). As a result, it appears that a new signature cannot be created from two old signatures. For the same reason, the active attack described in [4] will also not succeed.

9 Conclusions

A new public key cryptographic scheme based on elliptic curves over a ring Z_n has been proposed. The main advantage of the scheme is that it can be used on elliptic curves with arbitrary parameters. In addition, digital signatures can be produced that are of the same size as the message. Furthermore, the scheme does not appear to be prone to homomorphism attacks. Finally, the techniques used in this scheme can be employed to produce an elliptic curve analogue of the Pollard Rho method of factorisation.

10 Acknowledgement

The permission of the Director, Research, of Telecom Australia to publish this paper is hereby acknowledged.

11 References

[1] W. Diffie and M. Hellman, "New Directions in Cryptography", IEEE Transactions on Information Theory, Vol. 22, pp. 644-654, 1976.

[2] V.S. Miller, "Use of Elliptic Curves in Cryptography", Advances in Cryptology: Proceedings of CRYPTO '85, Lecture Notes in Computer Science, Vol. 218, pp. 417-426, Springer-Verlag, 1986.

[3] N. Koblitz, A Course in Number Theory and Cryptography, Spinger-Verlag, New York, 1987.

[4] K. Koyama, U.M. Maurer, T. Okamoto and S.A. Vanstone, "New Public-Key Schemes Based on Elliptic Curves over the Ring Zn", CRYPTO '91 Abstracts, Santa Barbara, CA, pp. 6-1 to 6-7, August 11-15, 1991.

[5] R.Schoof, "Elliptic Curves Over Finite Fields and the Computation of Square Roots mod p", Mathematics of Computation, Vol.44, No.170, pp. 483-494, 1985.

[6] A.K. Lenstra and H.W. Lenstra, Jnr., "Algorithms in Number theory", University of Chicago, Department of computer Science, Technical Report # 87-008, 1987.

[7] D.M. Bressoud, Factorisation and Primality Testing, Springer-Verlag, New York, 1989.

Weaknesses of a public-key cryptosystem based on factorizations of finite groups

Simon Blackburn * Sean Murphy
Information Security Group,
Royal Holloway and Bedford New College,
University of London,
Egham, Surrey TW20 0EX, U.K.

Jacques Stern
Laboratoire d'Informatique,
Ecole Normale Supérieure,
45, rue d'Ulm,
75230 Paris

Abstract. Recently, Qu and Vanstone have announced the construction of several new public-key cryptosystems based on group factorization. One of these was described at the last AUSCRYPT meeting [2]. We point out a serious weakness of this last system which makes it insecure. Our method only uses elementary algebra.

1 The proposed cryptosystem

Let G be a finite group. A *factorization* of G is a sequence $A_1, \cdots, A_s$ of subsets of G such that each element g of G can be expressed uniquely as a product

$$g = g_s g_{s-1} \cdots g_1$$

where $g_i \in A_i$.

The public-key cryptosystem described in [2] uses the additive group $G = Z_2^n$. Starting from a sequence $\alpha_1, \cdots, \alpha_n$ of generators, the authors build a sequence $G = G_0 > G_1 > \cdots > G_{n/2} = \{0\}$ of subgroups, where G_i is generated by $\{\alpha_{2i+1}, \alpha_{2i+2}, \alpha_{2i+3}, \cdots, \alpha_n\}$. Next, a complete set of coset representatives of G_i in G_{i-1}

$$\overline{A}_i = \{\overline{\alpha}[i,0], \overline{\alpha}[i,1], \overline{\alpha}[i,2], \overline{\alpha}[i,3]\}$$

is chosen, where

$$\overline{\alpha}[i,j] = j_1\alpha_{2i-1} + j_2\alpha_{2i} + \alpha_{i,j}$$

$j_1 j_2$ being the binary expansion of j and $\alpha_{i,j}$ a random element of G_i.

The family $\overline{A}_i$ is defined for $i \leq n/4 - 1$. For $n/4 \leq i \leq n/2 - 2$, a somehow similar construction is performed, with the difference that four complete sets

$$\overline{A}_{i,h} = \{\overline{\alpha}[i,0]_h, \overline{\alpha}[i,1]_h, \overline{\alpha}[i,2]_h, \overline{\alpha}[i,3]_h\}$$

are built instead of one.

* This author was supported by S.E.R.C. research grant GR/H23719

Finally, a one-one function f that maps the set $\{1, \cdots, n/4-1\}$ onto the set $\{n/4, \cdots, n/2-2\}$ is chosen and the public key $A_1, \cdots, A_{n/4-1}$ is defined where A_i is obtained by (randomly) ordering the set

$$\bigcup_{\substack{0 \le j \le 3 \\ 0 \le h \le 3}} \{\overline{\alpha}[i, h] + \overline{\alpha}[f(i), j]_h\}$$

as

$$\alpha[i, p] \qquad , \qquad 0 \le p \le 15$$

A message m of $n-4$ bits is encoded as follows: packing the bits 4 by 4, one obtains a sequence $m_1, \cdots, m_{n/4-1}$ of hexadecimal digits from which the ciphertext c is computed as:

$$c = \sum \alpha[i, m_i]$$

The secret data coming from the construction itself allow decoding: using the decomposition of c in the basis $\alpha_1, \cdots, \alpha_n$ successively gives, for each $i = 1 \cdots n/4-1$, the corresponding index h of the vector

$$\overline{\alpha}[i, h] + \overline{\alpha}[f(i), j]_h$$

chosen from A_i via m_i. But, once all the values h are known, the decomposition of c in the basis $\alpha_1, \cdots, \alpha_n$ also gives by an easy recursion on $k = n/4 \cdots n/2-2$ the missing part $\overline{\alpha}[k, j]_h$ of the vector

$$\overline{\alpha}[f^{-1}(k), h] + \overline{\alpha}[k, j]_h$$

chosen at level $f^{-1}(k)$. All this can be made quite efficient (see [2]). The value $n = 128$ is suggested for practical implementations.

2 Basic observations

In [2], it is stated that, although the cryptosystem is a kind of (modular) knapsack, methods using lattice reduction, such as the Lagarias-Odlyzko attack ([3]) do not apply. We agree with this opinion and we think it remains true even with recent improved versions of this attack such as [1]. Thus, we take another way.

Note that the group G used in section 1 is also a vector space: using Gaussian elimination over the field $Z/2Z$, it is easy to compute the dimension of a subspace of G generated by a given family X of vectors and to output a basis of this subspace.

Also, whenever a large family X is chosen in a subspace F of G, this family is quite likely to generate the entire subspace. For example, we have:

Theorem 1 *If P vectors are chosen independantly at random in a vector space of dimension K over the two element field, the probability that they do not generate the entire space is at most 2^{P-K}.*

The proof of this result is quite simple: given a subspace of dimension $k-1$, the probability that all choices remain in this subspace is at most $\frac{1}{2}^P$. But the number of such subspaces is exactly the number of non-zero linear functionals, i.e. $2^K - 1$.

From the public key of the above cryptosystem we define

$$A^i = \bigcup_{j>i} A_i$$

and we observe the following

Fact: ***With high probability, A^i generates the subspace G_i, provided i is not too large.***

Note that A^i contains $16(n/4 - i)$ vectors from the space G_i, which is of dimension $n-2i$. Even if the vectors from A^i are not really chosen independantly, the above theorem still gives a convincing estimate of the probability that A^i does not generate G_i, namely 2^{3n-14i}. For $n = 128$, this estimate remains below 2^{40} up to $i = 27$. This leaves out only four values.

Thus, it is fairly clear that some secret information leeks out. In the next section, we will see how to take advantage of this fact.

3 Cryptanalysis of the system

From section 2, we know that we can recover from the public data the sequence of subgroups G_i, for i not too large, say $i \leq i_0$. Our cryptanalysis include several steps.

Grouping the elements of A_i together, for i not too large. Although the elements of each A_i have been scrambled, it is possible to group together the elements

$$\overline{\alpha}[i,h] + \overline{\alpha}[f(i),j]_h$$

with the same h by using the equivalence relation

$$u \oplus v \in G_i$$

Since, the G_i's are known up to $i = i_0$, the grouping is properly recovered up to $i = i_0$ as well. We note that actually, whenever u and v are equivalent elements of A_i, the sum $u \oplus v$ belongs to $G_{n/4-1}$. This way, we can collect a fairly large family Y of elements of $G_{n/4-1}$.

Extending the method to the last few indices. For i between i_0 and $n/4 - 1$, we have not been able so far to compute accurately G_i because the sample A^i of elements of G_i was not large enough. Now, if we add to A^i the set Y that has been computed at the end of the last paragraph, then we see that we obtain a generating family for G_i. From this, we can also perform the correct grouping of A_i.

Recovering the secret permutation f. We work again with the equivalence relation on A_i defined above and, this time, we use the fact that, whenever u and v are equivalent elements of A_i, the sum $u \oplus v$ belongs to $G_{f(i)-1}$ (and not only to $G_{n/4-1}$ as was observed above). We let B_i be the set of sums $u \oplus v$ obtained from equivalent elements of A_i. Each B_i contains 24 elements. We define:

$$B^i = \bigcup_{j \neq i} B_i$$

fact:
i) if $f(i) = n/4$ then, with high probability, B^i generates the subspace $G_{n/4}$
ii) otherwise, with high probability, B^i generates the subspace $G_{n/4-1}$

This is because, we have, in each case a very large family of members of the corresponding space.

From the fact, it follows that we can recover both $f^{-1}(n/4)$ and $G_{n/4}$ by computing the dimension of all the spaces generated by the various families B^i. A recursive procedure will then achieve the same for $f^{-1}(n/4+j)$ and $G_{n/4+j}$: this procedure uses the same argument, the family B_i being restricted to those indices i for which $f(i)$ is not yet known (i.e. is $\geq n/4+j$).

Note that, at step j, we have $24(n/4-j-2)$ elements of a subspace of dimension $n/2-2j$ or $n/2-2j+2$. Using the estimate of theorem 1, we see that the probability of error remains quite small even for the last significant case ($j = n/4-3$), for which it is below 2^{-16}. Still, there is a slight chance that $f^{-1}(n/4+j)$ is not correctly computed for say the last two or three values. This issue will be addressed specifically when we turn to decoding.

Decoding. Given the ciphertext c, we first apply the following procedure:
for $i := 1$ *to* $n/4-1$ *do*
begin
pick any u *in* A_i *such that* $c \oplus u$ *belongs to* G_i
$u[i] := u$; *return* $u[i]$;
$c := c \oplus u$;
end;
return c

At the end of the procedure, we have reduced the posible choices of the unique element of A_i that contributes to the sum

$$c = \sum \alpha[i, m_i]$$

to a subset of each A_i consisting of the four elements equivalent to the vector $u[i]$ returned at step i. We denote this subset by $\tilde{A}_i$. Next, we apply the following.
for $i := n/4$ *to* $n/2-2$ *do*
begin
pick any v *in* $\tilde{A}_{f^{-1}(i)}$ *such that* $c \oplus u[i] \oplus v$ *belongs to* G_i
$v[i] := v$; *return* $v[i]$;
$c := c \oplus u$;
end

The value of $v[i]$ returned at step i is the vector $\alpha[i, m_i]$ of the sum

$$c = \sum \alpha[i, m_i]$$

This gives the plaintext m. Efficient implementations using decoding matrices can be implemented as in [2].

As observed above, there is a small chance that a mistake occurs for two or three values $i = f^{-1}(n/4-2)$, $i = f^{-1}(n/4-3)$, etc. This can be corrected by exhaustive search. Note that, since the mistake comes from the attack (and not from the ciphertext), the proper value of f can be recovered from a few decoding computations.

4 Conclusion

We have pointed out a serious weakness in the system proposed in [2]. Furthermore, we do not feel simple modifications of the system can restore its security. For example, it is quite possible to changer the order of the A_i's but the correct order can be recovered by computing dimensions with the same method we used to disclose f.

References

1. M. J. Coster, A. Joux, B. A. LaMacchia, A. M. Odlyzko, C. P. Schnorr and J. Stern, Improved low-density subset sum algorithms, *Computational Complexity*, to appear.
2. M.Qu, S. A. Vanstone, New public-key cryptosystems based on factorizations of finite groups, *AUSCRYPT'92*, preproceedings page 12.7-12.12.
3. J. C. Lagarias and A. M. Odlyzko, Solving low-density subset sum problems, *J. Assoc. Comp. Mach.* **32** (1985), 229-246.

Differentially uniform mappings for cryptography

KAISA NYBERG*
Institute of Computer Technology, Vienna Technical University

Abstract. This work is motivated by the observation that in DES-like ciphers it is possible to choose the round functions in such a way that every non-trivial one-round characteristic has small probability. This gives rise to the following definition. A mapping is called differentially uniform if for every non-zero input difference and any output difference the number of possible inputs has a uniform upper bound. The examples of differentially uniform mappings provided in this paper have also other desirable cryptographic properties: large distance from affine functions, high nonlinear order and efficient computability.

1. Introduction

The most successful and widely used block cipher has been the DES algorithm which was designed in the seventies. By the time, however, it has become too small, the key size being only 56 bits, and all attempts to increase the security by extending the DES by parallel or serial implementetations have failed.

In recent years the DES has been extensively analyzed in order to capture its properties of strength. Special attention has been focused on the nonlinearity properties of the round function, which is composed of permutations and eight small parallel substitution transformations, the S-boxes. It seems that the security can be increased only by increasing the size of the S-boxes or possibly by replacing the set of small parallel substitutions by one large transformation with desirable properties.

The necessary criteria for a substitution transformation or a round function of DES-like cipher include the following.

(i) High nonlinearity, large distance from linear functions;
(ii) High nonlinear order, the degrees of the outputbit functions are large;
(iii) Resistance against the differential cryptanalysis; and
(iv) Efficient construction and computability.

To satisfy requirement (iii) it is enough that for every fixed nonzero input difference to the function no output difference occurs with high probability. In other words, it is required that there is a uniform upperbound to the probability of the possible output differences. If this holds no strong characteristics for the success of the differential cryptanalysis exist as was proven in [7].

The purpose of this paper is to give examples of transformations of $GF(2^n)$ with properties (i) - (iv). Moreover, all these transformations are extendable in the sense that if parametrized by the length of the input, the complexity of the construction and implementation is polynomial but the security is exponential with key of linear length.

---

*Current address: Prinz Eugen-Straße 18/6, A-1040 Vienna, Austria
The work of the author on this project is supported by the MATINE Board, Finland

2. The resistance of DES-like ciphers against differential attacks

Let $(G, +)$ be a finite Abelian group, G' a subgroup of G, $F_i : G \to G'$ mappings, and $E_i : G' \to G$ such that $E_i - E_i(0)$ is an injective group homomorphism, $i = 1, 2, \ldots, r$. We define an r-round DES-like cipher over G as follows.

Given a plaintext $\mathbf{x} = (\mathbf{x}_L, \mathbf{x}_R) \in G' \times G'$ *and a key* $\mathbf{k} = (\mathbf{k}_1, \mathbf{k}_2, \ldots, \mathbf{k}_r) \in G^r$ *the ciphertext* $\mathbf{y} = (\mathbf{y}_L, \mathbf{y}_R)$ *is computed in r iterative rounds:*

Set $\mathbf{x}_L(0) = \mathbf{x}_L$ *and* $\mathbf{x}_R(0) = \mathbf{x}_R$ *and compute for* $i = 1, 2, \ldots, r$

$$\mathbf{x}(i) = (\mathbf{x}_L(i), \mathbf{x}_R(i)), \quad \textit{where}$$
$$\mathbf{x}_L(i) = \mathbf{x}_R(i-1) \quad \textit{and}$$
$$\mathbf{x}_R(i) = F_i(E_i(\mathbf{x}_R(i-1)) + \mathbf{k}_i) + \mathbf{x}_L(i-1).$$

Set $\mathbf{y}_L = \mathbf{x}_R(r)$ *and* $\mathbf{y}_R = \mathbf{x}_L(r)$.

An s-round *characteristic* $\chi = \chi(\boldsymbol{\alpha}(0), \ldots, \boldsymbol{\alpha}(s))$ of a DES-like cipher is a sequence of *differences* $\boldsymbol{\alpha}(i) = (\boldsymbol{\alpha}_L(i), \boldsymbol{\alpha}_R(i)) \in G' \times G'$, $i = 0, 1, \ldots, s$, such that $\boldsymbol{\alpha}_L(i) = \boldsymbol{\alpha}_R(i-1)$, $i = 1, 2, \ldots, s$.

Given $\mathbf{x}$, $\mathbf{x}^* \in G' \times G'$ and $\mathbf{k} = (\mathbf{k}_1, \ldots, \mathbf{k}_R) \in G^r$, $r \geq s$, we say that χ holds for $\mathbf{x}$ and $\mathbf{k}$ if $\mathbf{x}^* - \mathbf{x} = \boldsymbol{\alpha}(0)$ and

$$\mathbf{x}^*(i) - \mathbf{x}(i) = \boldsymbol{\alpha}(i),$$

for all $i = 1, 2, \ldots, s$, or what is the same,

$$\begin{aligned} &F_i(E_i(\mathbf{x}_R(i-1) + \boldsymbol{\alpha}_R(i-1)) + \mathbf{k}_i) - \\ &\qquad F_i(E_i(\mathbf{x}_R(i-1)) + \mathbf{k}_i) + \boldsymbol{\alpha}_L(i-1) = \boldsymbol{\alpha}_R(i) \end{aligned} \tag{1}$$

for all $i = 1, 2, \ldots, s$.

In what follows we find it convenient to consider a characteristic χ as a Boolean function of $\mathbf{x}$ and $\mathbf{k} = (\mathbf{k}_1, \ldots, \mathbf{k}_s)$ defined as follows

$$\chi(\mathbf{x}, \mathbf{k}) = \prod_{i=1}^{s} \chi_i(\mathbf{x}, \mathbf{k}_i)$$

where for $i = 1, 2, \ldots, s$ we set

$$\chi_i(\mathbf{x}, \mathbf{k}_i) = 1$$

if and only if (1) holds.

The differential cryptanalysis of iterated ciphers [1] [4] makes use of s-round characteristics to carry forward the information of a fixed input difference from the first round to the sth round *independently of the used key.* Given a plaintext pair $(\mathbf{x}, \mathbf{x}^*)$, *chosen by the cryptanalyst*, and the round keys $\mathbf{k}_1, \ldots, \mathbf{k}_s$, *unknown* *to the cryptanalyst*, a characteristic may or may not hold. The probability of the cryptanalyst's success, i.e., that a characteristic $\chi = \chi(\boldsymbol{\alpha}(0), \ldots, \boldsymbol{\alpha}(s))$ holds for a chosen plaintext pair $(\mathbf{x}, \mathbf{x}^*)$, $\mathbf{x} + \boldsymbol{\alpha}(0)$, is $P_K(\chi(\mathbf{x}, K) = 1)$, where the key $K = (K_i, \ldots, K_s) \in G^s$ is considered as a random variable. If the round keys $K_1, \ldots, K_s$ are independent then

$$P_K(\chi(\mathbf{x}, K) = 1) = \prod_{i=1}^{s} P_{K_i}(\chi_i(\mathbf{x}, K_i) = 1)$$

i.e., the probability of a characteristic is the product of the probabilities of its rounds.

To compute the one-round probabilities we assume that K_i is uniformly random. Let us denote

$$\boldsymbol{\alpha}'_R(i-1) = E_i(\boldsymbol{\alpha}_R(i-1)) - E_i(0).$$

Then

$$\begin{aligned} P_{K_i}(\chi(\mathbf{x}, K_i) = 1) =& P_{K_i}\{F_i(E_i(\mathbf{x}_R(i-1) + \boldsymbol{\alpha}_R(i-1)) + K_i) \\ & - F_i(E_i(\mathbf{x}_R(i-1)) + K_i) + \boldsymbol{\alpha}_L(i-1) = \boldsymbol{\alpha}_R(i)\} \\ =& P_{K_i}\{F_i(E_i(\mathbf{x}_R(i-1)) + K_i + \boldsymbol{\alpha}'_R(i-1)) \\ & - F_i(E_i(\mathbf{x}_R(i-1)) + K_i) + \boldsymbol{\alpha}_L(i-1) = \boldsymbol{\alpha}_R(i)\} \\ =& P_Z\{F_i(Z + \boldsymbol{\alpha}'_R(i-1)) - F_i(Z) + \boldsymbol{\alpha}_L(i-1) = \boldsymbol{\alpha}_R(i)\}, \end{aligned}$$

where $Z \in G$ is uniformly random. This shows, in particular, that the one-round probabilities are independent of $\mathbf{x}$. In the terminology of [4] this means that a DES-like cipher over an Abelian group is a Markov cipher with respect to the canonical difference.

Given $\mathbf{a}$, $\mathbf{b} \in G'$ let us denote

$$p_{F_i}(\mathbf{a}, \mathbf{b}) = P_Z\{F_i(Z + \mathbf{a}) - F_i(Z) = \mathbf{b}\}$$

where $Z \in G$ is uniformly random. Then

$$P_{K_i}(\chi(\mathbf{x}, K_i) = 1) = p_{F_i}(\boldsymbol{\alpha}'_R(i-1), \boldsymbol{\alpha}_R(i) - \boldsymbol{\alpha}_L(i-1)).$$

To prove resistance against differential cryptanalysis it suffices to show that for any given (or chosen) input pair $(\mathbf{x}, \mathbf{x}^*)$ the probability of guessing correctly the difference $\mathbf{x}^*(s) - \mathbf{x}(s)$, without any knowledge of the used key, is too small to be useful.

Given $\mathbf{x}$, $\boldsymbol{\alpha}$ and $\boldsymbol{\beta}$ the probability of getting $\mathbf{x}^*(s) - \mathbf{x}(s) = \boldsymbol{\beta}$ from $(\mathbf{x}^*, \mathbf{x})$ with $\mathbf{x}^* = \mathbf{x} + \boldsymbol{\alpha}$ is the sum of the probabilities of the different characteristics $\chi(\boldsymbol{\alpha}(0), \ldots, \boldsymbol{\alpha}(s))$ with $\boldsymbol{\alpha}(0) = \boldsymbol{\alpha}$, $\boldsymbol{\alpha}(s) = \boldsymbol{\beta}$, that is

$$\sum_{\substack{\boldsymbol{\alpha}(0) = \boldsymbol{\alpha} \\ \boldsymbol{\alpha}(s) = \boldsymbol{\beta}}} P_K\{\chi(\boldsymbol{\alpha}(0), \boldsymbol{\alpha}(1), \ldots, \boldsymbol{\alpha}(s-1), \boldsymbol{\alpha}(s))(\mathbf{x}, K) = 1\}.$$

Theorem 1 of [7] can be generalized to hold for a DES-like cipher over any Abelian group with different round functions. We have the following result.

THEOREM. *Let the round keys of a DES-like cipher be independent and uniformly random. Then for all* $\boldsymbol{\alpha}, \boldsymbol{\beta} \in G' \times G', \boldsymbol{\beta} \neq 0$ *and for all* $\mathbf{x} \in G' \times G'$

$$\sum_{\substack{\boldsymbol{\alpha}(0) = \boldsymbol{\alpha} \\ \boldsymbol{\alpha}(s) = \boldsymbol{\beta}}} P_K(\chi(\boldsymbol{\alpha}(0), \ldots, \boldsymbol{\alpha}(s))(\mathbf{x}, K) = 1) \leq 2(\max_{i,\mathbf{a},\mathbf{b},\mathbf{a}\neq 0} p_{F_i}(\mathbf{a}, \mathbf{b}))^2.$$

if $s \geq 4$. *If* $G' = G$ *and* F_i *is a permutation for all* $i = 1, 2, \ldots, r$, *then the estimate holds for* $s \geq 3$.

This motivates the following definition.

DEFINITION. *Let* G_1 *and* G_2 *be finite Abelian groups. A mapping* $F : G_1 \to G_2$ *is called differentially* δ*-uniform if for all* $\boldsymbol{\alpha} \in G_1$, $\boldsymbol{\alpha} \neq \mathbf{0}$, *and* $\boldsymbol{\beta} \in G_2$

$$|\{\mathbf{z} \in G_1 \mid F(\mathbf{z} + \boldsymbol{\alpha}) - F(\mathbf{z}) = \boldsymbol{\beta}\}| \leq \delta.$$

Now the result of the theorem can be stated as follows:

If the round functions of a DES-like cipher over G are differentially δ*-uniform and the round keys are independent and uniformly random then for every given input pair* $(\mathbf{x} + \boldsymbol{\alpha}, \mathbf{x})$, $\boldsymbol{\alpha} \neq 0$, *the average probability over the keys to obtain an output difference* $\boldsymbol{\beta} \neq 0$ *at the* s^{th} *round,* $s \geq 4$, *is less than or equal to* $2(\delta/|G|)^2$.

Examples of differentially 2-uniform mappings are the almost perfect nonlinear permutations of $GF(2^n)$ as defined in [7]. If $m - n$ output coordinates of a permutation of $GF(2^m)$ with property (P) (see [7]) are omitted the resulting mapping from $GF(2^m)$ to $GF(2^n)$ is differentially 2^{m-n+1}-uniform.

The purpose of this paper is to give other examples. The following two facts are useful.

PROPOSITION 1. *Let* $A : G_1 \to G_1$ *and* $B : G_2 \to G_2$ *be group isomorphisms and* $F : G_1 \to G_2$ *be differentially* δ*-uniform. Then* $B \circ F \circ A$ *is differentially* δ*-uniform.*

PROPOSITION 2. *Let* $F : G_1 \to G_2$ *be a differentially* δ*-uniform bijection. Then the inverse mapping of* F *is differentially* δ*-uniform.*

3. Power polynomials $F(\mathbf{x}) = \mathbf{x}^{2^k+1}$ in $GF(2^n)$ and their inverses

We shall first prove the following general results about the nonlinearity properties of power polynomial mappings.

PROPOSITION 3. *Let* $F(\mathbf{x}) = \mathbf{x}^{2^k+1}$ *be a power polynomial in* $GF(2^n)$ *and let* $s = \gcd(k, n)$. *Then* F *is differentially* 2^s*-uniform. If* $\frac{n}{s}$ *is odd, that is,* F *is a permutation, then the Hamming distance of the Boolean function* $f_{\boldsymbol{\omega}}(\mathbf{x}) = tr(\boldsymbol{\omega}F(\mathbf{x}))$ *from the set of linear Boolean functions is equal to* $2^{n-1} - 2^{\frac{n+s}{2}-1}$, *for all* $\boldsymbol{\omega} \in GF(2^n), \boldsymbol{\omega} \neq \mathbf{0}$.

PROOF: Given $\boldsymbol{\alpha}$, $\boldsymbol{\beta} \in GF(2^n)$, $\boldsymbol{\alpha} \neq \mathbf{0}$, the equation

$$(\mathbf{x} + \boldsymbol{\alpha})^{2^k+1} + \mathbf{x}^{2^k+1} = \boldsymbol{\beta} \tag{2}$$

has either zero or at least two solutions. Let $\mathbf{x}_1$ and $\mathbf{x}_2$ be two different solutions. Then

$$(\mathbf{x}_1 + \mathbf{x}_2)^{2^k} \boldsymbol{\alpha} + (\mathbf{x}_1 + \mathbf{x}_2)\boldsymbol{\alpha}^{2^k} = \mathbf{0}$$

or equivalently,

$$(\mathbf{x}_1 + \mathbf{x}_2)^{2^k-1} = \boldsymbol{\alpha}^{2^k-1}$$

from which it follows that

$$\mathbf{x}_1 + \mathbf{x}_2 \in \boldsymbol{\alpha}(G \setminus \{\mathbf{0}\})$$

where G is the subfield of $GF(2^n)$ of order 2^s. Hence given one solution $\mathbf{x}_0$ of (2) the set of all solutions is $\mathbf{x}_0 + \boldsymbol{\alpha}G$ of cardinality 2^s.

To prove the second part we make use of the technique of squared character sums. Let $\boldsymbol{\omega} \in GF(2^n)$, $\boldsymbol{\omega} \neq \mathbf{0}$ and denote the Walsh transform of $f_{\boldsymbol{\omega}}$ by $\widehat{F}_{\boldsymbol{\omega}}$. It suffices to show that

$$\max_{\mathbf{t} \in GF(2^n)} |\widehat{F}_{\boldsymbol{\omega}}(\mathbf{t})| = 2^{\frac{n+s}{2}}.$$

Let $\mathbf{t} \in GF(2^n)$. Then

$$\begin{aligned}(\widehat{F}_{\boldsymbol{\omega}}(\mathbf{t}))^2 &= \sum_{\mathbf{x} \in GF(2^n)} (-1)^{f_{\boldsymbol{\omega}}(\mathbf{x}) + \mathbf{t}\cdot\mathbf{x}} \sum_{\mathbf{y} \in GF(2^n)} (-1)^{f_{\boldsymbol{\omega}}(\mathbf{x}+\mathbf{y}) + \mathbf{t}\cdot(\mathbf{x}+\mathbf{y})} \\ &= \sum_{\mathbf{y} \in GF(2^n)} (-1)^{\mathbf{t}\cdot\mathbf{y}} \sum_{\mathbf{x} \in GF(2^n)} (-1)^{f_{\boldsymbol{\omega}}(\mathbf{x}+\mathbf{y}) + f_{\boldsymbol{\omega}}(\mathbf{x})}.\end{aligned}$$

Let $\mathbf{y} \neq \mathbf{0}$ and denote by $E_{\mathbf{y}}$ the range of the linear mapping

$$\mathbf{x} \mapsto F(\mathbf{x} + \mathbf{y}) + F(\mathbf{x}) + F(\mathbf{y}) = \mathbf{x}^{2^k}\mathbf{y} + \mathbf{y}^{2^k}\mathbf{x}.$$

Similarily as in the first part of the proof we see that the kernel of this linear mapping is $\mathbf{y}G$. Thus the dimension of the linear space $E_{\mathbf{y}}$ is $n - s$. For each $\mathbf{y} \neq \mathbf{0}$ either

$$tr(\boldsymbol{\omega}\boldsymbol{\beta}) = 0 \text{ for all } \boldsymbol{\beta} \in E_{\mathbf{y}}, \text{ or } \sum_{\boldsymbol{\beta} \in E_{\mathbf{y}}} (-1)^{tr(\boldsymbol{\omega}\boldsymbol{\beta})} = 0.$$

The vectors $\mathbf{y}$ for which $\mathrm{tr}(\boldsymbol{\omega}\boldsymbol{\beta}) = 0$ for all $\boldsymbol{\beta} \in E_{\mathbf{y}}$ or equivalently,

$$f_{\omega}(\mathbf{x}+\mathbf{y}) + f_{\omega}(\mathbf{x}) + f_{\omega}(\mathbf{y}) = \mathrm{tr}(\boldsymbol{\omega}(\mathbf{x}^{2^k}\mathbf{y} + \mathbf{x}\mathbf{y}^{2^k})) = 0$$

for all $\mathbf{x} \in GF(2^n)$, form a linear subspace Y of $GF(2^n)$. So we have

$$(\widehat{F}_{\boldsymbol{\omega}}(\mathbf{t}))^2 = 2^n + \sum_{\mathbf{y}\neq \mathbf{0}} (-1)^{\mathbf{t}\cdot\mathbf{y}+f_{\omega}(\mathbf{y})} 2^s \sum_{\boldsymbol{\beta}\in E_{\mathbf{y}}} (-1)^{\mathrm{tr}(\boldsymbol{\omega}\boldsymbol{\beta})}$$

$$= 2^n + 2^n \sum_{\mathbf{y}\in Y\setminus\{\mathbf{0}\}} (-1)^{\mathbf{t}\cdot\mathbf{y}+f_{\omega}(\mathbf{y})}$$

By definition of Y the function f_{ω} is linear on Y. Hence it remains to show that Y has 2^s elements.

Let $\mathbf{y} \in Y$. Then

$$\mathrm{tr}(\boldsymbol{\omega}\mathbf{y}\mathbf{x}^{2^k}) = \mathrm{tr}(\boldsymbol{\omega}\mathbf{y}^{2^k}\mathbf{x}) = \mathrm{tr}(\boldsymbol{\omega}^{2^k}\mathbf{y}^{2^{2k}}\mathbf{x}^{2^k})$$

for all $\mathbf{x} \in GF(2^n)$, which is equivalent to

$$\boldsymbol{\omega}\mathbf{y} = \boldsymbol{\omega}^{2^k}\mathbf{y}^{2^{2k}}$$

or, if $\mathbf{y} \neq \mathbf{0}$,

$$(\boldsymbol{\omega}F(\mathbf{y}))^{2^k-1} = \mathbf{1},$$

from which we get exactly $2^s - 1$ nonzero solutions $\mathbf{y}$, since F is assumed to be a permutation. This completes the proof.

If n is odd, $1 < k < n$ and $\gcd(n,k) = 1$, then the power polynomial $F(\mathbf{x}) = \mathbf{x}^{2^k+1}$ in $GF(2^n)$ is a differentially 2-uniform permutation. The public key cryptosystem C^* [5] is based on power polynomial permutations with $n = (2\ell+1)2^r$ and $k = b2^r$, $1 \leq b \leq \ell$. By Proposition 3 the coordinate functions of these polynomials are the more linear the larger r is. Finally notice that for $n = 2^m$ the polynomial $F(\mathbf{x}) = \mathbf{x}^{2^k+1}$ in $GF(2^n)$ is never a permutation.

The *degree* of a Boolean function f is the polynomial degree of the algebraic normal form of f and is denoted by $\deg(f)$. Let us denote by $\mathrm{w}_2(k)$ the 2-weight of a non-negative integer k. One proof of the following well-known result can be found in [2].

PROPOSITION 4. *Let $\boldsymbol{\omega} \in GF(2^n)$, $\boldsymbol{\omega} \neq \mathbf{0}$ and let $\mathbf{x} \mapsto \mathbf{x}^e$ be a permutation of $GF(2^n)$. Then*

$$\deg(\mathrm{tr}(\boldsymbol{\omega}\mathbf{x}^e)) = \mathrm{w}_2(e).$$

The permutations $\mathbf{x} \mapsto \mathbf{x}^{2^k+1}$ in $GF(2^n)$, n odd, satisfy properties (i), (iii) and (iv) but their output coordinate functions are only quadratic. Their inverses, however, have degrees linearly growing with n.

PROPOSITION 5. *Let n be odd, $gcd(n,k) = 1$ and $F(\mathbf{x}) = \mathbf{x}^{2^k+1}$. Then $F^{-1}(\mathbf{x}) = \mathbf{x}^{\ell}$, where*

$$\ell = \frac{2^{k(n+1)}-1}{2^{2k}-1} = \sum_{i=0}^{\frac{n-1}{2}} 2^{2ik} \mod (2^n-1)$$

with

$$w_2(\ell) = \frac{n+1}{2}.$$

PROOF:

$$\ell(2^k+1) = \sum_{i=0}^{\frac{n-1}{2}} 2^{(2i+1)k} + \sum_{i=0}^{\frac{n-1}{2}} 2^{2ik} \mod (2^n-1)$$

$$= \sum_{i=0}^{n} 2^{ik} \mod (2^n-1)$$

$$= \sum_{i=0}^{n} 2^{i} \mod (2^n-1)$$

$$= 2^{n+1}-1 \mod (2^n-1) = 1 \mod (2^n-1),$$

where the third equality follows from the fact that the mapping $i \mapsto ki$ permutes the integers modulo n if $\gcd(n,k) = 1$.

As a conclusion we list the following properties of the inverse of $F(\mathbf{x}) = \mathbf{x}^{2^k+1}$ in $GF(2^n)$ with n odd and $\gcd(n,k) = 1$.

(i) $\mathcal{N}(F^{-1}) = \min_{\boldsymbol{\omega}\neq\mathbf{0}} \min_{L \text{ lin.}} \min_{\mathbf{x}\in GF(2^n)} d(\mathrm{tr}(\boldsymbol{\omega}F^{-1}(\mathbf{x})), L(\mathbf{x})) = 2^{n-1} - 2^{\frac{n-1}{2}}$;

(ii) $\deg(\mathrm{tr}(\boldsymbol{\omega}F^{-1}(\mathbf{x}))) = w_2((2^k+1)^{-1} \mod (2^n-1)) = \frac{n+1}{2}$;

(iii) F^{-1} is differentially 2-uniform;

(iv) Using the fast exponentiation algorithm the computation of $F^{-1}(\mathbf{x})$ is of polynomial time requiring $\frac{n-1}{2}$ squarings and $\frac{n-1}{2}$ multiplications in $GF(2^n)$.

The first property follows from Theorem 1 of [6] which says that $\mathcal{N}(F^{-1}) = \mathcal{N}(F)$ and from Proposition 3.

4. The mapping $F(\mathbf{x}) = \mathbf{x}^{-1}$ in a finite field

Let $(\mathsf{F}, \cdot, +)$ be a finite field. Then the inversion mapping $F : \mathsf{F} \to \mathsf{F}$

$$F(\mathbf{x}) = \begin{cases} \mathbf{x}^{-1}, & \text{if } \mathbf{x} \neq \mathbf{0} \\ \mathbf{0}, & \text{if } \mathbf{x} = \mathbf{0} \end{cases}$$

is well defined.

PROPOSITION 6. *The inversion mapping is differentially 4-uniform in* $(\mathsf{F},+)$.

PROOF: Let $\boldsymbol{\alpha}$, $\boldsymbol{\beta} \in \mathsf{F}$ and $\boldsymbol{\alpha} \neq \mathbf{0}$ and consider the equation

$$(\mathbf{x}+\boldsymbol{\alpha})^{-1} - \mathbf{x}^{-1} = \boldsymbol{\beta}. \tag{3}$$

Assume that $\mathbf{x} \neq \mathbf{0}$ and $\mathbf{x} \neq -\boldsymbol{\alpha}$. Then (3) is equivalent to

$$\boldsymbol{\beta}\mathbf{x}^2 + \boldsymbol{\alpha\beta}\mathbf{x} - \boldsymbol{\alpha} = \mathbf{0}, \tag{4}$$

which has at most two solutions in F. If either $\mathbf{x} = \mathbf{0}$ or $\mathbf{x} = -\boldsymbol{\alpha}$ is solution to (3), then both of them are solutions and $\boldsymbol{\beta} = \boldsymbol{\alpha}^{-1}$. In that case (4) is equivalent to

$$\mathbf{x}^2 + \boldsymbol{\alpha}\mathbf{x} - \boldsymbol{\alpha}^2 = 0, \tag{5}$$

which may give two more solutions to (3).

Let us solve (5) in the special case $\mathsf{F} = GF(2^n)$. By squaring (5) and substituting $\mathbf{x}^2 = \boldsymbol{\alpha}\mathbf{x} + \boldsymbol{\alpha}^2$ we obtain

$$\mathbf{x}(\mathbf{x}^3 + \boldsymbol{\alpha}^3) = \mathbf{0},$$

which has no other solutions than $\mathbf{x} = \mathbf{0}$ or $\boldsymbol{\alpha}$ if $\gcd(3, 2^n - 1) = 1$, or equivalently, if n is odd. If n is even then 3 divides $2^n - 1$. Let $d = \frac{1}{3}(2^n - 1)$. Then there are two more solutions, $\mathbf{x} = \boldsymbol{\alpha}^{1+d}$ and $\mathbf{x} = \boldsymbol{\alpha}^{1+2d}$.

We list the following properties of the inversion mapping in $GF(2^n)$.

(i) $\mathcal{N}(F) = \min_{\boldsymbol{\omega}\neq\mathbf{0}} \min_{L \text{ lin.}} \min_{\mathbf{x}\in GF(2^n)} d(\mathrm{tr}(\boldsymbol{\omega}\mathbf{x}^{-1}), L(\mathbf{x})) \geq 2^{n-1} - 2^{\frac{n}{2}}$;

(ii) $\deg(\mathrm{tr}(\boldsymbol{\omega}\mathbf{x}^{-1})) = \mathrm{w}_2(2^n - 2) = n - 1$;

(iii) F is differentially 2-uniform if n is odd and it is differentially 4-uniform if n is even;

(iv) The Euclidean algorithm computes $\mathbf{x}^{-1}$ in polynomial time with respect to n.

Acknowledgements. The author's attention to the mapping $\mathbf{x} \mapsto \mathbf{x}^{-1}$ was drawn by C. Carlet. He observed that the high nonlinearity property (i) was actually proven in the work of Carlitz and Uchiyama **[3]**. L. R. Knudsen provided the author with examples demonstrating the difference between the odd and even case in (iii).

5. A mapping derived from the exponent mapping in a prime field

Let p be a prime and consider the Abelian group $G = \{0, 1, \ldots, p-1\}$ with the modulo p addition. Let $\mathbf{u}$ be an element of order q in the finite field $GF(p)$. We define a mapping $F : G \to G$ as follows:

$$F(x) = \mathbf{u}^x, \quad \text{for } x \in G,$$

where the exponentiation is computed in $GF(p)$.

Let $\alpha, \beta \in G$ and $\alpha \neq 0$. Then the equation

$$\mathbf{u}^{(x+\alpha) \bmod p} - \mathbf{u}^x = \beta \tag{6}$$

is equivalent to

$$\begin{cases} \mathbf{u}^{x+\alpha} - \mathbf{u}^x = \beta \text{ and } 0 \leq x \leq p - \alpha - 1 & (7) \\ \text{or} \\ \mathbf{u}^{x+\alpha-p} - \mathbf{u}^x = \beta \text{ and } p - \alpha \leq x \leq p - 1. & (8) \end{cases}$$

Since the solution x of

$$\mathbf{u}^{x+\alpha} - \mathbf{u}^x = \beta$$

is unique modulo q it follows that (7) has at most $\lceil \frac{p-\alpha}{q} \rceil$ solutions in G. Similarily equation (8) has at most $\lceil \frac{\alpha}{q} \rceil$ solutions in G. Consequently equation (6) has at most

$$\lceil \tfrac{p-\alpha}{q} \rceil + \lceil \tfrac{\alpha}{q} \rceil = \tfrac{p-1}{q} + 1$$

solutions in G. We have proved the following.

PROPOSITION 7. *Let F be the mapping from the set of integers modulo a prime p to itself as defined above using exponentiation and an element of order q in $GF(p)$. Then F is differentially $(\frac{p-1}{q} + 1)$-uniform.*

The mapping F defined in this section seems to be complex enough to be used as round function of a DES-like cipher over the integers modulo a prime with a small number of rounds. The computational complexity of such a cipher grows with the order of the base element $\mathbf{u}$. Proposition 7 shows the trade-off between the complexity of the enciphering (and deciphering) algorithm and the security against differential cryptanalysis.

6. Other security aspects.

Let us consider as an example a r-round DES-like cipher over $G = (GF(2^n), \oplus)$ with round functions $F_i(\mathbf{x}) = \mathbf{x}^{-1}$. From known plaintext-ciphertext pairs one gets polynomial equations of low degree (linear with the number of rounds) from which the round keys can be easily solved. The same is true if round functions $F_i(\mathbf{x}) = \mathbf{x}^3$ are used. Note the number of known plaintext-ciphertext pairs needed is constant with n. This number is at most linear with n if the inverses of $\mathbf{x} \mapsto \mathbf{x}^{2^k+1}$ are used as round functions.

However, the high nonlinear order of the inversion mapping and the inverses of $\mathbf{x}^{2^k+1}$ comes into effect if these mappings are combined with appropriately chosen linear or affine permutations which may vary from round to round and depend on the secret key. Hereby the virtues (i), (ii) and (iv) presented in §1 are not destroyed since they are linear invariants. By Proposition 1 the same is true for the differential uniformness that quarantees (iii).

An anonymous referee of this paper posed a natural question whether our approach is relevant to the situation where an attacker uses a notion of difference other than *xor* in his differential cryptanalysis attack. In our view, regarding DES-like ciphers, resistance against *xor*-differential analysis has no less crucial relevance as resistance against linear approximation.

Naturally, as well as a cryptanalyst may try any type of approximation he may try any type of differentials. Since all our examples of differentially uniform mappings in $GF(2^n)$ are multiplicative, we should consider differential cryptanalysis with respect to the multiplicative difference

$$\mathbf{x}^*\mathbf{x}^{-1}, \quad \text{for} \quad \mathbf{x}^*,\, \mathbf{x} \in GF(2^n).$$

Let us assume that F is a multiplicative permutation and A a linear permutation in $GF(2^n)$. Then for a DES-like cipher with $F \circ A$ as a round function, the probability of every one-round multiplicative differential with $\boldsymbol{\alpha} \neq \mathbf{1}$ is

$$\begin{aligned} &P_K\{F(A(\mathbf{x}\boldsymbol{\alpha} \oplus K))F(A(\mathbf{x} \oplus K))^{-1} = \boldsymbol{\beta}\} = \\ &P_K\{F((A(\mathbf{x}\boldsymbol{\alpha}) \oplus A(K))(A(\mathbf{x}) \oplus A(K))^{-1}) = \boldsymbol{\beta}\} = \\ &2^{-n}, \end{aligned}$$

since the mapping

$$\mathbf{z} \mapsto (\mathbf{a} \oplus \mathbf{z})(\mathbf{b} \oplus \mathbf{z})^{-1}$$

is a permutation in $GF(2^n)$ if $\mathbf{a} \neq \mathbf{b}$ and we set $(\mathbf{a} \oplus \mathbf{z})(\mathbf{b} \oplus \mathbf{z})^{-1} = \mathbf{1}$ for $\mathbf{z} = \mathbf{b}$.

Recent related work.

Some of the results of this paper were independently obtained by T. Beth and C. Ding. They present also more examples of almost perfect nonlinear permutations in their paper which is the next to follow in these proceedings.

References

1. E. Biham, A. Shamir, *Differential Cryptanalysis of DES-like Cryptosystems*, J. Cryptology **4** (1991).
2. C. Carlet, *Codes de Reed-Muller, codes de Kerdock et de Preparata, thesis*, Publication of LITP, Institut Blaise Pascal, Université Paris 6, **90.59** (1990).
3. L. Carlitz and S. Uchiyama, *Bounds for exponential sums*, Duke Math. J. **24** (1957), 37-41.
4. X. Lai, J. L. Massey and S. Murphy, *Markov Ciphers and Differential Cryptanalysis*, Advances in Cryptology - Eurocrypt '91. Lecture Notes in Computer Science **547**, Springer-Verlag (1992).
5. T. Matsumoto and H. Imai, *Public quadratic polynomial-tuples for efficient signature-verification and message-encryption*, Advances in Cryptology - Eurocrypt '88. Lecture Notes in Computer Science **330**, Springer-Verlag (1988).
6. K. Nyberg, *On the construction of highly nonlinear permutations*, Advances in Cryptology - Eurocrypt '92. Lecture Notes in Computer Science **658**, Springer-Verlag (1993).
7. K. Nyberg and L. R. Knudsen, *Provable Security Against Differential Cryptanalysis*, Proceedings of Crypto '92 (to appear).

On Almost Perfect Nonlinear Permutations

T. Beth and C. Ding

European Institute for System Security
University of Karlsruhe
P.O. Box 6980, Am Fasanengarten 5
D-W7500 Karlsruhe 1, Germany

abstract. In this paper basic properties of APN permutations, which can be used in an iterated secret-key block cipher as a round function to protect it from a differential cryptanalysis, are investigated. Several classes of almost perfect nonlinear permutations and other permutations in $GF(2)^n$ with good nonlinearity and high nonlinear order are presented. Included here are also three methods for constructing permutations with good nonlinearity.

1 Introduction

Many secret-key block ciphers are based on iterating a substitution function several times. Each iteration is called a round. Such a substitution function is refered to as the round function. The Security of such iterated block ciphers depends mainly on the "strength" of the round function. It is known that the nonlinearity of the round function is crutial for the security of an iterated block cipher.

There are two related concepts for nonlinearity of a substitution function: local nonlinearity and global nonlinearity. Let $f(x)$ be a substitution function of $GF(q)$, then

$$P_f(\alpha) = \max_{\gamma} P(f(X+\alpha) - f(X) = \gamma), \alpha \neq 0$$

is the measure of the local nonlinearity and

$$P_f = \max_{\alpha \neq 0} P_f(\alpha)$$

is that of the global nonlinearity of $f(X)$.

The linearity and nonlinearity of some cipher functions have been analyzed by Chaum and Evertse [3], Biham and Shamir [1, 2], Lai, Massey and Murphy [4], Nyberg and knudsen [5], Pieprzyk [6]. The differential cryptanalysis of DES-like functions introduced by Biham and Shamir [1,2] are closely related with the nonlinearity of cryptofunctions. As shown by Biham and Shamir [1,2], Lai, Massey and Murphy [4], Nyberg and Knudsen [5], to make an iterated block cipher immune to a differential cryptanalysis, it suffices to make the global nonlinearity P_f of the round function as small as enough.

For a round function f over $GF(2)^n$, the minimum value for P_f is 2^{1-n}. Permutations of $GF(2)^n$ with $P_f = 2^{1-n}$ were said to be almost perfect nonlinear(APN)[5]. In [5] Nyberg and Knudsen have given some results about quardatic APN permutations. In this paper basic properties of APN permutations in $GF(2)^n$ are developed in Section 2. Section 3 presents results about the relationships between permutations in $GF(2^n)$ and in $GF(2)^n$. Section 4 gives quadratic permutations with controllable nonlinearity. Section 5 provides with a class of permutations of order 3 with good nonlinearity. Section 6 presents a class of APN permutations in $GF(2)^n$ with maximum order $n-1$. Section 7 gives a class of permutations of order $n-2$ with controllable nonlinearity. Section 8 discusses the nonlinearity of the permutations X^d in $GF(2^n)$ with $d = 2^m - 1$.

2 Properties of APN Permutations

First of all, we would like to mention that the concept of nonlinearity is associated with a definite operation. In this paper we only discuss the nonlinearity of permutations in $GF(2^n)$ and in $GF(2)^n$ under the additions of them respectively.

From the definition of APN permutation, it is apparent that the following Lemma 1 holds:

Lemma 1 *Let $f(x)$ be a permutation of $GF(2)^n$ and $g(x,a) = f(x) + f(x+a)$. Then $f(x)$ is APN iff $g(x,a)$ takes exactly 2^{n-1} different nonzero vectors of $GF(2)^n$ and each of them two times when x runs over $GF(2)^n$ for each $a \neq 0$.*

It may be cryptographically beneficial to require that $g(x,a)$ takes each nonzero vector of $GF(2)^n$ equally likely, i.e. , $g(x,a)$ takes each vector of $GF(2)^n$ 2^n times when x runs over $GF(2)^n$ and a over $GF(2)^n - \{0\}$. We call such functions difference uniformly distributed (DUD). The $f(x)$ in the following Example 1 is APN, but not DUD. The permutation in Example 2 is APN and DUD.

Example 1 *Let $f(x) = (f_1, f_2, f_3)$ in $GF(2)^3$, where $f_1(x) = x_1 + x_2 + 1 + x_2x_3$, $f_2(x) = x_1 + x_3 + x_1(x_2 + x_3)$, $f_3(x) = x_2 + x_1x_3$.*

Example 2 *Let $f(x) = (f_1, f_2, f_3)$ in $GF(2)^3$, where $f_1(x) = x_1x_2 + x_1x_3 + x_1 + x_2$, $f_2(x) = x_1 + x_2x_3 + x_3$, $f_3(x) = 1 + x_1 + x_1x_2 + x_2x_3$.*

From Lemma 1 it follows that the following Theorem 1 holds:

Theorem 1 *Let $f(x) = (f_1(x), \cdots, f_n(x))$ be a permutation in $GF(2)^n$, then*
1) $f_i(x)$ is balanced, but not bent;
2) the order $ord(f_i) \leq n-1$;
3) if $f(x)$ is APN (DUD), then $h(x) = f(Ax+b)$ is also APN (DUD) for each nonsingular $n \times n$ matrix A over $GF(2)$ and each b in $GF(2)^n$.

Before to present a characterization of APN permutations, we need the following definitions and results:

Definition 1 *Let S be a subset of $GF(2)^n$-$\{0\}$. If for any $a \neq 0$, $b \neq 0$, $a+b \neq 0$, a, $b \in GF(2)^n$, there is at least one of the a, b, $a+b$, which belongs to S, then we call S a differential representation set of $GF(2)^n$. A differential representation set of $GF(2)^n$ such that $|S|$ is minimal, is called a differential basis of $GF(2)^n$, and $|S|$ is called the differential dimension (DD).*

Theorem 2 *A subset S of $GF(2)^n$ is a differential representation set of $GF(2)^n$ iff the difference of any two distinct elements of the set $GF(2)^n - S - \{0\}$ belongs to S.*

Theorem 3 *The differential dimension (DD) of $(GF(2)^n, +)$ is $2^{n-1}-1$ and the set $E = \{x : W_H(x) \text{ even}, x \in GF(2)^n, x \neq 0\}$ is a differential basis of $GF(2)^n$, where $W_H(x)$ denotes the Hamming weight of the vector x.*

Proof: Let S be any differential representation set of $GF(2)^n$, $|S| = k$, then $|S'| = |GF(2)^n - S - \{0\}| = 2^n - k - 1$. Suppose that $S' = \{s_1, \cdots, s_{2^n-k-1}\}$, then the elements s_1+s_2, s_1+s_3, $\cdots$, $s_1+s_{2^n-k-1}$, are distinct and all belongs to S, therefore we have $k \geq 2^n - k - 2$, which is equivalent to $k \geq 2^{n-1}-1$. It is clear that E is a representation set of $GF(2)^n$ and $|E| = 2^{n-1}-1$.

Theorem 4 *Let $D = \{d_1, \cdots, d_{2^{n-1}-1}\}$ be any differential basis of $GF(2)^n$, and $f(x)$ be a permutation in $GF(2)^n$. Then $f(x)$ is APN iff for each i, $g(x, d_i) = f(x) + f(x+d_i)$ takes exactly 2^{n-1} different nonzero vectors of $GF(2)^n$ and each two times.*

Proof: By definition the necessity is natural. What remains to be proved is the sufficiency. For any $a \neq 0$, $x \neq y$ and $x+y \neq a$, let $b = x+y$, then $b \neq a$, $b \neq 0$. Noticing that

$$\begin{aligned} d &= [f(x)+f(x+a)] + [f(y)+f(y+a)] \\ &= [f(x)+f(y)] + [F(x+a)+f(y+a)] \\ &= [f(x)+f(y+a)] + [f(x+a)+f(y)] \end{aligned}$$

and that there is at least one of the elements in $\{a, b, a+b\}$ that belongs to D, we have $d = 0$. This proves the sufficiency.

From Theorem 4 we see that $f(x)$ is APN iff for each nonzero vector e of even Hamming weight, $g(x, e)$ takes 2^{n-1} different nonzero vector of $GF(2)^n$. This result reduces largely the operation in searching for APN permutations.

Theorem 5 *Let $f(x) = (f_1(x), \cdots, f_n(x))$ be a APN permutation, then none of $f_1, \cdots, f_n$, is affine.*

Proof: Suppose that $f_1(x) = b_{1n}x_n + \cdots + b_{11}x_1 + b_0$, then

$$f_1(x) + f_1(x+c) = \sum_{i=1}^{n} b_{1i}c_i,$$

so we can find a vector $c \neq 0$ such that $f_1(x) + f_1(x+c) = 0$. whence

$$f(x) + f(x+c) = (0, f_2(x) + f_2(x+c), \cdots, f_n(x) + f_n(x+c)).$$

To ensure that $f(x)+f(x+c)$ takes 2^{n-1} distinct vectors of $GF(2)^n$, there must exist a vector x such that

$$f(x) + f(x+c) = (0, \cdots, 0).$$

This is contrary to the one-to-one property of $f(x)$. This completes the proof.

This theorem demonstrates that each component function of a APN permutation can not be affine. In what follows in this section, we shall discusss the nonlinear terms $x_i x_j (i \neq j)$ of APN permutations.

Theorem 6 *Let $f(x) = (f_1(x), \cdots, f_n(x))$ be a APN permutation of $GF(2)^n$. Then every quadratic term $x_i x_j (i \neq j)$ must appear in at least one of the component functions $f_1, \cdots, f_n$.*

Proof: For $c, x \in GF(2)^n$, let $x^c = 0$ when $x \neq c$, and $x^c = 1$ otherwise. Therefore $f(x)$ can be expressed as

$$\begin{aligned} f(x) &= \sum_{c \in GF(2)^n} x^c f(c) = \prod_{i=1}^{n} x_i \sum_{c \in GF(2)^n} f(c) \\ &+ \sum_{i=1}^{n-1} \sum_{1 \leq k_1 \leq \cdots \leq k_t \leq n} \Big(\prod_{j \neq k_1, \cdots, k_t} x_j \Big) \sum_{c} c'_{k_1} \cdots c'_{k_t} f(c) \end{aligned}$$

where $c'_i = 1 + c_i$. Without the loss of generality, we consider the cofficients of the term $x_{n-1} x_n$, which is

$$f(0 \cdots 001) + f(0 \cdots 000) + f(0 \cdots 010) + f(0 \cdots 011),$$

not equal to zero vector by the definition of APN permutations. This proves the theorem.

The nonlinear oder of a permutation $f(x) = (f_1(x), \cdots, f_n(x))$ is defined as

$$ord(f) = \max_{1 \leq i \leq n} ord(f_i),$$

where $ord(f_i)$ is the nonlinear order of $f_i(x)$. Theorem 1 means that the maximum nonlinear order of a APN permutation in $GF(2)^n$ is $n-1$. This upper bound is achievable (see Example 1 and 2, also Section 6). Theorem 6 tell us that any APN permutation must be depentent of all the quadratic terms, this may mean that the most important terms of a APN permutation are the quadratic ones.

3 The Nonlinearity of Permutations in $GF(2)^n$ and in $GF(2^n)$

If $f(x_1,\cdots,x_n) = (f_1(x),\cdots,f_n(x))$ is a permutation of $GF(2)^n$, let $B = \{\alpha_1,\cdots,\alpha_n\}$ be any basis of $GF(2^n)$ over $GF(2)$, then

$$F(X) = \sum_{i=1}^{n} f_i(x_1,\cdots,x_n)\alpha_i \tag{1}$$

is a permutation in $GF(2^n)$, and vice versa, where $X = \sum x_i\alpha_i \in GF(2^n)$. So there is an one-to-one correspondence between the permutations of $GF(2)^n$ and those of $GF(2^n)$ under a chosen basis of $GF(2^n)$ over $GF(2)$. We denote here and hereafter the permutation $f(x) = (f_1(x),\cdots,f_n(x))$ in (1) as $[F(X)]_B$.

For odd n, let $\{\alpha_1^*,\cdots,\alpha_n^*\}$ be the dual basis of B, then each component of $f(x)$ can be expressed as

$$f_i(x) = Tr(F(X)\alpha_i^*), \tag{2}$$

where $X = \sum x_i\alpha_i$.

The following result about the nonlinearity of the function $F(X)$ and $f(x)$ in (1) is obviously true, but is the theoretical foundation for constructing permutations in $GF(2)^n$ with good nonlinearity from those in $GF(2^n)$.

Theorem 7 *Let $B = \{\alpha_1,\cdots,\alpha_n\}$ be a basis of $GF(2^n)$ over $GF(2)$, $x = (x_1,\cdots,x_n)$, $y = (y_1,\cdots,y_n)$, $a = (a_1,\cdots,a_n)$, $b = (b_1,\cdots,b_n) \in GF(2)^n$, and $X = \sum x_i\alpha_i$, $Y = \sum y_i\alpha_i$, $A = \sum a_i\alpha_i$, $B = \sum b_i\alpha_i \in GF(2^n)$, then*
1) $P(F(X)+F(Y) = A \mid X+Y = B) = P(f(x)+f(y) = a \mid x+y = b)$;
2) $P_F(B) = P_f(b)$;
3) $P_F = P_f$;
4) $P_F = P_{F^{2^i}}$ for each integer i.

This theorem showes that the global and local nonlinearity of $F(X)$ and $f(x)$ is the same.

Theorem 8 *Let $f(x) = (f_1(x),\cdots,f_n(x))$ be a APN (DUD) permutation in $GF(2)^n$, then for each nonsingular $n \times n$ matrix A over $GF(2)$, $g(x) = (f_1(x),\cdots,f_n(x))A$ is also APN (DUD).*

The above Theorem 8 is useful in constructing APN permutations. Two permutations $f(x)$ and $g(x)$ in $GF(2)^n$ are said to be linearly equivalent if there are a nonsingular $n \times n$ matrix A over $GF(2)$ and a vector b in $GF(2)^n$ such that $f(x) = g(Ax+b)$.

Let $f(x) = [F(X)]_B$. For the changing of the basis, let $B' = \{\beta_1,\cdots,\beta_n\}$ be another basis of $GF(2^n)$ over $GF(2)$, $f'(x) = [F(X)]_{B'}$ and

$$(\beta_1,\cdots,\beta_n) = (\alpha_1,\cdots,\alpha_n)A^t, \tag{3}$$

then A is nonsingular and

$$f'(x) = (f_1(xA), \cdots, f_n(xA))A^{-1} \tag{4}$$

This result showes that the permutations deduced from a permutation in $GF(2^n)$ by changing the basis are usually not linear equivalent.

We now consider the conjugacy class of $Z^*_{2^n-1}$ mod $(2^n - 1)$. A conjugacy class C_k is the set $\{k2^i \bmod (2^n - 1), i = 0, 1, \cdots, \}$. Theorem 7 tell us that $P_F = P_{F^{2^i}}$ for any permutations in $GF(2^n)$, so we can construct a class of permutations with good nonlinearity, provided that we can construct one.

It is well known that X^d is a permutation of $GF(2^n)$ iff $\gcd(d, 2^n - 1) = 1$. In the following sections we investigate mainly the permutations X^d in $GF(2^n)$ with good nonlinearity. Before doing so, we need the following result about the nonlinear order of $[X^d]_B$, which was proved in [8] according to the citation of [7] (So we have deleted here our original proof).

Theorem 9 *Let B be a basis of $GF(2^n)$ over $GF(2)$, and d an integer, then $ord([X^d]_B) = W_H(d)$, where $W_H(d)$ is the Hamming weight of the binary representation of the integer d.*

4 Quadratic Permutations with Controllable Nonlinearity

In [5] Nyberg and Knudsen have studied the permutations f in $GF(2^m) = GF(2^d)^n$ which satisfy the property that every nonzero linear combination of the components of f is a balanced quadratic form x^tCx in n indeterminates over $GF(2^d)$ with $\text{rank}(C + C^t) = n - 1$. We now present a general result about the quadratic APN permutations.

Theorem 10 *Let $f(x) = (f_1, \cdots, f_n)$ be a permutation in $GF(2)^n$, where*

$$f_l(x) = \sum_{1 \leq i < j \leq n} a_{ij}^{(l)} x_i x_j + \sum_{i=1}^{n} b_i^{(l)} x_i + b_0^{(l)}, \quad 1 \leq l \leq n.$$

Setting the entries $a_{ij}^{(l)}$ of the matrix A_l as 0 when $i = j$, as $a_{\min\{i,j\}\max\{i,j\}}$ otherweise, then $f(x)$ is APN iff $\text{rank}(A_1 w^t, \cdots, A_n w^t) = n - 1$ for each $w \neq 0$.

Proof: Let

$$g_l(x, w) = f_l(x) + f_l(x + w) = xA_l w^t + \sum_{1 \leq i < j \leq n} a_{ij}^{(l)} w_i w_j + \sum_{i=1}^{n} b_i^{(l)} w_i$$

$$= xA_l w^t + f_l(w) + f_l(0).$$

For each $w \neq 0$, the set of linear equations

$$(g_1(x, w), \cdots, g_n(x, w)) = (d_1, \cdots, d_n) \neq 0 \tag{5}$$

has no solution or only two solutions iff $\text{rank}(A_1 w^t, \cdots, A_n w^t) = n - 1$ for each $w \neq 0$. This proves the theorem.

If we denote $f_l(x)$ as $f_l(x) = xC_l x^t + f_l(0)$, then $A_l = C_l + C_l^t$. Therefore the result presented here seems to be different from the one in [5]. From the forgoing proof it follows that the following Corollary 1 holds:

Corollary 1 *Let the symbols and notations as in Theorem 10. If*

$$\max_{w \neq 0} rank(A_1 w^t, \cdots, A_n w^t) = k,$$

then $P_f \leq 2^{-k}$.

Theorem 11 *Let* $d = 2^l(2^k + 1)$, $\gcd(d, 2^n - 1) = 1$ *and* $m = \gcd(2^n - 1, 2^k - 1)$, *B be any basis of* $GF(2^n)$ *over* $GF(2)$, $f(x) = [X^d]_B$, *then* $P_f \leq (m+1)/2^n$.

Proof: Because of Theorem 7 it suffices to prove the case $d = 2^k + 1$. Let

$$G(X, \beta) = X^d + (X + \beta)^d = X^{2^k}\beta + X\beta^{2^k} + \beta^{2^k+1} = \alpha \tag{6}$$

Noticing that $G(X, \beta)$ is a linearized function of X, we need only to consider the number of solutions of the equation

$$\beta X^{2^k} + \beta^{2^k} X = 0, \tag{7}$$

which is equivalent to $X = 0$ and $(X\beta^{-1})^{2^k - 1} = 1$.

Setting $H = \{x : x^u = 1, x \in GF(2^n)\}$, we see that H is a subgroup of the cyclic group $GF(2^n)^*$, so it is also cyclic, say $H = (h)$, then it is obvious that $h^m = 1$. Hence oder(h) divides m. It follows that the number of solutions of equation (7) is at most $m + 1$, so is that of equation (6). This proves the theorem.

Corollary 2 *Let* $\gcd(2^k + 1, 2^n - 1) = 1$, *then the permutation* $[X^{2^l(2^k+1)}]_B$ *is APN iff* $\gcd(k, n) = 1$.

Proof: The permutation $[X^{2^l(2^k+1)}]_B$ is APN iff $m = \gcd(2^k - 1, 2^n - 1) = 1$, which is equivalent to $\gcd(k, n) = 1$.

For odd n the result of Corollary 2 has been proved in [5]. We get here a general result without requiring n being odd. On the other hand, it is apparent that $ord([X^{2^l(2^k+1)}]_B) = 2$.

5 A Class of Permutations of Order 3 with Good Nonlinearity

For a quadratic APN permutation $f = (f_1, \cdots, f_n)$ in $GF(2)^n$, it is not difficult to see that each $f_i(x)$ has a linear structure, i.e. , there is a vector w such that $f_i(x) + f_i(x + w) = f_i(w) + f_i(0)$. This may be a cryptographical demerit. In this sense it is important to construct permutations which have good nonlinearity and high nonlinear order. In this section we present a class of permutations of order 3 with good nonlinearity.

Theorem 12 *Let* $\gcd(3, 2^n - 1) = 1$, $d = 2^{i+2} + 2^{i+1} + 2^i$, *and* $i \geq 0$, B *a basis of* $GF(2^n)$ *over* $GF(2)$ *and* $f(x) = [X^d]_B$, *then* $ord(f) = 3$ *and* $P_f = 2^{1-n}$ *or* $3 * 2^{1-n}$.

Proof: Because of Theorem 7 and $d = 7 * 2^i$, it suffices to prove the case $d = 7$. Let

$$G(X, \beta) = X^d + (X + \beta)^d, \beta \neq 0, \tag{8}$$

then $G(X, \beta) = \alpha$ is equivalent to

$$Y^d + (Y + 1)^d = \gamma, \tag{9}$$

where $Y = X/\beta$, $\gamma = \alpha\beta^{-d}$. If $\gamma = 1$, then equation (9) is equivalent to $Y(Y^6 - 1) = 0$. Noticing that $\gcd(6, 2^n - 1) = \gcd(3, 2^n - 1) = 1$, so (9) has only two solutions.

If $\gamma \neq 0$, assume that (9) has two solutions in $GF(2^n)$, say $Y_1, 1+Y_1$. Suppose it has another two solutions Y_2 and $1 + Y_2$ in $GF(2^n)$, let Y_3 and $1 + Y_3$ be the other two solutions of (9) in an extension field of $GF(2^n)$. By making use of the relationships between the cofficients and roots of equation (9), we get

$$Y_1 + Y_2 + Y_3 = 0 \text{ or } 1.$$

This means that $Y_3 \in GF(2^n)$. Whence $G(X, \beta) = \alpha$ has either no solution or two solutions or six solutions in $GF(2^n)$. This proves the first part of the theorem. Finally, it follows from Theorem 9 that $\text{ord}(f) = 3$.

In the following sections we will see that permutation $[X^7]_B$ of $GF(2)^5$ is APN. We now discuss when the $f(x)$ in Theorem 12 is APN. If (9) has more then two solutions in $GF(2^n)$, then it follows from the above proof that it has six solutions, say, Y_1, $1+Y_1$, Y_2, $1+Y_2$, Y_3, $1+Y_3$. By making use of the relations between the cofficients and roots of equation (9), we get

$$\begin{cases} (Y_1^2 + Y_1)^2 + (Y_2^2 + Y_2)^2 + (Y_1^2 + Y_1)(Y_2^2 + Y_2) = 1 \\ (Y_1^2 + Y_1)(Y_2^2 + Y_2)(Y_2^2 + Y_2 + Y_1^2 + Y_1) = r + 1 \end{cases}$$

Let $Y_1^2 + Y_1 = a$, $Y_2^2 + Y_2 = b$, then a, $b \in GF(2^n)$. Whence we obtain

$$\begin{cases} a^2 + b^2 + ab = 1 \\ ab(a + b) = r + 1, \end{cases}$$

which is equivalent to

$$\begin{cases} b^3 + b + r + 1 = 0 \\ a^3 + a + r + 1 = 0 \\ (a + b)^3 + a + b + r + 1 = 0, \end{cases}$$

because a, $b \neq 0$. This means that the equation

$$X^3 + X + r + 1 = 0 \tag{10}$$

has three solutions in $GF(2^n)$.

On the other hand, let

$$x^3 + X + r + 1 = (X + a)(X^2 + aX + c),$$

then we have

$$\begin{cases} a^2 + c = 1 \\ ac = r + 1. \end{cases}$$

Since $X^2 + aX + a^2 + 1 = a^2[(X/a)^2 + (X/a) + (a^2 + 1)/a^2]$ and it is known that polynomial $Y^2 + Y + d$ is reducible in $GF(2^n)$ iff $Tr(a) = Tr(a^{-1})$, where $d \in GF(2^n)$. Therefore $X^3+X+r+1$ has only one solution a iff $Tr(a) = Tr(a^{-1})$. Thus, if we can give a condition such that $\gcd(3, 2^n - 1) = 1$ and every solution Y of equation (9) satisfies $Tr(Y) = Tr(Y^{-1})$ in $GF(2^n)$, then the permutation f in Theorem 12 must be APN.

6 A Class of APN Permutations of Order $n - 1$ in $GF(2)^n$

It has been already been mentioned that constructing higher order permutations with good nonlinearity is cryptographically desirable. In this section we present a class of maximum order permutations in $GF(2)^n$.

Theorem 13 *Let* $\gcd(3, 2^n - 1) = 1$ *and* $d = 2^n - 2^i - 1$, $0 \leq i \leq n - 1$, B *a basis of* $GF(2^n)$ *over* $GF(2)$. *Then* $f(x) = [X^d]_B$ *is a maximum order APN permutation in* $GF(2)^n$.

Proof: We first consider the case $i = 0$. Then $F(X) = X^d = 0$ when $X = 0$, $F(X) = X^{-1}$ otherweise. Now we discuss the number of solutions of the equation

$$X^d + (X + \beta)^d = \alpha \tag{11}$$

If $\alpha = \beta^d$, then 0 and β are two solutions of (11) in $GF(2^n)$. Suppose that $X \neq 0, \beta$, is another solution of (11) in $GF(2^n)$, then we get from (11) that

$$X^2 + \beta X + \beta^2 = 0. \tag{12}$$

It follows that $X^3 = \beta^3$, which gives $X = \beta$, because that $\gcd(3, 2^n - 1) = 1$. A contridiction. Hence, in this case (11) has only two solutions.

If $\alpha \neq \beta^d$, then (11) has no solutions 0 and β. Whence (11) can be written as

$$G(X, \beta) = X^{-1} + (X + \beta)^{-1} = \beta / X(X + \beta) = \alpha, \tag{13}$$

which is equivalent to

$$X^2 + \beta X + \alpha^{-1}\beta = 0. \tag{14}$$

Obviously, (14) has at most two solutions for each $\alpha \neq \beta^d$, so has equation (13).

Summarizing the above results, we see that $[X^d]_B$ is APN. Noticing that $d = 2^n - 2$, we get $W_H(d) = n - 1$. Whence $\mathrm{ord}(f) = n - 1$. Finally, it follows from Theorem 7 that for each $d = 2^n - 1 - 2^i$, the conclusion of the theorem is true.

7 A Class of Permutaions of Order $n - 2$ in $GF(2)^n$ with Good Nonlinearity

This section presents a class of permutations of order $n - 2$ in $GF(2)$ with good nonlinearity.

Theorem 14 *Let* $\gcd(3, 2^n - 1) = 1$, $\gcd(7, 2^n - 1) = 1$ *and* $d = 2^n - 2^{i+1} - 2^i - 1$, $0 \leq i \leq n - 2$. *Then the permutation* $f(x) = [X^d]_B$ *has order n-2 and nonlinearity* $P_f = 2^{1-n}$ *or* $3 * 2^{1-n}$.

Proof: Because of Theorem 7 it suffices to prove the case $d = 2^n - 1 - 3$. Consider now the equation

$$G(X, \beta) = X^d + (X + \beta)^d = \alpha, \ \alpha \neq \beta^d, 0. \tag{15}$$

Apparently, (15) has no solutions 0 and β. Therefore (15) is equivalent to

$$X^6 + X^5\beta + X^4\beta^2 + X^3\beta^3 + X^2\alpha^{-1}\beta + X\alpha^{-1}\beta^2 + \alpha^{-1}\beta^3 = 0. \tag{16}$$

Similar to the proof of Theorem 12, we can prove that (15) has either no solution or two or six solutions in $GF(2^n)$.

What remains to be consided, is the equation

$$X^d + (X + \beta)^d = \beta^d. \tag{17}$$

Let $Y = X/\beta$, then (17) is equivalent to

$$Y^d + (1 + Y)^d = 1. \tag{18}$$

We conclude that (18) has only two solutions 0 and 1 in $GF(2^n)$. If not so, say that $Y_1 \neq 0, 1$, is another one in $GF(2^n)$. Then we get

$$1 + Y_1 + Y_1^2 + Y_1^3 + Y_1^4 + Y_1^5 + Y_1^6 = 0. \tag{19}$$

Whence $Y_1^7 = 1$. It follows that $Y_1 = 1$, a contridiction. Hence (18) has only two solutions in $GF(2^n)$.

By summarizing the above results, we see that $P_f = 2^{1-n}$ or $3 * 2^{1-n}$. It can be easily seen that ord(f)=$n - 2$.

8 On the Nonlinearity of the Permutations X^d in $GF(2^n)$ with $d = 2^m - 1$

For $d = 2^m + 1$ with $\gcd(m, n) = 1$, we have seen that X^d is APN in $GF(2^n)$. It is natural to ask whether the permutation X^{2^m-1} is APN. A simple example is that x^7 is APN in $GF(2^5)$, but not APN in $GF(2^4)$. Therefore X^{2^m-1} may be APN or not in $GF(2^n)$, it depents on the structure of the field $GF(2^n)$. To investigate the problem further. We need the following lemma:

Lemma 2 *Assume that n and $2^n - 1$ are two primes, then each nonzero conjugacy class of $Z^*_{2^n-1}$mod($2^n - 1$) has n elements, and there are $(2^n - 2)/n$ such conjugacy classes.*

Since $d = 2^m - 1$, we get

$$G(X, \beta) = X^d + (X + \beta)^d = \beta^d(Y^d + 1)/(Y + 1),\ X \neq \beta,$$

where $Y = X/\beta$. Therefore we need only to discuss the number of solutions of the equation

$$Y^{2^m-1} + 1 = r(Y + 1),\ r \neq 0, 1. \tag{20}$$

For the solutions of (20), we have the following conjecture:

Conjecture 1 *Assume that n and $2^n - 1$ are two primes, then for each $2 \leq i \leq n - 1$, equation (20) has at most two solutions other then 1 in $GF(2^n)$.*

In the case $m = n - 1$, the conclusion has already been proved in Theorem 13. If the conjecture is true, then every permutation $[X^{2^m-1}]_B (2 \leq m \leq n - 1)$ is APN in $GF(2)^n$.

9 Summary and Remarks

In this paper basic properties of APN permutations are presented. These results are useful in seeing the nature of the APN permutations and in constructing these permutations. By investigating mainly the permutations X^d in $GF(2^n)$, several classes of permutations in $GF(2)^n$ with good nonlinearity have been obtained. Some of them have high nonlinear order.

Included here are also three kinds of methods for constructing APN permutations in $GF(2)^n$. 1) Matrix Method: From an APN permutation $f = (f_1, \cdots, f_n)$ by multiplying a $n \times n$ nonsingular matrix A over $GF(2)$ to obtain another APN permutation $g(x) = (f_1, \cdots, f_n)A$; 2) Conjugacy Method: From a APN permutation $F(X)$ in $GF(2^n)$ to get $G(X) = F(X)^{2^i} = F(X^{2^i})$; 3) Basis Method: From a APN permutation $F(X)$ in $GF(2^n)$ to obtain different $[F(X)]_B$, by changing the basis B of $GF(2^n)$ over $GF(2)$.

After the simultaneous submissions of this paper and [7], we have found some overlaps between them, which should be made clear. Theorem 11 in this paper is an overlap with the first part of Proposition 3 in [7], and Theorem 13 here is a special case of of Proposition 6 in [7].

During Eurocrypt'93 Dr Nyberg has made some comments and suggestions for this paper. One is that the condition $\gcd(3, 2^n - 1) = 1$ in Theorem 12 and 13 is actually that n is odd. Another suggestion of Nyberg is a generalized definition of the linear equivalence for permutations, by which two permutations f and g over a field F are said to be linearly equivalent if there are two affine permutations A and B such that $f = B \circ g \circ A$, here $\circ$ denotes the composition operation of functions. By this definition all the functions x^d's for d's in the same conjugacy class are linearly equivalent.

Acknowlegements: The authors would like to thank Dr. Nyberg for the above helpful comments and suggestions, and the referees for pointing out a necessary condition of Corollary 2 and some typos of the original paper.

References

[1] E. Biham, A. Shamir. *Differential Cryptanalysis of DES-like Cryptosystems.* Advances in Cryptology, Proceedings of Crypt'90, Springer-Verlag, 1990.

[2] E. Biham, A. Shamir. *Differential Cryptanalysis of Snefru, Khafre, Redoc-II, Lokl and Lucifer.* Crypto'91, LNCS Vol.576, Springer-Verlag, 1991, pp.156-171.

[3] D. Chaum, J.H. Evertse. *Cryptanalysis of DES with a reduced number of rounds.* Advances in Cryptology, Proceedings of Crypto'85, Springer-Verlag, 1986, pp.192-211.

[4] X. Lai, J.L.Massey, S. Murphy. *Markov Ciphers and Differential Cryptanalysis.* Advances in Cryptology, Eurocrypto'91, LNCS Vol.547, Springer-Verlag, 1991, pp.17-38.

[5] K. Nyberg, L.R. Knudsen. *Provable Security against Differential Cryptanalysis.* Advances in Cryptology, Crypto'92.

[6] J. Pieprzyk. *Nonlinearity of Exponent Permutations*, Advances in Cryptology, Proc. Eurocrypt'89, Springer-Verlag, 1990.

[7] K. Nyberg. *Differentially uniform mappings for cryptography,* Eurocrypt'93, this proceedings.

[8] C. Carlet. *Codes de Reed-Muller, Codes de kerdock et de Preparata,* thesis, Publication of LITP, Institut Blaise Pascal, Université Paris 6, 90.59 (1990).

Two New Classes of Bent Functions

Claude Carlet

INRIA Rocquencourt, Domaine de Voluceau,
Bat 10, BP 105, 78153 Le Chesnay Cedex France
and
LAMIFA, Université de Picardie, France.

Abstract. We introduce a new class of bent functions on $(GF(2))^n$ (n even). We prove that this class is not included in one of the known classes of bent functions, and that, when n equals 6, it covers the whole set of bent functions of degree 3. This class is obtained by using a result from J.F. Dillon. We generalize this result and deduce a second new class of bent functions which we checked was not included in one of the preceding ones.

1. Introduction

Let $n = 2p$ ($p \in \mathbf{N}^*$) be an even positive integer.

The bent functions on $(GF(2))^n = \{0, 1\}^n$ are those boolean functions whose Hamming distance to the set of all affine functions on $(GF(2))^n$ (viewed as a vector space over the field GF(2)) is maximum. They play an important role in cryptography (in stream ciphers, for instance), as well as in error correcting coding (where they are used to define optimum codes such as the Kerdock codes and the Delsarte-Goethals codes). They have been studied by J. F. Dillon [5], [4] (in the wider framework of difference sets) and O. S. Rothaus [9] in the seventies. Since then, generalizations have been studied by several authors (cf. for instance [6], [8], and in another direction [3], see also the papers dealing with the covering radius of the Reed-Muller code of order 1 or with bent sequences), but very few papers lead to new results on the bent functions themselves (cf. [2]). In fact, no paper introducing new classes of bent functions has been published since 1975.

All quadratic bent functions are known (we say that a function is quadratic if the global degree of its algebraic normal form, cf. def. below, is at most 2, cf. [7] ch. 15). If n is at least 4, then any bent function has degree at most n/2 (cf [9]). Therefore, all bent functions on $(GF(2))^2$ and $(GF(2))^4$ are quadratic. Excepted these values, the only (even) value of n for which all bent functions are known is n = 6. In [9], O. S. Rothaus exhibits three classes of bent functions of degree 3 on $(GF(2))^6$ (the elements of a same class are equivalent each other up to an affine nonsingular transformation on the variable). But the problem of finding a simple characterization of the bent functions of degree 3 on $(GF(2))^6$ is still open.

Using a result from J.F. Dillon [4], we introduce (cf. corollary 1 and the definition which follows it) a new class of bent functions of degree $\frac{n}{2}$ on $(GF(2))^n$. The algebraic normal forms of the elements of this class are deduced from those of some of the elements of Maiorana-Mc Farland's class (whose definition will be recalled below) by adding a function whose support is an $\frac{n}{2}$- dimensional subspace of G. We call $\mathcal{D}$ the new class of bent functions. We check that it is not included in the completed versions of Maiorana-Mc Farland's class and Partial Spread class (cf. def. below). The size of class $\mathcal{D}$ has approximately same order as that of Maiorana-Mc Farland's class.

We prove that the bent functions of degree 3 on $(GF(2))^6$ all belong to class $\mathcal{D}$. That gives a simple characterization of these functions.

Generalizing Dillon's result, we obtain a theorem which characterizes the conditions under which, a bent function f and a flat E being chosen, the function $f + \phi_E$ is bent (where ϕ_E is the charateristic function of E). We deduce a second class of bent functions that we denote by $\mathcal{C}$. We check that this class is not included in the preceding ones.

We recount now with more details the definitions and known properties about bent functions.

Let F denote the Galois field GF(2), and G the F-space F^n (whose zero (0,…,0) will be simply denoted by 0). We denote by G' the space F^p. Clearly, G may be identified with G'^2.

The *dot product* on G is defined for any elements $x = (x_1, \dots, x_n)$ and $s = (s_1, \dots, s_n)$ of G by : $x \cdot s = x_1 s_1 + \dots + x_n s_n \in F$ (where the operation + is in F). We will use the same notation to denote the dot product on G'.

A well-known property which will often be used in this paper is the following :

if E is any F-linear subspace of G and a, b are any elements of G, the sum $\sum_{x \in b+E} (-1)^{a \cdot x}$ is equal to $|E| (-1)^{a \cdot b}$ (where |E| denotes the size of E) if a belongs to the dual of E (that is the linear space : $E^{\perp} = \{y \in G \,/\, \forall x \in E, x \cdot y = 0\}$), and to 0 otherwise.

We will call this property the character-sum property (it extends to more general character sums) and denote it by (1).

Let f be a boolean function on G. We denote by $\widehat{F}$ the Walsh (or Hadamard or discrete Fourier) transform of the real-valued function $(-1)^{f(x)}$:

$$\widehat{F}(s) = \sum_{x \in G} (-1)^{f(x) + x \cdot s} \quad .$$

It satisfies *Parseval's formula* (cf [7], p.416, corollary 3) : $\sum_{s \in G} (\widehat{F}(s))^2 = 2^{2n}$.

The boolean function f is called *bent* if (cf. [4], [5], [9]) for any element s of G, $\widehat{F}(s)$ is equal to: $\pm 2^p$. According to Parseval's formula and since $\widehat{F}(s)$ is related to the Hamming distance between f and the *affine function* $h_s : x \rightarrow s \cdot x + \varepsilon$ ($\varepsilon \in F$) by the relation : $(-1)^{\varepsilon} \widehat{F}(s) = 2^n - 2\, d(f, h_s)$, that is equivalent with the fact that f is at maximum distance from the set of all affine functions. Another equivalent definition is (cf. [4], [7]) : for any non-zero element s of G, the function on G: $x \rightarrow f(x) + f(x+s)$ is *balanced* (a boolean function g on G is called balanced if its support $\{x \in G \,/\, g(x) = 1\}$ has size 2^{n-1}, or equivalently if the sum $\sum_{x \in G} (-1)^{g(x)}$ equals 0).

The notion of bent function is invariant under any affine nonsingular transformation on the variable (or in other words under any linear nonsingular mapping, and any translation). If f is bent, then for any affine function g, the function f + g is bent. We shall say that a class of bent functions is *complete* if it is globally invariant under the addition of any affine function and the composition (on the right) with any nonsingular affine transformation.

If a boolean function f on G is bent, then the boolean function $\tilde{f}$ defined by :

$$\widehat{F}(s) = 2^p (-1)^{\tilde{f}(s)}$$

is bent itself. Following Dillon, we shall call it the *"Fourier" transform* of f. Its properties are (cf. [1] p. 55-59, [4]) :

- the mapping $f \rightarrow \tilde{f}$ is an isometry (i.e. the Hamming distance between two bent functions is equal to that of their "Fourier" transforms)
- if b is any element of G and ε any element of F, let g be the boolean function defined by : $g(x) = f(x) + b \cdot x + \varepsilon$ (respectively $g(x) = f(x+b) + \varepsilon$) then $\tilde{g}(x)$ is equal to : $\tilde{f}(x+b) + \varepsilon$ (respectively $\tilde{f}(x) + b \cdot x + \varepsilon$). (2)

Any boolean function on G admits an algebraic normal form, that is a polynomial expression by means of the coordinates $x_1, \ldots, x_n$, each coordinate appearing in any monomial with the degree 0 or 1 (cf [7], ch.13). If n is at least 4 and f is a bent function, then the (global) degree of its algebraic normal form is at most p (cf [9]).

Any quadratic function $f(x) = \sum_{1 \le i < j \le n} a_{i,j} x_i x_j + h(x)$ (h affine, $a_{i,j} \in F$) is bent if and only if one of the following equivalent properties is satisfied (cf [7], ch 15) :

- its associated symplectic form :

$$\varphi_f : (x, y) \to f(0) + f(x) + f(y) + f(x+y)$$

is non-degenerate

- the skew-symmetric matrix $(m_{i,j})_{i,j \in \{1,\ldots,n\}}$ over F , defined by : $m_{i,j} = a_{i,j}$ if i<j, $m_{i,j} = 0$ if i=j, and $m_{i,j} = a_{j,i}$ if i>j, is regular
- f(x) is equivalent, up to an affine nonsingular transformation of the variables, to the function on G:

$$x_1 x_2 + x_3 x_4 + \ldots + x_{n-1} x_n + \varepsilon \quad (\varepsilon \in F) \qquad (3)$$

A first general class of bent functions is the so-called Maiorana-Mc Farland's class (cf. [4] p. 90, [5]) denoted by $\mathcal{M}$:

we use the identification between G and G'^2, a general element of G being denoted by (x,y) (where x and y belong to G'), and we denote by "$\cdot$" the dot product on G'; the elements of class $\mathcal{M}$ are all the functions of the form :

$$f(x,y) = x \cdot \pi(y) + h(y)$$

where π is any permutation on G' and h any boolean function on G' . Notice that function (3) corresponds to the case π = id, h = ε modulo a permutation of the coordinates.

The "Fourier" transform $\tilde{f}(x,y)$ is then equal to :

$$y \cdot \pi^{-1}(x) + h(\pi^{-1}(x))$$

where π^{-1} denotes the inverse permutation of π.

Class $\mathcal{M}$ is not complete. We denote by $\mathcal{M}^{\#}$ its completed version.

A second important class of bent functions is that of Partial Spreads, denoted by $\mathcal{PS}$ (cf. [4] p. 95, [5]) :

$\mathcal{PS}$ is the disjoint union of two classes $\mathcal{PS}^-$ and $\mathcal{PS}^+$:

- the elements of $\mathcal{PS}^-$ are those functions whose supports are the unions of 2^{p-1} "disjoint" p-dimensional subspaces of G , less the point 0, "disjoint" meaning that any two of these spaces admit 0 as only common element, and therefore that their sum is direct and equal to G. In other words, they are the sums of 2^{p-1} characteristic functions of "disjoint" p-dimensional subspaces.
- the elements of $\mathcal{PS}^+$ are those functions whose supports are the unions of $2^{p-1}+1$ "disjoint" p-dimensional subspaces of G. They are the sums of $2^{p-1}+1$ characteristic functions of "disjoint" p-dimensional subspaces.

The "Fourier" transform of any function of $\mathcal{PS}$ is (very simply) deduced from the function itself by replacing the spaces by their duals.

This class is not complete. We obtain the completed version, that we denote by $\mathcal{PS}^{\#}$, by changing the subspaces into flats, two of them having a single (fixed) point in common, and by adding affine functions.

Classes $\mathcal{M}^{\#}$ and $\mathcal{PS}^{\#}$ are the only "effective" known classes of bent functions : there exist other classes of bent functions, but their definitions involve non-obvious conditions, so that none of them leads to an explicit description of bent functions. In fact, class $\mathcal{PS}$ is not really effective (the condition on the spaces which are involved in the definition is not simple, contrary to the condition on π which stands in the definition of class $\mathcal{M}$), but class $\mathcal{PS}$ contains subclasses (cf [4] p. 97...) which are more effective.

The generalized bent functions are defined as follows (cf [6], [8]) :

let n and q be any integers greater than 1. Let J_q and G be respectively the ring $\mathbf{Z}/q\mathbf{Z}$ of all integers modulo q, and the J_q-module $(J_q)^n$. Let $w = e^{2\pi i/q}$, then a function f from G to J_q is called bent if it satisfies one of the following equivalent properties :

1) for any element s of G, the sum $\sum_{x \in G} w^{f(x) - s \cdot x}$ (where "$\cdot$" denotes the usual dot product on G) has modulus $q^{n/2}$ (f is called regular if there exists a function $\tilde{f}$ from G to J_q such that, for any s, this sum is equal to $q^{n/2}\, w^{\tilde{f}(s)}$)

2) for any element s of $G\backslash\{0\}$, the sum $\sum_{x\in G} w^{f(x+s)-f(x)}$ is zero (ie the value of the autocorrelation function of f is zero on any nonzero element).

Class $\mathcal{M}$ generalizes to any q : if n is even and π is any permutation on $G'=(J_q)^{n/2}$, the function on $G = G'^2 : (x,y) \rightarrow x \cdot \pi(y)$ is regular-bent.

2. A New Class of Bent Functions

The idea which is the starting point of this work is the following : if we want to obtain new bent functions, a simple way would be to use known ones and to alter them without losing their property.

J.F. Dillon gives in [4, remark 6.2.15 p.82] a result which may be used in this sense. It may be stated as follows : *let f be a bent function on G; suppose its support contains a p-dimensional linear subspace E of G. Then, denoting by ϕ_E the boolean function of support E, the function $f + \phi_E$ is bent.*

Notice that, more generally, the condition : *E is contained in the support of f* may be replaced by: *the restriction of f to E is affine* . Indeed, if this restriction is equal to $a \cdot x + \varepsilon$, then E is included in the support of the (bent) function $f(x) + a \cdot x + \varepsilon + 1$, on which Dillon's remark may be applied.

We will see (cf. Corollary 1) that this result leads to new bent functions if we apply it to the elements of Maiorana-Mc Farland's class (it does not do so if we try to apply it to the elements of $\mathcal{PS}$). We also wish to determine the "Fourier" transforms of the bent functions that we obtain. It would be possible to deduce them from the proof given by Dillon, but it will be almost as simple and more convenient to give a direct proof of the whole result. To achieve it, a lemma will be useful, which is a slight generalization of [4, theorem 6.2.11 p.79]:

Lemma 1 *Let E be any linear subspace of G, f any bent function on G, and $\widetilde{f}$ its "Fourier" transform. Then for any elements a and b of G, we have :*

$$\sum_{x\in a+E} (-1)^{f(x)+b\cdot x} = 2^{\dim E - p} (-1)^{a\cdot b} \sum_{x\in b+E^{\perp}} (-1)^{\widetilde{f}(x)+a\cdot x} .$$

If E has dimension p and if the restriction of f(x) to E is 0 (respectively 1) then the restriction of $\tilde{f}$ *to* $E^{\perp}$ *is 0 (respectively 1).*

Proof:

According to the definition of the "Fourier" transform, we have :

$$\sum_{x \in b+E^{\perp}} (-1)^{\tilde{f}(x) + a \cdot x} = 2^{-p} \sum_{x \in b+E^{\perp}} \sum_{y \in G} (-1)^{f(y) + y \cdot x + a \cdot x} =$$

$$2^{-p} \sum_{y \in G} (-1)^{f(y)} \Big(\sum_{x \in b+E^{\perp}} (-1)^{(y+a) \cdot x} \Big).$$

According to the character-sum property (1), the sum $\Big(\sum_{x \in b+E^{\perp}} (-1)^{(y+a) \cdot x} \Big)$ is equal to $|E^{\perp}| (-1)^{(y+a) \cdot b}$ if y + a belongs to E, and to 0 otherwise. Therefore, we have :

$$\sum_{x \in b+E^{\perp}} (-1)^{\tilde{f}(x) + a \cdot x} = |E^{\perp}|\, 2^{-p} \sum_{y \in a+E} (-1)^{f(y) + (y+a) \cdot b}$$ and the first part of the lemma holds, since $|E^{\perp}|$ is equal to $2^{2p - \dim E}$.

If E has dimension p, the restriction of f(x) to E is 0 (respectively 1) if and only if $\sum_{x \in E} (-1)^{f(x)}$ is equal to 2^p (respectively -2^p), and the conclusion holds, applying the preceding equality with a = b = 0. □

Proposition 1 *Let E be a p-dimensional linear subspace of G and* ϕ_E *its characteristic function. Let f be a bent function on G whose restriction to E is affine. Then the function on G :*

$$f(x) + \phi_E(x)$$

is bent, and its "Fourier" transform is :

$$\tilde{f}(x) + \phi_{E^{\perp}}(x + a),$$

where a is any element of G such that the restriction to E of f(x) is equal to $a \cdot x + \varepsilon$ *(*$\varepsilon \in F$*).*

Proof:

Replacing f(x) by $f(x) + a \cdot x + \varepsilon$, and using property (2) (of section 1), we may without loss of generality assume : $a=0$, $\varepsilon=0$.

For any λ in G, we have :

$$\sum_{x\in G} (-1)^{f(x)+\phi_E(x) + \lambda\cdot x} =$$

$$\sum_{x\in G} (-1)^{f(x) + \lambda\cdot x} - 2 \sum_{x\in E} (-1)^{f(x) + \lambda\cdot x} .$$

The sum : $\sum_{x\in G} (-1)^{f(x) + \lambda\cdot x}$ is equal to : $2^p (-1)^{\tilde{f}(\lambda)}$, and the sum : $\sum_{x\in E} (-1)^{f(x) + \lambda\cdot x} = \sum_{x\in E} (-1)^{\lambda\cdot x}$ is equal to : $2^p \phi_{E^\perp}(\lambda)$, according to the character-sum property (1).

So, if λ does not belong to $E^\perp$, then the sum : $\sum_{x\in G} (-1)^{f(x)+\phi_E(x)+\lambda\cdot x}$ is equal to $2^p (-1)^{\tilde{f}(\lambda)}$, and if λ belongs to $E^\perp$, it is equal to $2^p (-1)^{\tilde{f}(\lambda)} - 2^{p+1}$ which is equal to $-2^p = 2^p (-1)^{\tilde{f}(\lambda)+1}$, since according to lemma 1, $\tilde{f}(\lambda)$ is equal to 0.

So, f is bent and $\sum_{x\in G} (-1)^{f(x)+\phi_E(x) + \lambda\cdot x}$ is equal to : $2^p (-1)^{\tilde{f}(\lambda) + \phi_{E^\perp}(\lambda)}$. □

In next corollary, we identify G with G'^2, so that we denote by (x,y) any element of G (x,y ∈ G').

Corollary 1 *Let E be a p-dimensional linear subspace of G and π a permutation on G' such that, for any (x, y) in E, the number : $x \cdot \pi(y)$ equals 0. Then the function defined on G as :*

$$x \cdot \pi(y) + \phi_E(x, y)$$

is bent, and its "Fourier" transform is the function:

$$y \cdot \pi^{-1}(x) + \phi_{E^\perp}(x, y).$$

Proof :

π being a permutation, the function f defined by :

$$f(x, y) = x \cdot \pi(y)$$

belongs to Maiorana-Mc Farland's class, and so is bent. Its "Fourier" transform is the function :

$$\tilde{f}(x,y) = y \cdot \pi^{-1}(x).$$

So, the result follows directly from proposition 1 with a = 0 and ε = 0. □

Remark

1) The class of bent functions that we obtain cannot be considered as an effective one since there is no simple description of all the subspaces and permutations satisfying the condition of Corollary 1. But there is a simple subcase : when E is equal to the cartesian product of two subspaces E_1 and E_2 of G' such that dim E_1 + dim E_2 = p and $\pi(E_2) = E_1^{\perp}$. This will lead to our first new class of bent functions (whose definition is below).

When E_1 is equal to the trivial space $\{0\}$, (and therefore E_2 = G'), the condition $\pi(E_2) = E_1^{\perp}$ is obviously satisfied. This special case leads to a subclass.

Of course, when E_1 = G' (and $E_2 = \{0\}$), the condition on π is empty too, but in that case, the function that we deduce belongs to Maiorana-Mc Farland's class, and we so obtain no new bent function.

2) Corollary 1 may be extended to some non-binary cases : let q be any positive even integer, let J_q, G' and G be respectively **Z**/q**Z**, $(J_q)^p$ and $(J_q)^n = G'^2$ (n =2p). Let E be any subgroup of order q^p of G and π any permutation on $(J_q)^p$. Suppose that, for any (x,y) in E : $x \cdot \pi(y) = 0$. Then the function :

$$(x,y) \rightarrow x \cdot \pi(y) + \frac{q}{2}\phi_E(x,y)$$

is bent :

let $w = e^{2\pi i/q}$, we have (since $w^{q/2} = -1$) :

$$\sum_{(x,y)\in G} w^{x\cdot\pi(y)+(q/2)\phi_E(x,y) - \lambda\cdot x - \mu\cdot y} =$$

$$\sum_{(x,y)\in G} w^{x\cdot\pi(y) - \lambda\cdot x - \mu\cdot y} - 2 \sum_{(x,y)\in E} w^{x\cdot\pi(y) - \lambda\cdot x - \mu\cdot y}.$$

The sum : $\sum_{(x,y)\in G} w^{x\cdot\pi(y) - \lambda\cdot x - \mu\cdot y}$ is equal to : $q^p\, w^{\mu\cdot\pi^{-1}(\lambda)}$ (cf.[6], p.100)

and the sum : $\sum_{(x,y)\in E} w^{- \lambda\cdot x - \mu\cdot y}$ is equal to : $q^p\, \phi_{E^{\perp}}(\lambda,\mu)$, (Lemma 1 generalizes).

That completes this sketch of proof.

Definition 1 *We call $\mathcal{D}$ the class of all the boolean functions of the form :*

$$(x, y) \in G \rightarrow \phi_E(x, y) + x \cdot \pi(y)$$

where E is a subspace of G equal to $E_1 \times E_2$, E_1 *and* E_2 *are subspaces of G' such that* $\dim E_1 + \dim E_2 = p$, *and* π *is any permutation on G' such that* $\pi(E_2) = E_1^{\perp}$.

We call $\mathcal{D}_o$ *the subclass of all the functions of the form :*

$$(x, y) \to \prod_{i=1}^{p} (x_i + 1) + x \cdot \pi(y) .$$

$\mathcal{D}_0$ corresponds to the case : $E = \{0\} \times G'$.

Example :

Assume $\pi(0) = 0$. Let z be any nonzero element of G'. Let E_1 be the linear hyperplane $\pi(z)^{\perp}$ and E_2 the line $\{0,z\}$. The function:

$$x \cdot \pi(y) + \phi_{E_1}(x)\, \phi_{E_2}(y) = x \cdot \pi(y) + (x \cdot \pi(z) + 1) \left(\prod_{i=1}^{p} (y_i+1) + \prod_{i=1}^{p} (y_i+z_i+1)\right)$$

belongs to class $\mathcal{D}$.

Remark

1) Both classes $\mathcal{M}^{\#}$ and $\mathcal{PS}^{\#}$ are invariant under the "Fourier" transform $f \to \tilde{f}$. According to Corollary 1, that is still the case of classes $\mathcal{D}_o^{\#}$ and $\mathcal{D}^{\#}$ (the completed classes of $\mathcal{D}_o$ and $\mathcal{D}$) .

2) The sizes of $\mathcal{D}$ and $\mathcal{M}$ have approximately same order since the number 2^{2^p} of boolean functions on F^p is small, compared with the number of permutations on the same space : $(2^p)!$.

We check now that class $\mathcal{D}$ is not included in class $\mathcal{M}^{\#}$. We shall obtain this result as a corollary of next proposition.

Proposition 2 *If* $p \geq 4$ *and if the restriction of permutation* π *to any linear hyperplane of G' is not affine, then the following function does not belong to class* $\mathcal{M}^{\#}$:

$$(x,y) \to \prod_{i=1}^{p} x_i + (x + 1) \cdot \pi(y)$$

(where 1 denotes the all-one word).

Proof :

We know that if a function f belongs to $\mathcal{M}^{\#}$, then there exists a p-dimensional subspace E of G, such that, for any elements (a,a') and (b,b') of E, the function :

$$x \rightarrow f(x,y) + f(x+a,y+a') + f(x+b,y+b') + f(x+a+b,y+a'+b')$$

is equal to 0 . Indeed, if f(x,y) is equal to $x \cdot \pi'(y) + h(y)$, where π' is a permutation, we may take E = G' x {0}, and any element of $\mathcal{M}^{\#}$ is equivalent to such a function, up to a nonsingular affine transformation on the variable .

Suppose that, two elements (a,a') and (b,b') being chosen in G, the function :

$$f(x,y) = \prod_{i=1}^{p} x_i + (x + \mathbb{1}) \cdot \pi(y)$$

satisfies the condition :

$$\forall (x,y) \in G,\ f(x,y) + f(x+a,y+a') + f(x+b,y+b') + f(x+a+b,y+a'+b') = 0.$$

That implies that the degree of the function :

$$\prod_{i=1}^{p} x_i + \prod_{i=1}^{p} (x_i+a_i) + \prod_{i=1}^{p} (x_i+b_i) + \prod_{i=1}^{p} (x_i+a_i+b_i)$$

is at most 1 (since the degree relative to x of : $(x+\mathbb{1}) \cdot \pi(y) + (x+a+\mathbb{1}) \cdot \pi(y+a') + (x+b+\mathbb{1}) \cdot \pi(y+b') + (x+a+b+\mathbb{1}) \cdot \pi(y+a'+b')$ is at most 1).

For any pair {i,j} of indices, the coefficient of $\prod_{k \neq i,j} x_k$ in that expression is :

$$a_i a_j + b_i b_j + (a_i+a_j)(b_i+b_j) = a_i b_j + a_j b_i$$

and must be equal to 0, since $p \geq 4$. So, any two elements (a,a') and (b,b') of E are such that a and b are linearly dependent (ie one of them is 0, or they are equal each other).We deduce that E is either equal to {0} x G' or to the direct sum of a line $\{0,\alpha\}$ (where α is a nonzero element of G') and of an hyperplane H of G'. In any case, there exists at least a linear hyperplane H of G' such that {0} x H is included in E. This hyperplane satisfies that for any elements a' and b' of H, and any elements x and y of G', we have :

$$(x + \mathbb{1}) \cdot (\pi(y) + \pi(y+a') + \pi(y+b') + \pi(y+a'+b')) = 0.$$

Since the restriction to H of at least one of the coordinate functions of π is not affine, we arrive to a contradiction. □

Corollary 2 *Classes $\mathcal{D}_0$ and $\mathcal{D}$ are in general not included in class $\mathcal{M}^{\#}$.*

Proof:

The function $\prod_{i=1}^{p} x_i + (x+1) \cdot \pi(y)$ is equivalent to the function $\prod_{i=1}^{p} (x_i+1) + x \cdot \pi(y)$ which belongs to class $\mathcal{D}_0$. So, all we need to prove is that there does exist in general a permutation π whose restriction to any linear hyperplane is not affine.

Let us identify G' with the Galois field of order 2^p. Let i be any integer prime to $2^p - 1$. The mapping on G' : $x \rightarrow x^i$ is a permutation on G'. Its retriction to a linear hyperplane $\{x \in G' / tr(ax) = 0\}$ (where tr is the trace function from G' to F and a is any nonzero element of G') is affine if and only if its restriction to the linear hyperplane $H_0 = \{x \in G' / tr(x) = 0\}$ is affine (since $\pi(ax) = a^i \pi(x)$).

H_0 being equal to the image of the linear mapping : $x \rightarrow x^2 + x$, that is true if and only if the mapping : $x \rightarrow (x^2 + x)^i$ is affine.

It is a simple matter to show that there exists in general i such that this last mapping is not affine. Take for instance $i = 1 + 2^j$. Suppose 2j is prime to p (and so, p is odd), then i is prime to 2^p-1. Suppose $2^{j+1}+2 < 2^p$. Then $(x^2 + x)^i$, equal to :$x^{2^{j+1}+2} + x^{2^{j+1}+1} + x^{2^j+2} + x^{2^j+1}$ cannot be affine since it is a non-affine polynomial of degree at most 2^p-1 (cf [7] p. 402). □

We now wish to prove that class $\mathcal{D}$ is not included in class $\mathcal{PS}^\#$. That is much more difficult since there does not seem to exist simple necessary conditions for a function to belong to $\mathcal{PS}^\#$. That is perhaps why it has never been proved until now that class $\mathcal{M}$ is not included in class $\mathcal{PS}^\#$ (J.F.Dillon has only proved in his thesis [4 p.53] that class $\mathcal{M}$ is not included in class $\mathcal{PS}$).

Proposition 3 *If p is any odd integer at least equal to 5, the function:*

$$(x,y) \in G \rightarrow \prod_{i=1}^{p} x_i + (x+1) \cdot y$$

does not belong to class $\mathcal{PS}^\#$.

Proof :

We have to prove that, for any affine function g on G, the function : $f(x,y) = \prod_{i=1}^{p} x_i + (x+1) \cdot y + g(x,y)$ is not equivalent, up to a nonsingular affine transformation, to a function of $\mathcal{PS}$.

Suppose first that it is equivalent to a function of $\mathcal{PS}^-$. There exist $k=2^{p-1}$ flats $H_1,\ldots, H_k$ of G such that any two of them intersect in a single (fixed) point (a,b), and that the support of f is their union less the point (a,b).

Let i be any element of $\{1,\ldots,k\}$. The intersection between H_i and the support of f is $H_i \setminus \{(a,b)\}$ and so has an odd number of elements. Therefore, denoting by h_i the boolean function of support H_i, the function f h_i (whose value in x is $f(x)h_i(x)$) has an odd weight, and so has degree 2p. Consequently, since $\prod_{i=1}^{p} x_i$ is the only monomial in the algebraic normal form of f (x,y) whose degree is p, and since h_i has degree p, the product of the function h_i with any function $\prod_{i=1}^{p} (x_i+\lambda_i)$ has degree 2p. Applying this result to $\lambda_i = a_i + 1$ (i=1...p), we deduce that the flat $\{a\}\times G'$ has an odd number of elements in commun with H_i. That means that it has the point (a,b) only in commun with H_i.

We deduce :

$\forall y \in G', f(a,y) = 0,$

$$\forall\, y, \prod_{i=1}^{p} a_i + (a+1) \cdot y + g(x,y) = 0 .$$

So, replacing g(x,y) by its value, we obtain :

$$f(x,y) = \prod_{i=1}^{p} x_i + \prod_{i=1}^{p} a_i + (x+a) \cdot y .$$

The translation $(x,y) \rightarrow (x+a, y+b)$ translates the point (a,b) in (0,0) and changes f (x,y) in:

$$\prod_{i=1}^{p} (x_i+a_i) + \prod_{i=1}^{p} a_i + x \cdot (y+b) .$$

The subspace $E = \{0\}xG'$ is disjoint from the support of f and so is "disjoint" from any of the H_i. Using an idea from J. F. Dillon [4, p.53], we may deduce that there exist linear mappings ϕ_i on G' such that, for any i, H_i is equal to the set : $\{(x,\phi_i(x)), x\in G'\}$. Since for any i, the set $H_i \setminus \{(0,0)\}$ is included in the support of f, we deduce :

$\forall$ i,j $\forall$ x $\in$ G', $f(x,\phi_i(x)) = f(x,\phi_j(x))$, and therefore :

$\forall$ i,j $\forall$ x $\in$ G', $x \cdot (\phi_i(x)+\phi_j(x)) = 0$.

That implies that the matrix of the linear mapping $\phi_i+\phi_j$ is skew-symmetric. If $i\neq j$, H_i and H_j admit 0 as only commun element. So, this matrix is regular. The dimension p being odd, that is impossible.

Suppose now that the function :

$$f : (x,y) \rightarrow \prod_{i=1}^{p} x_i + (x+\mathbb{1}) \cdot y + g\ (x,y)$$

is equivalent to a function of $\mathcal{PS}^+$. Its support is the union of $2^{p-1}+1$ p-dimensional flats, any two of them intersecting in a single fixed point (a,b).

We may suppose without loss of generality that g depends only on y (if $g(x,y) = u \cdot x + v \cdot y + \varepsilon$, change y in y + u). The restriction of f to the flat $H = \{(a,y), y\in G'\}$ is the function :

$$y \rightarrow \prod_{i=1}^{p} a_i + (a+\mathbb{1}) \cdot y + g\ (y)\ .$$

This function is affine. Suppose it is not the constant function 1, then its weight is at most 2^{p-1}, and at least two spaces H_i have (a,b) as only commun point with H. We can apply the translation of vector (a,b) and complete the proof as previously.

Otherwise, f(x,y) is equal to : $\prod_{i=1}^{p} x_i + (x+a) \cdot y + \prod_{i=1}^{p} a_i + 1$. Let us apply again the translation of vector (a,b), so that f(x,y) becomes : $\prod_{i=1}^{p} (x_i+a_i) + x \cdot (y+b) + \prod_{i=1}^{p} a_i + 1$.

Let α be an element of G' such that : $\prod_{i=1}^{p} (\alpha_i+a_i) + \alpha \cdot b + \prod_{i=1}^{p} a_i = 1$ (such an element exists since the function $\prod_{i=1}^{p} (x_i+a_i) + x \cdot b + \prod_{i=1}^{p} a_i$ is not the zero function).

Let E denote the linear hyperplane of G' : $\{x \in G' / \alpha \cdot x = 0\} = \alpha^{\perp}$.

The restriction of f to the space : $\{0,\alpha\} \times E$ is balanced since it is equal to 1 on $\{0\} \times E$ and to 0 on $\{\alpha\} \times E$. Therefore, there exist at least two spaces H_i (say H_1 and H_2) which are "disjoint" from the space $\{0,\alpha\} \times E$.

We shall now compose f on the right by an automorphism ψ of G which maps $\{0\} \times G'$ onto $\{0,\alpha\} \times E$, so that we can apply on $f \circ \psi$ the same technique as the one we applied previously on f. Let β be an element of G' such that $\alpha \cdot \beta = 1$, and $E' = \beta^{\perp}$.

We have : $G = (\{0,\alpha\} \oplus E') \times (\{0,\beta\} \oplus E)$.

Let ψ be the involutive isomorphism of G defined by :

$\forall\, \varepsilon, \eta \in F, \forall\, u \in E, \forall\, v \in E', \psi(\varepsilon\, \alpha + v, \eta\, \beta + u) = (\eta\, \alpha + v, \varepsilon\, \beta + u)$.

ψ maps $\{0\} \times G'$ onto $\{0,\alpha\} \times E$.

The function $\left(\prod_{i=1}^{p} (x_i+a_i)\right) \circ \psi$ has support : $\psi(\{a+\mathbb{1}\} \times G')$, since $\prod_{i=1}^{p} (x_i+a_i)$ has support $\{a+\mathbb{1}\} \times G'$ and $\psi^{-1} = \psi$. If γ and w are the elements of F and E' (respectively) such that $a+\mathbb{1}$ is equal to : $\gamma\, \alpha + w$, this support is the set : $\{(\varepsilon\, \alpha + w, \gamma\, \beta + u), \varepsilon \in F, u \in E\}$.

Remember that f(x,y) is equal to $\prod_{i=1}^{p} (x_i+a_i) + x \cdot (y+b) + \prod_{i=1}^{p} a_i + 1$. We deduce : $\forall\, \varepsilon, \eta \in F, \forall\, u \in E, \forall\, v \in E'$,

$$f \circ \psi\, (\varepsilon\, \alpha + v, \eta\, \beta + u) = (\eta+\gamma+1)\ \ \chi_w(v) + \eta\, \varepsilon + v \cdot u + (\eta\, \alpha + v) \cdot b + \prod_{i=1}^{p} a_i + 1$$

where $\chi_w(v) = \begin{cases} 1 \text{ if } v = w \\ 0 \text{ otherwise} \end{cases}$.

We know that there exist (at least) two linear subspaces $\psi(H_1)$ and $\psi(H_2)$ (since $\psi^{-1} = \psi$) of G which are "disjoint" from each other and "disjoint" from the space $\{0\} \times G'$ and which are included in the support of $f \circ \psi$. We deduce that there exist two linear mappings ϕ_1 and ϕ_2 from G' to itself such that :

$\forall\, \varepsilon \in F, \forall\, v \in E', f \circ \psi\, (\varepsilon\, \alpha + v, \phi_1(\varepsilon\, \alpha + v)) = f \circ \psi\, (\varepsilon\, \alpha + v, \phi_2(\varepsilon\, \alpha + v))$.

Let l_i and L_i (i=1, 2) be respectively the boolean function on G' and the linear mapping from G' onto E such that : $\phi_i(\varepsilon\alpha + v) = l_i(\varepsilon\alpha + v)\beta + L_i.(\varepsilon\alpha + v)$. We have :

$\forall \varepsilon \in F, \forall v \in E'$,

$(l_1 + l_2)(\varepsilon\alpha + v)\ \ \chi_w(v) + ((l_1 + l_2)(\varepsilon\alpha + v))\,\varepsilon + v \cdot [(L_1 + L_2)(\varepsilon\alpha + v)] +$

$[(l_1 + l_2)(\varepsilon\alpha + v)]\,[\alpha \cdot b] = 0$.

We shall prove that this is impossible.

If v is different from w, then for any ε, we have :

$((l_1 + l_2)(\varepsilon\alpha + v))\,\varepsilon + v \cdot [(L_1 + L_2)(\varepsilon\alpha + v)] + [(l_1 + l_2)(\varepsilon\alpha + v)]\,[\alpha \cdot b] = 0$.

That means that the function of the variable (ε,v) which is equal to the LHS of that equality has weight at most 2. But, this function is quadratic, and we know (cf. [7] ch 15) that if a function is quadratic, then either its weight is at least 2^{p-2} (ans so is at least 8) or it is the zero function. We deduce that it is the zero function. But this function is equal to : $x \cdot (\phi_1(x) + \phi_2(x))$, where $x = \varepsilon\alpha + v$.

Thus, the matrix of the linear mapping $\phi_1 + \phi_2$ is skew-symmetric and regular, a contradiction. □

Corollary 3 *Classes $\mathcal{D}_0$, $\mathcal{D}$ and $\mathcal{M}$ are not included in class $\mathcal{PS}^{\#}$.*

Proof :

It is straightforward, according to Proposition 3, since the function $(x,y) \in G \rightarrow \prod_{i=1}^{p} x_i + (x+1) \cdot y$ of Proposition 3 belongs to classes $\mathcal{D}_0^{\#}$, $\mathcal{D}^{\#}$, and $\mathcal{M}^{\#}$, . □

3. Generalization of Dillon's result

We shall now extend Dillon's result to cases where E is a flat whose dimension is not necessarily equal to p. That will lead us to new bent functions.

Before we state the theorem, we need some preliminary definition and lemmas.

Lemma 2 *Let f be any boolean function on G. Let E be any flat of G and k its dimension. Let ψ be any affine mapping from F^k to G such that $E = \psi(F^k)$. Then the degree of the boolean function $f \circ \psi$ on F^k does not depend on the choice of ψ.*

Proof:

Suppose that ψ_1 and ψ_2 are two affine mappings from F^p to G such that $\psi_1(F^k) = \psi_2(F^k) = E$. The boolean functions $f \circ \psi_1$ and $f \circ \psi_2$ are then equivalent and so have same degree (cf [4], p.39, [7] ch.13). □

Definition 2 *Let f be any boolean function on G and E any flat in G. We call degree of the restriction of f to E the degree of the function $f \circ \psi$ on F^k, where k is the dimension of E and ψ is any affine mapping from F^k to G such that $\psi(F^k) = E$.*

Lemma 3 *Let f be any boolean function on G and E any k-dimensional flat in G. If there exists an integer r such that, for any element a of G, the sum :*

$$\sum_{x \in E} (-1)^{f(x) + a \cdot x}$$

is divisible by 2^r, then the degree of the restriction of f to E is at most : $k - r + 1$.

Proof:

We just adapt the proof due to Rothaus [9] on the degrees of the bent functions.

We may suppose that E is equal to F^k (otherwise, we can compose by an appropriate affine nonsingular mapping). Let d be the degree of the restriction of f to E and $\prod_{i \in I} x_i$ one of its monomials of degree d (I is a subset of $\{1,\ldots,k\}$ of size d).

The sum : $\sum_{x \in E / x_i = 0, \forall i \notin I} (-1)^{f(x)}$ is equal to the size 2^d of the linear subspace $\{x \in E \,/\, x_i = 0, \forall i \notin I\}$ minus twice the weight of the restriction of f to this subspace, which is odd, since the degree of this function is equal to the dimension of this subspace (cf. [7], ch 13 or [9]) . Therefore, $\sum_{x \in E / x_i = 0, \forall i \notin I} (-1)^{f(x)}$ is divisible by 2 but not by 4 .

For any a in G, let $\lambda_a = \sum_{x \in E} (-1)^{f(x) + a \cdot x}$.

According to the inverse formula of the Walsh transform (cf. [7], p. 127), for any x in E, we have: $(-1)^{f(x)} = 2^{-k} \sum_{a \in E} \lambda_a (-1)^{a \cdot x}$, and therefore :

$$\sum_{x \in E / x_i = 0, \forall i \notin I} (-1)^{f(x)} = \sum_{x \in E / x_i = 0, \forall i \notin I} 2^{-k} \sum_{a \in E} \lambda_a (-1)^{a \cdot x} =$$

$$2^{-k} \sum_{a \in E} \lambda_a \Big(\sum_{x \in E / x_i = 0, \forall i \notin I} (-1)^{a \cdot x} \Big) = 2^{|I| - k} \sum_{a \in E / a_i = 0, \forall i \in I} \lambda_a$$ (according to the character-sum property (1)).

So, the sum $\sum_{x \in E / x_i = 0, \forall i \notin I} (-1)^{f(x)}$ is divisible by $2^{|I| - k + r} = 2^{d - k + r}$ and therefore, $d - k + r$ is at most 1. We so have proved : $d \leq k - r + 1$. □

Theorem *Let $E = b + E'$ be any flat in $G = F^{2p}$ (E', its direction, is a linear subspace of G). Let ϕ_E be the boolean function whose support is E and f(x) any bent function on G. Then the function $f^* = f + \phi_E$ is bent if and only if one of the following equivalent conditions is satisfied :*

1) for any x in $G \setminus E'$, the function:

$$y \rightarrow f(y) + f(x+y)$$

is balanced on E

2) for any λ in G, the restriction of the function $\widetilde{f}(x) + b \cdot x$ to the flat $\lambda + E'^{\perp}$ is either constant or balanced.

If one of these conditions is satisfied, then E has dimension at least p and the degree of the restriction of f to E is at most dim E - p + 1.

If E has dimension p, then this last condition is also sufficient and the function $\widetilde{f^*}(x)$ is equal to :

$$\widetilde{f}(x) + \phi_{a + E'^{\perp}}(x),$$

where a is any element of G such that for any x in E : $f(x) = a \cdot x + \varepsilon$.

Proof :

1) The function f* is bent if and only if, for any x in $G \setminus \{0\}$, the function $y \rightarrow f^*(y) + f^*(x+y)$ is balanced on G, that is :

$$\sum_{y \in G} (-1)^{f^*(y) + f^*(x+y)} = 0.$$

If x belongs to E', then we have for any y in G : f*(y) +f*(x+y) =f(y) +f(x+y) and therefore :

$$\sum_{y\in G} (-1)^{f^*(y) + f^*(x+y)} = \sum_{y\in G} (-1)^{f(y) + f(x+y)} = 0 \text{ (since f is bent).}$$

If x does not belong to E', then the flats E and x+E are disjoint, the function $\phi_E(y) + \phi_E(x+y)$ takes the value 1 on $E \cup (x+E)$ and :

$$\sum_{y\in G} (-1)^{f^*(y) + f^*(x+y)} =$$

$$\sum_{y\in G} (-1)^{f(y) + f(x+y)} - 2 \sum_{y\in E} (-1)^{f(y) + f(x+y)} - 2 \sum_{y\in x+E} (-1)^{f(y) + f(x+y)} =$$

$$- 4 \sum_{y\in E} (-1)^{f(y) + f(x+y)}.$$

We deduce that f* is bent if and only if, for any element x of G \ E', the function $y \to f(y) + f(x+y)$ is balanced on E.

2) We have :

$$\sum_{x\in G} (-1)^{f^*(x) + \lambda \cdot x} = \sum_{x\in G} (-1)^{f(x) + \lambda \cdot x} - 2 \sum_{x\in E} (-1)^{f(x) + \lambda \cdot x}$$

$$= 2^p (-1)^{\tilde{f}(\lambda)} - 2 \sum_{x\in E} (-1)^{f(x) + \lambda \cdot x}.$$

So, f* is bent if and only if for any λ in G, the sum $\sum_{x\in E} (-1)^{f(x) + \lambda \cdot x}$ is equal either to 0 or to $2^p (-1)^{\tilde{f}(\lambda)}$.

According to Lemma 1, we have :

$$\sum_{x\in E} (-1)^{f(x) + \lambda \cdot x} = |E|\, 2^{-p} (-1)^{\lambda \cdot b} \sum_{x\in \lambda+E'^{\perp}} (-1)^{\tilde{f}(x) + b \cdot x}.$$

That sum is equal to $2^p (-1)^{\tilde{f}(\lambda)}$ if and only if $\sum_{x\in \lambda+E'^{\perp}} (-1)^{\tilde{f}(x) + b \cdot x}$ is equal to: $\frac{2^{2p}}{|E|} (-1)^{\tilde{f}(\lambda) + \lambda \cdot b}$, that is if and only if $\tilde{f}(x) + b \cdot x$ is constant on $\lambda + E'^{\perp}$, since $\frac{2^{2p}}{|E|}$ is equal to the size of $E'^{\perp}$. This same sum is equal to zero if and only if $\tilde{f}(x) + b \cdot x$ is balanced on $\lambda + E'^{\perp}$. That completes the proof of part 2.

If f and f* are bent, then the degrees of their algebraic normal forms are at most p (cf. [9]), and therefore, ϕ_E has degree at most p. That is equivalent with the fact that the dimension of E is at least p.

For any λ in G, since f and f* are bent and since $\sum_{x\in G} (-1)^{f^*(x) + \lambda\cdot x}$ is equal to :

$\sum_{x\in G} (-1)^{f(x) + \lambda\cdot x} - 2\sum_{x\in E} (-1)^{f(x) + \lambda\cdot x}$, the number $2\sum_{x\in E} (-1)^{f(x) + \lambda\cdot x}$

is the difference between two numbers which are both equal to $\pm 2^p$. So, it is divisible by 2^{p+1} and lemma 3 (with r = p) may be applied. Thus the restriction of f to E has degree at most dim E - p + 1.

If E has dimension p, then the restriction of f to E is affine (that is the converse of Dillon's result). There exist a in G and ε in F such that, for any x in E: $f(x) = a \cdot x + \varepsilon$. Proposition 1 and property (2) complete the proof. □

Remark

1) If E is the whole space G, then conditions 1 and 2 in the theorem are obviously satisfied. That corresponds to the fact that for any bent function f, the function f+1 is bent.

If E is an hyperplane, then $\phi_E(x)$ is of the form : $a \cdot x + \varepsilon$ ($a \in G$, $\varepsilon \in F$). So conditions 1 and 2 must be satisfied (since if f(x) is any bent function then $f(x) + a \cdot x + \varepsilon$ is bent). It is a simple matter to check it. Notice that, in that case, $\tilde{f}^*(x)$ is equal to : $\tilde{f}(x+a) + \varepsilon$. We see that the expression of $\tilde{f}^*(x)$ by means of $\tilde{f}(x)$ may be quite different depending on whether E has dimension p or not.

2) The characterization by condition 1 of those bent functions f such that $f + \phi_E$ is bent generalizes to non-binary cases the following way :

let n and q be any integers greater than 1. Let J_q and G be respectively the groups **Z**/q**Z** and $(J_q)^n$, f a bent function from G to J_q, E' any subgroup of G, b any element of G, E the set b + E' and λ any element of J_q. Then the function $f^* = f + \lambda\,\phi_E$ is bent if and only if, for any x in G \ E', the element of J_q equal to:

$(w^{-\lambda} - 1) \sum_{y\in E} w^{f(x+y) - f(y)} + (w^{\lambda} - 1) \sum_{y\in -x+E} w^{f(x+y) - f(y)}$ is zero. Indeed, let x be any nonzero element of G, the sum :

$$\sum_{y \in G} w^{f^*(x+y) - f^*(y)}$$

is equal to : $\sum_{y \in G} w^{f(x+y) - f(y)} = 0$ if x belongs to E' (since x + E is then equal to E), and to:

$$\sum_{y \in G} w^{f(x+y) - f(y)} + (w^{-\lambda} - 1) \sum_{y \in E} w^{f(x+y) - f(y)} + (w^{\lambda} - 1) \sum_{y \in -x+E} w^{f(x+y) - f(y)}$$

otherwise.

Condition 2 may also be generalized to some non-binary cases, but only for regular-bent functions.

We deduce now the existence of another superclass of $\mathcal{D}_0$ whose elements are bent functions :

Corollary 4 *Let L be any linear subspace of G' = F^p and π any permutation on G' such that, for any element λ of G', the set $\pi^{-1}(\lambda + L)$ is a flat. Then the function on G:*

$$x \cdot \pi(y) + \phi_{L^\perp}(x)$$

is bent.

Proof:

Let E be the subspace of G : $L^\perp$ x G'.

The function f(x,y) = x · π(y) belongs to Maiorana-Mac Farland's class and so is bent. Its "Fourier" transform is $\tilde{f}(x,y) = y \cdot \pi^{-1}(x)$. Let (λ, μ) be any element of G. The size of the support of the restriction of $\tilde{f}(x,y)$ to the set $(\lambda, \mu) + E^\perp = (\lambda + L) \times \{\mu\}$ is equal to that of the support of the restriction of the function : $x \to \mu \cdot x$ to the flat $\pi^{-1}(\lambda + L)$, which is either balanced or constant, since this function is affine. So, condition 2 of the theorem is satisfied. □

Definition *We call* $\mathcal{C}$ *the class of all the functions of the form :*

$$x \cdot \pi(y) + \phi_{L^\perp}(x) ,$$

where L and π satisfy the conditions of the preceding corollary.

Class $\mathcal{C}$ contains $\mathcal{D}_0$ (which corresponds to the case L = G'), and so is not included in classes $\mathcal{M}^{\#}$ and $\mathcal{PS}^{\#}$.

Notice that class $\mathcal{C}$ is not included in class $\mathcal{D}^{\#}$, since it contains functions of degrees less than p.

4. A simple characterization of the bent functions on F^6

We shall deduce from the theorem a characterization of the bent functions of degree 3 on F^6.

Proposition 4 *Let f be any boolean function of degree 3 on F^6 of the form :*
$f(x_1,\dots,x_6) = x_1 x_2 x_3 + x_1 h_1(x_4,x_5,x_6) + x_2 h_2(x_4,x_5,x_6) + x_3 h_3(x_4,x_5,x_6) + g(x_4,x_5,x_6)$

where h_1, h_2, and h_3 are three (quadratic) functions on F^3, and g is a boolean function on F^3.

Then f is bent if and only if :

1) the mapping $(x_1,x_2,x_3) \to (h_1(x_1,x_2,x_3), h_2(x_1,x_2,x_3), h_3(x_1,x_2,x_3))$ is a permutation on F^3

2) the function $h_1 + h_2 + h_3 + g$ is affine.

Any bent function of degree 3 on F^6 is equivalent, up to a nonsingular affine transformation on the variables, to such a function .

Proof:

Suppose f is bent, then the functions on F^6 :

$f(x_1, x_2,\dots,x_6) + f(x_1+1, x_2,\dots,x_6) = x_2x_3 + h_1(x_4,x_5,x_6)$

$f(x_1, x_2, x_3,\dots,x_6) + f(x_1+1, x_2+1, x_3,\dots, x_6) = (x_1+x_2+1)x_3 + h_1(x_4, x_5, x_6) + h_2(x_4, x_5, x_6)$

$f(x_1, x_2, x_3, x_4, x_5, x_6) + f(x_1+1, x_2+1, x_3+1, x_4, x_5, x_6) =$
$x_1x_2+x_1x_3+x_2x_3+x_1+x_2+x_3+1 + h_1(x_4, x_5, x_6)+ h_2(x_4, x_5, x_6)+ h_3(x_4, x_5, x_6)=$
$(x_1+x_2+1)(x_1+x_3+1) + h_1(x_4, x_5, x_6)+ h_2(x_4, x_5, x_6)+ h_3(x_4, x_5, x_6)$

are balanced.

We have (cf [2]) : $\sum_{(x_2,\ldots,x_6)\in F^5} (-1)^{x_2x_3+h_1(x_4,x_5,x_6)} = 2 \sum_{(x_4,x_5,x_6)\in F^3} (-1)^{h_1(x_4,x_5,x_6)}$

Thus, h_1 is balanced. Similarly, h_2, h_3, $h_1 + h_2$, $h_1 + h_3$, $h_2 + h_3$, and $h_1 + h_2 + h_3$ are balanced.

It is then a simple matter to prove that (h_1, h_2, h_3) is a permutation on F^3 :

let us denote by h the real-valued function on F^3 whose value on any element (a_1,a_2,a_3) of F^3 is equal to the size of the set :

$\{(x_4,x_5,x_6) \in F^3 / (h_1(x_4,x_5,x_6) = a_1,\ h_2(x_4,x_5,x_6) = a_2$, and $h_3(x_4,x_5,x_6) = a_3 \}$.

The Walsh transform of h is the function :

$$(x_1, x_2, x_3) \rightarrow \sum_{(a_1,a_2,a_3)\in F^3} h(a_1, a_2, a_3)\ (-1)^{x_1a_1+x_2a_2+x_3a_3} = \sum_{(x_4,x_5,x_6)\in F^3} (-1)^{x_1h_1(x_4,x_5,x_6)+x_2h_2(x_4,x_5,x_6)+x_3h_3(x_4,x_5,x_6)}.$$

So, it is equal to 8 if $x_1=x_2=x_3=0$, and to 0 otherwise (since the functions h_1, h_2, h_3, $h_1 + h_2$, $h_1 + h_3$, $h_2 + h_3$, and $h_1 + h_2 + h_3$ are balanced). According to the inverse formula of the Walsh transform (cf.[7]), h is the constant function equal to 1, and the mapping :

$$(x_4,x_5,x_6) \in F^3 \rightarrow (h_1 (x_4,x_5,x_6) ,\ h_2 (x_4,x_5,x_6) ,\ h_3 (x_4,x_5,x_6)) \in F^3$$

is therefore a permutation.

So, the function $x_1\ h_1(x_4,x_5,x_6)+ x_2\ h_2(x_4,x_5,x_6)+ x_3\ h_3(x_4,x_5,x_6)+ g(x_4,x_5,x_6)$ belongs to class $\mathcal{M}$. It is equal to $f(x) + x_1x_2x_3$.

$x_1x_2x_3$ is the algebraic normal form of the 3-dimensional flat of equations $x_1=x_2=x_3=1$. According to the theorem, the restriction of f(x) to this flat must be affine, and so, $h_1 + h_2 + h_3 + g$ is affine. So, 1) and 2) are satisfied.

The converse is straightforward, according to the theorem.

Let f be now any bent function of degree 3 on F^6. We may without loss of generality suppose that its algebraic normal form contains the monomial $x_1x_2x_3$. Let $g_1(x_4,x_5,x_6)$, $g_2(x_4,x_5,x_6)$, and $g_3(x_4,x_5,x_6)$ be the factors in $f(x_1,\ldots,x_6)$ of respectively x_2x_3, x_1x_3, and x_1x_2. Then $f(x_1,\ldots,x_6)$ is equal to :

$x_1x_2x_3 + x_1x_2\ g_3(x_4,x_5,x_6) + x_1x_3\ g_2(x_4,x_5,x_6) + x_2x_3\ g_1(x_4,x_5,x_6)$ plus an expression whose (global) degree relative to x_1, x_2 and x_3 is at most 1. So, there exist boolean functions h_1, h_2, h_3, and g on F^3 such that :

$f(x_1,\ldots,x_6) =$

$(x_1+g_1(x_4,x_5,x_6))\ (x_2+g_2(x_4,x_5,x_6))\ (x_3+g_3(x_4,x_5,x_6)) + (x_1+g_1(x_4,x_5,x_6))$ $h_1(x_4,x_5,x_6)+ (x_2+ g_2(x_4,x_5,x_6))\ h_2(x_4,x_5,x_6)+ (x_3+g_3(x_4,x_5,x_6))\ h_3(x_4,x_5,x_6)+$ $g(x_4,x_5,x_6)$.

Thus, $f(x_1,\ldots,x_6)$ is equivalent to :

$$x_1\, x_2\, x_3 + x_1\, h_1(x_4,x_5,x_6)+ x_2\, h_2(x_4,x_5,x_6)+ x_3\, h_3(x_4,x_5,x_6)+ g(x_4,x_5,x_6)$$

up to the nonsingular affine transformation :

$$(x_1,\ldots,x_6) \rightarrow (x_1+g_1(x_4,x_5,x_6),\ x_2+g_2(x_4,x_5,x_6),\ x_3+g_3(x_4,x_5,x_6),\ x_4,\ x_5,\ x_6)\ .\ \square$$

Corollary 5 *The bent functions of degree 3 on F^6 all belong to class $\mathcal{D}_0^{\#}$.*

Conclusion

We have now twice more classes of bent functions than we had before.

We have also obtained new generalized bent functions, but the extension to non-binary cases has only been sketched in this paper. That gives a direction in which a research may be done.

Acknowledgement

We wish to thank J. Wolfmann for having drawn our attention to Dillon's remark.

References

[1] C. Carlet, *Codes de Reed-Muller, codes de Kerdock et de Preparata* , *thèse* , publication du LITP n° 90.59 (1990), Institut Blaise Pascal, Université Paris 6, 4 place Jussieu, 75005 Paris, France.

[2] C. Carlet, *A transformation on boolean functions, its consequences on some problems related to Reed-Muller codes* , EUROCODE '90, Lecture Notes in Computer Science 514, 42-50 (1991).

[3] C. Carlet, *Partially Bent Functions* , Designs, Codes and Cryptography, 3, 135-145 (1993) , presented at Crypto'92, Santa Barbara, USA.

[4] J. F. Dillon, *Elementary Hadamard Difference Sets*, Ph. D. Thesis, Univ. of Maryland (1974).

[5] J. F. Dillon, *Elementary Hadamard Difference Sets*, in Proc. Sixth S-E Conf. Comb. Graph Theory and Comp., p 237-249, F. Hoffman et al. (Eds), Winnipeg Utilitas Math (1975)

[6] P. V. Kumar, R. A. Scholtz, and L. R. Welch, *Generalized Bent Functions and their Properties*, Journal of Combinatorial Theory, Series A 40, 90-107 (1985)

[7] F. J. Mac Williams & N. J. A. Sloane, *The Theory of Error Correcting Codes*, North Holland 1977.

[8] Kaisa Nyberg, *Constructions of Bent Functions and Difference Sets*, EUROCRYPT'90, Lecture Notes in Computer Science 473, 151-160 (1991).

[9] O. S. Rothaus, *On Bent Functions* , J. Comb. Theory, 20A, 300- 305 (1976)

Boolean functions satisfying a higher order strict avalanche criterion

Thomas W. Cusick

Department of Mathematics
State University of New York at Buffalo
106 Diefendorf Hall
Buffalo, New York 14214
e—mail: V360EAKB@UBVMS.CC.BUFFALO.EDU

Abstract. The Strict Avalanche Criterion (SAC) for Boolean functions was introduced by Webster and Tavares in connection with a study of the design of S–boxes. Later Forré extended this notion by defining strict avalanche criteria of order k for Boolean functions of n variables, where $0 \leq k \leq n-2$; the case $k = 0$ is the original SAC. Recent work by Lloyd, Preneel and others has been concerned with the problem of counting the functions which satisfy SAC of various orders. If the order is $n-2$ or $n-3$, this problem has been completely solved; the work in these cases is made easier by the fact that only quadratic Boolean functions occur. In this paper, we give good estimates for the number of Boolean functions which satisfy the SAC of order $n-4$. We also give a detailed description of the functions which satisfy SAC of order $n-4$, so the actual construction of these functions for cryptographic applications is made easy.

1. Introduction

The Strict Avalanche Criterion (SAC) was introduced by Webster and Tavares [10] in connection with a study of the design of S–boxes; a Boolean function is said to satisfy the SAC if complementing a single input bit results in changing the output bit with probability one half. Forré [3] extended this concept by defining higher order strict avalanche criteria. A Boolean function on n variables satisfies the SAC of order k , $0 \leq k \leq n-2$, if whenever k input bits are fixed arbitrarily, the resulting function of $n-k$ variables satisfies the SAC. It is easy to see (Lloyd [5]) that if a function satisfies the SAC of order $k > 0$, then it also satisfies the SAC of order j for any $j = 0, 1, \ldots, k-1$. As is the case with any Boolean function criterion of cryptographic significance, it is of interest to count the functions which satisfy the criterion. A number of recent papers have dealt, wholly or in part, with counting functions that satisfy the SAC of various orders, for example, Lloyd [5, 6, 7] and Preneel et al. [8]. In all of these papers, when the number of variables is large only quadratic Boolean functions (that is, functions whose algebraic normal form contains only terms of degree ≤ 2) are counted. The simplest cases involve the functions satisfying the SAC of order $n-2$ or $n-3$; in these cases, no non–quadratic function can satisfy the criteria, so a complete count is obtained.

The problem of counting the functions which satisfy the SAC of order $\leq n-4$ is difficult, because many of the functions in these cases are non–quadratic. In this paper we apply some methods from group theory and combinatorics to give good estimates for the number of functions which satisfy the SAC of order $n-4$. It is known (see Lemma 2 below) that all such functions have degree ≤ 3. We also give a detailed characterization of the quadratic and cubic functions which can occur in this case, so the actual construction of such functions is made routine.

2. Preliminaries

We define the degree of a Boolean function $f(x_1,\dots,x_n)$ (notation: deg(f)) to be the maximum of the degrees of the terms which occur in the algebraic normal form

$$f(x_1,\dots,x_n) = a_0 \oplus \sum_{1\leq i\leq n} a_i x_i \oplus \sum_{1\leq i<j\leq n} a_{ij} x_i x_j \oplus \dots \oplus a_{12\dots n} x_1 x_2 \dots x_n$$

(Here $\oplus$ denotes addition modulo 2. Some writers use the term "nonlinear order" instead of degree.) We say the Boolean function f is linear if it has degree one with no constant term a_0 and we say f is affine if it has degree one. We say f is quadratic or cubic if it has degree 2 or 3, respectively.

We use the abbreviation SAC(k) for the Strict Avalanche Criterion of order k. Our first lemma states the simple result that in testing whether a Boolean function f satisfies SAC(k), we can discard the affine terms (if any) in f.

Lemma 1. If a Boolean function f of n variables satisfies SAC(k) for some $k, 0 \leq k \leq n-2$, then so does $f \oplus g$, where g is any affine function of n variables.

Two fundamental results on Boolean functions satisfying SAC(k) were given by Preneel et al. [8].

Lemma 2 (Preneel et al. [8, Th. 10, p. 169]) Suppose f is a Boolean function of $n \geq 2$ variables. If f satisfies $SAC(n-2)$, then f has degree 2. If f satisfies $SAC(k)$, $0 \leq k \leq n-3$, then $\deg(f) \leq n-k-1$.

Lemma 3 (Preneel et al. [8, Th. 11, p. 170]) Suppose f is a quadratic Boolean function of $n \geq 2$ variables. Then f satisfies $SAC(k)$, $0 \leq k \leq n-2$, if and only if every variable x_i occurs in at least $k+1$ second degree terms of the algebraic normal form.

If we define

$$(1) \qquad q(x_1,\dots,x_n) = \sum_{1 \leq i < j \leq n} x_i x_j ,$$

then each variable x_i occurs in exactly $n-1$ terms. It follows immediately from Lemmas 1 to 3 that any Boolean function on n variables which satisfies $SAC(n-2)$ has nonaffine part equal to $q(x_1,\dots,x_n)$, so we have:

Lemma 4. There are 2^{n+1} Boolean functions of $n \geq 2$ variables which satisfy $SAC(n-2)$; they are exactly the functions $q(x_1,\dots,x_n) \oplus g(x_1,\dots,x_n)$, where g is affine.

The count in Lemma 4 was first found in a less direct way by Lloyd [5].

It is possible to use Lemmas 1 to 3 to count the Boolean functions which satisfy $SAC(n-3)$:

Lemma 5. Define the sequence $\{w_i\}$ of integers by $w_1 = 1$, $w_2 = 2$, $w_n = w_{n-1} + (n-1)w_{n-2}$ for $n \geq 3$. The number of Boolean functions of $n \geq 3$ variables which satisfy $SAC(n-3)$ is $2^{n+1} w_n$.

Proof. By Lemma 1, it suffices to show that the number of functions which satisfy $SAC(n-3)$ and have no affine terms is w_n. By Lemmas 2 and 3, any such function is obtained by deleting zero or more terms from the sum in (1) in such a way that the remaining sum has the property that every variable x_i occurs in at least $n-2$ terms. Thus S is a set of terms which we are allowed to delete if and only if no subscript i occurs in a term $x_i x_j$ in S more than once. It is easy to find a recursion for the number w_n of such sets S: Obviously $w_1 = 1$ (the empty set) and $w_2 = 2$. Clearly any set of terms $T = \{x_i x_j\}$ which is counted in w_{n-1} is also a set which must be counted in w_n, and this includes all sets which do not contain any term $w_i w_n$. If we have any set T which includes $\leq n-2$ variables from $x_1,\ldots,x_{n-1}$, we may add a term $x_k x_n$ to T and get a set to be counted in w_n if and only if x_k does not already occur in a term in T. There are w_{n-2} such sets T, by our definitions, so we count w_{n-2} sets for each k, $1 \leq k \leq n-1$. Hence $w_n = w_{n-1} + (n-1)w_{n-2}$ and the lemma is proved.

The numbers w_n have been previously studied in a combinatorial setting because w_n is the number of permutations in the symmetric group S_n whose square is the identity. In particular, Chowla et al. [2, p. 333] gave the following asymptotic formula for w_n:

$$w_n \sim (e^{1/4}\sqrt{2})^{-1} e^{\sqrt{n}} (n/e)^{n/2} \quad \text{as } n \to \infty. \tag{2}$$

It is not difficult to use Lemmas 1 to 3 to deduce the expression

$$2^{n+1} \sum_{0 \leq i \leq n/2} \frac{n!}{(n-2i)!\,i!\,2^i} \tag{3}$$

for the number of Boolean functions of $n \geq 3$ variables which satisfy $SAC(n-3)$. This expression is more complicated to compute with than the recursion for w_n and it does not seem simple to deduce the nice asymptotic result (2) from (3). Formula (3) was found in a more complicated way by Lloyd [6, p. 171].

3. Orbits of Boolean functions of 4 variables

Our goal is to find good estimates for the number of Boolean functions of n variables which satisfy $SAC(n-4)$. By Lemma 1, we lose no generality in confining ourselves to functions which have no affine terms; for brevity, we shall sometimes call such a function "affineless". We define t_n for $n \geq 4$ by

t_n = number of affineless Boolean functions of n variables which satisfy SAC($n-4$) .

It is clear that the property of satisfying SAC($n-4$) is preserved if we apply any permutation of $x_1,\ldots,x_n$ to a Boolean function of n variables which has that property. Thus if we let the symmetric group S_n of permutations of $\{1,2,\ldots,n\}$ act on the set of Boolean functions of n variables in the natural way, either all or none of the functions in any orbit under this action will satisfy SAC($n-4$) . Therefore in estimating t_n it suffices to estimate the size and the number of the orbits which contain affineless Boolean functions of n variables satisfying SAC($n-4$) . Given such a function, if we fix any $n-4$ variables we have an affineless function of 4 variables which satisfies SAC . Our first theorem gives a complete description of all of the orbits of these functions of 4 variables. We abbreviate the terms $x_ix_jx_k$ and x_ix_j in such a function by ijk and ij , respectively.

Theorem 1. There are exactly 90 orbits under the action of S_4 on the set of the 2^{10} Boolean functions of 4 variables with no affine terms. The tables below give the following information about these orbits: a representative function for each orbit, the size of the orbit, and whether the functions in the orbit satisfy SAC . The tables group all orbits with representatives having a given number of third degree terms in the algebraic normal form of the Boolean function. Presence of a given second degree term in the representative is indicated by an "x".

Four third degree terms – 11 orbits containing 64 functions

Representative = 123 ⊕ 134 ⊕ 124 ⊕ 234 ⊕ indicated quadratic terms

Orbit	Quadratic terms included						Orbit	SAC
number	12	13	14	23	24	34	size	
1							1	yes
2	x						6	no
3	x					x	3	no
4	x			x			12	no
5	x	x		x			4	yes
6	x	x	x				4	no
7	x			x		x	12	no
8	x	x			x	x	3	yes
9	x	x		x		x	12	no
10	x	x	x	x	x		6	no
11	x	x	x	x	x	x	1	no

Three third degree terms – 20 orbits containing 256 functions

Representative = 123 ⊕ 124 ⊕ 234 ⊕ indicated quadratic terms

Orbit	Quadratic terms included						Orbit	SAC
number	12	13	14	23	24	34	size	
12							4	no
13	x						12	no
14		x					12	no
15	x					x	12	no
16	x			x			12	no
17		x	x				12	no
18	x	x					24	no
19	x			x	x		4	no
20		x	x			x	4	no
21	x	x		x			12	no
22	x	x	x				12	no
23		x	x		x		24	no
24			x	x	x		24	no
25	x	x			x	x	12	no
26		x	x		x	x	12	no
27	x	x		x	x		12	no
28	x	x		x		x	24	no
29	x		x	x	x	x	12	no
30	x	x	x		x	x	12	no
31	x	x	x	x	x	x	4	no

Two third degree terms – 28 orbits containing 384 functions

Representative = 123 ⊕ 234 ⊕ indicated quadratic terms

Orbit number	Quadratic terms included 12	13	14	23	24	34	Orbit size	SAC
32							6	no
33			x				6	no
34				x			6	no
35	x						24	no
36			x	x			6	no
37	x					x	12	no
38	x	x					12	no
39	x				x		12	no
40		x		x			24	no
41	x		x				24	no
42	x	x		x			12	no
43	x	x	x				12	yes
44	x		x		x		12	no
45	x			x	x		12	no
46	x			x		x	12	no
47	x		x			x	12	yes
48	x		x	x			24	no
49	x				x	x	24	no
50	x	x			x	x	6	no
51	x		x	x		x	12	yes
52	x	x	x	x			12	yes
53		x	x	x		x	12	no
54	x		x		x	x	24	no
55		x		x	x	x	24	no
56	x	x		x	x	x	6	no
57	x	x	x		x	x	6	no
58		x	x	x	x	x	24	no
59	x	x	x	x	x	x	6	no

No third degree terms – 11 orbits containing 64 functions

Representative = indicated quadratic terms

Orbit number	Quadratic terms included 12	13	14	23	24	34	Orbit size	SAC
80							1	no
81	x						6	no
82	x					x	3	yes
83	x			x			12	no
84	x	x		x			4	no
85	x	x	x				4	yes
86	x			x		x	12	yes
87	x	x			x	x	3	yes
88	x	x		x		x	12	yes
89	x	x	x	x	x		6	yes
90	x	x	x	x	x	x	1	yes

One third degree term – 20 orbits containing 256 functions

Representative = 134 ⊕ indicated quadratic terms

Orbit number	Quadratic terms included 12	13	14	23	24	34	Orbit size	SAC
60							4	no
61	x						12	no
62		x					12	no
63	x					x	12	no
64	x			x			12	no
65		x	x				12	no
66	x	x					24	no
67	x			x	x		4	yes
68		x	x			x	4	no
69	x	x		x			12	no
70	x	x	x				12	no
71		x	x		x		24	no
72			x	x	x		24	no
73	x	x			x	x	12	no
74		x	x		x	x	12	no
75	x	x		x	x		12	yes
76	x	x		x		x	24	no
77	x		x	x	x	x	12	yes
78	x	x	x		x	x	12	no
79	x	x	x	x	x	x	4	yes

Proof. The number of orbits can be predicted in advance from the well–known Burnside counting formula (for example, see Artin [1, p. 196]), which says that if a group G with $|G|$ elements acts on a set S then

$$\text{number of orbits} = |G|^{-1} \sum_{g \in G} f(g) , \tag{4}$$

where $f(g)$ = the number of elements of S fixed by the action of g. In our case $|G| = |S_4| = 24$ and a computation shows that the sum in (4) is 2160.

Now it is a matter of calculation (we used the Mathematica software system), using the properties of the group action in order to minimize the work, to enumerate the orbits and test whether SAC is satisfied. This proves the theorem.

For our later work, it will be important to have more conceptual descriptions of the 18 orbits containing functions satisfying SAC. These are given in the Corollary below. In every case, the description in the Corollary follows in a straightforward way from the information given in Theorem 1 for the orbit in question.

Corollary to Theorem 1. The 18 orbits in Theorem 1 which contain functions satisfying SAC can be described as follows. The orbit numbers are the same as in Theorem 1. The quadratic terms are described in terms of the digit set {1, 2, 3, 4} = {h, i, j, k} , where the letters can stand for any one of the four numbers.

Orbit number	Orbit size	No. of degree 3 terms	Description of quadratic terms
1	1	4	None
5	4	4	Three terms containing only 3 digits, each twice
8	3	4	Four terms, consisting of two disjoint pairs
43	12	2	hi, hj, hk; h occurs only once in the two cubic terms
47	12	2	hi, hk, jk; h and k occur only once in the two cubic terms
51	12	2	hk and ij and two disjoint pairs, where h and k occur only once in the two cubic terms
52	12	2	hk and three pairs from one of the cubic terms; h and k occur only once in the cubic terms
67	4	1	hi, hj, hk; h does not occur in the cubic term
75	12	1	Four terms, where the two missing terms do not contain the number not in the cubic term
77	12	1	Five terms; the missing term is one of the 3 not containing the number not in the cubic term
79	4	1	All six possible pairs
82	3	0	One disjoint pair, e.g. 12, 34
85	4	0	Three pairs hi, hj, hk
86	12	0	Three pairs hi, ij, jk
87	3	0	Four terms consisting of two disjoint pairs
88	12	0	Four terms, where the two missing terms are not disjoint
89	6	0	Any five pairs
90	1	0	All six possible pairs

4 Functions with no affine terms satisfying SAC(n − 4)

We have already defined t_n $(n \geq 4)$ to be the number of affineless Boolean functions satisfying SAC(n − 4). By Lemma 2, if f is any such function, then $\deg(f) \leq 3$. We let T(f) denote the set of triples ijk such that $x_i x_j x_k$ is a term in the algebraic normal form of f and we let P(f) denote the corresponding set of pairs ij. Our next two lemmas give conditions that T(f) must satisfy if f satisfies SAC(n − 4).

Lemma 6. If $f(x_1,\dots,x_n)$ is a Boolean function satisfying SAC(n − 4), then no more than two triples in T(f) can have exactly two integers in common.

Proof. We give a proof by contradiction. Suppose f satisfies SAC(n − 4) and T(f) contains three triples 23i, 23j, 23k. We can assume with no loss of generality that the triples are 123, 234, 23k. If g is the function obtained when all variables except x_i $(1 \leq i \leq 4)$ are set equal to zero, then from Theorem 1 Corollary we see that if g satisfies SAC then P(g) must be one of the eight sets

$$\begin{array}{llll} \{12, 13, 14\}, & \{12, 14, 34\}, & \{13, 14, 24\}, & \{14, 24, 34\}, \\ \{12, 13, 14, 23\}, & \{12, 14, 23, 34\}, & \{13, 14, 23, 24\}, & \{14, 23, 24, 34\} \end{array} \quad (5)$$

(notice 14 is in every set and the last four sets are just the first four sets with 23 added).

First suppose 12 and 13 occur in P(g) (an analogous argument deals with the case when 24 and 34 occur). If g_1 is the function obtained when all variables except x_2, x_3, x_4 and x_k are set equal to zero, then $P(g_1)$ (which is in the list (5) with 1 replaced by k) must contain both 2k and 3k (otherwise either 24 or 34 is in $P(g_1)$, so P(g) contains 12, 13, 14 and at least one of 24, 34; this contradicts the fact that P(g) is in the list (5)). If g_2 is the function obtained when all variables except x_1, x_2, x_3, x_k are set equal to zero, we now have that $P(g_2)$ (which is in the list (5) with 4 replaced by k) contains at least 1k, 12, 13, 2k and 3k; this contradicts the fact that $P(g_2)$ can have no more than 4 elements.

Now suppose 12 and 34 occur in $P(g)$ (an analogous argument deals with the case when 13 and 24 occur). It follows that $P(g_1)$ must contain both 2k and 4k (otherwise 24 occurs in $P(g_1)$ and so in $P(g)$, contradicting the fact that $P(g)$ is in the list (5)). This means $P(g_2)$ must contain both 12 and 2k, which is impossible since $P(g_2)$ is in the list (5) with 4 replaced by k. This completes the proof.

Lemma 7. If $f(x_1,...,x_n)$ is a Boolean function satisfying $SAC(n-4)$, then no two triples in $T(f)$ can have exactly one integer in common.

Proof. We give a proof by contradiction. Let f satisfy $SAC(n-4)$. Suppose $T(f)$ contains two triples 123 and 14k, $k > 4$, and does not contain any of the triples 124, 134, 234. If g is the function obtained when all variables except x_1, x_2, x_3, x_4 are set equal to zero, then Theorem 1 Corollary shows that if g satisfies SAC, then $P(g)$ must be $\{14, 24, 34\}$ plus any subset of $\{12, 13, 23\}$. If g_1 is the function obtained when all variables except x_1, x_2, x_3, x_4, x_k are set equal to zero and $x_k = 1$, then $T(g_1) = \{123\}$ and $P(g_1)$ does not contain 14. This means g_1 does not satisfy SAC, so f cannot satisfy $SAC(n-4)$.

Now suppose $T(f)$ contains 123, 14k and 234. With g defined as above, we saw in the proof of Lemma 6 that $P(g)$ must be one of the sets in (5). Since 14 is in every set in (5), looking at g_1 defined above again gives a contradiction. A similar argument gives a contradiction in the case where $T(f)$ contains 123, 14k and one of 124, 134.

Finally suppose $T(f)$ contains 123, 14k, 124 and 234 (the case 123, 14k, 134, 234 is similar and other choices of two or more triples from $\{124, 134, 234\}$ are ruled out by Lemma 6). If g is defined as above, then $T(g)$ has three triples and so by Theorem 1 g does not satisfy SAC and f does not satisfy $SAC(n-4)$. This completes the proof of the lemma.

We turn to the quadratic functions which satisfy $SAC(n-4)$. By Lemma 3, such a function is characterized by the property that every variable x_i occurs in at least $n-3$ of the terms $x_i x_j$ in the algebraic normal form. Thus every such function with no affine terms is obtained from the sum in (1) by deleting a set S of zero or more terms in which no subscript i occurs more than twice; hence we can count the functions by counting the corresponding sets S (the same idea was already used in the proof of Lemma 5). In our next lemma we give a proof of an asymptotic formula for the number of these sets. By analogy with the proof of Lemma 5, we define

$$v_n = \text{the number of quadratic functions } \Sigma x_i x_j \text{ of } n \text{ variables such that no variable } x_i \text{ occurs more than twice}$$

Lemma 8. The number of quadratic Boolean functions of $n \geq 4$ variables, with no affine terms, which satisfy SAC$(n-4)$ is v_n. We have

$$v_n \sim (2^{5/2}\pi)^{-1/2}(n+1)^{-3/4}\exp((\tfrac{1}{2}(n+1))^{1/2})n! \tag{6}$$

as $n \to \infty$.

Proof. The first sentence of the lemma follows from the remarks in the paragraph preceding the lemma. The number v_n is clearly the number of graphs (no multiple edges) on the vertex set $\{1,2,\ldots,n\}$ such that every component is an isolated vertex, an edge, a cycle (of length at least 3) or a path. By well-known combinatorial arguments (see [9, Example 6.5, p. 134] for a similar problem), the exponential generating function is

$$\sum_{n=0}^{\infty} v_n x^n/n! = \exp(-x^2/4 + x/(2(1-x)))(1-x)^{-1/2}$$

This function is "admissible" in the sense of Hayman [4] and it follows from his Theorem I, Corollary II that (6) holds as $n \to \infty$.

Lemma 8 Corollary. The number v_n satisfies

$$\log v_n \sim n \log n \text{ as } n \to \infty.$$

Proof. This follows from (6) and Stirling's formula $n! \sim \sqrt{2\pi n}\ (n/e)^n$.

We are now ready to give our estimates for t_n .

Theorem 2. The following inequalities hold for the number t_n of Boolean functions of n variables which satisfy SAC(n − 4) and have no affine terms:

$$n!g(n) < t_n < 3.2\ n^3\ n!\ g(n-4)$$

where

$$g(n) \sim C(n+1)^{-3/4} \exp((\tfrac{1}{2}(n+1))^{1/2})$$

as $n \to \infty$, with $C = (2^{5/2}\pi)^{-1/2} = .237\ldots$.

Proof. We see from the tables after Theorem 1 that there are no orbits which contain functions satisfying SAC and which have three third degree terms. It follows from this and Lemmas 6 and 7 that if f is an affineless Boolean function satisfying SAC(n − 4) , then the set of triples T(f) is either made up of disjoint triples or is made up of exactly two triples with a common pair of elements, and possibly some further disjoint triples.

We define

$$g(n) = v_n/n! , \tag{7}$$

where v_n is defined above Lemma 8.

We first consider the case where T(f) contains a single triple, which we may take to be 123. Suppose now that we fix all of the variables except those with subscripts 1, 2, 3 and j for some j , $4 \le j \le n$. The resulting function of 4 variables satisfies SAC and so by Theorem 1 it lies in one of four orbits which contain functions with a single term of degree 3. If 123 is the term of degree 3, then by the Corollary to Theorem 1 the set of pairs for the function of the variables 1,2,3,j is

1j, 2j, 3j plus some subset of {12, 13, 23} ;

all of the eight possible subsets can occur. It follows from this that the pair set P(f) for the function f of n variables must be some subset of {12, 13, 23} plus 14, 24, 34, 15, 25, 35,... ,

1n, 2n, 3n plus more terms involving $x_4,...,x_n$.

Let $r(x_1,...,x_n)$ denote the quadratic function of n variables formed from the terms 15, 25, 35 and all the subsequent terms in the above list. If we fix variables x_1,x_2,x_3 in some way, since our original function f satisfies SAC(n − 4) , the resulting quadratic function r of n − 3 variables must satisfy SAC of order n − 7 = (n − 3) − 4 . Since there is no choice for the affine terms in such a function r , by Lemma 8 the number of such functions n is $\leq v_{n-3}$. Thus the total number of functions f satisfying SAC(n − 4) and with T(f) containing a single triple is less than

$$8\binom{n}{3}g(n-3)(n-3)!\ ;$$

here the binomial coefficient gives the number of ways of choosing the single triple, say ijk ; 8 is the number of ways of choosing the associated subset of {ij, ik, jk} ; and (7) gives the final factors.

Now we suppose the set of triples T(f) contains t disjoint triples. A straightforward extension of the argument in the previous paragraph now gives the upper bound

$$8^t\binom{n}{3t}g(n-3t)(n-3t)!$$

for the number of such functions f which satisfy SAC(n − 4) . Summing our estimates, we find that the number of functions f satisfying SAC(n − 4) and having T(f) made up of disjoint triples is less than

$$g(n-3)\sum_{t=1}^{[n/3]} 8^t\binom{n}{3t}(n-3t)! < (e^2-1)n!\,g(n-3)$$

since

$$\sum_{t=1}^{[n/3]} 8^t/(3t)! < \sum_{t=1}^{\infty} 8^{t/3}/t! = e^2-1\ .$$

Next we consider the case where $T(f)$ contains two triples with a common pair, which we may take to be 123, 234. Suppose now that we fix all of the variables except those with subscripts 1,2,3,4. The resulting function of 4 variables satisfies SAC and so by Theorem 1 it lies in one of four orbits which contain functions with two terms of degree 3. If we fix x_4 we have a function of $n-1$ variables satisfying $SAC(n-5)$ with triple set 123 only. Thus our above analysis for the case of a singleton triple set applies (with $n-1$ in place of n), and so the number of such functions of $n-1$ variables is less than

$$8\binom{n-1}{3} g(n-4)(n-4)!\,.$$

To return to functions of n variables with two–element triple set made up of two triples with some pair ij in common we must multiply this bound by

$$\binom{n}{3} 3\,(n-3)\,;$$

here the binomial coefficient gives the number of ways of choosing the first triple, 3 is the number of ways of choosing ij from the first triple and $n-3$ is the number of ways of choosing the third element in the second triple.

As in our previous work, we can extend the above argument to the case where the triple set is made up of two triples with a common pair plus t more disjoint triples. This gives the following upper bound for the number of functions f satisfying $SAC(n-4)$ and having triple set $T(f)$ containing two triples with a common pair:

$$3(n-3)\binom{n}{3} \sum_{t=1}^{[(n-4)/3]} 8^t \binom{n-1}{3t} g(n-1-3t)(n-1-3t)!$$

$$< \frac{1}{2}(n-1)(n-2)(n-3)n!\, g(n-4) \sum_{t=1}^{[(n-4)/3]} 8t/(3t)!$$

$$< \frac{1}{2}(e^2-1)n^3\, n!\, g(n-4)\,.$$

Since $\frac{1}{2}(e^2-1) = 3.195\ldots$, putting together our estimates and using Lemma 8 gives the upper bound in Theorem 2. The lower bound follows immediately from (7) and Lemma 8.

References

[1] M. Artin, *Algebra*, Prentice–Hall, 1992.

[2] S. Chowla, I. N. Herstein and K. Moore, On recursions connected with symmetric groups I, *Canadian J. Math.* 3 (1951), 328–334.

[3] R. Forré, The strict avalanche criterion: spectral properties of Boolean functions and an extended definition, *Advances in Cryptology – Crypto '88*, Lect. Notes Comp. Sci. 403, Springer–Verlag, 1990, pp. 450–468.

[4] W. K. Hayman, A generalisation of Stirling's formula, J. reine agnew. Math. 196 (1956), 67–95.

[5] S. Lloyd, Counting functions satisfying a higher order strict avalanche criterion, *Advances in Cryptology – Eurocrypt '89*, Lect. Notes Comp. Sci. 434, Springer– Verlag, 1990, pp. 63–74.

[6] S. Lloyd, Characterising and counting functions satisfying the strict avalanche criterion of order $(n-3)$, in *Cryptography and Coding II*, Clarendon Press, Oxford, 1992, pp. 165–172.

[7] S. Lloyd, Counting binary functions with certain cryptographic properties, *J. Cryptology* 5 (1992), 107–131.

[8] B. Preneel, W. Van Leekwijck, L. Van Linden, R. Govaerts and J. Vandewalle, Propagation characteristics of Boolean functions, *Advances in Cryptology – Eurocrypt '90*, Lect. Notes Comp. Sci. 473, Springer–Verlag, 1991, pp. 161–173.

[9] R. P. Stanley, Generating functions, in *Studies in Combinatorics*, MAA Studies 17, Math. Assoc. America, 1978, pp. 100–141.

[10] A. F. Webster and S. E. Tavares, On the design of S–boxes, *Advances in Cryptology – Crypto '85*, Lect. Notes Comp. Sci. 218, Springer–Verlag, 1986, pp. 523–534.

Size of Shares and Probability of Cheating in Threshold Schemes*

Marco Carpentieri Alfredo De Santis Ugo Vaccaro

Dipartimento di Informatica ed Applicazioni, Università di Salerno, 84081 Baronissi (SA), Italy

Abstract

In this paper we study the amount of secret information that must be given to participants in any secret sharing scheme that is secure against coalitions of dishonest participants in the model of Tompa and Woll [20]. We show that any (k, n) threshold secret sharing algorithm in which any coalition of less than k participants has probability of successful cheating less than some $\epsilon > 0$ it must give to each participant shares whose sizes are at least the size of the secret plus $\log \frac{1}{\epsilon}$.

1 Introduction

In 1979 Blakley [2] and Shamir [15] gave protocols to solve the following problem: divide a secret s in n shares $d_1, \ldots, d_n$ in such a way that:

i) the knowledge of k or more d_i's makes s computable,

ii) the knowledge of $k - 1$ or less d_i's leaves s *completely* indeterminate.

This problem, known in the literature as "(k, n) Threshold Secret Sharing", has received considerable attention in the last few years because of its many applications to several fields, as data security, secure computation and others [10]. For an extensive bibliography and illustration of the main results in the area we refer the reader to [17] and [18].

Let $\mathcal{P} = \{P_1, \ldots, P_n\}$ be a set of participants, S be the set of secrets and $D_1, \ldots, D_n$ be the sets in which the shares to participants $P_1, \ldots, P_n$ are taken. Any probability distribution $\{p(s)\}_{s \in S}$ on the set of secrets S and a sharing algorithm for secrets in S (both known by each participant) naturally induce a probability distribution $\{p(d_1, \ldots, d_n)\}_{d_1 \in D_1, \ldots, d_n \in D_n}$ on the joint space $D_1 \times \ldots \times D_n$ of the possible values of the shares. Therefore, we will consider each D_i as a random variable. Formally, a (k, n) Threshold Secret Sharing Scheme is a method to distribute shares to the n participants such that:

*Partially supported by Italian Ministry of University and Research (M.U.R.S.T.) and by National Council for Research (C.N.R.).

1. for any k-tuple of distinct indexes $i_1, \ldots, i_k$, $1 \leq i_j \leq n$, for each $(d_1, \ldots, d_k) \in D_{i_1} \times \ldots \times D_{i_k}$ such that $p(d_1, \ldots, d_k) > 0$ there exists an unique secret $s \in S$ such that $p(s \mid d_1, \ldots, d_k) = 1$,

2. for any $j < k$, for any j-tuple of distinct indexes $i_1, \ldots, i_j$, $1 \leq i_j \leq n$, for each $d_1, \ldots, d_j \in D_{i_1} \times \ldots \times D_{i_j}$, such that $p(d_1, \ldots, d_j) > 0$, for each $s \in S$ it holds $p(s \mid d_1, \ldots, d_j) = p(s)$.

The first property implies that the shares held by any group of k participants univocally determines the secret $s \in S$. Notice that the second property means that the probability that the secret is equal to s given that the shares held by any group of $j < k$ participants are $d_1, \ldots, d_j$ is the same as the *a priori* probability that the secret is s. Therefore, no amount of knowledge of shares of less than k participants enables a Bayesian opponent to modify an *a priori* guess regarding which the secret is.

Using the information theoretic concepts of entropy (see Appendix for definitions and properties) the two previous conditions can be stated as follows [12], [8], [3]:

1′. for any k-tuple of distinct indexes $i_1, \ldots, i_k$, $1 \leq i_j \leq n$, it holds $H(S \mid D_{i_1}, \ldots, D_{i_k}) = 0$,

2′. for any $j < k$, for any j-tuple of distinct indexes $i_1, \ldots, i_j$, $1 \leq i_j \leq n$, it holds $H(S \mid D_{i_1}, \ldots, D_{i_j}) = H(S)$.

Tompa and Woll [20] considered the following scenario: let us suppose that $k-1$ participants $P_{i_1}, \ldots, P_{i_{k-1}}$ want to cheat a k-th participant P_{i_k}. Let $d_1, \ldots, d_{k-1}, d_k$ be the shares held by participants $P_{i_1}, \ldots, P_{i_k}$ and s be the correct secret, that is, the secret the participants would reconstruct if they pooled together their shares. The $k-1$ cheaters $P_{i_1}, \ldots, P_{i_{k-1}}$, not knowing d_k, could return $d'_1, \ldots, d'_{k-1}$ forged shares in a tentative to force the k-th participant P_{i_k} to reconstruct a secret $s' \neq s$. Tompa and Woll showed that Shamir's scheme [15] is insecure against this attack, in the sense that even a single participant, with high probability, can deceive other $k-1$ honest participants. Tompa and Woll, however, modified Shamir's scheme to make it secure against cheating. Briefly, they proposed a sharing algorithm that specifies a subset S_{legal} of the set S of possible secrets. A secret will be accepted as authentic only if it is an element of S_{legal}. If a set of k participants calculate the secret to be an element of $S_{illegal} = S - S_{legal}$, then they realize that at least one of them is cheating. In other words S_{legal} is the set of legal secrets that each participant would expect to reconstruct. $S_{illegal}$ is a set of illegal secrets that is introduced only to reveal cheating. Other papers that addressed the problem of coping with cheaters in secret sharing schemes are [1], [7], [14], [16] and [20].

An important issue in the implementation of secret sharing schemes is the size of shares distributed to participants since the security of a system degrades as the amount of the information that must be kept secret increases. Recently, several papers studied this topic and both upper bounds and lower bounds on the size of the shares have been provided [3], [4], [5], [6], [8], [19]. In this paper we study the amount of secret information that must be given to participants in terms of the probability that the previously described attack be successful. Our motivations are based on the observation that the Tompa and Woll secret sharing scheme requires that each participant must receive an amount of secret information that grows with the level of security one imposes against dishonest coalitions. We show that this phenomenon is unavoidable, in the sense that in any secret sharing scheme that has probability of successful cheating less than some $\epsilon > 0$, it must give to each participant shares whose size is at least the size of the secret plus $\log \frac{1}{\epsilon}$.[1]. The security of the schemes presented in this paper is unconditional, since they are not based on any computational assumption.

2 Robust Secret Sharing Schemes

Tompa and Woll [20] defined the cheating probability as "*the probability that from $k-1$ forged shares $d'_1, \ldots, d'_{k-1}$ and any d_k the secret s' reconstructed is legal, but not a correct one*". In order to formally define the problem let us introduce some notations. For each k-tuple of distinct indexes $i_1, \ldots, i_k$, $1 \leq i_j \leq n$, and for any $(d_1, \ldots, d_k) \in D_{i_1} \times \ldots \times D_{i_k}$ such that $p(d_1, \ldots, d_k) > 0$ (i.e., for any k-tuple of shares that the secret sharing algorithm can possibly give to participants $P_{i_1}, \ldots, P_{i_k}$ to share a particular secret $s \in S$), let us denote by $(d_1, \ldots, d_k) \to s$ the fact that the values $d_1, \ldots, d_k$ force participants $P_{i_1}, \ldots, P_{i_k}$ to reconstruct the secret $s \in S$. Since the sharing algorithm and the probability distribution $\{p(s)\}_{s \in S}$ are known to all participants, it follows that the probability distribution $\{p(d_1, \ldots, d_n)\}_{d_1 \in D_1, \ldots, d_n \in D_n}$ is also known. Assume that the $k-1$ cheaters $P_{i_1}, \ldots, P_{i_{k-1}}$ know the correct secret s and their legal shares $d_1, \ldots, d_{k-1}$. From Decision Theory it is well known (see for example [11]) that the decision rule that minimizes the probability of error is the Bayesian decision rule that chooses the hypothesis with largest "a posteriori" probability. Therefore, the best strategy the $k-1$ cheaters $P_{i_1}, \ldots, P_{i_{k-1}}$ can follow to cheat a k-th participant P_{i_k} is to give him forged values $d'_1, \ldots, d'_{k-1}$ that maximize the following quantity

$$\sum_{d_k \in D_{i_k} : (d'_1, \ldots, d'_{k-1}, d_k) \to S - \{s\}} p(d_k \mid d_1, \ldots, d_{k-1}, s) .$$

Averaging on all secrets in S and on the possible shares the sharing algorithm could give, we have that the maximum average probability $P(Cheat \mid D_{i_1}, \ldots, D_{i_{k-1}}, S)$ that $k-1$ participants $P_{i_1}, \ldots, P_{i_{k-1}}$, knowing the correct secret, succedees in cheating the k-th participant P_{i_k} is:

---

[1] All logarithms in this paper are of base 2

$$P(Cheat \mid D_{i_1}, \ldots, D_{i_{k-1}}, S)$$

$$= \sum_{d_1 \in D_{i_1}, \ldots, d_{k-1} \in D_{i_{k-1}}, s \in S} p(d_1, \ldots, d_{k-1}, s) \max_{d'_1, \ldots, d'_{k-1}} \left[\sum_{d_k \in D_{i_k} : (d'_1, \ldots, d'_{k-1}, d_k) \to S - \{s\}} p(d_k \mid d_1, \ldots, d_{k-1}, s) \right]$$

For a fixed ϵ, $1 \geq \epsilon > 0$, we define a (k, n, ϵ) Robust Secret Sharing Scheme as secret sharing scheme that satisfies the following properties:

P1) for any k-tuple of distinct indexes $i_1, \ldots, i_k$, it holds $H(S \mid D_{i_1}, \ldots, D_{i_k}) = 0$,

P2) for any j-tuple of indexes $i_1, \ldots, i_j$, $1 \leq j < k$, it holds $H(S \mid D_{i_1}, \ldots, D_{i_j}) = H(S)$,

P3) $P(Cheat \mid D_{i_1}, \ldots, D_{i_{k-1}}, S) \leq \epsilon$.

Properties P1 and P2 are those of a (k, n) threshold scheme. The property P3 assures that any cheating tentative has arbitrarily small probability of succeeding, even though the cheaters know the correct secret.

Note that the condition $\epsilon > 0$ for (k, n, ϵ) Robust Secret Sharing Schemes is necessary, since the probability of cheating $P(Cheat \mid D_{i_1}, \ldots, D_{i_{k-1}}, S)$ cannot be 0 in any (k, n) Threshold Secret Sharing Scheme. For a proof of this fact see the remark following Lemma 2.

A quantity that will play an important role to derive our result is the probability that $k-1$ participants $P_{i_1}, \ldots, P_{i_{k-1}}$ can guess the share of the k-th participant P_{i_k}, given that they know the correct secret s besides their own shares $d_1, \ldots, d_{k-1}$. Again, the best strategy that the $k-1$ participants can follow is to choose the value $\underline{d_k} \in D_{i_k}$ that maximizes the conditional probability

$$p(d_k \mid d_1, \ldots, d_{k-1}, s).$$

Averaging on all the possible shares and secrets it follows that the maximum average probability $P(GuessD_{i_k} \mid D_{i_1}, \ldots, D_{i_{k-1}}, S)$ that $k-1$ participants $P_{i_1}, \ldots, P_{i_{k-1}}$, knowing the correct secret, succeeds in guessing the value of the share of the k-th participant P_{i_k} is:

$$P(GuessD_{i_k} \mid D_{i_1}, \ldots, D_{i_{k-1}}, S)$$

$$= \sum_{d_1 \in D_{i_1}, \ldots, d_{k-1} \in D_{i_{k-1}}, s \in S} p(d_1, \ldots, d_{k-1}, s) \max_{d_k \in D_{i_k}} p(d_k \mid d_1, \ldots, d_{k-1}, s).$$

The following lemma holds.

Lemma 1 *In any (k, n) Threshold Secret Sharing Schemes one has*

$$H(D_{i_k} \mid D_{i_1}, \ldots, D_{i_{k-1}}, S) \geq \log \frac{1}{P(GuessD_{i_k} \mid D_{i_1}, \ldots, D_{i_{k-1}}, S)}.$$

Proof. Since the function $\log x$ is convex, the lemma follows from Jensen inequality [13]. Indeed, we have

$$
\begin{aligned}
-H(D_{i_k} \mid D_{i_1},\ldots,D_{i_{k-1}},S) &= \sum_{d_1,\ldots,d_k,s} p(d_1,\ldots,d_k,s)\log p(d_k \mid d_1,\ldots,d_{k-1},s) \\
&\le \log\left[\sum_{d_1,\ldots,d_k,s} p(d_1,\ldots,d_k,s)\, p(d_k \mid d_1,\ldots,d_{k-1},s)\right] \\
&\qquad\qquad\text{(from Jensen inequality)} \\
&\le \log\left[\sum_{d_1,\ldots,d_{k-1},s} p(d_1,\ldots,d_{k-1},s)\max_{d_k} p(d_k \mid d_1,\ldots,d_{k-1},s)\right] \\
&= \log\ P(GuessD_{i_k} \mid D_{i_1},\ldots,D_{i_{k-1}},S).
\end{aligned}
$$

□

The relationship between the cheating probability $P(Cheat \mid D_{i_1},\ldots,D_{i_{k-1}},S)$ and the guess probability $P(GuessD_{i_k} \mid D_{i_1},\ldots,D_{i_{k-1}},S)$ is stated by the following lemma:

Lemma 2 *In any (k,n) Threshold Secret Sharing Schemes one has*

$$P(GuessD_{i_k} \mid D_{i_1},\ldots,D_{i_{k-1}},S) \le P(Cheat \mid D_{i_1},\ldots,D_{i_{k-1}},S)\ .$$

Proof. The proof follows from the following argument. For all $d_1 \in D_{i_1},\ldots,d_{k-1} \in D_{i_{k-1}}$ and $s \in S$, in correspondence of a value $\underline{d}_k \in D_{i_k}$ that maximizes the guess probability, that is, for which $\max_{d_k \in D_{i_k}} p(d_k \mid d_1 \ldots d_{k-1}s) = p(\underline{d}_k \mid d_1 \ldots d_{k-1}s)$, it must exist a choice of $k-1$ shares $d'_1 \in D_{i_1},\ldots,d'_{k-1} \in D_{i_{k-1}}$ such that $(d'_1,\ldots,d'_{k-1},\underline{d}_k) \rightarrow s'$, for some $s' \in S-\{s\}$. In the opposite case, the value $\underline{d}_k$ would univocally identify a secret s, in the sense that the k-th participant P_{i_k} knowing only the share $\underline{d}_k$, could reconstruct the correct secret s. This contradicts Property P2. Therefore,

$$\max_{d_k \in D_{i_k}} p(d_k \mid d_1,\ldots,d_{k-1},s) \le \max_{d'_1,\ldots,d'_{k-1}}\left[\sum_{d_k \in D_{i_k}:(d'_1,\ldots,d'_{k-1},d_k)\rightarrow S-\{s\}} p(d_k \mid d_1,\ldots,d_{k-1},s)\right]$$

and the lemma follows. □

Remark. It is clear that the probability of guessing $P(GuessD_{i_k} \mid D_{i_1},\ldots,D_{i_{k-1}},S)$ is positive. Therefore, from Lemma 2 it follows that $P(Cheat \mid D_{i_1},\ldots,D_{i_{k-1}},S) > 0$.

The following theorem represents our main result.

Theorem 1 *In any (k,n,ϵ) Robust Secret Sharing Scheme, for any k-tuple of distinct indexes $i_1,\ldots,i_k$ it holds*

$$H(D_{i_k} \mid D_{i_1},\ldots,D_{i_{k-1}}) \geq H(S) + \log \tfrac{1}{\epsilon}.$$

Proof. From (3) of Appendix we have:

$$\begin{aligned} H(D_{i_k}, S \mid D_{i_1},\ldots,D_{i_{k-1}}) &= H(S \mid D_{i_1},\ldots,D_{i_{k-1}}) + H(D_{i_k} \mid D_{i_1},\ldots,D_{i_{k-1}}, S) \\ &= H(S \mid D_{i_1},\ldots,D_{i_k}) + H(D_{i_k} \mid D_{i_1},\ldots,D_{i_{k-1}}) . \end{aligned}$$

As a consequence of Properties P1 and P2 it follows that

$$H(D_{i_k} \mid D_{i_1},\ldots,D_{i_{k-1}}) = H(S) + H(D_{i_k} \mid D_{i_1},\ldots,D_{i_{k-1}}, S).$$

From Lemmas 1 and 2, and Property P3 of (k,n,ϵ) Robust Secret Sharing Scheme, the theorem follows. Indeed,

$$\begin{aligned} H(D_{i_k} \mid D_{i_1},\ldots,D_{i_{k-1}}) &= H(S) + H(D_{i_k} \mid D_{i_1},\ldots,D_{i_{k-1}}, S) \\ &\geq H(S) + \log \frac{1}{P(GuessD_{i_k} \mid D_{i_1},\ldots,D_{i_{k-1}}, S)} \quad \text{(by Lemma 1)} \\ &\geq H(S) + \log \frac{1}{P(Cheat \mid D_{i_1},\ldots,D_{i_{k-1}}, S)} \quad \text{(by Lemma 2)} \\ &\geq H(S) + \log \frac{1}{\epsilon} \quad \text{(by Property P3).} \end{aligned}$$

□

Corollary 1 *In any (k,n,ϵ) Robust Secret Sharing Scheme, if the secret is uniformly chosen it holds*

$$\log |D_i| \geq \log |S| + \log \tfrac{1}{\epsilon}.$$

Proof. The proof is immediate from Theorem 1 and properties (1) and (2) of the entropy in the Appendix. □

The corollary shows that in a (k,n,ϵ) Robust Secret Sharing Scheme the size of the shares given to participants — measured as the number of bits necessary to their representation — necessarily grows as ϵ decreases. For completeness, we recall that in the Tompa and Woll algorithm the size of shares $\log |D_i|$ satisfies the bound

$$2\log\left(\frac{(|S|-1)(k-1)}{\epsilon} + k\right) < \log |D_i| < 2\log\left(\frac{(|S|-1)(k-1)}{\epsilon} + k\right) + 1.$$

We have proved a tradeoff between the size of the shares and the probability of successful cheating in perfect (i.e., properties 1. and 2. hold) threshold schemes. The same technique can be used for non-perfect schemes, such as, for example, ramp schemes.

References

[1] M. Ben-Or, T. Rabin *Verifiable Secret Sharing and Multiparty Protocols with Honest Majority*, Proc. 21st ACM Symposium on Theory of Computing, pp. 73-85, 1989.

[2] G. R. Blakley, *Safeguarding Cryptographic Keys*, Proceedings AFIPS 1979 National Computer Conference, pp. 313-317, 1979.

[3] C. Blundo, A. De Santis, L. Gargano, and U. Vaccaro, *On the Information Rate of Secret Sharing Schemes*, "Advances in Cryptology - CRYPTO 92", Ed. E. Brickell, "Lecture Notes in Computer Science", Springer-Verlag, (to appear).

[4] C. Blundo, A. De Santis, D. R. Stinson, and U. Vaccaro, *Graph Decomposition and Secret Sharing Schemes*, "Advances in Cryptology - EUROCRYPT 92", Ed. R. Rueppel, "Lecture Notes in Computer Science", Springer-Verlag, (to appear).

[5] E. F. Brickell and D. M. Davenport, *On the Classification of Ideal Secret Sharing Schemes,* J. Cryptology, Vol.**4**, 123–134, 1991.

[6] E. F. Brickell and D. R. Stinson, *Some Improved Bounds on the Information Rate of Perfect Secret Sharing Schemes*, J. Cryptology, Vol. **5**, pp. 153–156, 1992.

[7] E. F. Brickell, D. R. Stinson, *The Detection of Cheaters in Threshold Schemes*, SIAM J. Disc. Math, Vol. **4**, pp. 502-510, 1991.

[8] R. M. Capocelli, A. De Santis, L. Gargano, U. Vaccaro, *On the Size of Shares for Secret Sharing Schemes*, Advances in Cryptology – CRYPTO '91, J. Feigenbaum (Ed.), Lectures Notes in Computer Science, Vol. **576**, pp. 101–113, 1992, Springer-Verlag. Also to appear in Journal of Cryptology.

[9] I. Csiszár and J. Körner, *Information Theory. Coding Theorems for Discrete Memoryless Systems,* Academic Press, 1981.

[10] D. Denning, *Cryptography and Data Security,* Addison–Wesley, Reading, MA, 1983.

[11] T.S. Ferguson, *Mathematical Statistics*, Academic Press, New York, 1967.

[12] E. D. Karnin, J. W. Greene, and M. E. Hellman, *On Secret Sharing Systems,* IEEE Trans. on Inform. Theory, Vol. **IT-29**, pp. 35–41, 1983.

[13] A.W. Marshall, I. Olkin, *Inequalities: Theory of Majorization and Its Applications*, Academic Press, New York, 1979.

[14] R. J. McEliece, D. V. Sarwate, *On Sharing Secrets and Reed-Solomon Codes*, Communications of the ACM, Vol. **24**, pp. 583-584, 1981.

[15] A. Shamir, *How to Share a Secret*, Communication of the ACM, Vol. **22**, pp. 612-613, 1979.

[16] G. Simmons, *Robust Shared Secret Schemes or "How to be Sure You Have the Right Answer Even Though You Do Not Know the Question"*, Congr. Numer., Vol. **68**, pp. 215-248, 1989.

[17] G. J. Simmons, *An Introduction to Shared Secret and/or Shared Control Schemes and Their Application*, Contemporary Cryptology, IEEE Press, pp. 441–497, 1991.

[18] D. R. Stinson, *An Explication of Secret Sharing Schemes*, Design, Codes and Cryptography, Vol. **2**, pp. 357–390, 1992.

[19] D. R. Stinson, *Decomposition Constructions for Secret Sharing Schemes*, Technical Report UNL-CSE-92-020, Department of Computer Science and Engineering, University of Nebraska, September 1992.

[20] M. Tompa, H. Woll, *How to Share a Secret with Cheaters*, Journal of Cryptology, Vol. 1, pp. 133-139, 1988.

3 Appendix

In this section we shall review the information theoretic concepts we used in the paper. For a complete treatment of the subject we refer the reader to [9].

Given a probability distribution $\{p(x)\}_{x\in X}$ on a finite set X, define the *entropy* of X, $H(X)$, as:

$$H(X) = -\sum_{x\in X} p(x)\log p(x).$$

The entropy $H(X)$ is a measure of the average information content of the elements in X or, equivalently, a measure of the average uncertainty one has about which element of the set X has been chosen when the choices of the elements from X are made according to the probability distribution $\{p(x)\}_{x\in X}$. The entropy $H(X)$ enjoys the following property:

$$0 \le H(X) \le \log|X|, \tag{1}$$

where $H(X) = 0$ if and only if there exists $x_0 \in X$ such that $p(x_0) = 1$; $H(X) = \log|X|$ if and only if $p(x) = 1/|X|$, for each $x \in X$.

Given two sets X and Y and a joint probability distribution $\{p(x,y)\}_{x\in X, y\in Y}$ on their cartesian product, the *conditional entropy* $H(X|Y)$ of X given Y is defined as:

$$H(X|Y) = -\sum_{y\in Y}\sum_{x\in X} p(y)p(x|y)\log p(x|y).$$

The conditional entropy satisfies the following inequalities

$$H(X) \ge H(X|Y) \ge 0. \tag{2}$$

The entropy of the joint space XY satisfies:

$$H(XY) = H(X) + H(Y|X) = H(Y) + H(X|Y).$$

Analogously, the conditional entropy of XY given Z satisfies:

$$H(XY|Z) = H(X|Z) + H(Y|XZ) = H(Y|Z) + H(X|YZ). \tag{3}$$

Nonperfect Secret Sharing Schemes and Matroids

Kaoru KUROSAWA, Koji OKADA, Keiichi SAKANO,
Wakaha OGATA, Shigeo TSUJII

Department of Electrical and Electronic Engineering,
Faculty of Engineering, Tokyo Institute of Technology
2-12-1 O-okayama, Meguro-ku, Tokyo 152, Japan
E-mail: kkurosaw@ss.titech.ac.jp

Abstract. This paper shows that nonperfect secret sharing schemes (NSS) have matroid structures and presents a direct link between the secret sharing matroids and entropy for both perfect and nonperfect schemes. We define natural classes of NSS and derive a lower bound of $|V_i|$ for those classes. "Ideal" nonperfect schemes are defined based on this lower bound. We prove that every such ideal secret sharing scheme has a matroid structure. The rank function of the matroid is given by the entropy divided by some constant. It satisfies a simple equation which represents the access level of each subset of participants.

1 Introduction

Secret sharing schemes are defined by using entropy such as follows. The inputs to a secret sharing scheme are a secret S and a random number R. The outputs of the scheme are V_1 through V_n, which are called shares. Each V_i is given to a party P_i. We assume that S and R are uniformly distributed. Then, V_i becomes a random variable with a certain distribution. We denote the entropy as $H(V_i)$. In a "perfect" secret sharing scheme, any subset of parties is an access set or a non-access set. If A is an access set, A can recover S. The conditional entropy is that $H(S|A) = 0$. If B is a non-access set, B has absolutely no information on S. That is, $H(S|B) = H(S)$, which equals the bit length of S (denoted by $|S|$) because S is assumed to be uniformly distributed. No subset is allowed in between.

Many researchers have investigated perfect secret sharing schemes extensively so far [1]~[16]. Let's review the history of perfect secret sharing schemes. An access structure Γ is defined as the family of all access sets.

1. First, (k, n) threshold schemes were proposed by Shamir and Blakley [1][2].
2. Later, more general access structures were considered. It was shown that Γ is an access structure of a perfect secret sharing scheme if and only if Γ is monotone [3].

The meaning of monotone is as follows. If A can recover S, then any set A' which contains A can also recover S. Formally, Γ is monotone if A belongs to Γ and A' contains A, then A' also belongs to Γ.

Further, it was proved that $|V_i| \geq |S|$ for any V_i [6][7]. This lower bound was obtained by using entropy. Recently, more tight lower bounds of V_i were shown for some access structures [6][8][9][11][12].

We call a scheme ideal if $|V_i| = |S|$. Brickell and Davenport showed that every ideal perfect scheme has a matroid structure by using a combinatorial argument [5]. Matroids play a central role in many combinatorial problems [17]. Many subjects can be more clearly understood by using the matroids. No relation is known between the entropy and the secret sharing matroids.

The size of V_i should be as small as possible. As we saw, in any perfect scheme, $|V_i| \geq |S|$. Therefore, if $|V_i| < |S|$, the scheme must be "nonperfect".

A nonperfect scheme consists of not only access sets and non-access sets but also semi-access sets. If C is a semi-access set, C has some information on S but can not recover S. $H(S|C)$ takes a value between 0 and $|S|$. (d, k, n) ramp schemes shown by Blakley and Meadows which are an extension of (k, n) threshold schemes, are such an example [16]. However, only a little effort has been paid for nonperfect schemes.

Let Γ_1 denote the family of access sets, Γ_2 denote the family of semi-access sets and Γ_3 denote that of non-access sets.

In [18], we showed the following results.

Result 1. $(\Gamma_1, \Gamma_2, \Gamma_3)$ has a nonperfect secret sharing scheme if and only if Γ_1 is monotone and $\Gamma_1 \cup \Gamma_2$ is monotone.

Result 2. $\max |V_i| \geq |S| / \sharp(A \backslash C)$, for any access set A in Γ_1 and any non-access set C in Γ_3, where $\sharp(A \setminus C)$ denotes the cardinality of A set minus C.

Result 2 shows a possibility that V_i can be smaller by the factor of $\sharp(A \setminus C)$ than $|S|$.

In this paper, we will show that nonperfect schemes also have matroid structures. We will also present a direct connection between the secret sharing matroids and the entropy for both perfect and nonperfect schemes.

We define natural classes of NSS and derive a lower bound of $|V_i|$ for those classes. "Ideal" nonperfect schemes are defined based on this lower bound. We prove that every such ideal nonperfect secret sharing scheme has a matroid structure. The rank function of the matroid is given by the entropy divided by some constant. It satisfies a simple equation which represents the access level of each subset of the participants in the NSS.

$H(X)$ denotes the entropy of X (see [19] or Appendix). $\sharp X$ denotes the cardinality of a finite set X. $|X| \triangleq \log_2 \sharp X$. $A \setminus B \triangleq \{x | x \in A \text{ } but \text{ } x \notin B\}$. 2^P denotes the family of all subsets of P. $\mathcal{Z}$ denotes the set of nonnegative integers. Γ^- denotes the family of minimal sets of a family Γ.

2 Perfect and Nonperfect Secret Sharing Scheme

1. $P = \{P_1, \cdots, P_n\}$ denotes a set of participants.
2. s denotes a secret uniformly distributed over a finite set S $(H(S) = |S|)$.
3. v_i is the share of P_i distributed over a finite set V_i. $V \triangleq \{V_1, \cdots, V_n\}$.

Usually, access structures are defined as a subset of 2^P. For convenience, we define them as a subset of 2^V. We use P_i and V_i interchangeably such as follows. $\tilde{\Gamma}_i$ denotes a subset of 2^P. Γ_i denotes a subset of 2^V. $(V_{i1}, \cdots, V_{ik}) \in \Gamma_i$ iff $(P_{i1}, \cdots, P_{ik}) \in \tilde{\Gamma}_i$. (The index set in $\tilde{\Gamma}_i$ and that in Γ_i are the same.)

Definition 1. (Π, S, V) is a secret sharing scheme (SS) if Π is a mapping: $S \times R \rightarrow V_1 \times V_2 \times \cdots \times V_n$, where R is a set of random inputs.

Definition 2. Let $\Gamma \subseteq 2^V$. We say that an SS is a perfect SS (PSS) on Γ if

(1) $H(S|A) = 0 \quad$ for $\forall A \in \Gamma$.
(2) $H(S|C) = H(S) \quad$ for $\forall C \notin \Gamma$.

Remark.

1. A is called an access subset. (1) means that A can recover S.
2. C is called a non-access subset. (2) means that C obtains absolutely no information on S.

Definition 3. A family Γ is said to be monotone if $A \in \Gamma, A \subseteq A' \Rightarrow A' \in \Gamma$.

Proposition 4. *[3][4] There exists a PSS on Γ if and only if Γ is monotone.*

Proposition 5. *[6] $|V_i| \geq |S|$ for any i in PSSs if $V_i \in \exists A \in \Gamma^-$.*

Definition 6. Suppose that $\Gamma_1 \subseteq 2^V, \Gamma_2 \subseteq 2^V, \Gamma_1 \cap \Gamma_2 = \phi$. We say that an SS is a nonperfect SS (NSS) on $(\Gamma_1,\ \Gamma_2)$ if

(1) $H(S|A) = 0 \quad$ for $\forall A \in \Gamma_1$.
(2) $0 < H(S|B) < H(S) \quad$ for $\forall B \in \Gamma_2$.
(3) $H(S|C) = H(S) \quad$ otherwise.

The authors showed the following results in [18].

Proposition 7. *[18] Suppose that $\sharp S$ is not a prime. There exists an NSS on $(\Gamma_1,\ \Gamma_2)$ if and only if Γ_1 is monotone and $\Gamma_1 \cup \Gamma_2$ is monotone.*

Proposition 8. *[18]*

$$\max_i |V_i| \geq |S| / \sharp(A \setminus C), \forall A \in \Gamma_1, \forall C \in \Gamma_3 \ ,$$

where $\Gamma_3 \triangleq 2^V \setminus (\Gamma_1 \cup \Gamma_2)$.

Proposition 8 shows a possibility that $|V_i|$ can be smaller by the factor $1/\sharp(A \setminus C)$ than $|S|$.

3 Matroid

A matroid $M = (W, \mathcal{I})$ is a finite set W and a collection $\mathcal{I}$ of subsets of W such that $(I1) \sim (I3)$ are satisfied [17].

(I1) $\phi \in \mathcal{I}$.
(I2) If $X \in \mathcal{I}$ and $Y \subseteq X$, then $Y \in \mathcal{I}$.
(I3) If X and Y are members of $\mathcal{I}$ with $\sharp X = \sharp Y + 1$, then there exists $x \in X \setminus Y$ such that $Y \cup \{x\} \in \mathcal{I}$.

We show an example. Let W be a finite vector space and let $\mathcal{I}$ be the collection of linearly independent subsets of vectors of W. Then, such a pair of W and $\mathcal{I}$ is a matroid.

The elements of W are called the points of the matroid and the sets $\mathcal{I}$ are called independent sets. A base of M is a maximal independent subset of W. The rank function of a matroid is a function $\rho : 2^W \to \mathcal{Z}$ defined by $\rho(A) = \max(\sharp X : X \subseteq A, X \in \mathcal{I})$. The rank of matroid, denoted by $\rho(M)$, is the rank of the set W.

There exists an equivalent axiom of a matroid based on the rank function.

Proposition 9. *A function ρ is the rank function of a matroid on W if and only if for $X \subseteq W, y, z \in W$,*

(R0) $\rho(X)$ *takes a value of a non-negative integer.*
(R1) $\rho(\phi) = 0$.
(R2) $\rho(X) \leq \rho(X \cup y) \leq \rho(X) + 1$.
(R3) *If* $\rho(X \cup y) = \rho(X \cup z) = \rho(X)$, *then* $\rho(X \cup y \cup z) = \rho(X)$.

4 Overview

4.1 Background

The background of our problem is summarized as follows. In a perfect scheme, it is known that $|V_i| \geq |S|$ [6][7]. This was proved by using entropy. If they are equal for all i, the scheme is called ideal. On the other hand, an ideal perfect scheme has a matroid structure [5]. No relation between the matroid and the entropy is known.

Now, we ask

(1) Do the matroids have any relation with the entropy ?
(2) Suppose that Π_1 and Π_2 are two ideal perfect schemes for the same access structure Γ. Then each Π_i has a matroid structure. What is common between the two matroids?
(3) Does an ideal nonperfect scheme also have a matroid structure (if "ideal" is properly defined for nonperfect schemes) ?

This paper gives answers to these questions.

4.2 Perfect SS

Our observation is as follows.

In a PSS, from Definition 2,

$$H(S|A) = H(SA) - H(A) = \begin{cases} 0 & \text{if } A \in \Gamma \\ H(S) & \text{if } A \notin \Gamma \end{cases} . \tag{1}$$

Define $\hat{\rho}(A)$ as

$$\hat{\rho}(A) \triangleq \frac{H(A)}{H(S)} . \tag{2}$$

Then, from eq.(1), we obtain that

$$\hat{\rho}(SA) - \hat{\rho}(A) = \begin{cases} 0 & \text{if } A \in \Gamma \\ 1 & \text{if } A \notin \Gamma \end{cases} . \tag{3}$$

We will prove that, in an ideal PSS, $\hat{\rho}(A)$ so defined is the rank function of a matroid.

Note that eq.(2) gives a direct connection between the secret sharing matroid and the entropy. This is an answer to our problem 1.

Also note that eq.(3) depends only on Γ, not on each scheme. Thus, this is an answer to our problem 2.

It will be proved that our $\hat{\rho}$ satisfies the conditions (R0)∼(R3) of Proposition 9. The proof will be given in Section 6 in a more general form.

4.3 Nonperfect SS

In a nonperfect scheme, $H(S|A)$ can take a value between 0 and $|S|$. As an example, let's assume that

$$H(S|A) = H(SA) - H(A) = 0,\ H(S)/3,\ 2H(S)/3 \text{ or } H(S) .$$

Let

$$\hat{\rho}(A) \triangleq \frac{H(A)}{H(S)/3} .$$

Then, we have

$$\hat{\rho}(SA) - \hat{\rho}(A) = 0, 1, 2, \text{ or } 3 .$$

We will prove that in an ideal nonperfect scheme, $\hat{\rho}(A)$ so defined is the rank function of a matroid.

This is an answer for our problem 3 if "ideal nonperfect" is defined. However, we have not yet defined "ideal nonperfect". In Section 5, we will give a definition of "ideal nonperfect".

5 "Ideal" Nonperfect Secret Sharing Schemes

5.1 Access Hierarchy

In this subsection, we will define a natural class of nonperfect schemes.

Definition 10. Let d be a positive integer. We say that an SS (Π, S, V) has a level d access hierarchy $(\Sigma_0, \Sigma_1, \cdots, \Sigma_d)$ if

$$\bigcup_{i=0}^{d} \Sigma_i = 2^V, \quad \Sigma_i \cap \Sigma_j = \phi \quad (i \neq j) \text{ and}$$

$$H(S|A) = (k/d)H(S) \quad \textit{for } \forall A \in \Sigma_k \ .$$

Theorem 11. *Suppose that* $\sharp S = q^d$ *for some positive integer* q*. There exists an SS which has a level* d *access hierarchy* $(\Sigma_0, \Sigma_1, \cdots, \Sigma_d)$ *if and only if* $\Delta_k \triangleq \bigcup_{i=0}^{k} \Sigma_i$ *is monotone for* $0 \leq \forall k \leq d-1$.

Proof. "Only if" part is clear. We prove "if" part. The secret s can be expressed as $(s_0, \cdots, s_{d-1})$ such that $s_i \in \{0, \cdots, q-1\}$. From Proposition 4, there exists a PSS T_k on each Δ_k. Apply T_k to s_k for $0 \leq \forall k \leq d-1$, independently. Then, it is easy to see that the above scheme has a level d access hierarchy. □

5.2 Lower Bound of $|V_i|$

This subsection will derive a lower bound of $|V_i|$ (Note that Proposition 8 gives a lower bound of the "max"$|V_i|$).

Theorem 12. *If an SS has a level* d *access hierarchy* $(\Sigma_0, \Sigma_1, \cdots, \Sigma_d)$ *and if* $V_i \in A \in \Sigma_k^-$ *for some* A *and some* k $(\leq d-1)$*, then*

$$|V_i| \geq H(V_i) \geq H(S)/d \ .$$

Proof.

$$\begin{aligned} H(V_i) &\geq H(V_i|A \setminus \{V_i\}) \\ &\geq I(S; V_i|A \setminus \{V_i\}) \\ &= H(S|A \setminus \{V_i\}) - H(S|A) \\ &\geq (k+1)/d \times H(S) - k/d \times H(S) \\ &= H(S)/d \ . \end{aligned}$$

□

5.3 Definition of "Ideal"

Based on Theorem 12, we will define "ideal" as follows.

Definition 13. We say that an SS of a level d access hierarchy is ideal if

$$|V_i| = H(V_i) = H(S)/d, \quad \forall V_i \in V \ .$$

Theorem 14. *If an SS has a level d access hierarchy* $(\Sigma_0, \Sigma_1, \cdots, \Sigma_d)$ *and if the SS is ideal, then for* $\forall A \in \Sigma_i, \forall C \in \Sigma_j$,

$$\sharp(A \setminus C) \geq j - i \quad (j > i) \ .$$

Proof.

(1) First we assume that $B = (A \setminus C)$. Then,

$$\begin{aligned} I(S; B|C) &= H(S|C) - H(S|CB) \\ &= H(B|C) - H(B|SC) \\ &\leq H(B|C) \leq H(B) \leq \sum_{V_i \in B} H(V_i) \ . \end{aligned}$$

Therefore,

$$\begin{aligned} \sharp(A \setminus C)H(S)/d &= \sum_{V_i \in B} H(V_i) \\ &\geq H(S|C) - H(S|A) \\ &= (j - i)H(S)/d \ . \end{aligned}$$

Hence,

$$\sharp(A \setminus C) \geq j - i \ .$$

(2) Next we assume that $C \not\subset A$. Let $A' \triangleq C \cup A, A' \in \Sigma_k$. It is clear that $k \leq i$. Then, from (1) of this proof,

$$\sharp(A \setminus C) = \sharp(A' \setminus C) \geq j - k \geq j - i \ .$$

□

5.4 Mixed Access Hierarchy

Now, we will define a slight variation of Definition 10.

Definition 15. Suppose that $S = S_1 \circ S_2 \circ \cdots \circ S_d$ and $|S_i| = |S|/d$ for all i ($\circ$ means concatenation). Let $W \triangleq \{S_1, \cdots, S_d, V_1, \cdots, V_n\}$. We say that an SS (Π, S, V) has a level d mixed access hierarchy $(\hat{\Sigma}_0, \hat{\Sigma}_1, \cdots, \hat{\Sigma}_d)$ if

$$\bigcup_{i=0}^{d} \hat{\Sigma}_i = 2^W, \quad \hat{\Sigma}_i \cap \hat{\Sigma}_j = \phi \ (i \neq j) \quad \text{and}$$

$$H(S|A) = (k/d)H(S) \quad \mathit{for} \ \forall A \in \hat{\Sigma}_k \ .$$

Remark.

1. Many examples of NSS in [16] have mixed access hierarchies.
2. A PSS has a level 1 mixed access hierarchy.

The following theorem clearly holds.

Theorem 16. *If an SS has a level d mixed access hierarchy* $(\hat{\Sigma}_0, \hat{\Sigma}_1, \cdots, \hat{\Sigma}_d)$, *it has a level d access hierarchy* $(\Sigma_0, \Sigma_1, \cdots, \Sigma_d)$ *such that* $\Sigma_k = \hat{\Sigma}_k \cap 2^V$.

Therefore, Theorem 12 also holds for an SS of a level d mixed access hierarchy.

Definition 17. We say that an SS of a level d mixed access hierarchy is ideal if

$$|a| = H(a) = H(S)/d, \quad \forall a \in W .$$

Theorem 18. *If an SS has a level d mixed access hierarchy* $(\hat{\Sigma}_0, \hat{\Sigma}_1, \cdots, \hat{\Sigma}_d)$ *and if the SS is ideal, then for* $\forall A \in \hat{\Sigma}_i, \forall C \in \hat{\Sigma}_j$,

$$\sharp(A \setminus C) \geq j - i \quad (j > i) .$$

The proof is similar to Theorem 14.

6 Ideal NSS and Matroid

In this section, we will show that each ideal nonperfect SS (in the sense of Definition 17) has a matroid structure. The rank function of the matroid is given by the entropy divided by some constant. It satisfies a simple equation which represents the access level of the subset. This property also holds for ideal perfect SSs.

6.1 Ideal NSS and Matroid

Theorem 19. *Suppose that*

1. *An SS has a level d mixed access hierarchy* $(\hat{\Sigma}_0, \hat{\Sigma}_1, \cdots, \hat{\Sigma}_d)$ *and the SS is ideal.*
2. *For* $\forall a \in V$ *such that* $\{a\} \in \hat{\Sigma}_d$, *there exists* $B \in \hat{\Sigma}_{d-1}^-$ *such that* $a \in B$.

Then, there exists a matroid on $W \triangleq \{S_1, \cdots, S_d, V_1, \cdots, V_n\}$ *with a rank function* ρ *such that*

(N1) $\rho(S_1 \cdots S_d) = d$.
(N2) $\rho(S_1 \cdots S_d X) - \rho(X) = k$ *if* $X \in \Sigma_k$, *where* $\Sigma_k = \hat{\Sigma}_k \cap 2^V$.

To prove the Theorem, we define

$$\hat{\rho}(X) \triangleq \begin{cases} 0 & \text{if } X = \phi \\ H(X) \times (d/|S|) & \text{otherwise} \end{cases} .$$

We will prove that $\hat{\rho}$ is the desired rank function. We have to show that $\hat{\rho}$ satisfies (R0)~(R3) of Proposition 9 and (N1), (N2) of Theorem 19. The proof of (R0) will be given in the next subsection.

Lemma 20. *$\hat{\rho}$ satisfies (R0)∼(R3), (N1) and (N2).*

Proof. (R1) and (N1) are clear.

(R2) $H(X) \leq H(X \cup y) \leq H(X) + H(y) = H(X) + |S|/d$. Hence,

$$dH(X)/|S| \leq dH(X \cup y)/|S| \leq dH(X)/|S| + 1 \ .$$

(R3) $H(X \cup y \cup z) = H(X) + H(y|X) + H(z|yX)$.

Suppose that

$$H(X \cup y) = H(X \cup z) = H(X) \ .$$

Then,

$$H(y|X) = H(X \cup y) - H(X) = 0 \ .$$

Similarly,

$$H(z|X) = 0 \ .$$

Since $0 \leq H(z|yX) \leq H(z|X) = 0$,

$$H(z|yX) = 0 \ .$$

(N2) If $X \in \Sigma_k$,

$$\begin{aligned}(k/d)|S| &= H(S|X) \\ &= H(SX) - H(X) \\ &= H(S_1 \cdots S_d X) - H(X) \ .\end{aligned}$$

□

As a special case of Theorem 19, we have the following corollary.

Corollary 21. *For a perfect ideal SS, there exists a matroid on $\{S, V_1, \cdots, V_n\}$ with a rank function ρ such that*

1. $\rho(S) = 1$.
2. $\rho(SX) - \rho(X) = \begin{cases} 0 \text{ if } X \text{ is an access subset} \\ 1 \text{ if } X \text{ is a non-access subset.} \end{cases}$

6.2 $H(X) = (|S|/d) \times$Integer

Lemma 22. *If $X \in \hat{\Sigma}_{i+1}$ and $(X \cup y) \in \hat{\Sigma}_i$, then $H(y|X) = |S|/d, H(y|XS) = 0$.*

Proof.

$$\begin{aligned}I(y; S|X) &= H(S|X) - H(S|Xy) \\ &= ((i+1)/d)H(S) - (i/d)H(S) \\ &= H(S)/d \ .\end{aligned}$$

On the other hand,

$$I(y; S|X) = H(y|X) - H(y|XS) \ .$$

Then,

$$0 \leq H(y|XS) = H(y|X) - H(S)/d \leq H(y) - H(S)/d = 0 \ .$$

Therefore,

$$H(y|XS) = 0 \ .$$

Hence,

$$H(y|X) = H(S)/d = |S|/d \ .$$

□

Lemma 23. $\forall A \in \hat{\Sigma}_i,\ \forall C \in \hat{\Sigma}_{i+2},\ \sharp(A \setminus C) \geq 2.$

Proof. It is clear from Theorem 18. □

Lemma 24. *For* $0 \leq \forall i \leq d-1$, *if* $a \in B \subseteq A \in \hat{\Sigma}_i$, $(A \setminus \{a\}) \in \hat{\Sigma}_i$ *and* $B \in \hat{\Sigma}_i^-$, *then* $H(a|(A \setminus \{a\})) = 0$.

Proof. Choose $C \subseteq (A \setminus \{a\})$ such that $C \in \hat{\Sigma}_i^-$. Let $D \triangleq (B \setminus \{a\})$.
Since $C \subseteq C \cup D \subseteq C \cup B \subseteq A$ and $C \in \hat{\Sigma}_i^-, A \in \hat{\Sigma}_i$, then $C \cup D \in \hat{\Sigma}_i, C \cup B \in \hat{\Sigma}_i$.
Therefore,

$$H(S|CD) = H(S|CB) \ .$$

On the other hand,

$$H(aS|CD) = H(a|CD) + H(S|CB) = H(S|CD) + H(a|SCD) \ .$$

Then,

$$0 \leq H(a|(A \setminus \{a\})) \leq H(a|CD) = H(a|SCD) \leq H(a|SD) = 0$$

(from Lemma 22).

□

Lemma 25. *For* $\forall X \in \hat{\Sigma}_d$, $H(X) = (|S|/d) \times$ *integer.*

Proof. Let X be a minimal set such that

$$X \in \hat{\Sigma}_d \quad \textit{and} \quad H(X) \neq (|S|/d) \times \textit{integer} \ .$$

Claim 26. $\forall y \in X, \quad H(X \setminus \{y\}) = (\sharp X - 1)|S|/d.$

Proof. Let $X \setminus \{y\} = \{a_1, \cdots, a_l\}$. From the minimality of X,

$$q_i \triangleq H(a_1 \cdots a_i) = (|S|/d) \times \textit{integer} \ .$$

Therefore,

$$t_i \triangleq H(a_i|a_1 \cdots a_{i-1}) = q_i - q_{i-1} = (|S|/d) \times \textit{integer} \ .$$

On the other hand,

$$0 \le t_i \le H(a_i) = |S|/d \ .$$

Hence.

$$t_i = 0 \ or \ |S|/d \ .$$

If $t_i = 0$,

$$H(a_i|X \setminus \{a_i\}) = 0$$

because

$$0 \le H(a_i|X \setminus \{a_i\}) \le H(a_i|a_1 \cdots a_{i-1}) = 0 \ .$$

Then,

$$H(X) = H(X \setminus \{a_i\}) + H(a_i|X \setminus \{a_i\}) = H(X \setminus \{a_i\}) \ .$$

This contradicts the minimality of X. Therefore,

$$t_i = |S|/d \quad \text{for } 1 \le i \le l \ .$$

Hence,

$$H(X \setminus \{y\}) = H(a_1) + t_2 + \cdots + t_l = (\sharp X - 1)|S|/d \ .$$

□

Claim 27. *There exists* $Y = \{y_1, \cdots, y_k\} \in \hat{\Sigma}_d$ *such that* $(X \cup Y) \in \hat{\Sigma}_{d-1}$ *and* $(X \cup Y) \setminus \{\forall y_i\} \in \hat{\Sigma}_d$.

Proof. From the assumption of Theorem 19,

$$\forall a \in X, \ \exists B \in \hat{\Sigma}^{-}_{d-1}, \ s.t. \ \ a \in B \ .$$

Clearly, $B \setminus X \in \hat{\Sigma}_d$. Let $Y \subseteq (B \setminus X)$ be a minimal set such that $(X \cup Y) \in \hat{\Sigma}_{d-1}$.

□

Claim 28. $\forall Z \subseteq X, \quad H(Z \cup Y) = H(Z) + \sharp Y|S|/d.$

Proof. Let

$$u_i \triangleq H(y_i|Z \cup \{y_1, \cdots, y_{i-1}\}) \ .$$

Then,

$$u_i \le H(y_i) = |S|/d \ .$$

On the other hand,

$$u_i \ge H(y_i|(X \cup Y) \setminus \{y_i\}) = |S|/d \ .$$

The equality comes from Lemma 22. Therefore,

$$u_i = |S|/d \ .$$

Hence

$$H(Z \cup Y) = H(Z) + u_1 + \cdots + u_k = H(Z) + \sharp Y|S|/d \ .$$

□

Claim 29. $H(X \cup Y) \neq |S|/d \times \mathit{integer}$.

Proof. From Claim 3,

$$H(X \cup Y) = H(X) + \sharp Y|S|/d \ .$$

□

Claim 30. $\forall a \in X, \quad (X \cup Y) \setminus \{a\} \in \hat{\Sigma}_{d-1}$.

Proof. Suppose that

$$\exists a \in X, \quad (X \cup Y) \setminus \{a\} \in \hat{\Sigma}_d \ .$$

Then, from Lemma 22,

$$H(a|(X \cup Y) \setminus \{a\}) = |S|/d \ .$$

Therefore,

$$\begin{aligned} H(X \cup Y) &= H((X \cup Y) \setminus \{a\}) + H(a|(X \cup Y) \setminus \{a\}) \\ &= H(X \setminus \{a\}) + \sharp Y|S|/d + |S|/d \\ &= ((\sharp X - 1) + \sharp Y + 1)|S|/d = (\sharp X + \sharp Y)|S|/d \ . \end{aligned}$$

The second line comes from Claim 3. The third line comes from Claim 1. This is against Claim 4. □

(Proof of Lemma 25). Choose $B \in \hat{\Sigma}^{-}_{d-1}$ such that $B \subseteq (X \cup Y)$. Let $a \in (B \cap X)$. From Claim 5 and Lemma 24,

$$H(a|(X \cup Y) \setminus \{a\}) = 0 \ .$$

Then, from Claim 3 and Claim 1,

$$\begin{aligned} H(X \cup Y) &= H((X \cup Y) \setminus \{a\}) + H(a|(X \cup Y) \setminus \{a\}) \\ &= H(X \setminus \{a\}) + \sharp Y|S|/d \\ &= (\sharp X - 1 + \sharp Y)|S|/d \ . \end{aligned}$$

This is against Claim 4. □

Theorem 31. *For* $0 \leq \forall k \leq d$,

$$\forall A \in \hat{\Sigma}_k, \quad H(A) = |S|/d \times \mathit{integer} \ . \tag{4}$$

Proof. We will prove by induction on k. When $k = d$, (4) holds from Lemma 25. Suppose that (4) holds for $k \geq i + 1$. Let A be a minimal set such that

$$A \in \hat{\Sigma}_i, \quad H(A) \neq (|S|/d) \times \mathit{integer}.$$

(1) Assume that

$$\exists a \in A, \ \ A \setminus \{a\} \notin \hat{\Sigma}_i \ .$$

From Lemma 23,

$$A \setminus \{a\} \in \hat{\Sigma}_{i+1} \ .$$

Then, from Lemma 22,

$$H(a|A \setminus \{a\}) = |S|/d \ .$$

Hence

$$H(A) = H(A \setminus \{a\}) + H(a|A \setminus \{a\}) = H(A \setminus \{a\}) + |S|/d \ .$$

From the hypothesis of the induction,

$$H(A \setminus \{a\}) = |S|/d \times integer \ .$$

This is a contradiction.

(2) Assume that

$$\forall a \in A, \ \ A \setminus \{a\} \in \hat{\Sigma}_i \ .$$

Choose $B \in \hat{\Sigma}_i^-$ such that $B \subseteq A$. Let $b \in B$. From Lemma 24,

$$H(b|A \setminus \{b\}) = 0 \ .$$

Then,

$$H(A) = H(A \setminus \{b\}) + H(b|A \setminus \{b\}) = H(A \setminus \{b\}) \ .$$

This contradicts the minimality of A.

Therefore,

$$\forall A \in \hat{\Sigma}_i, \ \ H(A) = |S|/d \times integer \ .$$

□

6.3 Other Theorems

Theorem 32. *Under the assumption of Theorem 19, let Y be any maximal independent set contained in X. Then, $X \in \hat{\Sigma}_i$ if and only if $Y \in \hat{\Sigma}_i$.*

Proof. Let $X = Y \cup Z$. Because Y be a maximal independent set,

$$H(X) = H(Y) \ .$$

On the other hand,

$$H(X) = H(Y) + H(Z|Y) \ .$$

Therefore,

$$H(Z|Y) = 0 \ .$$

Here,

$$0 \leq H(Z|YS) \leq H(Z|Y) = 0 \ .$$

Hence,

$$H(Z|YS) = 0 .$$

Then,

$$I(S;Z|Y) = H(Z|Y) - H(Z|YS) = 0 = H(S|Y) - H(S|YZ) .$$

Now, we have

$$H(S|Y) = H(S|YZ) = H(S|X) .$$

□

Theorem 33. *If there exists a representable matroid over a finite field $GF(q)$ on W which satisfies (N1) and (N2), there exists an SS which has a level d mixed access hierarchy $(\hat{\Sigma}_0, \hat{\Sigma}_1, \cdots, \hat{\Sigma}_d)$ and is ideal.*

Proof. There exist a vector space D over $GF(q)$ and a mapping $\phi : W \to D$, which preserves rank. Let $\phi(S_i) = \alpha_i$ and $\phi(V_i) = \beta_i$. α_i and β_i are column vectors. For a secret $s = (s_1, \cdots, s_d)$ $(s_i \in GF(q))$, choose a vector γ such that

$$s_i = \alpha_i' \cdot \gamma \quad (1 \leq i \leq d)$$

at random, where $\cdot$ means inner product. We can do this because the rank of $\{\alpha_1, \cdots, \alpha_d\}$ equals d. Then, compute each share v_i as

$$v_i = \beta_i' \cdot \gamma \quad (1 \leq i \leq n) .$$

It is easy to see that the above scheme satisfies the desired condition. □

Remark. Let $E \triangleq \{x_1, x_2, \cdots, x_n\}$, where x_i is a random variable. It is known that (E, H) is a polymatroid [20]. The rank function of a polymatroid takes a value in nonnegative real numbers. It doesn't have to be integer valued, while the rank function of a matroid must be integer valued. Generally, $H(X)$ is not integer valued. Our contribution is to show that $H(S)$ is integer valued in ideal secret sharing schemes (for both perfect and nonperfect.)

7 Summary

This paper has shown that nonperfect secret sharing schemes (NSS) have matroid structures and has presented a direct link between the secret sharing matroids and entropy for both perfect and nonperfect schemes. We have defined natural classes of NSS and have derived a lower bound of $|V_i|$ for those classes. "Ideal" nonperfect schemes are defined based on this lower bound. We have proved that every such ideal secret sharing scheme has a matroid structure. The rank function of the matroid has been given by the entropy divided by some constant. It satisfies a simple equation which represents the access level of each subset of participants.

Acknowledgement

We would like to thank Prof. S.Ueno of Tokyo Institute of Technology for useful discussion.

References

1. G.R.Blakley : Safeguarding cryptographic keys. Proc. of the AFIPS 1979 National Computer Conference, vol.48, pp.313-317 (1979)
2. A.Shamir : How to share a secret. Communications of the ACM, 22, (11), pp.612-613 (1979)
3. M.Itoh, A.Saito, T.Nishizeki : Secret sharing scheme realizing general access structure. Proc. of IEEE Globecom '87, Tokyo, pp.99-102 (1987)
4. J.C.Benaloh, J.Leichter : Generalized secret sharing and monotone functions. Crypto'88, pp.27-36 (1990)
5. E.F.Brickell, D.M.Davenport : On the classification of ideal secret sharing schemes. Journal of Cryptology, vol.4, No.2, pp.123-134 (1991)
6. R.M.Capocelli, A.De Santis, L.Gargano, U.Vaccaro : On the size of shares for secret sharing schemes. Crypto'91, pp.101-113 (1991)
7. E.D.Karnin, J.W.Green, M.E.Hellman : On secret sharing systems. IEEE Trans. IT-29, No.1, pp.35-41 (1982)
8. E.F.Brickell, D.R.Stinson : Some improved bounds on the information rate of perfect secret sharing schemes. Crypto'90, pp.242-252 (1990)
9. C.Blund, A.De Santis, D.R.Stinson, U.Vaccaro : Graph decomposition and secret sharing schemes. Eurocrypt'92, pp.1-20 (1992)
10. Y.Frankel, Y.Desmedt : Classification of ideal homomorphic threshold schemes over finite Abelian groups. Eurocrypt'92, pp.21-29 (1992)
11. C.Blund, A.De Santis, L.Gargano, U.Vaccaro : On the information rate of secret sharing schemes. Crypto'92 (1992)
12. D.R.Stinson : New general bounds on the information rate of secret sharing schemes. Crypto'92 (1992)
13. A.Beimel, B.Chor : Universally ideal secret sharing schemes. Crypt'92 (1992)
14. W.A.Jackson, K.M.Martin : Cumulative arrays and geometric secret sharing schemes. Auscrypt'92 (1992)
15. M.Bertilsson, I.Ingemarsson : A construction of practical secret sharing schemes using linear block codes. Auscrypt'92 (1992)
16. G.R.Blakley, C.Meadows : Security of ramp schemes. Crypto'84, pp.242-268 (1984)
17. D.J.A.Welsh : Matroid theory. Academic Press (1976)
18. W.Ogata, K.Kurosawa, S.Tsujii : Nonperfect secret sharing schemes. Auscrypt'92 (1992)
19. R.G.Gallager : Information Theory and Reliable Communications. John Wiley & Sons, New York, NY, (1968)
20. S.Fujishige : Polymatroidal dependence structure of a set of random variables. Information and Control 39, pp.55-72, (1978)

Appendix

Given a probability distribution $\{p(x)\}_{x \in X}$, the *entropy* of X is defined as

$$H(X) \triangleq -\sum_{x \in X} p(x) \log_2 p(x) .$$

It holds that

$$0 \leq H(X) \leq \log_2 \sharp X = |X| ,$$

where $H(X) = 0$ if and only if there exists $x \in X$ such that $p(x) = 1$; $H(X) = |X|$ if and only if $p(x) = 1/\sharp X$, for $\forall x \in X$.

Given two sets X and Y and a joint probability distribution $\{p(x,y)\}_{x \in X, y \in Y}$ on their Cartesian product, the *conditional entropy* $H(X|Y)$ is defined as

$$H(X|Y) \triangleq -\sum_{y \in Y} \sum_{x \in X} p(x,y) \log_2 p(x|y) .$$

From the definition of conditional entropy, it is easy to see that

$$H(X|Y) \geq 0 .$$

The entropy of the joint space XY satisfies

$$H(XY) = H(X) + H(Y|X) = H(Y) + H(X|Y) .$$

The *mutual information* between X and Y is defined by

$$I(X;Y) \triangleq H(X) - H(X|Y) .$$

The mutual information has the following properties:

$$I(X;Y) = I(Y;X) ,$$
$$I(X;Y) \geq 0 .$$

From the above inequality, one gets

$$H(X) \geq H(X|Y) .$$

The *conditional mutual information* is defined by

$$I(X;Y|Z) \triangleq H(X|Z) - H(X|YZ) .$$

$I(X;Y|Z)$ satisfies the following properties.

$$I(X;Y|Z) \geq 0 ,$$
$$I(X;Y|Z) = I(Y;X|Z) ,$$
$$I(X;YZ) = I(X;Z) + I(X;Y|Z) .$$

From the memoirs of a Norwegian cryptologist

Ernst S. Selmer
Idrettsvegen 20, N-1400 Ski, Norway

Norwegian cryptology was first organized, in the 30's, by then Capt. R. A. Roscher Lund. He set up a "Cryptology club", recruited partly from amateurs, partly from mathematicians. Many members came from a bridge club with the appropriate name "Forcing". Around the outbreak of the war a "Defense information office" was established with some (very) few cryptologists.

Personally, I was too young then, and first met cipher work during the war. My friend *Nils Stordahl* was strongly involved in the Norwegian underground movement, which needed secure communication with Stockholm. Stordahl recruited me and some other students of mathematics for a cipher service. Under the circumstances, only a hand cipher on letters was possible. The actual types are a transposition (letter permutation) or a substitution cipher. Our underground system in Norway was based on transposition. The letters of a key word (or key phrase) are numbered from their position in the alphabet, for instance

S	E	L	M	E	R
6	1	3	4	2	5
M	A	R	Y	X	H
A	D	X	A	X	L
I	T	T	L	E	X
L	A	M	B		

(X between words could be dropped.) The cipher text is then read off column by column in the numbered order.

The result is a single transposition, which is reasonably simple to break. We therefore used a *double transposition*, with a new key word on the result of the first step. I have read later that one should use key words of length 20–30, and messages of several hundred letters. Further, the text below the key numbering should not fill complete rectangles.

It is very easy to make mistakes in the transposition. My first encounter with cipher was extremely tedious and boring! But of course, we were glad to be of some use during the occupation.

Late in 1943, the Germans closed the University of Oslo and arrested the students they got hold of. I managed to escape to Sweden. There was my friend Stordahl already, and he saw to it that I got more cipher training before I was sent to London (where I arrived in the Spring of 1944 together with the first flying bomb, V1).

The already mentioned Roscher Lund had organized a Norwegian cipher network abroad, especially between Stockholm, London and U.S.A. The communication was mainly on the *Hagelin* cipher machine. This was a great success all over the world. At the end of the war, the U.S. army had 140,000 machines, and the producers, the Swedish Hagelin family, had become very wealthy.

In addition to some purely mechanical versions, there was also an electrical model with keyboard, which we were using in London. I was on a course in the Hagelin factory in Stockholm, and had just taken an electrical machine into parts and pieces, when Hagelin senior passed by. Addressed to me, he said that "it is easy to take it apart, more complicated to reassemble it, and much more difficult to make it work afterwards". But the next day, I could show him a working machine. The unforeseen result was that later, the complete technical responsibility for all Norwegian military and diplomatic cipher machines in London was put on top of my other duties.

The actual ciphering work on the Hagelin machines was only slightly less boring than the earlier double transposition. In the beginning, it was very exciting with all the military and diplomatic secrets you could read about. Very soon, however, one becomes completely blasé.

Roscher Lund's cryptologists had suspected that the Hagelin cipher might be broken, at least from corresponding plaintext and cipher, that is, from pure key. We therefore shuffled different parts of each message, and used a strange mixture of — often abbreviated — Norwegian and English. After the war, we learned from German archives that they had not been able to break our Hagelin cipher.

When the war ended, I was sent to Tromsø in Northern Norway as a cipher officer. But there was no need for cipher, so I spent a fabulous Summer under the midnight sun. I managed to get hold of some of the famous German *Enigma* cipher machines, and took them with me to the military headquarters in Oslo. But if I had been smart, I should have kept them for myself. Since the story of the Enigma breaking became publicly known in the 70's, the prices of old Enigmas have exploded. I have heard mentioned auction prices of $15,000 for a unit.

In 1946, Stordahl became head of the military cipher office, a position he held until 1983. He died much too early, in 1984. I always considered him my best friend. One of Stordahl's early initiatives was to engage me as a consultant. My first big task, and my most fantastic cryptological experience, was to establish a (hopefully) safe communication system for the Norwegian equivalent of MI5 and Scotland Yard Special Branch ("Overvåkningspolitiet" = "watch-over police"). We based it on the German Siemens teleprinter ("Fernschreiber"="remote writer"), with an additional unit for encryption/decryption (*"Geheimschreiber"*="secret writer"). There were many of them in Norway, and the Norwegian (public) Telephone and Telegraph Company had collected them to dismantle the cipher unit and use them as ordinary teleprinters. We had to prevent this, and could not let it pass official channels. I cooperated with another of Stordahl's men, a young electrical engineer named Asbjørn Mathisen. Some time in 1946, we were supplied with two large military trucks and 20 German prisoners of war. We commanded them in our school-German, and drove to the Telecompany store just outside Oslo. The few attendants protested vigorously, but did not stand a chance against us and the Germans. Including a wooden case, each G-Schreiber weighed 180 kg, so the POW's were really needed. We got everything onto the trucks and drove it to a safe hiding place.

Mathisen used wires and torch bulbs to reconstruct the coupling diagram of the G-Schreibers, including 20 relays. I knew absolutely nothing about relays then, but had to find out how everything worked, from a maze of unsystematized wires.

The same reconstruction has been performed much later by other people. The Norwegian Technical Museum in Oslo had managed to get hold of 3 G-Schreibers. Two of them were given — for exchange purposes — to museums in London and Munich, and these were analyzed by Donald W. Davies in *Cryptologia* 6,7 (1982,1983). Some more historical information was supplied by Wolfgang Mache in *Cryptologia* 10 (1986).

The cipher unit contained 10 large notched wheels, stepped by a pawn mechanism. The number of steps in a revolution varied from 47 to 73, all pairwise coprime. The cam pattern was the same for all machines. The cams activated two contacts for each wheel. Over 20 relays, one contact set controlled an irregular stepping of the cam wheels, while the other set performed the ciphering of each 5 bit teleprinter symbol, in two rounds: one bitwise addition key, and one permutation (only 32 out of the 5!=120 were used). Details can be found in the Cryptologia papers.

I did not want to use the German outfit exactly as it was, so I made some modifications which Mathisen implemented. I could not change the notched wheels (with contact sets), but I fiddled with the above-mentioned relay control of the cam wheel stepping. I thought I made it more complicated, but had nobody to check whether the outfit had become cryptologically stronger.

The G-Schreibers came in successive versions 52 a/b, c, d, e. The first version was broken by the Swedish mathematician Arne Beurling (cf. *Kahn: The Codebreakers*) from the traffic passing Sweden between Norway and Germany. Later versions were "occasionally, but not routinely" broken by the British cipher office at Bletchley. — All this has become known later; in my case, I just had to hope for the best.

In my later designs of cipher machines (for NATO), I always said like the waiter: "Not my table", about all questions regarding protocols, key distribution, initiating and closing routines, etc. But for the police cipher, I had to work this out all by myself. In 1948, we summoned 20 police officers from different stations to Oslo, where I drilled them in the use of the G-Schreiber for a 3 weeks course. And the system was used from 1949 until around 1960.

I spent 1951 and the first half of 1952 in U.S.A. with a Rockefeller grant, to study computers, primarily von Neumann's famous Princeton machine. It became operative in January 1952, and I got the opportunity to run some of my number theoretical problems (indeterminate equations) on it. My programs were in fact the first ones to go through on the machine without any programming errors.

The machine had no printer, so we read off the numerical results from a display of lamps, arranged hexadecimally. In this connection, von Neumann made one of his very few errors: He wrote that because of the computers, humanity should be prepared to switch from the decimal system to numbers in base 8 or 16.

While in U.S.A., I was asked by a medium sized electronic company — on von Neumann's recommendation — to perform the logical design of a commercial computer. I undertook the job, which was finished after my return to Norway. I actually did the design down to every single tube (no transistors existed then). The company wanted a completely decimal machine, with a magnetic drum as its main memory. After a while, the company got economic problems and was swallowed by Burroughs, who entered the computer race with "my" machine *Burroughs 205*. In the late 50's, this was the most serious competitor to the famous *IBM 650*. My machine was perhaps the only, and certainly the last, larger computer where the complete logical design was a "one man job".

Returning home in 1952, I told Stordahl that we needed a military computer in Norway. The next day, he came back to me and said that he had 1 million Norwegian kroner at his disposal! The whole affair was extremely hush-hush, with money paid secretly from U.S. to Norwegian intelligence, for Russian communication intercepted in Northern Norway. Anyway, the result for us was a Ferranti Mercury computer, which after some delay was installed at our Defense Research Institute in 1957.

Already in London during the war, I had been working on how to break the Hagelin machine cipher, at least with "normal" plaintext (and not distorted as we used in London). I continued with this after the war, and it was quite clear that only manual calculations would be too time-consuming. In 1952, I still did not have an electronic computer at hand, but now I did at least know very much about computer design. I used this to draw a special purpose relay computer, earmarked for breaking the Hagelin cipher. It was built by Finn Didriksen in Stordahl's office, and was used for several years breaking the diplomatic cipher of some foreign countries.

It is unnecessary for me to go into details with the Hagelin machine, which is described over 60 pages in *Beker & Piper: Cipher Systems*, including the breaking both with and without known plaintext. They refer to two papers from the late 70's. I have not studied these; it sufficed for me that I did the same job 25 years earlier.

Let me mention an interesting incidence from the 50's. The Norwegian Standard Telephone and Cable Company (STK), then a subsidiary of ITT, planned to build a one time cipher system based on 5 hole teleprinter tape. But how, in those early days, to generate a completely random binary sequence? A young Norwegian officer, Bjørn Rørholt, got the idea to use radiation from a radioactive source (cobalt). The project was very successful, and about 2000 one time cipher units were delivered before paper tape became obsolete. A "radioactive factory" produced the tons and tons of one time paper tape necessary to support the units. The most famous connection was the *"Hot Line"* between Kremlin and the White House (but the type of equipment for this line was not disclosed until many years later).

To generate a random binary sequence, the Germans had tried to read off a very quickly oscillating circuit with pulses from a slow one. We were two young Norwegians who went down to Germany to look at this concept. The result was negative, except for one experience: We met Hitler's famous (notorious?) chief of intelligence, *Gehlen*, who after the war had been recruited by U.S. intelligence.

In 1957, I left the University of Oslo to take over the position as a (full) professor of pure mathematics at the University of Bergen. Here my two NATO ciphers were "born". In the late 50's, I was asked to study the theory of *linear shift registers*, to use them for a cipher system which Standard was to build for the Norwegian forces. The subject was new to me, but I quickly caught up with the existing theory. The word "cipher" was hardly mentioned in the literature (although shift registers must have been used in cipher machines before 1960). I had to discover myself that with pure key known — which is always assumed today — breaking a system of purely linear registers amounts to solving a set of linear equations. So nonlinearity would be necessary somewhere in the system. But I also discovered what today is well known, that nonlinear feedbacks may lead to very short periods. What I ended up with was a design which today would be characterized as completely conventional: A series of binary linear registers with maximal sequences guaranteed a very large period. On the output of these came some nonlinear components, a system which today is usually called "feed forward". — You may perhaps say that my work then was a part of forming today's conventions.

Since Norway is a NATO country, my cipher machine needed acceptance by National Security Agency (NSA) in Washington. Fortunately, it passed the test without comments or corrections. It was then produced by Standard under the name *"Cryptel"*, and was used for many years. The engineers in charge of the project at Standard were Kaare Meisingseth and Per Abrahamsen.

The theory of linear shift registers was so interesting that I went on with it, both at home and during a sabbatical year 1964/65 in Cambridge, England, where I lectured on the subject. The lectures appeared in duplicated form at the University of Bergen in 1966, under the name *"Linear Recurrence Relations over Finite Fields"*. Since then, a series of (unaltered) editions have been produced, to meet the demand. My lectures must hold a world record for the number of copies of a duplicated monograph. Many of my cryptology friends call it their "bible" of linear recurrence — or at least the Old Testament. If we study Ch. 5 in Beker & Piper, we will see that they have landed exactly on my notation.

As already mentioned, I had to find out myself about the nonlinear parts of the Cryptel cipher machine, and use them in a feed forward system. Life would be much easier for a cryptologist if someone could establish a comprehensive theory about periods, cycles etc. for nonlinear registers. In an attempt, I asked my student Johannes Mykkeltveit to look more into this. From my cryptological point of view, the result was negative. However, the project led to Mykkeltveit's well known proof of Golomb's famous conjecture about the maximum number of cycles generated by a linear or nonlinear register. (The maximum is attained for the pure cycling register.)

In my book, I was only able to solve completely one particular aspect of nonlinearity: *Multiplication of the output* of two linear sequences in a particular case. If the sequences have minimal polynomials $f(x)$ and $g(x)$, it is easily seen that any product sequence is generated by what I called $f \S g$, the polynomial whose roots are all the products of one root from $f(x)$ with one from $g(x)$. My main result, over an arbitrary finite field, was that $f \S g$ is irreducible if and only if $f(x)$ and $g(x)$ are irreducible polynomials of coprime degrees. This was the start of a long line of papers on such multiplication problems.

I have also joined the design of another NATO-approved cipher. This time, it was no one man job, but a teamwork where my part was more modest.

An officer at the naval base in Bergen, Cato Seeberg, had sent a suggestion for a machine cipher to Stordahl's office. His cryptologists turned the thumb down, but as a comfort, they asked him to contact me, since I was working in Bergen. I realized that the concept was new and original, and might be developed into a strong cipher.

Seeberg used a directed graph, with flip-flops and binary adders as components. The input flip-flop is triggered continuously. Each pulse makes the unit flip from one side to the other, and the incoming pulse passes out of the flip-flop on the side determined by its state. In the drawing, it is assumed that both flip-flops (nodes) A and B start in the "left" state.

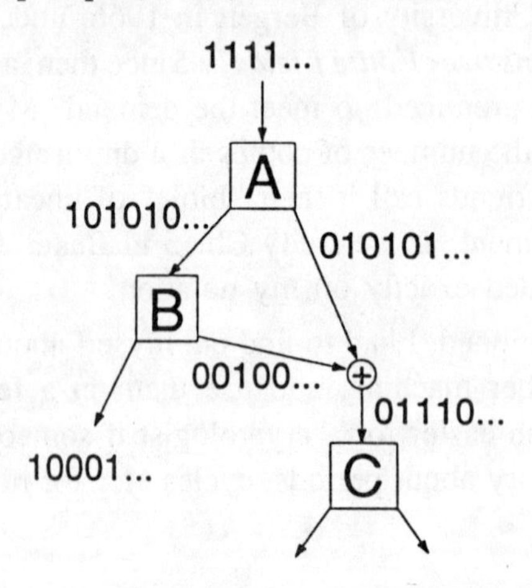

We do not want the graph spreading out continuously. On the contrary, some pulse paths are collected in adders. In particular, a final adder gives just one output of the graph.

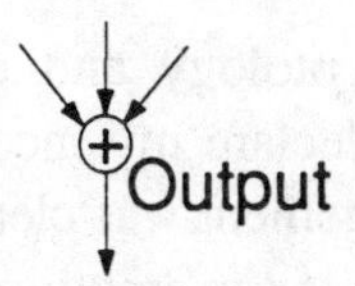

The inputs may come from different "levels". The final output can then be used to construct a key sequence. If there are n flip-flops, the number of different states of the graph is 2^n, so the output must be periodic.

There are two paths to the node C above, of lengths 1 and 2. If the longest path to the output has length l, then the period of the output sequence is at most 2^l. We want this to be independent of the initial state of the graph when the pulsing starts. A sufficient condition for this is that among the paths to any node, there is always just one of greatest length. In this case we can call 2^l the period of the graph.

After my positive reaction, Stordahl's cryptologists Kjell Kjeldsen and Ben Johnsen started working on Seeberg's concept. It was clear that the period length 2^l was too short, and it was increased as follows: Each flip-flop was replaced by a circulating binary sequence, which steps once for each incoming pulse. The 1's and 0's in this sequence then replace the sequence 1010... of flip-flop states.

Johnsen and particularly Kjeldsen made an extremely sophisticated analysis of the output from a modified Seeberg graph. Three papers were published in *Information and Control*; one by Kjeldsen (1976), one by Johnsen (1974), and a first common paper in 1973.

The periodic output sequence should obviously have a distribution of 1's and 0's as even as possible. The most important condition for this is that the output from the binary "skeleton" (Seeberg's original design, with only flip-flops) should have a completely even distribution. We had not obtained this when Kjeldsen, Johnsen and I went to Washington in 1971 to get an approval for NATO use from the NSA. One evening, the other two went to a movie (Marx Brothers), but I stayed in my hotel room to look more at the distribution. And suddenly it struck me that inclusion of just one more flip-flop in our suggested graph gave an even distribution. It was one of those aha-experiences which you never forget.

There are of course many other details which I cannot go into. The cipher machine got its NATO approval, and the electronic company *Lehmkuhl* took over the production of the so-called *"Omnicoder"*. Up to now, they have sold about 2000 units.

Until fairly recently, cryptology and cryptologists were very hush-hush. I was not allowed to declare my income from the consultant job in Stordahl's office! (The arrangement was cleared with the Auditor-General.) I did not earn more from the tax exemption, since my salary was adjusted for this.

As I remarked earlier, the word *cipher* did not turn up in the early literature on shift registers, not even in my textbook from 1966. Let me use the Omnicoder to illustrate how things suddenly became more relaxed.

I mentioned the papers by Kjeldsen and Johnsen. As mathematics, they were excellent, but many readers might ask what it is all about.

Kjeldsen added three more papers on cascade coupled sequences, and used this to get his Dr.philos. degree at the University of Bergen. Our degree is different from the American Ph.D. We have no oral examinations, but the requirements for the publications are higher. During the disputation, two opponents are dissecting the candidate's contributions, asking and criticizing.

Kjeldsen's disputation was in 1978, and I was one of the opponents. Since his 1976 paper, the secrecy of the *word* "cipher" had suddenly disappeared. Kjeldsen could now tell freely that he was working at the Defense cipher office, and in my introduction as opponent, I could explain his graphs in the same way that I have done above.

As you all know, the relaxing trend has accelerated. Just think of these international crypto seminars.

On the Linear Complexity of Products of Shift-Register Sequences

Rainer Göttfert and Harald Niederreiter

Institute for Information Processing
Austrian Academy of Sciences
Sonnenfelsgasse 19
A-1010 Vienna, Austria

E-mail: goet@qiinfo.oeaw.ac.at
nied@qiinfo.oeaw.ac.at

Abstract: In the theory of stream ciphers the termwise product of shift-register sequences plays a crucial role. In this paper we improve and generalize earlier results on the linear complexity of a termwise product of two shift-register sequences and we also provide information on the minimal polynomial of such a product.

1 Introduction

A fundamental problem in the theory of stream ciphers is the determination of the linear complexity of keystreams. A detailed account of this problem can be found in the survey article of Rueppel [8]. In practical implementations, the basic ingredients of typical algorithms for keystream generation are shift-register sequences and Boolean combining functions. It is essential for the analysis of such algorithms to investigate the behavior of shift-register sequences under elementary operations such as termwise addition and multiplication. If this behavior is known, then the effect of general combining functions can also be predicted. Since the sum of shift-register sequences is comparatively easy to analyze, research has focused on the linear complexity of the product of shift-register sequences. It suffices, of course, to treat the product of two shift-register sequences, as we can then proceed by induction to obtain information about the product of any finite number of shift-register sequences. For prior work on the linear complexity of products of shift-register sequences we refer e.g. to [1], [3], [9], [10]. The basic result in this area says that the linear complexity of a product of shift-register sequences is bounded from above by the product of the linear complexities of the individual sequences (this result is implicit in [10] and stated explicitly in [3, Corollary 3.5]). In this paper we improve and generalize earlier results on the linear complexity of a product of two shift-register sequences and we also provide information on the minimal polynomial of such a product.

For an arbitrary field F and a nonconstant monic polynomial $f \in F[x]$ let $S_F(f)$ be the set of all (linear feedback) shift-register sequences in F with characteristic polynomial f. For the constant polynomial $f = 1$, $S_F(f)$ consists by definition of the zero sequence, i.e., the sequence all of whose terms are 0. If $f \in F[x]$ is any monic polynomial, then $S_F(f)$ is a vector space over F of dimension $\deg(f)$ under termwise operations on sequences. If a shift-register sequence σ in F belongs to $S_F(f)$, but to no space $S_F(g)$ with g being a proper factor of f, then f is called the *minimal polynomial* of σ and its degree $\deg(f)$ is the *linear complexity* of σ. We write $M_F(f)$ for the subset of $S_F(f)$ consisting of the sequences with minimal polynomial f.

For arbitrary sequences $\sigma_1, \ldots, \sigma_k$ of elements of F we define $\sigma_1 + \cdots + \sigma_k$ to be the sequence which is the termwise sum, and for arbitrary sequences σ and τ of elements of F we let $\sigma\tau$ be the sequence which is the termwise product. Thus, for instance, if $\sigma = (s_n)_{n=0}^{\infty}$ and $\tau = (t_n)_{n=0}^{\infty}$, then $\sigma\tau = (s_n t_n)_{n=0}^{\infty}$, where all $s_n, t_n \in F$.

2 Results and Proofs

For cryptologic purposes, the only fields F of interest are finite fields, but it should be pointed out that most of our results have analogs for arbitrary fields (see Göttfert [2]). Throughout this paper, $\mathbb{F}_q$ denotes a fixed finite field of order q and characteristic p. If $F = \mathbb{F}_q$, we write $S(f)$ and $M(f)$ for $S_F(f)$ and $M_F(f)$, respectively. We refer to [4, Chapter 6] for background information on shift-register sequences in finite fields. The following theorem is our main result. There is no serious loss of generality in considering only spaces $S(f)$ with $f(0) \neq 0$ (compare with [4, p. 222]).

Theorem 1. *Let $f, g \in \mathbb{F}_q[x]$ be nonconstant monic polynomials with $f(0) \neq 0$ and $g(0) \neq 0$. Let $\alpha_1, \ldots, \alpha_r$ be the distinct roots of f in the splitting field E of fg over $\mathbb{F}_q$, with corresponding multiplicities $a_1, \ldots, a_r$. Similarly, let $\beta_1, \ldots, \beta_s \in E$ be the distinct roots of g with multiplicities $b_1, \ldots, b_s$, respectively. Suppose that the following two conditions hold:*

(i) *the rs elements $\alpha_i\beta_j \in E$, $1 \leq i \leq r$, $1 \leq j \leq s$, are distinct;*

(ii) $\binom{a_i + b_j - 2}{a_i - 1} \not\equiv 0 \bmod p$ *for $1 \leq i \leq r$, $1 \leq j \leq s$.*

Then for any shift-register sequences σ and τ in $\mathbb{F}_q$ with $\sigma \in M(f)$ and $\tau \in M(g)$, the product sequence $\sigma\tau$ is a shift-register sequence in $\mathbb{F}_q$ with minimal polynomial

$$\prod_{\substack{1 \leq i \leq r \\ 1 \leq j \leq s}} (x - \alpha_i\beta_j)^{a_i + b_j - 1} \in \mathbb{F}_q[x].$$

Corollary 1. *Under the conditions of Theorem 1, the sequence $\sigma\tau$ has linear complexity $L = s\deg(f) + r\deg(g) - rs$, which can be written as*

$$L = \deg(f)\deg(g) - (\deg(f) - r)(\deg(g) - s).$$

For the proof of Theorem 1 we need three auxiliary results. Lemma 2 below is taken from [4, Theorem 6.57].

Lemma 1. *Let F be an arbitrary field and α a nonzero element of F. Then for all integers $c, t \geq 0$ we have*

$$\left(\binom{n+c}{t}\alpha^n\right)_{n=0}^{\infty} \in M_F((x-\alpha)^{t+1}).$$

Proof. We assign to the sequence $\left(\binom{n+c}{t}\alpha^n\right)_{n=0}^{\infty}$ its generating function

$$G(x) = \sum_{n=0}^{\infty}\binom{n+c}{t}\alpha^n x^{-n-1}$$

in the sense of Niederreiter [7]. Then

$$\begin{aligned} G(x) &= \sum_{n=\max(c,t)}^{\infty}\binom{n}{t}\alpha^{n-c}x^{-n+c-1} \\ &= \sum_{n=t}^{\infty}\binom{n}{t}\alpha^{n-c}x^{-n+c-1} - \sum_{n=t}^{\max(c,t)-1}\binom{n}{t}\alpha^{n-c}x^{-n+c-1} \\ &= \frac{\alpha^{t-c}x^c}{(x-\alpha)^{t+1}} - h(x) \end{aligned}$$

for some polynomial $h \in F[x]$. Thus $G(x)$ is a rational function which in its reduced form has the denominator $(x-\alpha)^{t+1}$, and so the result follows from [7, Lemma 2]. □

Lemma 2. *Let F be a finite field. If $\sigma_i \in M_F(h_i)$ for $1 \leq i \leq k$ and the monic polynomials $h_1, \ldots, h_k \in F[x]$ are pairwise relatively prime, then $\sigma_1 + \cdots + \sigma_k \in M_F(h_1 \cdots h_k)$.*

Lemma 3. *Let F be an arbitrary field of characteristic p, let $\alpha, \beta \in F$ with $\alpha\beta \neq 0$, and let a and b be positive integers. Let σ and τ be shift-register sequences in F with minimal polynomials $(x-\alpha)^a$ and $(x-\beta)^b$, respectively. Then $\sigma\tau$ is always in $S_F((x-\alpha\beta)^{a+b-1})$, and $\sigma\tau \in M_F((x-\alpha\beta)^{a+b-1})$ if and only if $\binom{a+b-2}{a-1} \not\equiv 0 \bmod p$.*

Proof. By Lemma 1,

$$\psi_i := \left(\binom{n}{i}\alpha^n\right)_{n=0}^{\infty} \in M_F((x-\alpha)^{i+1}) \qquad \text{for } 0 \leq i \leq a-1.$$

It follows that the sequences $\psi_0, \ldots, \psi_{a-1}$ are linearly independent over F and thus form a basis of the a-dimensional vector space $S_F((x-\alpha)^a)$. Hence every $\sigma \in S_F((x-\alpha)^a)$ can be uniquely represented in the form

$$\sigma = c_0\psi_0 + \cdots + c_{a-1}\psi_{a-1} \qquad \text{with all } c_i \in F, \tag{1}$$

and we have $\sigma \in M_F((x-\alpha)^a)$ if and only if $c_{a-1} \neq 0$. Again by Lemma 1,

$$\omega_j := \left(\binom{n+j}{j}\beta^n\right)_{n=0}^{\infty} \in M_F((x-\beta)^{j+1}) \qquad \text{for } 0 \leq j \leq b-1.$$

Thus, the sequences $\omega_0, \ldots, \omega_{b-1}$ form a basis of $S_F((x-\beta)^b)$, and every $\tau \in M_F((x-\beta)^b)$ has a unique representation

$$\tau = d_0\omega_0 + \cdots + d_{b-1}\omega_{b-1} \qquad \text{with all } d_j \in F \text{ and } d_{b-1} \neq 0. \tag{2}$$

If $\sigma \in M_F((x-\alpha)^a)$ and $\tau \in M_F((x-\beta)^b)$, then from (1) and (2) we get

$$\sigma\tau = \sum_{\substack{0 \leq i \leq a-1 \\ 0 \leq j \leq b-1}} c_i d_j \psi_i \omega_j, \tag{3}$$

and so $\sigma\tau$ is a linear combination of product sequences of the form $\psi_i\omega_j$. The terms of $\psi_i\omega_j$ are

$$\binom{n}{i}\binom{n+j}{j}\alpha^n\beta^n, \qquad n = 0, 1, \ldots.$$

Now for all $i, j, n \geq 0$ we have

$$\binom{n}{i}\binom{n+j}{j} = \binom{i+j}{i}\binom{n+j}{i+j}. \tag{4}$$

Thus, we can write the terms of $\psi_i\omega_j$ as

$$\binom{i+j}{i}\binom{n+j}{i+j}(\alpha\beta)^n, \qquad n = 0, 1, \ldots,$$

where the first factor is independent of n. Consequently, $\psi_i\omega_j$ is either the zero sequence, namely if $\binom{i+j}{i} \equiv 0 \bmod p$, or by Lemma 1 we have $\psi_i\omega_j \in M_F((x-\alpha\beta)^{i+j+1})$ if $\binom{i+j}{i} \not\equiv 0 \bmod p$. From (3) it follows that $\sigma\tau \in S_F((x-\alpha\beta)^{a+b-1})$. Furthermore, since $c_{a-1}d_{b-1} \neq 0$, we have $\sigma\tau \in M_F((x-\alpha\beta)^{a+b-1})$ if and only if $\psi_{a-1}\omega_{b-1}$ is not the zero sequence, i.e., if and only if $\binom{a+b-2}{a-1} \not\equiv 0 \bmod p$. □

Proof of Theorem 1. Let E again be the splitting field of fg over $\mathbb{F}_q$. Then the sequences $\sigma \in M(f)$ and $\tau \in M(g)$ can be viewed as shift-register sequences in E. They have representations

$$\sigma = \sigma_1 + \cdots + \sigma_r,$$
$$\tau = \tau_1 + \cdots + \tau_s,$$

where the σ_i and τ_j are shift-register sequences in E with $\sigma_i \in M_E((x-\alpha_i)^{a_i})$ for $1 \le i \le r$ and $\tau_j \in M_E((x-\beta_j)^{b_j})$ for $1 \le j \le s$. It follows that

$$\sigma\tau = \sum_{\substack{1\le i\le r\\ 1\le j\le s}} \sigma_i\tau_j.$$

By Lemma 3 and condition (ii) we have $\sigma_i\tau_j \in M_E((x-\alpha_i\beta_j)^{a_i+b_j-1})$. Since the $\alpha_i\beta_j$ are distinct by condition (i), we obtain from Lemma 2 that $\sigma\tau$, considered as a shift-register sequence in E, has the minimal polynomial

$$h(x) = \prod_{\substack{1\le i\le r\\ 1\le j\le s}} (x-\alpha_i\beta_j)^{a_i+b_j-1}.$$

But from $f, g \in \mathbb{F}_q[x]$ we get

$$h(x)^q = \prod_{\substack{1\le i\le r\\ 1\le j\le s}} (x^q-\alpha_i^q\beta_j^q)^{a_i+b_j-1} = \prod_{\substack{1\le i\le r\\ 1\le j\le s}} (x^q-\alpha_i\beta_j)^{a_i+b_j-1} = h(x^q),$$

and so $h \in \mathbb{F}_q[x]$. Consequently, h is the minimal polynomial of $\sigma\tau$ as a shift-register sequence in $\mathbb{F}_q$. □

Remark 1. Consider the special case of Theorem 1 in which one of f and g has only simple roots. We then have $\deg(f) - r = 0$ or $\deg(g) - s = 0$. We assume again (i), and we note that (ii) is automatically satisfied since $\binom{a_i+b_j-2}{a_i-1} = 1$ for $1 \le i \le r$, $1 \le j \le s$. It follows from Corollary 1 that the linear complexity L of the product sequence $\sigma\tau$ is then $L = \deg(f)\deg(g)$. Thus, this special case of Theorem 1 yields a result of Herlestam [3] which says that under the indicated conditions the linear complexity of $\sigma\tau$ is equal to the product of the linear complexities of σ and τ (compare with [3, Corollary 3.5]).

Remark 2. The congruence of Lucas [5] for binomial coefficients (see also [6] for a recent proof) implies that

$$\binom{a+b-2}{a-1} \not\equiv 0 \bmod p \Leftrightarrow A_m + B_m < p \quad \text{for all } m,$$

where the A_m and B_m are the digits in the p-ary expansion of $a-1$ and $b-1$, respectively, i.e.,

$$a-1 = \sum_{m\ge 0} A_m p^m \quad \text{and} \quad b-1 = \sum_{m\ge 0} B_m p^m$$

with $0 \le A_m, B_m < p$ for all m. The connection with the operation $\vee$ introduced by Zierler and Mills [10] for positive integers a and b is given by

$$\binom{a+b-2}{a-1} \not\equiv 0 \bmod p \Leftrightarrow a \vee b = a+b-1.$$

The fact that the condition (ii) in Theorem 1 is needed can be shown by the following simple example. Take $\alpha, \beta \in \mathbb{F}_q$ with $\alpha\beta \neq 0$ and integers $a, b \geq 1$ with $\binom{a+b-2}{a-1} \equiv 0 \bmod p$. Then Lemma 3 shows that the conclusion of Theorem 1 is not valid for $\sigma \in M((x-\alpha)^a)$ and $\tau \in M((x-\beta)^b)$. In extreme cases such as the one considered in Theorem 2 below, the product sequence $\sigma\tau$ can even become the zero sequence.

Theorem 2. *Let $f, g \in \mathbb{F}_q[x]$ be nonconstant monic polynomials with $f(0) \neq 0$ and $g(0) \neq 0$ and with canonical factorizations*

$$f = \prod_{i=1}^{k} f_i^{a_i} \quad \text{and} \quad g = \prod_{j=1}^{l} g_j^{b_j}$$

in $\mathbb{F}_q[x]$. Suppose the exponents a_i and b_j are such that $\binom{a_i+b_j-2}{a_i-1} \equiv 0 \bmod p$ for $1 \leq i \leq k$, $1 \leq j \leq l$. Then there exist shift-register sequences $\sigma \in M(f)$ and $\tau \in M(g)$ in $\mathbb{F}_q$ such that $\sigma\tau$ is the zero sequence.

Proof. For a finite extension field F of $K = \mathbb{F}_q$ let $\mathrm{Tr}_{F/K}$ be the trace function from F onto K (compare with [4, Definition 2.22]). For $1 \leq i \leq k$ let σ_i be the sequence with terms

$$\binom{n}{a_i - 1} \mathrm{Tr}_{F_i/K}(\alpha_i^n), \qquad n = 0, 1, \ldots,$$

where α_i is a root of f_i and $F_i = \mathbb{F}_q(\alpha_i)$. It follows from Lemma 1 and Lemma 2 that $\sigma_i \in M(f_i^{a_i})$ for $1 \leq i \leq k$, and so $\sigma := \sigma_1 + \cdots + \sigma_k \in M(f)$ by Lemma 2. Similarly, we define for $1 \leq j \leq l$ the sequence τ_j with terms

$$\binom{n + b_j - 1}{b_j - 1} \mathrm{Tr}_{E_j/K}(\beta_j^n), \qquad n = 0, 1, \ldots,$$

where β_j is a root of g_j and $E_j = \mathbb{F}_q(\beta_j)$. Then $\tau_j \in M(g_j^{b_j})$ for $1 \leq j \leq l$ and $\tau := \tau_1 + \cdots + \tau_l \in M(g)$. By (4) and the given hypothesis, for all $1 \leq i \leq k$, $1 \leq j \leq l$ we have

$$\binom{n}{a_i - 1}\binom{n + b_j - 1}{b_j - 1} = \binom{a_i + b_j - 2}{a_i - 1}\binom{n + b_j - 1}{a_i + b_j - 2} \equiv 0 \bmod p,$$

and so $\sigma_i\tau_j$ is the zero sequence. This implies that $\sigma\tau = \sum_{i=1}^{k}\sum_{j=1}^{l} \sigma_i\tau_j$ is the zero sequence. □

We recall a definition from Zierler and Mills [10]. If f and g are nonconstant monic polynomials over $\mathbb{F}_q$ with simple roots, then $f \vee g \in \mathbb{F}_q[x]$ is the monic polynomial whose roots are exactly the distinct elements of the form $\alpha\beta$, where α is a root of f and β a root of g in the splitting field of fg over $\mathbb{F}_q$. The following result can be viewed as an extension of Theorem 2.

Theorem 3. *Let $f, g \in \mathbb{F}_q[x]$ be nonconstant monic polynomials with $f(0) \neq 0$ and $g(0) \neq 0$ and with canonical factorizations*

$$f = \prod_{i=1}^{k} f_i^{a_i} \quad \text{and} \quad g = \prod_{j=1}^{l} g_j^{b_j}$$

in $\mathbb{F}_q[x]$. Then there exist shift-register sequences $\sigma \in M(f)$ and $\tau \in M(g)$ in $\mathbb{F}_q$ such that $\sigma\tau$ has

$$\operatorname{lcm}\left\{ (f_i \vee g_j)^{a_i+b_j-1} : 1 \leq i \leq k,\ 1 \leq j \leq l,\ \binom{a_i + b_j - 2}{a_i - 1} \not\equiv 0 \bmod p \right\}$$

as a characteristic polynomial.

Proof. For $1 \leq i \leq k$ and $1 \leq j \leq l$ define the sequences σ_i and τ_j as in the proof of Theorem 2, and put $\sigma = \sigma_1 + \cdots + \sigma_k \in M(f)$ and $\tau = \tau_1 + \cdots + \tau_l \in M(g)$. Then

$$\sigma\tau = \sum_{\substack{1 \leq i \leq k \\ 1 \leq j \leq l}} \sigma_i \tau_j .$$

In the proof of Theorem 2 it was shown that $\sigma_i\tau_j$ is the zero sequence for all (i, j) with $\binom{a_i+b_j-2}{a_i-1} \equiv 0 \bmod p$. Thus we can write

$$\sigma\tau = \sum_{(i,j) \in I} \sigma_i \tau_j \tag{5}$$

with $I = \left\{(i, j) : 1 \leq i \leq k,\ 1 \leq j \leq l,\ \binom{a_i+b_j-2}{a_i-1} \not\equiv 0 \bmod p\right\}$. Since $\sigma_i \in S(f_i^{a_i})$ and $\tau_j \in S(g_j^{b_j})$, it follows from Zierler and Mills [10] that $\sigma_i\tau_j$ has $(f_i \vee g_j)^{a_i \vee b_j}$ as a characteristic polynomial. By Remark 2 we have $a_i \vee b_j = a_i + b_j - 1$ for $(i, j) \in I$. Hence we obtain from (5) that a characteristic polynomial of $\sigma\tau$ is given by the least common multiple of the polynomials $(f_i \vee g_j)^{a_i+b_j-1}$, $(i, j) \in I$. □

Finally, we combine Theorem 1 with a result of Rueppel and Staffelbach [9].

Corollary 2. *Let $f \in \mathbb{F}_q[x]$ be a monic irreducible polynomial with $f(0) \neq 0$ and let $g \in \mathbb{F}_q[x]$ be a nonconstant monic polynomial with $g(0) \neq 0$ and simple roots. Define $R = \operatorname{ord}(f)/\gcd(\operatorname{ord}(f), \operatorname{ord}(g))$. Suppose the multiplicative order of q modulo R equals $\deg(f)$. Let a and b be two positive integers with $\binom{a+b-2}{a-1} \not\equiv 0 \bmod p$. Then for any shift-register sequences σ and τ in $\mathbb{F}_q$ with $\sigma \in M(f^a)$ and $\tau \in M(g^b)$, the product sequence $\sigma\tau$ has the minimal polynomial $(f \vee g)^{a+b-1}$ and so the linear complexity $(a + b - 1)\deg(f)\deg(g)$.*

Proof. In [9] Rueppel and Staffelbach have shown that under the conditions in the corollary the products of a root of f and a root of g are distinct. Therefore the polynomials f^a and g^b satisfy the hypotheses of Theorem 1. □

References

[1] Golić, J. Dj.: On the linear complexity of functions of periodic GF(q) sequences, *IEEE Trans. Inform. Theory* **35**, 69–75 (1989).

[2] Göttfert, R.: Produkte von Schieberegisterfolgen, Ph.D. Dissertation, Univ. of Vienna, 1993.

[3] Herlestam, T.: On functions of linear shift register sequences, *Advances in Cryptology — EUROCRYPT '85* (F. Pichler, ed.), Lecture Notes in Computer Science, vol. **219**, pp. 119–129, Springer-Verlag, Berlin, 1986.

[4] Lidl, R., and Niederreiter, H.: *Introduction to Finite Fields and Their Applications*, Cambridge University Press, Cambridge, 1986.

[5] Lucas, E.: Sur les congruences des nombres eulériens et des coefficients différentiels des fonctions trigonométriques, suivant un module premier, *Bull. Soc. Math. France* **6**, 49–54 (1878).

[6] McIntosh, R. J.: A generalization of a congruential property of Lucas, *Amer. Math. Monthly* **99**, 231–238 (1992).

[7] Niederreiter, H.: Sequences with almost perfect linear complexity profile, *Advances in Cryptology — EUROCRYPT '87* (D. Chaum and W. L. Price, eds.), Lecture Notes in Computer Science, vol. **304**, pp. 37–51, Springer-Verlag, Berlin, 1988.

[8] Rueppel, R. A.: Stream ciphers, *Contemporary Cryptology: The Science of Information Integrity* (G. J. Simmons, ed.), pp. 65–134, IEEE Press, New York, 1992.

[9] Rueppel, R. A., and Staffelbach, O. J.: Products of linear recurring sequences with maximum complexity, *IEEE Trans. Inform. Theory* **33**, 124–131 (1987).

[10] Zierler, N., and Mills, W. H.: Products of linear recurring sequences, *J. Algebra* **27**, 147–157 (1973).

Resynchronization Weaknesses in Synchronous Stream Ciphers

Joan Daemen, René Govaerts and Joos Vandewalle

Katholieke Universiteit Leuven, Laboratorium ESAT
Kardinaal Mercierlaan 94, B-3001 Heverlee, Belgium
email: joan.daemen@esat.kuleuven.ac.be

Abstract. In some applications for synchronous stream ciphers, the risk of loss of synchronization cannot be eliminated completely. In these cases frequent resynchronization or resynchronization upon request may be necessary. In the paper it is shown that this can lead to significant deterioration of the cryptographic security. A powerful general attack on nonlinearly filtered linear (over $\mathbb{Z}_2$) systems is presented. This attack is further refined to efficiently cryptanalyze a linear system with a multiplexer as output function.

1 Introduction

Synchronous stream ciphers have the advantage that digit-value errors in the ciphertext only affect the corresponding digits in the plaintext, i.e. there is no error propagation due to the decryption process. The price paid for this property is that perfect synchronization is required between sender and receiver. If synchronization is lost, the output of the decryptor is unintelligible for the receiver and synchronization has to be regained.

If the plaintext has enough redundancy to identify correctness of decryption, the resynchronization (resync) can be performed by the receiver without the intervention of the sender. This process involves trying all possible offsets between the sender's and the receiver's clock. In applications where high-speed online decryption is needed, automatic synchronization recovery can be technologically infeasible. The risk of synchronization loss (SL) can be reduced somewhat at the cost of inserting synchronization patterns in the ciphertext. On the other hand the consequences of SL have to be limited. This can be achieved by performing resync at fixed timesteps (*fixed resync*). For instance, if resync occurs every M digits, SL will on the average result in the loss of M/2 digits. The appropriate frequency of resync depends on the probability of SL and the nature of the application. If there is a channel from the receiver to the sender, the receiver can request resync (*requested resync*) to the sender upon detection of SL.

The aim of resync is to ensure that encryptor and decryptor have the same internal state at a certain time. An internal state different from all previous resync states has to be chosen, to prevent the re-use of keysequences. For stream ciphers whose security is (partly) based on the secrecy of the internal state, the resync state may not be communicated in the clear at the time of resync. Since the reason for resync is the possible unavailability of the stream encryption, this new internal state has to be encrypted with an additional cipher. This introduces extra complexity and cost into the system.

A more elegant (and cheap) solution is to specify the new internal state in terms of information that is only known to the sender and receiver. In this case sender and receiver calculate the new internal state from the original internal state (or key) and a public parameter (e.g. the time at the moment of resync) using a publicly known algorithm. At the time of resync no confidential information has to be transmitted, hence no additional cipher system is required. The subject of this paper is the impact of this form of resync on the cryptographic security of practical stream ciphers.

In the following section we will define the different components of a stream cipher system with resynchronization. The term *key secure* is introduced and defined with respect to several systems that differ in their resync abilities. Section 3 treats the cryptographic security of nonlinearly filtered linear systems. It contains a general attack and an attack that is specific for a multiplexer output function.

2 The Cryptographic Setting

Stream encryption is performed by encrypting the plaintext digit by digit. The plaintext digit at time t denoted by x^t is transformed into a ciphertext digit y^t by a transformation $\mathrm{f_e}()$. This transformation depends on a keystream digit z^t and must be invertible with respect to x^t. In all practical systems the ciphertext, plaintext and keystream digits belong to the same alphabet, denoted by $\mathcal{A}$. We have

$$y^t = \mathrm{f_e}(x^t, z^t) \quad \text{and} \quad x^t = \mathrm{f_d}(y^t, z^t) \tag{1}$$
$$\text{with} \qquad x = \mathrm{f_d}(\mathrm{f_e}(x, z), z),\ \forall x, z \in \mathcal{A}$$

For the majority of designs the digits are bits. In this case (1) can be simplified to (with $\oplus$ denoting XOR)

$$y^t = x^t \oplus z^t$$

In most other designs the plaintext, keysequence and ciphertext bits are lumped into m-bit digits and the encryption can be described by

$$y_i^t = x_i^t \oplus z_i^t \tag{2}$$

where $x^t = (x_1^t, x_2^t \ldots x_\mathrm{m}^t)$, $y^t = (y_1^t, y_2^t \ldots y_\mathrm{m}^t)$, $z^t = (z_1^t, z_2^t \ldots z_\mathrm{m}^t)$ and $\oplus$ denotes bitwise XOR.

2.1 The Synchronous Stream Encryptor/Decryptor

In *synchronous* stream ciphers the keystream digits z_i are generated independently of the message stream by a pseudorandom sequence generator(PSG). This is a finite state machine whose operation is governed by the two rules

$$s^{t+1} = \mathrm{F_s}(s^t, p) \tag{3}$$

$$z^t = \mathrm{f_o}(s^t, p) \tag{4}$$

with s^t the internal state at time t and p an optional parameter that is fixed during pseudorandom sequence generation. The finite state machine is initialized by loading the *initialization vector* v into the stateregister and by loading the parameter p. In hardware this loading mechanism requires additional circuitry. Therefore the initialization is sometimes integrated into the state-transition function:

$$s^{t+1} = \mathrm{F_s}(s^t, p, e) \tag{5}$$

where the e is an external input that is used to 'load' the initial state η digits at a time. Initialization is performed by resetting the internal state to a fixed value and subsequently iterating the finite state machine a specified number of times. With this mechanism the initialization vector v is defined as the array of η-digit e-inputs during the initialization process. During pseudorandom sequence generation this e-input must be constant and the state-transition can be modeled by (4). The time origin is defined by the initialization, hence the keysequence z looks like $z^0 z^1 z^2 \ldots$ and the initial state is s^0.

The cryptographic security of the stream cipher can be based entirely on the secrecy of (part of) the initial state, entirely on the secrecy of the parameter p or on the secrecy of both. This secret information is generally referred to as the *key* K. The dependence of the initialization vector v and the parameter p on the key K can be formally expressed by a function $\mathrm{F_k}()$:

$$(v, p) = \mathrm{F_k}(K) \tag{6}$$

In this paper the specification of a PSG consists by definition of the three functions $\mathrm{F_s}(), \mathrm{f_o}()$ and $\mathrm{F_k}()$.

In general the cryptographic security that is required from a cipher system depends on the application. Therefore, *cryptographic security of a cipher system* can best be defined as security in the worst possible circumstances. Clearly, a cipher that is claimed to cryptographically secure by this definition, is claimed to be secure in all applications. We introduce the term *key secure* to denote cryptographic systems whose security is *indicated* by the entropy of the key K.

Definition 1. A pseudorandom sequence generator is *key secure* if for any a priori distribution Ω of K, the cheapest way of using any known part of the keysequence to obtain knowledge about the other part of the keysequence includes complete determination of the specific value of K by exhaustive key search (with respect to Ω).

This definition implies that the cryptanalist has only access to the output of the PSG. The term *cheap* must be seen in the context of the real world where hardware, software, manpower etc. are for sale.

2.2 The Resync Mechanism

The initialization vector v and the parameter p for a given resync depend not only on the key but also on a publicly known randomization variable ρ by a publicly known *randomization function* $\mathrm{F_r}()$:

$$(v_i, p_i) = \mathrm{F_r}(K, \rho) \tag{7}$$

In most designs the parameter p is independent of ρ. If resync is performed on fixed timesteps, ρ stands for the serial number of the resync. For requested resync ρ can be the time given by a commonly available clock or a string that is sent by the receiver together with the request. The set of all possible ρ inputs is denoted by $\mathcal{R}$. If resync is performed it can be advantageous to do a number (say μ) of 'blank' iterations of the PSG after the loading of the initial state. During these iterations no pseudorandom digits are given at the output. This can be modeled by removing the first μ digits of the keystream: $z = z^{\mu} z^{\mu+1} z^{\mu+2}$. The resync mechanism (RM) consists of the randomization function $\mathrm{F_r}()$ and the constant μ.

2.3 Fixed Resync Setting

In a practical situation the security of an encryption system has to be evaluated in its totality. In a fixed resync application the stream encryption system consists basically of 3 components: the PSG, the randomization function $\mathrm{F_r}()$ and the total number of resyncs, denoted by ℓ.

In this setting the keysequence is no longer the result of iterating the PSG starting from a single initial state. After every resync a 'new' keysequence is started. Since these keysequences are the result of initialization vectors v_i (and parameters p) that are computed from the same key, they are not independent. It will be advantageous to describe the keysequences (that are in fact successive in time) as emerging simultaneously from a number of finite state machines. The initialization vector and parameter of each of these machines is given by the RM. For the i-th machine this is expressed as

$$(v_i, p_i) = \mathrm{F_r}(K, \rho_i) \Longrightarrow z_i : z_i^{\mu} z_i^{\mu+1} z_i^{\mu+2} \ldots \tag{8}$$

The definition of key security of a PSG can be adapted in a natural way to this system:

Definition 2. A fixed resync stream encryption system is *key secure* if for any a priori distribution Ω of K, the cheapest way of using any known part of the keysequences z_i to obtain knowledge about the other part of the keysequences z_i includes complete determination of the specific value of K by exhaustive key search (with respect to Ω).

The parallel representation emphasizes that not one but ℓ digits per clockcycle are available to the cryptanalist. This can possibly be exploited in a key-reconstruction algorithm. On the other hand, possible correlations between the different keysequences can be exploited to predict some sequences from other ones.

2.4 Requested Resync Setting

To investigate the cryptographic security of an encryption system in a requested resync application only the PSG and the RM are considered. The number ℓ is no longer given in advance. If the cryptanalist can impersonate as the receiver, he (or she) can request resync with a *chosen* ρ. Say the set of all possible keysequences for a given key is denoted by Λ_k. We have

$$\Lambda_k = \{z : \mathrm{F_r}(K, \rho) \Longrightarrow z \text{ with } \rho \in \mathcal{R}\} \tag{9}$$

We define the key security of a requested resync system as follows:

Definition 3. A requested resync stream encryption system is *key secure* if for any a priori distribution Ω of K, the cheapest way of using any known part of all $z_i \in \Lambda_k$ to obtain knowledge about the other part of all $z_i \in \Lambda_k$ includes complete determination of the specific value of K by exhaustive key search (with respect to Ω).

It is clear that a system that is key secure in a requested resync application, can be used to make a key secure fixed resync application.

2.5 Resync Security of a PSG

Stream ciphers proposals in the literature usually contain a specification of the PSG and no specification of a RM. If one is confronted with the problem of choosing a PSG as a component of a fixed or requested resync stream cipher system, the function $\mathrm{F_k}()$ has to be replaced by an appropriate RM.

A RM consists of a randomization function $F_r(K, \rho)$ and a positive integer μ. The function $F_r(K, \rho)$ has to fulfill the following two criteria to be regarded as a valid randomization function.

- $\rho_1 \neq \rho_2 \Rightarrow F_r(K, \rho_1) \neq F_r(K, \rho_1)$: different public arguments must not give rise to equal initialization vectors v *and* parameters p.
- $k_1 \neq k_2 \Rightarrow F_r(K_1, \rho) \neq F_r(K_2, \rho)$: different keys must not give rise to equal initialization vectors v *and* parameters p.

At the best, the specific choice of the RM has no influence on the security of the system. At the worst, no secure system can be built around the PSG without choosing a randomization function that has certain cryptographic properties on its own. For the complexity of the overall system it is advantageous to have a RM as simple as the PSG allows.

For a given PSG classes in the space of all possible RM can be identified that give rise to loss of security. The result of this process can be a belief or claim that a PSG is *resync key secure* with respect to a subclass Ξ of all possible RM, as defined in

Definition 4. A PSG is *resync key secure* with respect to Ξ if the requested resync system composed of the PSG and all possible RM in Ξ are key secure.

The statement that a PSG is key secure according to Def.1 is equivalent to the statement that it is resync key secure with respect to the resync mechanism where $F_r()$ is defined by $F_k()$ and $\mathcal{R} = \emptyset$. Because the randomization function has no public argument, no resync can be applied in this case.

2.6 Packet encryption with Resync Key Secure PSGs

The ease and security of resync in a resync key secure PSG allows the use of synchronous stream encryption even for small data packets. Every time a new packet is encrypted the PSG is resynchronized with a unique parameter such as the packet identifier (that is not encrypted) as public parameter. If the PSG is in fact resync key secure, the security of the system is guaranteed by the size of the keyspace.

3 Attacks on Nonlinearly Filtered Linear Systems

As an illustration the effect of resync on a popular class of PSGs is examined. In this class the state-transition function is linear over $\mathbb{Z}_2$ and the nonlinear output function is fixed. The cryptographic security is based on the inability to reconstruct the internal state from the output of the nonlinear output function. We suppose that a linear randomization function is used. In the parallel representation introduced in Sect. 2.5 the operation of the i-th system can be described by matrix equations:

$$s_i^0 = AK \oplus R_i \tag{10}$$

$$s_i^{t+1} = F s_i^t \tag{11}$$

Where A and F are binary matrices. Say the number of statebits is n. It is supposed that F is nonsingular (as it is in all practical proposals). R_i is a vector that is fixed by ρ_i in (8). The attack focuses on the reconstruction of the entire internal state for a certain PSG at a certain moment in time: s_p^τ. From this internal state, s_p^0 can be computed and subsequently all other initial states s_i^0. This knowledge is sufficient to calculate all keysequences z_i.

$$s_p^0 = F^{-\tau} s_p^\tau \tag{12}$$

$$s_i^0 = s_p^0 \oplus R_i \oplus R_p \tag{13}$$

The difference modulo 2 of any two simultaneous states s_i^t and s_j^t can be calculated:

$$s_i^t \oplus s_j^t = F^t(R_i \oplus R_j) \ . \tag{14}$$

The output function computes the keysequence digit from the internal state. Since in most practical designs the keysequence digits are bits, we will consider only binary output functions. If m-bit keysequence digits are produced, the output function can be decomposed in its binary components.

In most practical proposals, the keysequence bit only depends on a subset of statebits. Suppose the values of these statebits are collected in the binary vector u with φ components. The vector u is a projection of the internal state s. This can be expressed in a matrix equation

$$u_i^t = \mathrm{G} s_i^t \; . \tag{15}$$

The linear relations between the internal states can be exploited to strengthen existing attacks or to construct new attacks.

3.1 A Simple General Resync Attack

Suppose the vector u_p^τ can be completely reconstructed for a certain τ. This gives the cryptanalist φ bits of s_p^τ. Since s_p^τ depends on s_p^0 by $\mathrm{F}^\tau s_p^0 = s_p^\tau$ this results in φ linear equations for the bits of s_p^0. If a number m of vectors u_p^t can be reconstructed, this results in $m\varphi$ (not necessarily linearly independent) linear relations for the bits of s_p^0. If n independent linear relations are found s_p^0 can be fixed. Therefore reconstructing $\lceil \frac{\mathrm{n}}{\varphi} \rceil$ u_p^t vectors is sufficient if all relations are independent. In the rare case that there are too much linear dependencies, more u-vectors can be reconstructed until n independent linear relations are found. The resulting set of equations can be efficiently solved by methods of linear algebra.

Now we are left with the problem of reconstructing u_p^t for some t. (In the following of this section we will omit the superscript t.) Suppose z_p is known to the cryptanalist. The correct u_p has to fulfill $\mathrm{f_o}(u_p) = z_p$. For every z_i that is known to the cryptanalist there is such a relation. We have $\mathrm{f_o}(u_i) = z_i$ and $u_i = u_p \oplus \mathrm{G}(\mathrm{R}_i \oplus \mathrm{R}_p)$. Substitution yields

$$\mathrm{f_o}(u_p \oplus \mathrm{GF}^t(\mathrm{R}_i \oplus \mathrm{R}_p)) = z_i \tag{16}$$

for any z_i that is known to the cryptanalist. This is a set of nonlinear boolean equations with φ variables. If the number of equations is larger than φ, the number of solutions is expected to converge to 1. If φ is small enough, the solution can be found by performing exhaustive search over the space of all possible u-vectors. This involves on the average 2^φ evaluations of $\mathrm{f_o}()$. If a certain bit of u only appears in the function $\mathrm{f_o}()$ in linear combination with another bit, the absolute values of these bits can not be deduced from the above equations. However, their XOR can be deduced, yielding a linear relation. This can be converted to a linear relation for the initial state without a problem. Since the output function only depends on the XOR of the two bits and not on the bitvalues itself, only $2^{\varphi-1}$ inputs have to be tested.

For this attack to be possible, the number of resyncs has to be larger than φ. The cryptanalist must know at least φ keysequence bits for a number of $\lceil \frac{\mathrm{n}}{\varphi} \rceil$ timesteps. The expected workload is

$$\left\lceil \frac{\mathrm{n}}{\varphi} \right\rceil 2^\varphi \tag{17}$$

evaluations of $f_o()$ and some additional linear algebra computations. The attack can easily be parallelized by distributing the search over a number of different processors. It can be observed that the complexity of this attack only grows linearly with the size of the internal state. For this general attack to be infeasible, it is essential that the output function depends on a large number of statebits.

Say we have a PSG with a linear state-transition and a nonlinear output function. The evaluation of the resync security of such a PSG for linear randomization functions boils down to the analysis of the output function. The cryptanalytic problem is constructing an algorithm that can extract a vector u in feasible time and space if $f_o(u \oplus e)$ can be asked for all possible binary vectors e. From the definition exhaustive key search must be the most efficient attack for key secure stream ciphers, hence the complexity of this algorithm imposes an upper limit on the size of the keyspace for a key secure system.

The general attack works independent of the specific properties of the output function. In the following section we will show that a significant speedup can be attained in cases where the structure of the $f_o()$ can be exploited.

3.2 A Simple Specific Resync Attack

Consider the multiplexer generator [1]. This is a linear finite state machine with a multiplexer as output function. A multiplexer with β address inputs has 2^β data inputs. If the address inputs are denoted by $a^0 a^1 \ldots a^{\beta-1}$, the data inputs by $d^0 d^1 \ldots d^{2^\beta - 1}$ and the output by z, the operation of the multiplexer can be described by

$$z = d^a \text{ with } a = \sum a_i 2^i \; . \tag{18}$$

Here a is the interpretation of $a^{\beta-1} \ldots a^1 a^0$ as an integer. The vector u_p can be split up in a_p and d_p. Since $u_i \oplus u_j$ is completely known for all i, j , so are $a_i \oplus a_j$ and $d_i \oplus d_j$. Suppose the keysequence bits z_p and z_i are known and that $a_p \oplus a_i = 0$. If $z_p = z_i$ the integer a_p (and equivalently a_i) must select a bit in d such that $d_p^{a_p} \oplus d_i^{a_p} = 0$. If $z_p \neq z_i$ a similar argument applies. In both cases about half of the possible values for a_p are eliminated. This can be repeated for every two keystream bits z_e, z_f where $a_e \oplus a_f = 0$. The correct value for a_p will be found after investigating β pairs (or a few more). If the correct value of a_p is found in this way, the remaining $2^\beta - \beta$ values of d_p can systematically be scanned by considering all keysequence bits z_{g_i} where $a_{g_i} = a_p \oplus i$ for all $0 \leq i < 2^\beta$.

The complete process of fixing the u-vector takes about $\beta + 2^\beta$ evaluations of the multiplexer function. This has to be repeated $\lceil \frac{n}{\varphi} \rceil$ times where $\varphi = \beta + 2^\beta$. The complexity of the complete attack can be approximated by

$$\left\lceil \frac{n}{\varphi} \right\rceil \varphi \approx n \tag{19}$$

output function evaluations and some additional linear algebra computations. In fact, the linear algebra will be responsible for the largest part of the workload in realistic cases.

The applicability of this attack is somewhat limited by the demands for the difference patterns in a. However, if the total number of resyncs is not significantly smaller than 2^{β} this is no problem. A multiplexer system that can be attacked by this scheme is currently recommended in [2] for video encryption. This system does 256 resyncs and has $\beta = 5$. In [3] we present a ciphertext-only attack that can be implemented in hardware for on-line decryption of the video component without the key.

4 Conclusions

It is pointed out that the application of resynchronization can jeopardize the security of a synchronous stream cipher. This is illustrated by a general attack on the class of nonlinearly filtered linear (over $\mathbb{Z}_2$) systems that comprises a large number of practical proposals. A specific attack is presented for linear systems where the output function is a multiplexer. Both attacks are conceptually very simple and show that applying resynchronization can result in total loss of security for a large number of published stream cipher proposals.

References

[1] R. A. Rueppel, 'Stream Ciphers', in *Contemporary Cryptology*, Gustavus J. Simmons Ed. IEEE Press, New York.

[2] 'Specification of the systems of the MAC/packet family'. EBU Technical Document 3258–E, Oct 1986.

[3] J. Daemen, R. Govaerts, J. Vandewalle, Cryptanalysis of MUX-LFSR Based Scramblers, in *Proceedings of SPRC '93*, 15–16 February, Roma. (to appear)

Blind Synchronization of m-Sequences with Even Span

Richard A. Games and Joseph J. Rushanan

The MITRE Corporation, Bedford, MA, 01730, USA

Abstract. The problem of recovering the phase on a known binary m-sequence that is corrupted by a binary noise source is considered. This problem arises in the cryptanalysis of stream ciphers formed from a nonlinear combination of m-sequences. A synchronization procedure is developed for even span n. The procedure obtains a reliable estimate of the phase of an m-sequence of span n from unreliable estimates of the phases of a small number of shifts of a fixed m-sequence of span $n/2$. These latter estimates can be obtained from a variety of methods available in the literature. The procedure results in a reduction of complexity but requires observing on the order of the square root of the m-sequence's period.

1 Introduction

In this paper we focus on the problem of recovering the phase on a known binary m-sequence that is corrupted by a binary noise source. This problem arises in cryptanalysis and a number of methods have been suggested for its solution ([CS], [MS], [S], [ZH]). More precisely, we assume that we observe some terms of the binary sequence $\mathbf{r} = E^t\mathbf{s} + \mathbf{n}$, where $\mathbf{s}$ is the normal form of a known m-sequence of span n, E is the sequence shift-left operator (that is, $(E\mathbf{s})_i = (\mathbf{s})_{i+1}$), $\mathbf{n}$ is a sequence of independent and identically distributed binary random variables with probability p that $n_i = 0$, and the addition is modulo 2. We wish to recover the unknown phase t. We call this the blind synchronization problem to distinguish it from the usual synchronization process that takes advantage of approximate knowledge of the correct phase t.

We develop a blind synchronization procedure applicable when the span n of the m-sequence is even. We show how a reliable estimate of the phase of an m-sequence of span n can be obtained from unreliable estimates of the phases of the shifts of a fixed m-sequence of span $n/2$. A variety of approaches can be applied to obtain these latter estimates [CS], [MS], [S], [ZH]. The decrease in span from n to $n/2$ can result in a dramatic decrease in overall complexity. However, this complexity reduction is achieved at the expense of having to observe on the order of the square root of the m-sequence's period.

Section 2 reviews the array properties of m-sequences that we need, including the definition and properties of the shift sequence. Section 3 describes the new array blind synchronization procedure. Section 4 gives a performance analysis for the technique, and section 5 is the conclusion.

2 Shift Sequences of m-Sequences

Let f be a primitive polynomial of degree n, m an integer dividing n, $v = (2^n-1)/(2^m-1)$, and α a root of f. Then α is a primitive element of GF(2^n) and $\beta = \alpha^v$ is a primitive element of GF(2^m). Let Tr^n_m denote the trace mapping from GF(2^n) to GF(2^m). The shift sequence $\mathbf{e} = (e_0, e_1, \ldots, e_{2^n-2})$ of f for m dividing n is defined by

$$e_k = \begin{cases} \infty, & \text{if } \mathrm{Tr}^n_m(\alpha^k) = 0 \\ e, & \text{if } \mathrm{Tr}^n_m(\alpha^k) = \beta^e \end{cases}$$

The sequence $\mathbf{e}$ is called the shift sequence because when the m-sequence $\mathbf{s} = (\mathrm{Tr}^n_1(\alpha^i))$ is arranged in a 2^m-1 by v array, the nonzero columns of this array are comprised of shifts of the column m-sequence $\mathbf{c} = (\mathrm{Tr}^m_1(\beta^i))$. In particular, for $k = 0, 1, \ldots, v-1$, column k of this array is identically zero if $e_k = \infty$ and otherwise is equal to $E^{e_k}\mathbf{c}$.

The finite elements of the shift sequence can be viewed as elements of the integers modulo $2^m - 1$, denoted by $\mathbf{Z}_{2^m-1}$. By convention we have for any $e \in \mathbf{Z}_{2^m-1}$, $e \pm \infty = \infty$ and $\infty \pm \infty = \infty$. The shift sequence can be extended periodically, that is, the subscripts of $\mathbf{e}$ can be regarded modulo $2^n - 1$.

The shift sequence satisfies the following facts. The first two follow easily from the definition; the third is proved in [G].

FACT 1. *The first v terms of the shift sequence determine the remaining terms: For $a = 1, 2, \ldots, 2^m - 2$,*

$$(e_{av}, e_{av+1}, \ldots, e_{av+v-1}) = (e_0 + a, e_1 + a, \ldots, e_{v-1} + a).$$

FACT 2. *For $n = 2m$, $e_0 = \infty$ corresponds to the single zero column.*

FACT 3. *The shift sequence satisfies a uniform modular difference property: For a fixed integer difference offset $k \not\equiv 0 \pmod v$, the list of differences $\Delta_k = (e_{j+k} - e_j \pmod{2^m-1} : j = 0, 1, \ldots, v-1)$ contains each element of $\mathbf{Z}_{2^m-1}$ exactly 2^{n-2m} times.*

3 Array Blind Synchronization

Let $\mathbf{s} = (s_i) = (\mathrm{Tr}^n_1(\alpha^i))$ be the normal form of a known m-sequence of span n, with $n = 2m$. We assume that an unknown phase t of $\mathbf{s}$ is transmitted and that $\mathbf{r} = E^t\mathbf{s}+\mathbf{n}$ is observed. We also assume the existence of a blind synchronization procedure for m-sequences of span n, denoted by BS(n, N), where N denotes the number of observed terms used. See, for example, the procedures described in [CS], [MS], [S], and [ZH]. In this section we describe a procedure that uses multiple applications of BS($n/2, l$), for some integer l, to produce an estimate for t.

The array blind synchronization procedure begins by collecting lv terms of $\mathbf{r}$, where $v = (2^{2m}-1)/(2^m-1) = 2^m+1$, and by forming the $l \times v$ array

$$\mathbf{A}_l(\mathbf{r}) = \begin{bmatrix} r_0 & r_1 & \cdots & r_j & \cdots & r_{v-1} \\ r_v & r_{v+1} & \cdots & r_{v+j} & \cdots & r_{2v-1} \\ \vdots & \vdots & & \vdots & & \vdots \\ r_{(l-1)v} & r_{(l-1)v+1} & \cdots & r_{(l-1)v+j} & \cdots & r_{lv-1} \end{bmatrix}.$$

We assume that the array $\mathbf{A}_l(\mathbf{s})$ for the underlying m-sequence $\mathbf{s}$ of span n has the following form

$$\mathbf{A}_l(\mathbf{s}) = \begin{bmatrix} s_{uv+w} & s_{uv+w+1} & \cdots & s_{uv+w+j} & \cdots & s_{uv+w+v-1} \\ s_{(u+1)v+w} & s_{(u+1)v+w+1} & \cdots & s_{(u+1)v+w+j} & \cdots & s_{(u+1)v+w+v-1} \\ \vdots & \vdots & & \vdots & & \vdots \\ s_{(u+l-1)v+w} & s_{(u+l-1)v+w+1} & \cdots & s_{(u+l-1)v+w+j} & \cdots & s_{(u+l-1)v+w+v-1} \end{bmatrix},$$

where the integers u and w satisfy $0 \leq w \leq v-1$, $0 \leq u \leq 2^m - 2$, and $t = uv + w$. We wish to determine reliable estimates for the unknown sequence offsets u and w.

The nonzero columns of $\mathbf{A}_l(\mathbf{s})$ correspond to shifts of the column m-sequence $\mathbf{c} = (\mathrm{Tr}_1^m(\alpha^{vi}))$, where the shift sequence is denoted by $\mathbf{e} = (e_0, e_1, \ldots, e_{2^n-2})$. For $n = 2m$, shift-sequence fact 2 implies there is a single zero column corresponding to $e_0 = \infty$, but the position of this zero column is unknown in the array $\mathbf{A}_l(\mathbf{s})$. Also, for $n = 2m$, shift-sequence fact 3 implies that the modular differences in Δ_k are distinct for each difference offset k. Thus each Δ_k corresponds to a vector of length v with exactly two ∞ entries and with the remaining $2^m - 1$ entries corresponding to the distinct elements of $\mathbf{Z}_{2^m-1}$.

Since the column sequence $\mathbf{c}$ is known, the blind synchronization procedure $\mathrm{BS}(m,l)$ can be applied to the first two columns of $\mathbf{A}_l(\mathbf{r})$ to derive estimates $\hat{e}_0$ and $\hat{e}_1$ of the shifts e_{uv+w} and e_{uv+w+1}, respectively. The difference $\hat{d}_{10} \equiv \hat{e}_1 - \hat{e}_0 \pmod{2^m - 1}$ occurs in some unique position in the sequence of first differences Δ_1 of $\mathbf{e}$. If the column sequence estimates are both correct, then by shift-sequence fact 1, $\hat{d}_{10} \equiv \hat{e}_1 - \hat{e}_0 \equiv e_{uv+w+1} - e_{uv+w} \equiv e_{w+1} + u - (e_w + u) \equiv e_{w+1} - e_w \pmod{2^m - 1}$, and the unique position determined by $\hat{d}_{10}$ reveals the sequence offset w, as well as $w+1$. It then follows that $u \equiv \hat{e}_1 - e_w \pmod{2^m - 1}$, and the sequence can be synchronized.

However, the estimates $\hat{e}_0$ and $\hat{e}_1$ are subject to error. If at least one of the estimates is in error, then the position P determined by $\hat{d}_{10}$ will be some other arbitrary value in the range of possible positions. To resolve this possibility, the blind synchronization procedure $\mathrm{BS}(m,l)$ is applied to the third column of $\mathbf{A}_l(\mathbf{r})$ to obtain the estimate $\hat{e}_2$ of the third column shift e_{uv+w+2}. If the estimates $\hat{e}_1$ and $\hat{e}_2$ are correct, the difference $\hat{d}_{21} \equiv \hat{e}_2 - \hat{e}_1 \pmod{2^m - 1}$ reveals the positions $w+1$ and $w+2$ in Δ_1. It is the consistent determination of position $w+1$ by the differences $\hat{d}_{10}$ and $\hat{d}_{21}$ that signals that the estimates are most likely correct, and the computed sequence offsets are accurate.

If the positions determined by $\hat{d}_{10}$ and $\hat{d}_{21}$ are inconsistent, then at least one of the estimates $\hat{e}_0$, $\hat{e}_1$, and $\hat{e}_2$ is in error. It could be that only the second estimate is wrong, implying that the difference $\hat{d}_{20} \equiv \hat{e}_2 - \hat{e}_0 \pmod{2^m - 1}$ would correspond to the correct shift-sequence positions w and $w+2$ in the sequence of unique second differences Δ_2. To verify this would require computing $\hat{e}_3$ using the blind synchronization procedure $\mathrm{BS}(m,l)$ on the fourth column of $\mathbf{A}_l(\mathbf{r})$. Then the differences $\hat{d}_{32}$, $\hat{d}_{31}$, and $\hat{d}_{30}$ would each yield a pair of positions to check.

In summary, the array blind synchronization procedure continues to estimate column shifts using $\mathrm{BS}(m,l)$ and then uses differences to obtain corresponding positions from the list of unique first differences Δ_1, second differences Δ_2, third differences Δ_3, etc. We assume that differences involving incorrect estimates will yield corresponding positions that are uniformly spread over the range of possible positions. Note that as m increases it becomes less likely that two such randomly determined positions will collide. On the other hand, differences from correctly estimated column shifts will cluster and produce collisions at the correct sequence offset. A collision will occur as soon as three correct column estimates (or three consistent incorrect column estimates!) are obtained. When the first collision of positions occurs, we stop the process and use the three estimates involved to determine the estimates of the sequence offsets.

Before giving the precise algorithm for array blind synchronization, we construct the necessary data structures. Given integers k and d, $1 \leq k \leq 2^m - 1$ and $0 \leq d \leq 2^m - 2$, by the distinct modular difference property there is a unique pair of positions $\{P, P+k\}$ in the shift sequence, with $1 \leq P \leq 2^m$, such that $d \equiv e_{P+k} - e_P \pmod{2^m - 1}$. The position P corresponds to the position that d occurs at in the list of differences Δ_k. A $(2^m-1) \times (2^m-1)$ **position** array is defined by $\mathbf{position}(k, d) = P$. Since we exclude the difference ∞, the values of $\mathbf{position}(k, d)$ range from 1 to 2^m, except for position $2^m + 1 - k$, which would be paired with $e_{2^m+1} = \infty$. The algorithm creates for the jth estimated column shift $\hat{e}_j$ a list of up to j distinct positions (exactly j positions if the algorithm does not stop at the jth estimate). This list, denoted by $\mathbf{list}(j)$, contains the right-hand positions, that is, $P+k$, obtained from the j differences involving the jth estimate. The value $P+k$ is reduced modulo v in step 4 to account for the row length of $\mathbf{A}(\mathbf{s})$.

Array blind synchronization procedure:

1. Collect lv terms of $\mathbf{r}$ and form $\mathbf{A}_l(\mathbf{r})$;
2. Set $j = 0$, $\mathbf{list}(0) = \emptyset$, and $wrap = 0$;
3. Apply the blind synchronization procedure BS(m, l) to the jth column of $\mathbf{A}_l(\mathbf{r})$ to obtain the estimate $\hat{e}_j$ of the shift of the column m-sequence $\mathbf{c}$; if $j = 0$, then increment j by 1 and repeat step 3;
4. For each difference offset k, $1 \leq k \leq j$, compute the difference $\hat{d}_{j,j-k} \equiv \hat{e}_j - \hat{e}_{j-k} \pmod{2^m - 1}$; if $\mathbf{position}(k, \hat{d}_{j,j-k})$ occurs in $\mathbf{list}(j-k)$, then go to step 6; else add $\mathbf{position}(k, \hat{d}_{j,j-k}) + k \pmod{v}$ to $\mathbf{list}(j)$;
5. Increment j by 1 and go to step 3;
6. Set $P = \mathbf{position}(k, \hat{d}_{j,j-k})$.
7. The estimate for w is $\hat{w} = P + k - j$; if $\hat{w} < 0$, then add $v = 2^m + 1$ to the estimate $\hat{w}$ and set $wrap = 1$;
8. The estimate for u is $\hat{u} \equiv \hat{e}_j - e_{P+k} - wrap \pmod{2^m - 1}$;
9. Stop.

When step 6 is reached, $\mathbf{position}(k, \hat{d}_{j,j-k})$ corresponds to a position in the shift sequence corresponding to two consistent differences: adding k to it gives the right-most position of the second consistent difference corresponding to the last estimate $\hat{e}_j$; subtracting j yields the estimate $\hat{w}$ of the position w involved in the first estimate: $\hat{e}_0 = e_{\hat{u}v+\hat{w}}$. The condition $\hat{w} < 0$ in step 7 is true when one of the estimated column shifts corresponds to the constant column. Adding v to the estimate produces the correct non-negative integer value of $\hat{w}$. The estimate for u can then be obtained as in step 8, with the variable *wrap* set to 1 in the case where one of the column shifts corresponds to the constant column.

Before giving an example of the algorithm, we prove that after step 4 is completed, **list**(j) contains distinct entries.

PROPOSITION 1. *If all j values of k in step 4 are processed without satisfying the stopping condition, then* **list**(j) *contains j distinct entries.*

PROOF: We show that if **list**(j) contains a repeated entry, then the algorithm would have already stopped at the first occurrence of that entry. Suppose for some j, there exist k_1 and k_2, $1 \leq k_1 < k_2 \leq j$, such that $\mathbf{position}(k_1, \hat{d}_{j,j-k_1}) + k_1 \equiv P \equiv \mathbf{position}(k_2, \hat{d}_{j,j-k_2}) + k_2 \pmod{v}$. Suppose $\hat{e}_j \equiv e_P + u \pmod{2^m - 1}$. Then $\hat{d}_{j,j-k_1} \equiv \hat{e}_j - \hat{e}_{j-k_1} \equiv e_P - e_{P-k_1} \pmod{2^m - 1}$ implies $\hat{e}_{j-k_1} \equiv e_{P-k_1} + u \pmod{2^m - 1}$. Similarly, $\hat{e}_{j-k_2} \equiv e_{P-k_2} + u \pmod{2^m - 1}$. Thus, $\hat{d}_{j-k_1, j-k_1-(k_2-k_1)} = \hat{d}_{j-k_1, j-k_2} \equiv \hat{e}_{j-k_1} - \hat{e}_{j-k_2} \equiv e_{P-k_1} - e_{P-k_2} \pmod{2^m - 1}$, and so $\mathbf{position}(k_2 - k_1, \hat{d}_{j-k_1, j-k_1-(k_2-k_1)}) + k_2 - k_1 \equiv P - k_1 \pmod{v}$ would have been added to **list**$(j-k_1)$ when the estimated shift $\hat{e}_{j-k_1}$ was processed. But then $\mathbf{position}(k_1, \hat{d}_{j,j-k_1}) \equiv P - k_1 \pmod{v}$ occurs in **list**$(j - k_1)$, and the algorithm would have stopped at k_1 before completing all j values of k. This is a contradiction.

Example: Let s be generated by the primitive polynomial $f(x) = x^8 + x^4 + x^3 + x^2 + 1$. Then $v = 17$, and the column sequence is $\mathbf{c} = (000100110101111)$, generated by $g(x) = x^4 + x + 1$. The shift sequence is $\mathbf{e} = (\infty, 2, 4, 2, 8, 12, 4, 0, 1, 9, 9, 14, 8, 5, 0, 3, 2)$. The first five difference offsets are:

j:	0	1	2	3	4	5	6	7	8	9	10	11	12	13	14	15	16
e_j :	∞	2	4	2	8	12	4	0	1	9	9	14	8	5	0	3	2
$d_{j+1,j}$:	∞	2	13	6	4	7	11	1	8	0	5	9	12	10	3	14	∞
$d_{j+2,j}$:	∞	0	4	10	11	3	12	9	8	5	14	6	7	13	2	∞	1
$d_{j+3,j}$:	∞	6	8	2	7	4	5	9	13	14	11	1	10	12	∞	0	3
$d_{j+4,j}$:	∞	10	0	13	8	12	5	14	7	11	6	4	9	∞	3	2	1
$d_{j+5,j}$:	∞	2	11	14	1	12	10	8	4	6	9	3	∞	13	5	0	7

Note that offsets $k = 2, 3, 4$, and 5 use $(e_{18}, e_{19}, e_{20}, e_{21}, e_{22}) = (e_1 + 1, e_2 + 1, e_3 + 1, e_4 + 1, e_5 + 1) = (3, 5, 3, 9, 13)$.

The **position** array is:

difference:	0	1	2	3	4	5	6	7	8	9	10	11	12	13	14
$k = 1$:	9	7	1	14	4	10	3	5	8	11	13	6	12	2	15
$k = 2$:	1	16	14	5	2	9	11	12	8	7	3	4	6	13	10
$k = 3$:	15	11	3	16	5	6	1	4	2	7	12	10	13	8	9
$k = 4$:	2	16	15	14	11	6	10	8	4	12	1	9	5	3	7
$k = 5$:	15	4	1	11	8	14	9	16	7	10	6	2	5	13	3

Suppose $(\hat{e}_0, \hat{e}_1, \hat{e}_2, \hat{e}_3) = (14, 6, 0, 3)$. The **list** array after iteration $j = 3$ and $k = 1$ is:

list(0)	$\emptyset$
list(1)	6
list(2)	12 1
list(3)	15

Then for $j = 3$ and $k = 2$, $\hat{d}_{31} = 3 - 6 = 12$ and **position**$(2, 12) = 6$, which appears in **list**(1), and the algorithm jumps to step 6. Step 7 computes the estimate $\hat{w} = \textbf{position}(k, \hat{d}_{j,j-k}) + k - j = 6 + 2 - 3 = 5$ and step 8 computes the estimate $\hat{u} = \hat{e}_3 - e_8 = 3 - 1 = 2$. Note that $(\hat{e}_0 - 2, \hat{e}_1 - 2, \hat{e}_3 - 2) = (12, 4, 1) = (e_5, e_6, e_8)$.

Next suppose $(\hat{e}_0, \hat{e}_1, \hat{e}_2, \hat{e}_3, \hat{e}_4, \hat{e}_5) = (2, 0, 8, 5, 4, 9)$. The **list** array after iteration $j = 5$ and $k = 1$ is:

list(0)	$\emptyset$			
list(1)	3			
list(2)	9	13		
list(3)	13	11	2	
list(4)	16	6	8	2
list(5)	11			

Then for $j = 5$ and $k = 2$, $\hat{d}_{53} = 9 - 5 = 4$ and **position**$(2, 4) = 2$, which appears in **list**(3), and the algorithm jumps to step 6. Step 7 computes the estimate $\hat{w} = \textbf{position}(k, \hat{d}_{j,j-k}) + k - j = 2 + 2 - 5 = -1$. Since $\hat{w} < 0$, the estimate is modified to $\hat{w} = -1 + 17 = 16$ and *wrap* is set to 1. Step 8 computes the estimate $\hat{u} = \hat{e}_5 - e_4 - 1 = 9 - 8 - 1 = 0$. Note that $(\hat{e}_0, \hat{e}_3, \hat{e}_5) = (2, 5, 9) = (e_{16}, e_{19}, e_{21}) = (e_{16}, e_2 + 1, e_4 + 1)$.

4 Performance Analysis

In this section we give a performance analysis for the array blind synchronization procedure. We assume that a given blind synchronization procedure BS(m, l) produces an estimate of the unknown phase of the jth column of the array $\mathbf{A}_l(\mathbf{r})$ with a probability of error $P_{\text{col}}(m, l, p)$ and a computational complexity $C_{P_{\text{col}}}(m, l, p)$, where recall p is the agreement probability that $r_i = s_{i+t}$. Our goal is to determine the probability of error P_{array} of the array blind synchronization procedure and its computational complexity. We begin with a noise only case that provides an upper bound on the expected number of applications of BS(m, l) required.

4.1 Uniformly Determined Shift Sequence Positions

Suppose that we are given a sequence of estimates $\hat{e}_0, \hat{e}_1, \ldots$ whose differences correspond to positions in Δ_k that are drawn uniformly and independently from the range 1 to 2^m, inclusive. We model the behavior of the array blind synchronization procedure in this situation using a multiple component version of the classic birthday repetition problem [F, page 31]. We use this model to determine how long we can expect the algorithm to run before we receive three consistent shifts.

Let **stop**(j) be the event that the array blind synchronization procedure stops while processing $\hat{e}_j$, and $\overline{\textbf{stop}}(j)$ be the event that the algorithm does not stop. Define the probabilities

$$q_j = P(\overline{\textbf{stop}}(j) \mid \overline{\textbf{stop}}(j-1))$$
$$1 - q_j = P(\textbf{stop}(j) \mid \overline{\textbf{stop}}(j-1)).$$

Since three terms are required before stopping, $P(\textbf{stop}(0))$ and $P(\textbf{stop}(1))$ are both zero.

The array blind synchronization procedure can be described by the state diagram depicted in figure 1. Here open circles denote that the algorithm has not stopped yet, while closed circles are where the algorithm halts. The q_j are the transition probabilities. The probability of reaching any given state is the sum over all paths from the starting state to the ending state of the probability of following that path. (In figure 1 there is a unique path from the starting state to any given ending state.) The probability of a path is just the product of the transition probabilities multiplied by the probability of the starting state. Thus we see that for $j \geq 2$:

$$P(\overline{\mathbf{stop}(j)}) = q_2 q_3 \cdots q_j$$
$$P(\mathbf{stop}(j)) = q_2 q_3 \cdots q_{j-1}(1 - q_j)$$

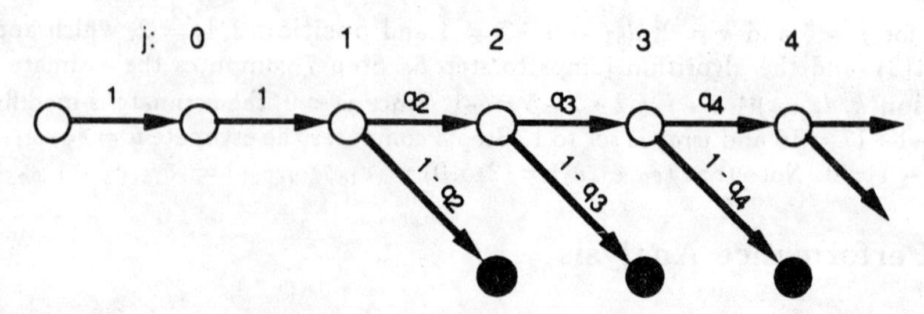

Figure 1. State Diagram for the Array Blind Synchronization Procedure

We wish to evaluate the q_j. Suppose that we do not stop at the jth estimated shift. Then for each difference offset k, $1 \leq k \leq j$ in step 4 of the array blind synchronization procedure, the difference $\hat{d}_{j,j-k} \equiv \hat{e}_j - \hat{e}_{j-k} \pmod{2^m - 1}$ determines a $\mathbf{position}(k, \hat{d}_{j,j-k})$ that does not occur in $\mathbf{list}(j-k)$, which by proposition 1 has $j-k$ distinct entries. If we assume that the new positions are determined uniformly in the range 1 to 2^m, then the probability that the new position is not in $\mathbf{list}(j-k)$ is

$$\frac{M-(j-k)}{M},$$

where $M = 2^m - 1$ (there are 2^m possible positions, but one value is excluded because of the ∞ term).

Assuming independence, we take

$$q_j = \prod_{k=1}^{j} \frac{M-(j-k)}{M} = \frac{M-(j-1)}{M} \cdot \frac{M-(j-2)}{M} \cdots \frac{M-1}{M} \cdot \frac{M}{M}.$$

However, this value is not accurate because we have not forced the elements of the form $\mathbf{position}(k, \hat{d}_{j,j-k}) + k \pmod{v}$ in $\mathbf{list}(j)$ to be distinct as required by proposition 1. To accurately model this extra requirement, the probability of adding the second element to $\mathbf{list}(j)$ would have to depend on which value was chosen for the first element and so forth. This soon leads to a large number of cases. For large values of M that we are interested in, simulations indicate that it is safe to ignore this extra requirement. This is because, as we shall see, the array blind synchronization process halts before it is likely that a repeated element is encountered in the formation of **list**(j).

Figure 2 shows a graphical comparison between the experimentally determined probability of stopping and our simple model. The simulation, unlike the analysis, forced the elements of **list**(j) to be distinct. The solid curve in the figure is the theoretical prediction and the dashed curve is the probability distribution derived from 250,000 experimental trials for $M = 2^{16} - 1$ corresponding to span $n = 32$. The close agreement between the curves in figure 2 suggests that the assumptions used in the calculation of the q_j are accurate. (The rise at the end of the empirical curve is because lengths greater than or equal to 150 were accumulated in one bin.)

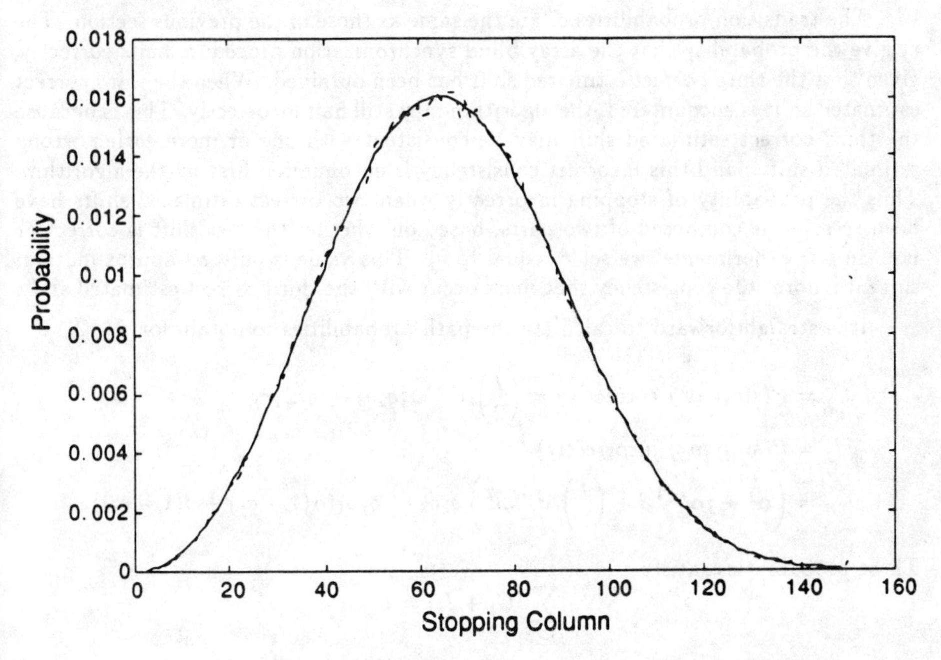

Figure 2. Probability Distribution for Stopping at Column j in Array Blind Synchronization Procedure ($n = 32$) (Theoretical Curve Solid; Empirical Curve Dashed)

The expected number of columns processed by the array blind synchronization procedure can be calculated using this simple model. These means represent an upper bound on the expected number of applications of the column blind synchronization procedure BS(m, l, p) when there is a sequence present. These means are listed below for some representative values of the span n:

span n	Mean
16	11.8
20	17.9
24	27.5
28	42.7
32	66.9

4.2 Probability of Error of Array Blind Synchronization

We next examine the case that the column blind synchronization procedure produces an estimate $\hat{e}_j$ that has a probability of error $P_{col}(m,l,p) = \alpha$ and thus is correct with probability $\beta = 1 - \alpha$. Figure 3 shows a transition diagram that distinguishes stopping correctly or incorrectly. The open circles represent non-terminal states. The closed circles represent stopping incorrectly due to three consistent shifts. Obtaining a correct estimated shift corresponds to moving down one row in the diagram. Finally, the last row of closed squares represent stopping correctly, that is, three correct estimated shifts have been obtained.

The transition probabilities q_j are the same as those in the previous section. The r_j give the probability that the array blind synchronization procedure halts correctly, given that the third correct estimated shift has been obtained. When the third correct estimated shift is encountered, the algorithm may still halt incorrectly. This is because the third correct estimated shift may be consistent with one or more earlier wrong estimated shifts, and this incorrect consistency is encountered first by the algorithm. Thus the probability of stopping incorrectly when two correct estimated shifts have been received is composed of two parts, based on whether the last shift is correct or not. In our experiments, we set r_j equal to q_j. This value is only an approximation, since it ignores the consistency that must occur with the third correct estimated shift.

It is straightforward to calculate the path probabilities to obtain for $j \geq 2$:

$$\begin{aligned} C_j &= P(\text{stop at } j \text{ correctly}) = \binom{j}{2}\alpha^{j-3}\beta^3 q_2 q_3 \cdots q_{j-1} r_j \\ I_j &= P(\text{stop at } j \text{ incorrectly}) \\ &= \left(\alpha^j + j\alpha^{j-1}\beta + \binom{j}{2}\alpha^{j-2}\beta^2\right) q_2 q_3 \cdots q_{j-1}(\alpha(1-q_j) + \beta(1-r_j)). \end{aligned}$$

These probabilities satisfy

$$\sum_{j\geq 2}(C_j + I_j) = 1.$$

The probability of error for the array blind synchronization procedure is

$$P_{\text{array}}(n, P_{\text{col}}(m,l,p)) = \sum_{j\geq 2} I_j.$$

The probability of stopping correctly is $1 - P_{\text{array}}$, which is the sum of the C_j. Note that changing the value of r_j will change both C_j and I_j, but the sum $C_j + I_j$, which is the probability of stopping at the jth shift, remains unchanged.

To test our analytic results, we applied the array blind synchronization procedure to three actual m-sequences of spans 16, 24, and 32. The primitive polynomials were respectively $x^{16}+x^5+x^3+x^2+1$, $x^{24}+x^4+x^3+x+1$, and $x^{32}+x^{22}+x^2+x+1$. For each probability of column shift error α considered, an estimated shift sequence was derived from the actual shift sequence by selecting a correct shift with probability β; otherwise an incorrect shift was selected uniformly from the remaining $2^m - 2$ possibilities. The array blind synchronization procedure was then applied. This process was repeated 1000 times and statistics on the number of estimated shifts used and the number of correct phases obtained were tallied.

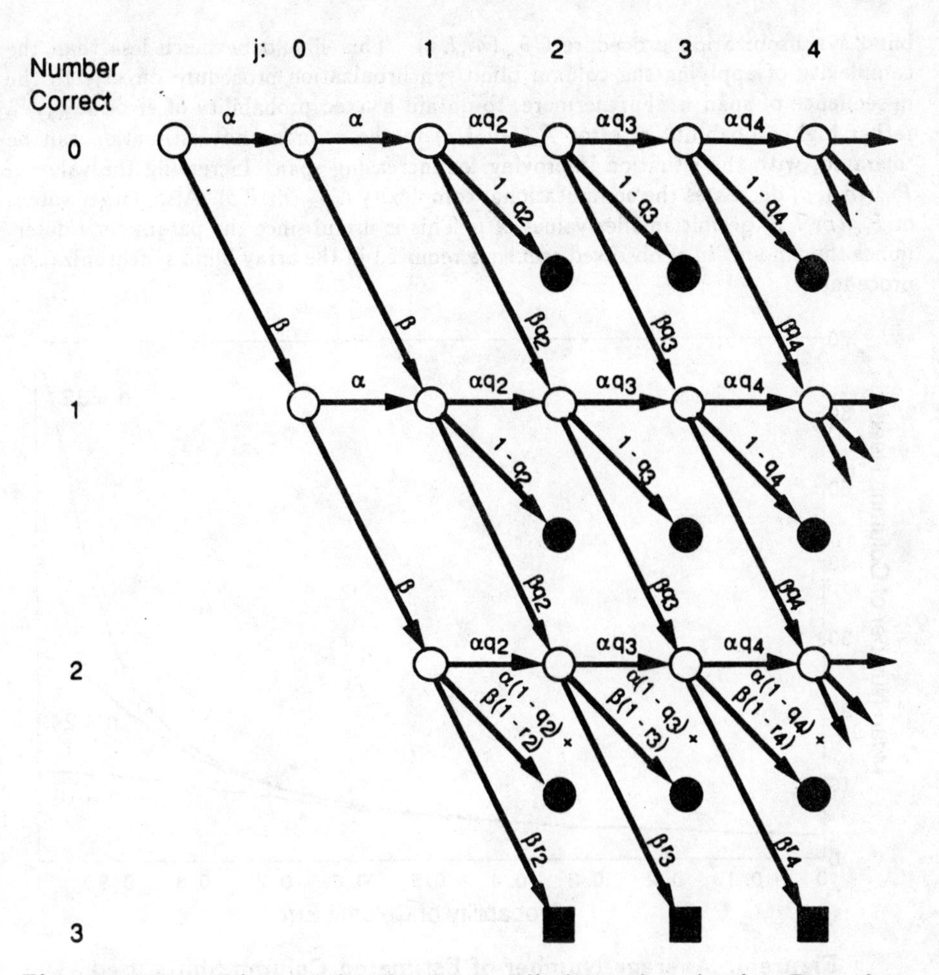

Figure 3. State Diagram for the Array Blind Synchronization Procedure

Figure 4 is a comparison of the analytic and simulation results for the expected number of columns processed by the array blind synchronization procedure as a function of the probability that the column estimates are in error. There is very good agreement between the analytic and simulation results. Also, as the column error probability approaches 1, the expected number of columns processed approaches the upper bound computed previously. Figure 5 is a comparison of the analytic and simulation results for the probability of error for the array blind synchronization procedure as a function of the probability that the column estimates are in error. Here we see that the agreement is not quite as good, due to our choice of r_j. Note that the closer fit between the experimental and theoretical curves in figure 4 can be explained by the fact that the expected number of columns processed is independent of the value of r_j, depending instead on the $C_j + I_j$.

Two facts are clear from these results. The modest expected number of columns processed means that the expected computational complexity of the array blind synchronization procedure is dominated by the computational complexity of the column blind synchronization procedure $C_{P_{col}}(m,l,p)$. This should be much less than the complexity of applying the column blind synchronization procedure directly to the m-sequence of span n. Furthermore, to obtain a fixed probability of error P_{array}, a rather high probability of error $P_{col}(m,l,p)$ in the column shift estimation can be tolerated, with the situation improving for increasing span. Increasing the value of $P_{col}(m,l,p)$ decreases the computational complexity $C_{P_{col}}(m,l,p)$. Also, larger values of $P_{col}(m,l,p)$ permit smaller values of l. This is useful since the parameter l determines the amount lv of observed sequence required in the array blind synchronization procedure.

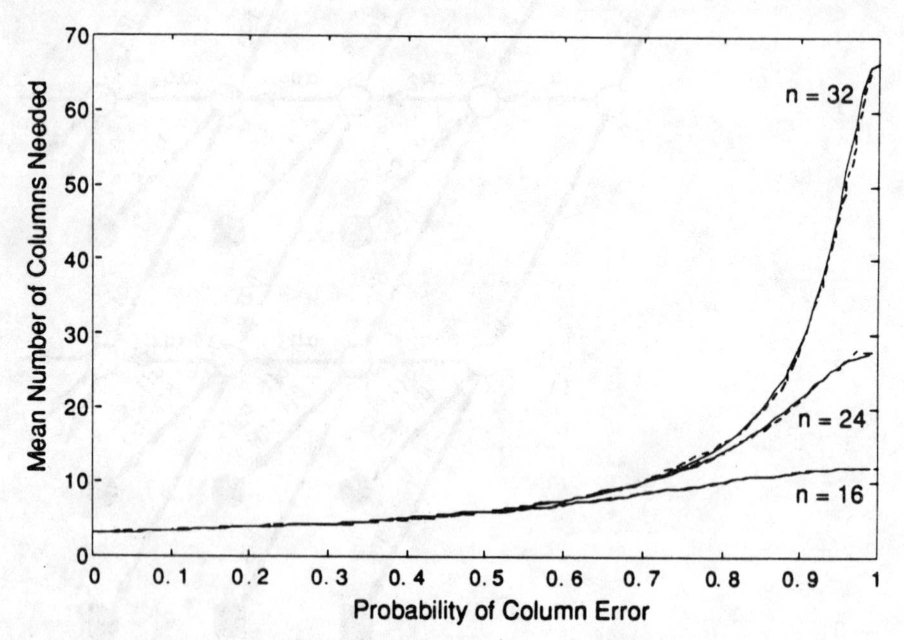

Figure 4. Average Number of Estimated Column Shifts used by Array Blind Synchronization as a Function of Probability of Error of Column Shift Estimation (Theoretical Curves Solid; Empirical Curves Dashed)

5 Conclusion

In this paper we developed a blind synchronization procedure applicable to m-sequences with even span n. In particular we showed how a reliable estimate of the phase of an m-sequence of span n can be obtained from unreliable estimates of the phases of a relatively small number of shifts of a fixed m-sequence of span $n/2$. The computational complexity of the procedure is dominated by the complexity of determining the phases of the smaller m-sequence of span $n/2$. The decrease in span from n to $n/2$ should result in a dramatic drop in complexity.

The procedure requires observing on the order of the square root of the period of the sequence. The observed terms are arranged in an array containing $2^{n/2}+1$

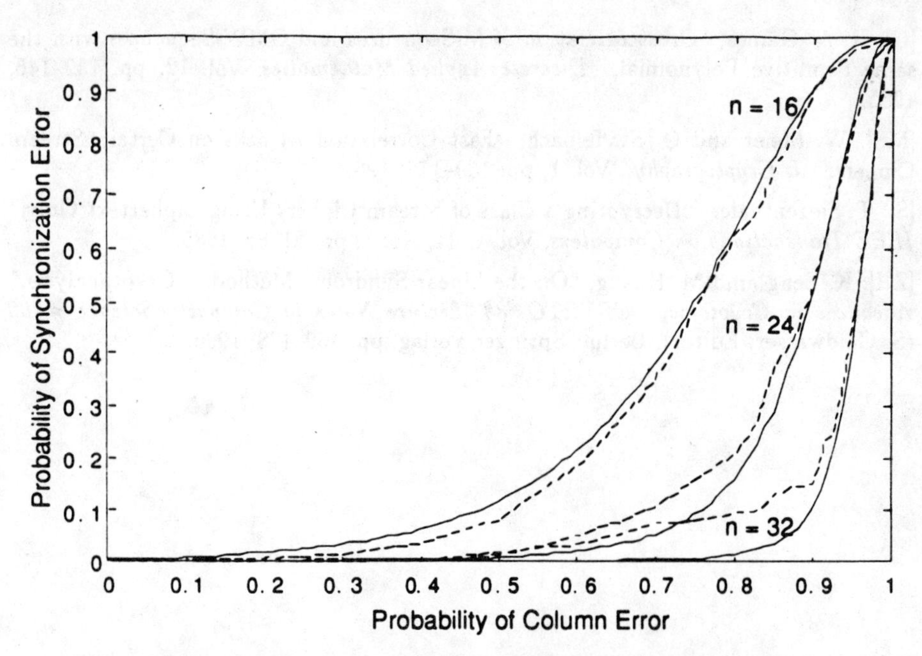

Figure 5. Probability of Error for Array Blind Synchronization as a Function of Probability of Error of Column Shift Estimation (Theoretical Curves Solid; Empirical Curves Dashed)

columns. The number of rows of this array can be minimized using the fact that the estimates of the phases of the shifts of the fixed m-sequence of span $n/2$ can have a high probability of error. Only a very small percentage of these terms are actually used in the procedure, and the required m-sequences of span $n/2$ can be collected directly using decimations of $2^{n/2}+1$. In the future, specific performance gains for the published blind synchronization procedures could be explored ([CS], [MS], [S], [ZH]), although we note that precise performance analyses of these techniques may have to be derived first.

The array blind synchronization procedure can be generalized to other factorizations of the span n, but the case $n = 2m$ is the most practical. This is mainly because for other factorizations, the long dimension of the array grows. It is also the case that the modular differences no longer are distinct when $n/m \neq 2$, which would further complicate the procedure.

References

[CS]. V. Chepyzhov and B. Smeets, "On a Fast Correlation Attack on Certain Stream Ciphers," *Advances in Cryptology—EUROCRYPT '91, Lecture Notes in Computer Science #547* (D. W. Davies, Editor), Berlin: Springer-Verlag, pp. 176–185, 1991.

[F]. W. Feller, *An Introduction to Probability Theory and Its Applications, Volume I*, New York: John Wiley & Sons, 1957.

[G]. R. A. Games, "Crosscorrelation of M-Sequences and GMW-Sequences with the same Primitive Polynomial," *Discrete Applied Mathematics*, Vol. 12, pp. 139–146, 1985.

[MS]. W. Meier and O. Staffelbach, "Fast Correlation Attacks on Certain Stream Ciphers," *J. Cryptography* , Vol. 1, pp. 159–176, 1989.

[S]. T. Siegenthaler, "Decrypting a Class of Stream Ciphers Using Ciphertext Only," *IEEE Transactions on Computers*, Vol. C-34, No. 1, pp. 81–85, 1985.

[ZH]. K. Zeng amd M. Huang, "On the Linear Syndrome Method in Cryptanalysis," *Advances in Cryptology—CRYPTO '88, Lecture Notes in Computer Science #403* (S. Goldwasser, Editor), Berlin: Springer-Verlag, pp. 469–478, 1990.

On Constructions and Nonlinearity of Correlation Immune Functions (Extended Abstract)

Jennifer Seberry *, Xian-Mo Zhang ** and Yuliang Zheng ***

Department of Computer Science, The University of Wollongong
Wollongong, NSW 2522, AUSTRALIA
E-mail: {jennie,xianmo,yuliang}@cs.uow.edu.au

Abstract. A Boolean function is said to be correlation immune if its output leaks no information about its input values. Such functions have many applications in computer security practices including the construction of key stream generators from a set of shift registers. Finding methods for easy construction of correlation immune functions has been an active research area since the introduction of the notion by Siegenthaler. In this paper we study balanced correlation immune functions using the theory of Hadamard matrices. First we present a simple method for directly constructing balanced correlation immune functions of any order. Then we prove that our method generates exactly the same set of functions as that obtained using a method by Camion, Carlet, Charpin and Sendrier. Advantages of our method over Camion et al's include (1) it allows us to calculate the nonlinearity, which is a crucial criterion for cryptographically strong functions, of the functions obtained, and (2) it enables us to discuss the propagation characteristics of the functions. Two examples are given to illustrate our construction method. Finally, we investigate methods for obtaining new correlation immune functions from known correlation immune functions. These methods provide us with a new avenue towards understanding correlation immune functions.

1 Introduction

The main component of a stream cipher is a key stream generator which produces from a random seed a sequence of pseudo-random bits. These pseudo-random bits are added modulo 2 to bits in a plaintext and the resulting stream, a ciphertext, is sent to a receiver. The receiver can recover the plaintext by adding modulo 2 to the ciphertext the output of the stream generator with the same seed.

---

* Supported in part by the Australian Research Council under the reference numbers A49130102, A9030136, A49131885 and A49232172.

** Supported in part by the Australian Research Council 16 under the reference number A49130102.

*** Supported in part by the Australian Research Council under the reference number A49232172.

A common method for obtaining key stream generators is to combine a set of shift registers with a nonlinear function. Blaser and Heinzmann [1] observed that if the combining function leaks information about its component functions, then the work needed in attacking the cryptosystem can be significantly reduced. This idea was further developed by Siegenthaler in [8] where a new concept called correlation immune functions was introduced. Since then the topic has been an active research area and correlation immunity has become one of the central design criteria for stream ciphers based on shift registers [4, 5].

For practical applications, finding methods for easy construction of correlation immune functions is of most importance. In [8] Siegenthaler presented the first method for constructing (balanced) correlation immune functions. His method is recursive in nature and hence not very satisfactory in practical applications. Camion et al studied correlation immune functions from the point view of algebraic coding theory, and presented a method for constructing correlation immune functions of any order [2].

In this paper we study correlation immune functions using the theory of Hadamard matrices. First we present a method for directly constructing balanced correlation immune functions of any order. We then prove that our method generates exactly the same set of correlation immune functions as that obtained using Camion et al's method. Advantages of our method over Camion et al's include that, in addition to their orders of correlation immunity and algebraic degrees, it gives the nonlinearity and propagation characteristics of the functions obtained. We also study methods for constructing correlation immune functions on a higher dimensional space by combining known correlation immune functions on a lower dimensional space. The nonlinearity of functions thus constructed is also investigated.

The organization of the rest of the paper is as follows. Section 2 introduces notations and definitions that are needed in the paper. Section 3 reviews the previous construction methods for correlation immune functions. Our new construction method is described in Section 4. In the same section we also prove that the new construction method generates exactly the same set of correlation immune functions as that by Camion et al's method. Section 5 discusses the algebraic degree, nonlinearity and propagation characteristics of functions obtained using the new method. Two examples are shown in the same section. Section 6 is devoted to the combination of known correlation immune functions. Three combination methods are shown in the section, among which the first one can be viewed as an extension of the new construction method described in Section 4. The paper concludes with some remarks in Section 7.

2 Preliminaries

We consider V_m, the vector space of m tuples of elements from $GF(2)$. Note that there is a natural one to one correspondence between vectors in V_m and integers in $[0, 2^m - 1]$. This allows us to order the vectors according to their corresponding integer values. For convenience, we denote by α_i the vector in V_m whose integer representation is i.

Let f be a function from V_m to $GF(2)$ (or simply a function on V_m). Since f can be expressed as a unique polynomial in m coordinates $x_1, x_2, \ldots, x_m$, we will identify f with its unique multi-variable polynomial $f(x)$ where $x = (x_1, x_2, \ldots, x_m)$. To distinguish between a vector of coordinates and an individual coordinate, the former will be strictly denoted by w, x, y or z, while the later strictly by w_i, x_i, y_i, z_i or u, where i is an index. The algebraic degree of f is defined as the number of coordinates in its longest term when it is represented in the algebraic normal form. f is called an *affine function* if it takes the form of $f(x) = a_1x_1 \oplus \cdots \oplus a_mx_m \oplus c$, where a_j, $c \in GF(2)$. In particular, f is called a *linear* function if $c = 0$.

The *sequence* of f on V_m is a $(1, -1)$-sequence defined by $((-1)^{f(\alpha_0)}, (-1)^{f(\alpha_1)}, \ldots, (-1)^{f(\alpha_{2^m-1})})$, and the *truth table* of f is a $(0\ ,\ 1)$-sequence defined by $(f(\alpha_0), f(\alpha_1), \ldots, f(\alpha_{2^m-1}))$. f is said to be *balanced* if the truth table of f has 2^{m-1} zeros (ones).

The following notation will be used in this paper. Let $\alpha = (a_1, \cdots, a_m)$ and $\beta = (b_1, \cdots, b_m)$ be two vectors (or sequences), the *scalar product* of α and β, denoted by $\langle\alpha, \beta\rangle$, is defined as the sum of the component-wise multiplications. In particular, when α and β are from V_m, $\langle\alpha, \beta\rangle = a_1b_1 \oplus \cdots \oplus a_mb_m$, where the addition and multiplication are over $GF(2)$, and when α and β are $(1, -1)$-sequences, $\langle\alpha, \beta\rangle = \sum_{i=1}^{m} a_ib_i$, where the addition and multiplication are over the reals.

Now we introduce the concept of *correlation immune functions*, the central topic treated in this paper. Let f be a function on V_m. Let X be a random variable taking on values $x \in V_m$ with uniform probability 2^{-m}, let X_i be the random variable corresponding to the ith coordinate value $x_i \in GF(2)$, and let Y be the random variable produced by the function f, i.e., $Y = f(X)$. f is said to be a *kth-order correlation immune function* if the random variable Y is statistically independent of any subset $X_{i_1}, X_{i_2}, \ldots, X_{i_k}$ of k coordinates [8].

Xiao and Massey gave an equivalent definition for correlation immunity in terms of *Walsh transformations* [3]. The Walsh transformation $\hat{f}$ of a function f on V_m is defined as the real-valued function

$$\hat{f}(\beta) = \sum_{x \in V_m} f(x)(-1)^{\langle\beta, x\rangle},$$

where $\beta \in V_m$. Note that in the sum, $f(x)$ and $\langle\beta, x\rangle$ are regarded as real-valued functions.

Definition 1. Let f be a function on V_m. f is a *kth-order correlation immune function* if its Walsh transformation satisfies $\hat{f}(\beta) = 0$ for all $\beta \in V_m$ with $1 \leq W(\beta) \leq k$, where $W(\beta)$ indicates the Hamming weight of, i.e., the number of the nonzero components in, a vector β.

A relevant topic, correlation immune functions with memory, was studied in [4]. The next lemma is useful for constructing correlation immune functions with a view to using Hadamard matrices.

Lemma 2. *Let g be a function on V_m and let η be its sequence. Also let $x = (x_1, x_2, \ldots, x_m)$. Then g is a* kth-order correlation immune function *if and only if $\langle \eta, \ell \rangle = 0$ for any ℓ, where ℓ is the sequence of a linear function $h(x) = \langle \alpha, x \rangle$ on V_m constrained by $1 \leq W(\alpha) \leq k$.*

Proof. Note that

$$\begin{aligned}\langle \eta, \ell \rangle &= \sum_{x \in V_m} (-1)^{g(x)}(-1)^{h(x)} = \sum_{x \in V_m} (-1)^{g(x)+\langle \alpha, x \rangle} \\ &= \sum_{x \in V_m} (-1)^{\langle \alpha, x \rangle} - 2 \sum_{x \in V_m} g(x)(-1)^{\langle \alpha, x \rangle} \\ &= -2\hat{g}(\alpha).\end{aligned}$$

Thus $\langle \eta, \ell \rangle = 0$ if and only if $\hat{g}(\alpha) = 0$ (See also Section 4.2, [2]). □

The order k of correlation immunity of a function on V_m and its algebraic degree d are constrained by the relation $k+d \leq m$. The only functions on V_m that achieve the maximum $(m-1)$th-order correlation immunity are $g(x_1, \ldots, x_m) = x_1 \oplus \cdots \oplus x_m$ and $g(x_1, \ldots, x_m) = x_1 \oplus \cdots \oplus x_m \oplus 1$, both of which are affine. For balanced functions, if $k \neq 0$ or $m-1$, the relation becomes $k + d \leq m - 1$ [8].

Next we introduce a fundamental combinatorial structure, the *Hadamard matrix*. Properties of Hadamard matrices will be very useful in our constructions of correlation immune functions. A $(1, -1)$-matrix H of order m is called a Hadamard matrix if $HH^T = mI_m$, where H^T indicates the transpose of H and I_m is the identity matrix of order m. It is well known that the order m of an Hadamard matrix is 1, 2 or divisible by 4 [9, 6]. In this paper we will use a special kind of Hadamard matrices called *Sylvester-Hadamard matrices* or *Walsh-Hadamard matrices*. A Sylvester-Hadamard matrix (or Walsh-Hadamard matrix) of order 2^m, denoted by H_m, is generated by the following recursive relation

$$H_0 = 1, H_m = \begin{bmatrix} 1 & 1 \\ 1 & -1 \end{bmatrix} \otimes H_{m-1}, m = 1, 2, \ldots$$

where $\otimes$ denotes the Kronecker product. Note that H_m can be written as $H_m = H_s \otimes H_t$ for any nonnegative integers s and t with $s+t = m$. Sylvester-Hadamard matrices are closely related to linear functions, as is shown in the following lemma.

Lemma 3. *Write* $H_m = \begin{bmatrix} \ell_0 \\ \ell_1 \\ \vdots \\ \ell_{2^m-1} \end{bmatrix}$ *where ℓ_i is a row of H_m. Then ℓ_i is the sequence of a linear function $h_i = \langle \alpha_i, x \rangle$, where $x = (x_1, \ldots, x_m)$ and α_i is a vector in V_m as defined in the first paragraph of this Section. Conversely the sequence of any linear function on V_m is a row of H_m.*

A proof for the first half of the lemma can be found in [7]. The second half is true by noting the fact that H_m has 2^m distinct rows and that there are exactly 2^m distinct linear functions on V_m. Thus the rows of $\pm H_m$ comprise all the *affine* sequences of length 2^m.

Next we introduce a notation which is used throughout the rest of the paper. Given any vector $\delta = (i_1, \ldots, i_s) \in V_s$, we define a function on V_s by

$$D_\delta(y) = (y_1 \oplus \bar{i_1}) \cdots (y_s \oplus \bar{i_s})$$

where $y = (y_1, \ldots, y_s)$ and $\bar{i} = 1 \oplus i$ indicates the binary complement of i. Note that since $D_\delta(y) = 1$ if and only if $y = \delta$, a function f on V_{s+t} can be expressed as

$$f(y, x) = \bigoplus_{\delta \in V_s} D_\delta(y) f(\delta, x)$$

where $x = (x_1, \ldots, x_t)$.

Lemma 4. *Let* $f(y, x) = \bigoplus_{\delta \in V_s} D_\delta(y) f_\delta(x)$ *and* $g(y, x) = \bigoplus_{\delta \in V_s} D_\delta(y) g_\delta(x)$ *where* $y = (y_1, \ldots, y_s)$, *and* $x = (x_1, \ldots, x_t)$. *Then* $f = g$ *if and only if* $f_\delta = g_\delta$ *for all* $\delta \in V_s$.

Proof. $f = g$ if and only if $f(\delta, x) = g(\delta, x)$ for all $\delta \in V_s$. Note that since $D_\delta(y) = 1$ if and only if $y = \delta$, we have $f(\delta, x) = f_\delta(x)$ and $g(\delta, x) = g_\delta(x)$ for all $\delta \in V_s$. □

The following lemma can be found in [7].

Lemma 5. *Let* $\xi_{i_1 \cdots i_p}$, $(i_1, \ldots, i_p) \in V_p$, *be the sequence of a function* $f_{i_1 \cdots i_p}(x_1, \ldots, x_q)$ *on* V_q. *Let* ξ *be the concatenation of* $\xi_{0 \cdots 00}$, $\xi_{0 \cdots 01}$, $\ldots$, $\xi_{1 \cdots 11}$, *namely,* $\xi = (\xi_{0 \cdots 00}, \xi_{0 \cdots 01}, \ldots, \xi_{1 \cdots 11})$. *Then* ξ *is the sequence of a function on* V_{q+p} *given by*

$$f(y_1, \ldots, y_p, x_1, \ldots, x_q) = \bigoplus_{(i_1 \cdots i_p) \in V_p} D_{i_1 \cdots i_p}(y_1, \ldots, y_p) f_{i_1 \cdots i_p}(x_1, \ldots, x_q).$$

Let $\alpha = (a_1, a_2, \ldots, a_n) \in V_n$ and $\beta = (b_1, b_2, \ldots, b_m) \in V_m$. The *Kronecker product* of α and β, denoted by $\alpha \otimes \beta$, is defined as $\alpha \otimes \beta = (a_1\beta, a_2\beta, \ldots, a_m\beta)$. The following lemma will be used in the rest of the paper.

Lemma 6. *Let* ξ *be the sequence (or truth table) of a function* f *on* V_n *and* η *be the sequence (or truth table) of a function* g *on* V_m. *Then* $\xi \otimes \eta$ *is the sequence (or truth table) of the function* $\varphi(y, x) = f(y) \oplus g(x)$ *on* V_{n+m}.

Proof. For any fixed $y = \alpha \in V_n$, we have $\varphi(\alpha, x) = f(\alpha) \oplus g(x)$. □

The propagation characteristic is another nonlinearity measure for cryptographic functions. A function satisfies the propagation criterion of order k if complementing k or less input coordinates results in the output being complemented half the times over all input vectors. The formal definition for the propagation criterion follows.

Definition 7. Let f be a function on V_n. We say that f satisfies

1. the *propagation criterion with respect to a non-zero vector α in V_n* if $f(x) \oplus f(x \oplus \alpha)$ is a balanced function.
2. the *propagation criterion of degree k* if it satisfies the propagation criterion with respect to all $\alpha \in V_n$ with $1 \leq W(\alpha) \leq k$.

3 Previous Constructions

Siegenthaler presented a recursive construction in his pioneering work [8]. Let f_1 and f_2 be kth-order correlation immune functions on V_m. Then the concatenation of their sequences results in a new correlation immune function, namely,

$$f(u, x) = (u \oplus 1)f_1(x) \oplus uf_2(x) \tag{1}$$

is a kth-order correlation immune function on V_{m+1}, where u is a variable on $GF(2)$ and $x = (x_1, x_2, \ldots, x_m)$.

Camion et al [2] observed that in Siegenthaler's construction, if the Walsh transformations of f_1 and f_2 satisfy the condition

$$\hat{f}_1(\lambda) + \hat{f}_2(\lambda) = 0, \text{for all } \lambda \in V_m \text{ with } W(\lambda) = k,$$

then the order of the correlation immunity of f is improved to $k+1$. In particular, they show the following two pairs of functions satisfy the condition:

1. $g(x)$ and $1 \oplus g(x)$;
2. $g(x)$ and $g(\bar{x})$, where $\bar{x} = (1 \oplus x_1, 1 \oplus x_2, \ldots, 1 \oplus x_m)$;

where g is a kth-order correlation immune function on V_m. Note that $1 \oplus g(x)$ complements the output, while $g(\bar{x})$ complements the input. Therefore, both

$$f(x) = (u \oplus 1)g(x) \oplus u(1 \oplus g(x)) = u \oplus g(x) \tag{2}$$

and

$$f(x) = (u \oplus 1)g(x) \oplus ug(\bar{x}) = g(x) \oplus u(g(x) \oplus g(\bar{x})) \tag{3}$$

are $(k+1)$th-order correlation immune functions on V_{m+1}.

In the same paper, Camion et al also discovered a method for direct construction of correlation immune functions. Let m and n be positive integers with $m > n$. Let r and p_j, $j = 1, 2, \ldots, n$ be arbitrary functions on V_{m-n}. Also let $x = (x_1, x_2, \ldots, x_n)$ and $y = (y_1, y_2, \ldots, y_{m-n})$. Set

$$f(y, x) = \bigoplus_{j=1}^{n} x_j p_j(y) \oplus r(y). \tag{4}$$

Then the function f defined in (4) is a balanced kth-order correlation immune function on V_m, where k is an integer satisfying $k \geq \min\{W(P(y)) | y \in V_{m-n}\} - 1$, and $P(y) = (p_1(y), p_2(y), \ldots, p_n(y))$.

4 A New Construction

Let m and n be positive integers with $m > n$. Suppose that $\Phi_{m,n} = \{\varphi_{0\cdots 0}, \varphi_{0\cdots 1}, \ldots, \varphi_{1\cdots 1}\}$ is a set containing 2^{m-n} linear functions on V_n, each is indexed by a vector in V_{m-n}. $\Phi_{m,n}$ can be a multi-set and hence a linear function is allowed to appear more than once in $\Phi_{m,n}$. Let $x = (x_1, x_2, \ldots, x_n)$, $y = (y_1, y_2, \ldots, y_{m-n})$ and r be an arbitrary function on V_{m-n}. Set

$$g(y, x) = \bigoplus_{\delta \in V_{m-n}} D_\delta(y)\varphi_\delta(x) \oplus r(y) \tag{5}$$

The following corollary is a consequence of Theorem 9 and Corollary 10 to be stated below, though it can be proved directly.

Corollary 8. *The function g defined in (5) is a balanced kth-order correlation immune function on V_m, where k is an integer satisfying $k \geq \min\{W(\gamma_\delta) | \delta \in V_{m-n}\} - 1$, $\varphi_\delta(x) = \langle \gamma_\delta, x \rangle \in \Phi_{m,n}$ and $\gamma_\delta \in V_n$.*

Theorem 9. *The constructions (4) and (5) express the same set of functions.*

Proof. Let S_1 be the set of functions generated by (4) and S_2 the set of functions generated by (5).

First we prove that $S_1 \subseteq S_2$ by showing that a function obtained by (4) can always be represented in the form of (5). Let

$$f(y, x) = \bigoplus_{j=1}^{n} x_j p_j(y) \oplus r(y)$$

be a function in S_1. For any $\delta \in V_{m-n}$ we have

$$f(\delta, x) = \bigoplus_{j=1}^{n} x_j p_j(\delta) \oplus r(\delta).$$

Since $p_j(\delta) \in GF(2)$, $j = 1, \ldots, n$, $\bigoplus_{j=1}^{n} x_j p_j(\delta)$ is a linear function on V_n. Now let

$$\varphi_\delta(x) = \bigoplus_{j=1}^{n} x_j p_j(\delta),$$

and let

$$g(y, x) = \bigoplus_{\delta \in V_{m-n}} D_\delta(y)\varphi_\delta(x) \oplus r(y).$$

Note that $D_\delta(y) = 1$ if and only if $y = \delta$. Thus we have

$$g(\delta, x) = \varphi_\delta(x) \oplus r(\delta) = f(\delta, x).$$

Since δ is arbitrary, by Lemma 4 we have

$$f(y, x) = g(y, x).$$

Consequently, $f(y,x)$ can be represented in the form of (5). This means that $S_1 \subseteq S_2$.

Next we show that a function obtained by (5) can be represented in the form of (4). This will prove that $S_2 \subseteq S_1$. Let

$$g(y,x) = \bigoplus_{\delta \in V_{m-n}} D_\delta(y)\varphi_\delta(x) \oplus r(y)$$

be a function in S_2. Let δ be an arbitrary vector in V_{m-n}, and let

$$\varphi_\delta(x) = a_{\delta,1}x_1 \oplus \cdots \oplus a_{\delta,n}x_n \tag{6}$$

Now let p_j, $j = 1, 2, \ldots, n$, be a function on V_{m-n} such that

$$p_j(\delta) = a_{\delta,j}$$

for all $\delta \in V_{m-n}$. Also let $P = (p_1, \ldots, p_n)$ be a mapping from V_{m-n} to V_n such that

$$P(\delta) = (p_1(\delta), \ldots, p_n(\delta)) \tag{7}$$

for all $\delta \in V_{m-n}$. Now we define a function on V_m in the following way

$$f(y,x) = \bigoplus_{j=1}^{n} x_j p_j(y) \oplus r(y).$$

Again since $D_\delta(y) = 1$ if and only if $y = \delta$, we have

$$g(\delta, x) = \varphi_\delta(x) \oplus r(\delta).$$

By (6) and (7) we have

$$f(\delta,x) = \bigoplus_{j=1}^{n} x_j p_j(\delta) \oplus r(\delta) = \bigoplus_{j=1}^{n} x_j a_{\delta,j} \oplus r(\delta) = \varphi_\delta(x) \oplus r(\delta) = g(\delta,x).$$

Since δ is arbitrary, by Lemma 4 we have

$$g(y,x) = f(y,x).$$

This implies that $g(y,x)$ can be presented in the form of (4) and thus $S_2 \subseteq S_1$. This completes the proof that $S_1 = S_2$. □

Corollary 10. *In the proof of Theorem 9*

$$\min\{W(P(y))|y \in V_{m-n}\} - 1 = \min\{W(\gamma_\delta)|\delta \in V_{m-n}\} - 1.$$

where $\varphi_\delta(x) = \langle \gamma_\delta, x\rangle = a_{\delta,1}x_1 \oplus \cdots \oplus a_{\delta,n}x_n$ *and* $\gamma_\delta = (a_{\delta,1}, \ldots, a_{\delta,n})$ *are the same as in the proof of Theorem 9.*

Proof. From (7) we have $P(\delta) = (a_{\delta,1}, \ldots, a_{\delta,n})$, and from (6) we have $\varphi_\delta(x) = a_{\delta,1}x_1 \oplus \cdots \oplus a_{\delta,n}x_n, = \langle \gamma_\delta, x\rangle$. Thus we have $P(\delta) = \gamma_\delta$ and hence $\min\{W(P(y))|y \in V_{m-n}\} - 1 = \min\{W(\gamma_\delta)|\delta \in V_{m-n}\} - 1$. □

5 Applying the New Construction

For integers k and n with $0 \leq k < n$, let $\Omega_{k,n}$ denote the set of linear functions on V_n that have $k+1$ or more non-zero coefficients, namely

$$\Omega_{k,n} = \{\varphi | \varphi(x) = \langle \beta, x \rangle, \beta \in V_n, W(\beta) \geq k+1\} \qquad (8)$$

where $x = (x_1, \ldots, x_n)$. This set of functions will be used in our constructions of correlation immune functions.

5.1 Balanced Functions with Given Immunity

Given two integers m and k with $m \geq 3$ and $1 \leq k < m-1$, balanced kth-order correlation immune functions on V_m can be constructed in the following way.

1. Fix an integer n such that $k < n < m$.
2. Create a set $\Phi_{m,n}$ by selecting linear functions strictly from $\Omega_{k,n}$. Note that the size of $\Phi_{m,n}$ is 2^{m-n}, and repetition is permitted in the selection.
3. Construct a function by using the method (5).

By Corollary 8, we have

Theorem 11. *A function constructed according to the above three steps is a balanced kth-order correlation immune function on V_m.*

5.2 Algebraic Degrees

Let k and m be integers with $k \geq 1$ and $m \geq k+2$. As mentioned in Section 2, the algebraic degree of a balanced kth-order immune correlation functions on V_m is at most $m-k-1$. We are interested in constructing balanced kth-order correlation immune functions having the maximum algebraic degree $m-k-1$.

In order to discuss their algebraic degrees, we construct functions in the following three steps.

1. Fix an integer n such that $m > n \geq k+2$.
2. Choose a multi-set $\Phi_{m,n} = \{\varphi_\delta : V_n \to GF(2) | \delta \in V_{m-n}\}$ of linear functions in such a way that it satisfies the following three conditions:
   (C1) If $\varphi \in \Phi_{m,n}$ then $\varphi \in \Omega_{k,n}$, where $\Omega_{k,n}$ is defined in (8),
   (C2) $\Phi_{m,n}$ contains at least two distinct functions,
   (C3) there is a variable x_j that appears in an odd number of functions in $\Phi_{m,n}$. Note that the repetition of functions is counted by the number of appearance.
3. Employ the set $\Phi_{m,n}$ in the construction (5).

Since $\Phi_{m,n}$ is a multi-set, the condition (C1) can be satisfied. On the other hand, since $n \geq k+2$ and $\Omega_{k,n}$ contains more than two functions, the condition (C2) can also be readily satisfied.

Once the conditions (C1) and (C2) are satisfied, we check $\Phi_{m,n}$ to see if it satisfies the condition (C3). If not, we modify $\Phi_{m,n}$ in the following way. Since $\Phi_{m,n}$ satisfies the condition (C2), there are two distinct functions $\varphi_{\delta_1}(x), \varphi_{\delta_2}(x) \in \Phi_{m,n}$. Thus there exists some x_j that appears in $\varphi_{\delta_1}(x)$ but not in $\varphi_{\delta_2}(x)$. Now we replace $\varphi_{\delta_2}(x)$ by $\varphi_{\delta_1}(x)$. In this way we can modify the function set $\Phi_{m,n}$ so that it satisfies the condition (C3). When the condition (C3) is satisfied, there is a term $y_1 \cdots y_{m-n} x_j$ that appears an odd number of times in a function g constructed according to the above three steps. This term survives in the final algebraic normal form representation of g. In other words, the algebraic degree of g is $m - n + 1$.

From Theorem 11 and the above discussions, we know that g is a balanced kth-order correlation immune function of algebraic degree $m - n + 1$. Thus we have proved

Theorem 12. *Let k, n and m be integers with $k \geq 1$ and $m > n \geq k + 2$. Then a function constructed according to the above three steps is a balanced kth-order correlation immune function on V_m of algebraic degree $m - n + 1$. When n is chosen as $n = k+2$, the function achieve the maximum algebraic degree $m-k-1$.*

5.3 Nonlinearity

Given two functions f and g on V_m, the *Hamming distance* between f and g is defined as $d(f,g) = W(f(x) \oplus g(x))$. The *nonlinearity* of g is defined as $N_f = \min_{i=0,1,\ldots,2^{m+1}-1} d(f,\varphi)$ where $\varphi_0, \varphi_1, \ldots, \varphi_{2^{m+1}-1}$ comprise all the affine functions on V_m. It has been proved that $N_f \leq 2^{m-1} - 2^{\frac{m}{2}-1}$ for any function f on V_m [7]. Nonlinearity is an crucial criterion for cryptographic functions and it measures the ability of a cryptographic system using the functions to resist being expressed as a set of linear equations. If the system could be expressed as linear equations, it would be easily breakable by various attacks.

Let f_1 and f_2 be functions on V_m, ξ_1 and ξ_2 be the sequences of f_1 and f_2 respectively. Then $\langle \xi_f, \xi_g \rangle = \sum_{f(x)=g(x)} 1 - \sum_{f(x)\neq g(x)} 1 = 2^m - 2\sum_{f(x)\neq g(x)} 1$ $= 2^m - 2d(f,g)$. This proves the following result which is very useful in the study of the nonlinearity of functions.

Lemma 13. *Let f and g be functions on V_m whose sequences are ξ_f and ξ_g respectively. Then the distance between f and g can be calculated by $d(f,g) = 2^{m-1} - \frac{1}{2}\langle \xi_f, \xi_g \rangle$.*

Now we calculate the nonlinearity of correlation immune functions constructed by (5).

Theorem 14. *Let m and n be integers with $m > n > 2$, and let g be a function constructed by (5). Denote by t_δ the number of times a linear function φ_δ appears in $\Phi_{m,n}$, and let $t = \max\{t_\delta | \delta \in V_{m-n}\}$. Then the nonlinearity of g satisfies $N_g \geq 2^{m-1} - t2^{n-1}$.*

Proof. For convenience a vector $\delta \in V_{m-n}$ will be denoted by its corresponding integer between 0 and $2^{m-n}-1$. In this way, a linear function $\varphi_\delta \in \Phi_{m,n}$ indexed by δ is rewritten as φ_j and t_δ is rewritten as t_j, where t_δ is the number of times φ_δ appears in $\Phi_{m,n}$ and j is the integer representation of δ. We first consider the case when $r(y) = 0$ in the construction (5), namely

$$g(y,x) = D_{0\cdots 0}(y)\varphi_0(x) \oplus \cdots \oplus D_{1\cdots 1}(y)\varphi_{2^{m-n}-1}(x) \tag{9}$$

where $\varphi_j \in \Omega_{k,n}$, $y = (y_1, \ldots, y_{m-n})$, $x = (x_1, \ldots, x_n)$, and $D_{j_1\cdots j_{m-n}}$ is defined in Section 2.

Let h be any affine function on V_m. By Lemma 3, the sequence of h, denoted by L, is a row of $\pm H_m$. Since $H_m = H_{m-n} \otimes H_n$, L can be expressed as $L = \pm\ell' \otimes \ell''$, the Kronecker product of ℓ' and ℓ'', where ℓ' is a row of H_{m-n} while ℓ'' is a row of H_n. Write ℓ' as $\ell' = (c_0, c_1, \ldots, c_{2^{m-n}-1})$. Then L can be rewritten as $L = (c_0\ell'', c_1\ell'', \ldots, c_{2^{m-n}-1}\ell'')$. Note that by Lemma 3, ℓ'' is the sequence of a linear function. We denote the linear function by φ''.

Now let ζ_j be the sequence of φ_j, $j = 0, 1, \ldots, 2^{m-n}-1$. By Lemma 5, $\eta = (\zeta_0, \zeta_1, \ldots, \zeta_{2^{m-n}-1})$ is the sequence of g defined in (9). On the other hand, since the rows of an Hadamard matrix are mutually orthogonal, we have the following result:

$$\langle \zeta_j, \ell'' \rangle = \begin{cases} 2^n, & \text{if } \varphi_j = \varphi'' \\ 0, & \text{otherwise.} \end{cases}$$

Now we discuss $\langle \eta, L \rangle$ in the following two cases:

Case 1: there exists a j such that $\varphi_j = \varphi''$; since φ_j appears t_j times in $\Phi_{m,n}$, the total number of times when $\varphi_j = \varphi''$ is also t_j. Thus $|\langle \eta, L \rangle| \le t_j 2^n$.

Case 2: there exists no j such that $\varphi_j = \varphi''$; in this case we have $|\langle \eta, L \rangle| = 0$.

Summarizing Cases 1 and 2, we have $|\langle \eta, L \rangle| \le t2^n$. By Lemma 13, $d(g,h) \ge 2^{m-1} - t2^{n-1}$. Since h is arbitrary, we have $N_g \ge 2^{m-1} - t2^{n-1}$.

Now consider the more general case when $r(y) \ne 0$ in the construction (5). Since r is a function of y but not x, the sequence of g takes the form of $\eta = (e_0\zeta_0, e_1\zeta_1, \ldots, e_{2^{m-n}-1}\zeta_{2^{m-n}-1})$, where $e_i = (-1)^{r(\alpha_i)}$ and α_i is a vector in V_{m-n} whose integer representation is i. By a similar discussion to the case when $r(y) = 0$, we have $|\langle \eta, L \rangle| \le t2^n$ for any affine sequence L, and hence $N_g \ge 2^{m-1} - t2^{n-1}$. □

5.4 Propagation Characteristics

This section discusses the propagation characteristics of functions obtained by (5). For convenience, the construction method is repeated here:

$$g(y,x) = \bigoplus_{\delta \in V_{m-n}} D_\delta(y)\varphi_\delta(x) \oplus r(y)$$

In the following discussion, we assume that all linear functions φ_δ in the construction are distinct.

It is easy to prove that

$$D_\delta(y \oplus \beta) = D_{\delta\oplus\beta}(y).$$

Let $z = (y, x)$. Also let $\beta \in V_{m-n}$, $\alpha \in V_n$ and $\gamma = (\beta, \alpha)$. Then

$$\begin{aligned} g(z \oplus \gamma) &= \bigoplus_{\delta \in V_{m-n}} D_\delta(y \oplus \beta)\varphi_\delta(x \oplus \alpha) \oplus r(y \oplus \beta) \\ &= \bigoplus_{\delta \in V_{m-n}} (y) D_{\delta\oplus\beta}(y)\varphi_\delta(x \oplus \alpha) \oplus r(y \oplus \beta) \\ &= \bigoplus_{\delta\oplus\beta \in V_{m-n}} D_{\delta\oplus\beta}(y)\varphi_\delta(x \oplus \alpha) \oplus r(y \oplus \beta) \end{aligned}$$

Set $\sigma = \delta \oplus \beta$, we have

$$g(z \oplus \gamma) = \bigoplus_{\sigma \in V_{m-n}} D_\sigma(y)\varphi_{\sigma\oplus\beta}(x \oplus \alpha) \oplus r(y \oplus \beta)$$

and hence

$$g(z) \oplus g(z \oplus \gamma) = \bigoplus_{\sigma \in V_{m-n}} D_\sigma(y)(\varphi_\sigma(x) \oplus \varphi_{\sigma\oplus\beta}(x \oplus \alpha)) \oplus r(y) \oplus r(y \oplus \beta).$$

Note that for any fixed $y = \sigma$

$$(g(z) \oplus g(z \oplus \gamma))|_{y=\delta} = \varphi_\sigma(x) \oplus \varphi_{\sigma\oplus\beta}(x \oplus \alpha) \oplus r(\sigma) \oplus r(\sigma \oplus \beta).$$

Consider the case when $\beta \neq (0, \ldots, 0)$. By assumption $\varphi_\sigma(x)$ and $\varphi_{\sigma\oplus\beta}(x)$ are distinct linear functions. Hence $\varphi_\sigma(x) \oplus \varphi_{\sigma\oplus\beta}(x \oplus \alpha)) = \varphi_\sigma(x) \oplus \varphi_{\sigma\oplus\beta}(x) \oplus \varphi_{\delta\oplus\beta}(\alpha)$ is a non-constant affine function which is balanced. This shows that $g(z) \oplus g(z \oplus \gamma)$ is balanced for any $\gamma = (\beta, \alpha)$ with $\beta \neq (0, \ldots, 0)$. Thus we have proved

Theorem 15. *In the construction (5), if all φ_δ are distinct linear functions on V_n, then g satisfies the propagation criterion with respect to all γ with $\gamma = (\beta, \alpha)$, $\beta \in V_{m-n}$, $\alpha \in V_n$ and $\beta \neq 0$.*

Note that there are $2^{m-n} - 1$ choices for $\beta \neq 0$ and 2^n choices for all $\alpha \in V_n$. Therefore the total number of vectors with respect to which the function g satisfies the propagation criterion is at least $(2^{m-n} - 1)2^n = 2^m - 2^n$.

5.5 Examples

Theorem 12 gives us a general method to construct balanced correlation immune functions having any given immunity. The construction method allows us to easily calculate the algebraic degree and the nonlinearity of the functions, which is very desirable in designing cryptographic systems. Two concrete examples follow.

Let $n = 4$ and $k = 2$. Then

$$\begin{aligned}\Omega_{2,4} &= \{\varphi|\varphi(x) = \langle\beta, x\rangle, \beta \in V_4, W(\beta) \geq 3\} \\ &= \{x_1 \oplus x_2 \oplus x_3,\ x_1 \oplus x_2 \oplus x_4,\ x_1 \oplus x_3 \oplus x_4,\ x_2 \oplus x_3 \oplus x_4,\ x_1 \oplus x_2 \oplus x_3 \oplus x_4\}.\end{aligned}$$

where $x = (x_1, x_2, x_3, x_4)$.

Example 1. We construct a balanced 2nd-order immune function f on V_7, which achieves the maximum algebraic degree of 4. We also calculate the nonlinearity of the function.

Set

$$\begin{aligned}\varphi_1(x) &= x_1 \oplus x_2 \oplus x_3,\ \varphi_5(x) = \varphi_1(x) \\ \varphi_2(x) &= x_1 \oplus x_2 \oplus x_4,\ \varphi_6(x) = \varphi_2(x) \\ \varphi_3(x) &= x_1 \oplus x_3 \oplus x_4,\ \varphi_7(x) = \varphi_3(x) \\ \varphi_4(x) &= x_2 \oplus x_3 \oplus x_4,\ \varphi_8(x) = \varphi_3(x)\end{aligned}$$

and

$$\Phi_{7,4} = \{\varphi_1, \varphi_2, \varphi_3, \varphi_4, \varphi_5, \varphi_6, \varphi_7, \varphi_8\}.$$

$\Phi_{7,4}$ is a multi-set whose elements are all taken from $\Omega_{2,4}$. In addition, it contains four different functions, and x_1 appears in seven functions. Thus the three conditions (C1), (C2) and (C3) are all satisfied.

To complete the construction, let

$$\begin{aligned}f(y, x) = {} & D_{000}(y)\varphi_1(x) \oplus D_{001}(y)\varphi_2(x) \oplus D_{010}(y)\varphi_3(x) \oplus D_{011}(y)\varphi_4(x) \oplus \\ & D_{100}(y)\varphi_5(x) \oplus D_{101}(y)\varphi_6(x) \oplus D_{110}(y)\varphi_7(x) \oplus D_{111}(y)\varphi_8(x) \\ = {} & (1 \oplus y_2y_3 \oplus y_1y_2y_3)x_1 \oplus (1 \oplus y_2 \oplus y_2y_3 \oplus y_1y_2y_3)x_2 \oplus \\ & (1 \oplus y_3 \oplus y_2y_3)x_3 \oplus (y_2 \oplus y_3 \oplus y_2y_3)x_4\end{aligned}$$

where $y = (y_1, y_2, y_3)$ and $x = (x_1, x_2, x_3, x_4)$.

By Theorem 12, f is a balanced 2nd-order correlation immune function on V_7 of algebraic degree 4. To calculate the nonlinearity of the function, note that $\varphi_3 = \varphi_7 = \varphi_8$ and hence $t = \max\{t_j|j = 1, \ldots, 8\} = 3$. By Theorem 14, we have $N_f \geq 2^{7-1} - 3 \cdot 2^{4-1} = 40$. Note that the upper bound of the nonlinearity of balanced functions on V_7 is 56 (see Corollary 17 of [7]).

Example 2. In this example, we construct a balanced 2nd-order immune function g on V_6. Let

$$\begin{aligned}\varphi_1(x) &= x_1 \oplus x_2 \oplus x_3, \\ \varphi_2(x) &= x_1 \oplus x_2 \oplus x_4, \\ \varphi_3(x) &= x_1 \oplus x_3 \oplus x_4, \\ \varphi_4(x) &= x_1 \oplus x_2 \oplus x_3 \oplus x_4,\end{aligned}$$

and

$$\Phi_{6,4} = \{\varphi_1, \varphi_2, \varphi_3, \varphi_4\}.$$

Obviously $\Phi_{6,4}$ satisfies the three conditions (C1), (C2) and (C3).

Let

$$\begin{aligned} g(y,x) &= D_{00}(y)\varphi_1(x) \oplus D_{01}(y)\varphi_2(x) \oplus D_{10}(y)\varphi_3(x) \oplus D_{11}(y)\varphi_4(x) \\ &= x_1 \oplus (1 \oplus y_1 \oplus y_1y_2)x_2 \oplus \\ &\quad (1 \oplus y_2 \oplus y_1y_2)x_3 \oplus (y_1 \oplus y_2 \oplus y_1y_2)x_4 \end{aligned}$$

where $y = (y_1, y_2)$ and $x = (x_1, x_2, x_3, x_4)$.

g is a balanced 2nd-order correlation immune function on V_6. It satisfies the propagation criterion with respect to all $\alpha = (a_1, a_2, a_3, a_4, a_5, a_6) \in V_6$ with $a_1 \neq 0$ or $a_2 \neq 0$. The algebraic degree of g is 3 and the nonlinearity of g is $N_g \geq 2^{6-1} - 2^{4-1} = 24$. For comparison, note that the upper bound for the nonlinearity of balanced functions on V_6 is 26 (see [7]).

6 Combination of Correlation Immune Functions

The construction (5) described in Section 4 presents a method for directly constructing correlation immune functions of any order. In this section we discuss three methods for constructing correlation immune functions on a higher dimensional space from existing such functions on a lower dimensional space.

6.1 An Extension of the New Construction

The construction (5) can be extended. Let m, n, k and s be positive integers, where $m > n > k$, and let $w = (y, x, z)$, $y = (y_1, \dots, y_{m-n})$, $x = (x_1, \dots, x_n)$ and $z = (z_1, \dots, z_s)$. Also let $\Phi_{m,n} = \{\varphi_0, \dots, \varphi_{2^{m-n}-1}\}$ be a set of linear functions on V_n, each of which is selected from $\Omega_{k,n}$. Repetition is permitted in selecting the linear functions. Set

$$g_1(y,x) = D_{0\cdots0}(y)\varphi_0(x) \oplus \cdots \oplus D_{1\cdots1}(y)\varphi_{2^{m-n}-1}(x) \oplus r_1(y) \tag{10}$$

where r_1 is an arbitrary function on V_{m-n}. By Corollary 8, g_1 is a balanced kth-order correlation immune functions on V_m.

Now let $\{f_0, \dots, f_{2^{m-n}-1}\}$ be a set of pth-order correlation immune functions on V_s. Functions in the set need not be mutually distinct. Set

$$g_2(y,z) = D_{0\cdots0}(y)f_0(z) \oplus \cdots \oplus D_{1\cdots1}(y)f_{2^{m-n}-1}(z) \oplus r_2(y) \tag{11}$$

where r_2 is an arbitrary function on V_{m-n}. We further set

$$g(y,x,z) = g_1(y,x) \oplus g_2(y,z) \tag{12}$$

Theorem 16. *The function $g(y,x,z) = g_1(y,x) \oplus g_2(y,z)$ is a balanced $(k+p+1)$th-order correlation immune function on V_{m+s}. The nonlinearity of g satisfies*

$$N_g \geq 2^{m-1} - t \cdot 2^n(2^{s-1} - N)$$

where $t = \max\{t_j | j = 0,1,\ldots,2^{m-n}-1\}$, t_j denotes the number of times that φ_j appears in $\Phi_{m,n}$, and $N = \min\{N_{f_j} | j = 0,1,\ldots,2^{m-n}-1\}$.

Proof. We first consider the case when $r(y) = r_1(y) \oplus r_2(y) = 0$. Note that

$$g(y,x,z) = D_{0\cdots 0}(y)(\varphi_0(x) \oplus f_0(z)) \oplus \cdots \oplus D_{1\cdots 1}(y)(\varphi_{2^{m-n}-1}(x) \oplus f_{2^{m-n}-1}(z)).$$

Since each φ_j is balanced, each $\varphi_j(x) \oplus f_j(z)$ is also balanced (see Lemma 20 of [7]). Hence $g(y,x,z)$ is balanced.

Now we show that g is a $(k+p+1)$th-order correlation immune function. Let ζ_j and ξ_j be the sequences of φ_j and f_j respectively, $j = 0,1,\ldots,2^{m-n}-1$. By Lemma 6 $\zeta_j \otimes \xi_j$ is the sequence of $\varphi_j(x) \oplus f_j(z)$, and $\eta = (\zeta_0 \otimes \xi_0, \ldots, \zeta_{2^{m-n}-1} \otimes \xi_{2^{m-n}-1})$ is the sequence of $g(y,x,z)$ (see Lemma 5).

Let h be a linear function on V_{m+s}. By Lemma 3, the sequence of h, denoted by L, is a row of H_{m+s}. Since $H_{m+s} = H_{m-n} \otimes H_n \otimes H_s$, L can be expressed as $L = \ell_1 \otimes \ell_2 \otimes \ell_3$, where ℓ_1 is a row of H_{m-n}, ℓ_2 is a row of H_n, and ℓ_3 is a row of H_s. Write $\ell_1 = (c_0, c_1, \ldots, c_{2^{m-n}-1})$. Then L can be rewritten as $L = (c_0\ell_2 \otimes \ell_3, \ldots, c_{2^{m-n}-1}\ell_2 \otimes \ell_3)$. Let η be the sequence of g. Then

$$\begin{aligned}\langle \eta, L\rangle &= c_0\langle \zeta_0 \otimes \xi_0, \ell_2 \otimes \ell_3\rangle + \cdots + c_{2^{m-n}-1}\langle \zeta_{2^{m-n}-1} \otimes \xi_{2^{m-n}-1}, \ell_2 \otimes \ell_3\rangle \\ &= c_0\langle \zeta_0, \ell_2\rangle\langle \xi_0, \ell_3\rangle + \cdots + c_{2^{m-n}-1}\langle \zeta_{2^{m-n}-1}, \ell_2\rangle\langle \xi_{2^{m-n}-1}, \ell_3\rangle.\end{aligned}$$

Write $h(w) = \langle \gamma, w\rangle = \langle \beta, y\rangle \oplus \langle \alpha, x\rangle \oplus \langle \sigma, z\rangle$, where $\gamma = (\beta, \alpha, \sigma)$, $\beta \in V_{m-n}$, $\alpha \in V_n$ and $\sigma \in V_s$. By the definition of the sequence of a function, ℓ_1, ℓ_2 and ℓ_3 are the sequences of $\langle \beta, y\rangle$, $\langle \alpha, x\rangle$ and $\langle \sigma, z\rangle$ respectively.

Suppose that $W(\gamma) \leq k+p+1$. Since $W(\gamma) = W(\beta) + W(\alpha) + W(\sigma)$, we have $W(\alpha) + W(\sigma) \leq k+p+1$, which implies that either $W(\alpha) \leq k$ or $W(\sigma) \leq p$. Recall that $\varphi_j \in \Omega_{k,n}$. If $W(\alpha) \leq k$, ζ_j and ℓ_2 must be orthogonal, and hence $\langle \zeta_j, \ell_2\rangle = 0$. Otherwise if $W(\sigma) \leq p$, $\langle \xi_j, \ell_3\rangle = 0$, since each f_j is a pth-order correlation immune function. Thus $\langle \eta, L\rangle = 0$. By Lemma 2, $g(y,x,z)$ is a $(k+p+1)th$-order correlation immune function on V_{m+s}.

To obtain the nonlinearity of the function g, we assume that in the above discussion h is an arbitrary affine function on V_{m+s}. Then L, the sequence of h, can be expressed as $L = \pm\ell_1 \otimes \ell_2 \otimes \ell_3$, and hence

$$\langle \eta, L\rangle = \pm(c_0\langle \zeta_0, \ell_2\rangle\langle \xi_0, \ell_3\rangle + \cdots + c_{2^{m-n}-1}\langle \zeta_{2^{m-n}-1}, \ell_2\rangle\langle \xi_{2^{m-n}-1}, \ell_3\rangle).$$

By Lemma 5

$$\langle \xi_j, \ell_3\rangle \leq 2^s - 2N_{f_j} \leq 2^s - 2N.$$

On the other hand, since the rows of an Hadamard matrix are mutually orthogonal, we have the following result:

$$\langle \zeta_j, \ell_2\rangle = \begin{cases} 2^n \text{ if } \zeta_j = \ell_2, \\ 0 \ \text{ otherwise.}\end{cases}$$

When there is a j such that $\zeta_j = \ell_2$, we have $|\langle \eta, L\rangle| \le t \cdot 2^n(2^s - 2N)$. Otherwise if there is no j such that $\zeta_j = \ell_2$, $|\langle \eta, L\rangle| = 0$. In summary, we have $|\langle \eta, L\rangle| \le t \cdot 2^n(2^s - 2N)$. By Lemma 5, $d(g,h) \ge 2^{m-1} - t \cdot 2^n(2^{s-1} - N)$. Since h is arbitrary, $N_g \ge 2^{m-1} - t \cdot 2^n(2^{s-1} - N)$.

By a similar discussion as in the last part of the proof of Theorem 14, the theorem is true for the more general case when $r(y) = r_1(y) \oplus r_2(y) \neq 0$. □

The construction (12) can be considered as an extension of the construction (5), in the sense that if $s = 0$ and each function f_j is defined as a constant, the former is reduced to the latter.

6.2 Direct Sum of Two Correlation Immune Functions

Lemma 17. *Let f_1 be a k_1th-order correlation immune function on V_{n_1}, f_2 be a k_2th-order correlation immune function on V_{n_2}. Then $g(x,y) = f_1(x) \oplus f_2(y)$ is a $(k_1 + k_2 + 1)$th-order correlation immune function on $V_{n_1+n_2}$, where $x = (x_1, x_2, \ldots, x_{n_1})$ and $y = (y_1, y_2, \ldots, y_{n_2})$.*

Proof. Let ξ_1 and ξ_2 be the sequences of f_1 and f_2 respectively. Then by Lemma 6, $\eta = \xi_1 \otimes \xi_2$ is the sequence of g.

Let φ be a linear function on $V_{n_1+n_2}$. Then φ can be written as $\varphi = \langle \gamma, z\rangle = \langle \alpha, x\rangle \oplus \langle \beta, y\rangle$, where $z = (x,y), \gamma = (\alpha, \beta) \in V_{n_1+n_2}$, $\alpha \in V_{n_1}$ and $\beta \in V_{n_2}$. Now let L be the sequence of φ. By Lemma 3, L is a row of $H_{n_1+n_2}$. Since $H_{n_1+n_2} = H_{n_1} \otimes H_{n_2}$, L can be expressed as $L = \ell_1 \otimes \ell_2$, where ℓ_1 is a row of H_{n_1} and ℓ_2 is a row of H_{n_2}.

Now we show that ℓ_1 matches the sequence of $\langle \alpha, x\rangle$, and ℓ_2 matches the sequence of $\langle \beta, y\rangle$. Assume that ℓ_1' is the sequence of $\langle \alpha, x\rangle$, and ℓ_2' is the sequence of $\langle \beta, y\rangle$. By Lemma 6, $\ell_1' \otimes \ell_2'$ is the sequence of φ. Thus $L = \ell_1 \otimes \ell_2 = \ell_1' \otimes \ell_2'$. By Lemma 3, ℓ_1' is a row of H_{n_1} and ℓ_2' is a row of H_{n_2}. This means that $\ell_1 = \ell_1'$ and $\ell_2 = \ell_2'$. Put it in another way, ℓ_1 is the sequence of $\langle \alpha, x\rangle$, and ℓ_2 is the sequence of $\langle \beta, y\rangle$.

Now consider γ with $W(\gamma) \le k_1 + k_2 + 1$. In this case we have either $W(\alpha) \le k_1$ or $W(\beta) \le k_2$. Thus

$$\langle \eta, L\rangle = \langle \xi_1 \otimes \xi_2, \ell_1 \otimes \ell_2\rangle = \langle \xi_1, \ell_1\rangle\langle \xi_2, \ell_2\rangle = 0.$$

By Lemma 2, g is indeed a $(k_1 + k_2 + 1)$th-order correlation immune function on $V_{n_1+n_2}$. □

Lemma 18. *Let f_1 be a function on V_{n_1} and f_2 be a function on V_{n_2}. Suppose that their nonlinearities are $N_{f_1} = d_1$ and $N_{f_2} = d_2$ respectively. Then the nonlinearity of $g(x,y) = f_1(x) \oplus f_2(y)$ satisfies $N_g \ge d_1 2^{n_2} + d_2 2^{n_1} - 2d_1 d_2$.*

Proof. Let ξ_1, ξ_2, η, L, ℓ_1, ℓ_2, φ be the same as in the proof of Lemma 17. Let $\varphi_1 = \langle \alpha, x\rangle$ and $\varphi_2 = \langle \beta, y\rangle$.

By Lemma 13, we have

$$d_1 = N_{f_1} \le d(f_1, \varphi_1) = 2^{n_1 - 1} - \frac{1}{2}\langle \xi_1, \ell_1\rangle.$$

Thus

$$\langle \xi_1, \ell_1 \rangle \leq 2^{n_1} - 2d_1. \tag{13}$$

Similarly

$$\langle \xi_2, \ell_2 \rangle \leq 2^{n_2} - 2d_2. \tag{14}$$

Note that the right sides of (13) and (14) are both positive. Thus

$$\langle \eta, L \rangle = \langle \xi_1 \otimes \xi_2, \ell_1 \otimes \ell_2 \rangle = \langle \xi_1, \ell_1 \rangle \langle \xi_2, \ell_2 \rangle \leq (2^{n_1} - 2d_1)(2^{n_2} - 2d_2). \tag{15}$$

Again by Lemma 13,

$$d(g, \varphi) = 2^{n_1+n_2-1} - \frac{1}{2}\langle \eta, L \rangle \geq d_1 2^{n_2} + d_2 2^{n_1} - 2d_1 d_2.$$

It is easy to see that the right side of (15) is also positive. Thus if L is an *affine* sequence (i.e. φ is an affine function) (15) still holds. Since φ is an arbitrary affine function we have

$$N_g \geq d_1 2^{n_2} + d_2 2^{n_1} - 2d_1 d_2.$$

Therefore the lemma is true. □

Combining Lemmas 17 and 18 and using Lemma 20 of [7] we have

Theorem 19. *Let f_1 be a k_1th-order correlation immune function on V_{n_1} and f_2 be a k_2th-order correlation immune function on V_{n_2}. Also suppose that $N_{f_1} = d_1$ and $N_{f_2} = d_2$. Then $g(x, y) = f_1(x) \oplus f_2(y)$ is a $(k_1 + k_2 + 1)$th-order correlation immune function on $V_{n_1+n_2}$ whose nonlinearity satisfies*

$$N_g \geq d_1 2^{n_2} + d_2 2^{n_1} - 2d_1 d_2,$$

where $x = (x_1, x_2, \ldots, x_{n_1})$ and $y = (y_1, y_2, \ldots, y_{n_2})$. In particular g is balanced if either f_1 or f_2 is balanced.

6.3 Combination of Four Correlation Immune Functions

This section show that from four correlation immune functions, we can obtain a new functions that achieves a higher order of correlation immunity.

Theorem 20. *Let f_1 and f_2 be pth-order correlation immune functions on V_m, and let h_1 and h_2 be qth-order correlation immune functions on V_n. Let ξ_1, ξ_2, η_1 and η_2 be the sequences of f_1, f_2, h_1 and h_2 respectively. Let ζ be a $(1, -1)$-sequence obtained from ξ_1, ξ_2, η_1 and η_2 in the following way:*

$$\zeta = \frac{1}{2}(\xi_1 + \xi_2) \otimes \eta_1 + \frac{1}{2}(\xi_1 - \xi_2) \otimes \eta_2 \tag{16}$$

where $+$ denotes the component-wise integer addition and $\otimes$ denotes the Kronecker product. Then the function corresponding to ζ is a $(p + q + 1)$th-order correlation immune function on V_{m+n}.

Proof. Similarly to the proof of Lemma 17, we let φ be a linear function on V_{m+n} and L be the sequence of φ. By Lemma 3, L is a row of H_{m+n}. In addition, φ can be written as $\varphi = \langle\gamma, z\rangle = \langle\alpha, x\rangle \oplus \langle\beta, y\rangle$, where $\gamma = (\alpha, \beta) \in V_{m+n}$, $\alpha \in V_m$, $\beta \in V_n$, $z = (x_1, \ldots, x_m, y_1 \ldots, y_n)$, $x = (x_1, x_2, \ldots, x_m)$ and $y = (y_1, y_2, \ldots, y_n)$. Since $H_{m+n} = H_m \otimes H_n$ L can be expressed as $L = \ell_1 \otimes \ell_2$, where ℓ_1 is a row of H_m, and ℓ_2 is a row of H_n. By the same reasoning as in the proof of Lemma 17, it can be shown that ℓ_1 is the sequence of $\langle\alpha, x\rangle$, and ℓ_2 is the sequence of $\langle\beta, y\rangle$. Thus we have

$$\begin{aligned}\langle\zeta, L\rangle &= \frac{1}{2}\langle(\xi_1 + \xi_2) \otimes \eta_1, \ell_1 \otimes \ell_2\rangle + \frac{1}{2}\langle(\xi_1 - \xi_2) \otimes \eta_2, \ell_1 \otimes \ell_2\rangle \\ &= \frac{1}{2}\langle(\xi_1 + \xi_2), \ell_1\rangle\langle\eta_1, \ell_2\rangle + \frac{1}{2}\langle(\xi_1 - \xi_2), \ell_1\rangle\langle\eta_2, \ell_2\rangle. \end{aligned} \tag{17}$$

For $\gamma \in V_{m+n}$ with $W(\gamma) \leq p+q+1$, we have either $W(\alpha) \leq p$ or $W(\beta) \leq q$. This implies that either of the following two situations occurs: (1) $\langle\xi_1, \ell_1\rangle = 0$ and $\langle\xi_2, \ell_1\rangle = 0$, and (2) $\langle\eta_1, \ell_2\rangle = 0$ and $\langle\eta_2, \ell_2\rangle = 0$. As a consequence, we have $\langle\zeta, L\rangle = 0$. □

Note that a similar technique to the construction (16) has been used in obtaining higher order Hadamard matrices from lower order Hadamard matrices [6].

7 Conclusion

We have studied correlation immune functions using the theory of Hadamard matrices. In particular, we have presented a new method for directly constructing correlation immune functions. It is shown that the method generates the same set of functions as that by a method of Camion et al. The new method is more convenient for use in practice since it allows one to calculate the nonlinearity of functions obtained and to discuss the algebraic degrees and propagation characteristics of the functions. Three methods for obtaining correlation immune functions on a higher dimensional space from known correlation immune functions on a lower dimensional space are also presented. We believe that these various methods of generating correlation immune functions, by direct construction or by combining known correlation immune functions, will find a wide range of applications in computer security.

References

1. W. Blaser and P. Heinzmann. New cryptographic device with high security using public key distribution. In *Proceedings of IEEE Student Paper Contest 1979-1980*, pages 145–153, 1982.
2. P. Camion, C. Carlet, P. Charpin, and N. Sendrier. On correlation-immune functions. In *Advances in Cryptology: Crypto'91 Proceeding*, volume 576, Lecture Notes in Computer Science, pages 87–100. Springer-Verlag, Berlin-Heidelberg-New York, 1991.

3. Xiao Guo-zhen and J. L. Massey. A spectral characterization of correlation-immune combining functions. *IEEE Transactions on Information Theory*, 34 No. 3:569–571, 1988.
4. R. A. Rueppel. *Analysis and Design of Stream Ciphers*. Springer-Verlag, Berlin, Heidelberg, New York, London, Paris, Tokyo, Berlin, Heidelberg, New York, London, Paris, Tokyo, 1986. In Communications and Control Engineering Series, Editors: A. Fettweis, J. L. Massey and M. Thoma.
5. R. A. Rueppel. Stream ciphers. In G. J. Simmons, editor, *Contemporary Cryptography: the Science of Information Integrity*, chapter 2, pages 65–134. IEEE Press, New York, 1992.
6. J. Seberry and M. Yamada. Hadamard matrices, sequences, and block designs. In J. H. Dinitz and D. R. Stinson, editors, *Contemporary Design Theory: A Collection of Surveys*, chapter 11, pages 431–559. John Wiley & Sons, Inc, 1992.
7. J. Seberry and X. M. Zhang. Highly nonlinear 0-1 balanced functions satisfying strict avalanche criterion. Presented at AUSCRYPT'92, 1992.
8. T. Siegenthaler. Correlation-immunity of nonlinear combining functions for cryptographic applications. *IEEE Transactions on Information Theory*, IT-30 No. 5:776–779, 1984.
9. W. D. Wallis, A. Penfold Street, and J. Seberry Wallis. *Combinatorics: Room Squares, sum-free sets, Hadamard Matrices*, volume 292 of Lecture Notes in Mathematics. Springer-Verlag, Berlin-Heidelberg-New York, 1972.

Practical and Provably Secure Release of a Secret and Exchange of Signatures

Ivan Bjerre Damgård

Aarhus University, Mathematical Institute

Abstract. We present a protocol that allows a sender to gradually and verifiably release a secret to a receiver. We argue that the protocol can be efficiently applied to exchange secrets in many cases, for example when the secret is a digital signature. This includes Rabin, low-public-exponent RSA, and El Gamal signatures. In these cases, the protocol requires an interactive 3-pass initial phase, after which each bit (or block of bits) of the signature can be released non-interactively (i.e. by sending 1 message). The necessary computations can be done in a few seconds on an up-to-date PC. The protocol is statistical zero-knowledge, and therefore releases a negligible amount of side information in the Shannon sense to the receiver. The sender is unable to cheat, if he cannot factor a large composite number before the protocol is completed.
We also point out a simple method by which any type of signatures can be applied to fair contract signing using only one signature.

1 Introduction

1.1 The Basic Problem

Suppose parties A and B each possess a secret, s_A and s_B resp. Suppose further that both secrets represent some value to the other party, and that they are therefore willing to "trade" the secrets against each other. For example, s_A might be A's digital signature on a commitment to deliver some kind of service to B, while s_B could be a bank's signature on some digital cash. But if the parties do not trust each other, it is clear that none of them are willing to go first in releasing the secret - once one of them has done this, he may never get anything in return.

If the two secrets are represented as bit strings of the same length, this can be solved by exchanging the secrets bit by bit; if this is done honestly, no party will be more than one bit ahead of the other; put another way: if at some point, A can compute s_B in time T, then B can compute s_A in at most time $2T$ by guessing the bit he may be missing (this assumes, of course, that a bit of s_B tells A just as much as a bit of s_A tells B - we'll get back to this problem in Section 2.2).

However, this "solution" has created another problem: one party may have given away his secret, only to find in the final stage that in return he has been given garbage instead of bits of a genuine secret. Hence what we need is a way to release the secrets in small parts, such that the receiver can *verify* for each part that he has been given correct information. The alternative, namely to assume a trusted third party, is not attractive and probably not very realistic either.

Thus we can distill a basic primitive (introduced in [4]) which we will call *gradual and verifiable release of a secret*, the intuitive meaning of which should be clear from the above. It should also be clear that a gradual and verifiable release protocol can be used to implement an exchange of secrets between any number of parties. In order for such an exchange to be *fair*, the secrets involved have to satisfy certain conditions (see Section 2.2 and 2.3). In addition, the concept of a release protocol makes sense in its own right, and might be useful for other purposes than implementing exchanges of secrets.

1.2 Comparison with Earlier Work

Exchange and release of secrets has attracted a lot of attention in the past, and a large body of literature exists on the subject [4, 5, 6, 7, 9, 10, 16, 17, 18, 19, 20, 22, 24]. However, since the discovery of zero-knowledge proofs and arguments for any NP-language [3, 12], the problem has lost most of its theoretical interest because these techniques can be used to construct a release protocol that is as secure as the bit-commitment scheme used in the zero-knowledge proof. Thus existence of any one-way function is a sufficient assumption to implement a secure gradual release. This is relatively trivial to see for methods that release specific bits of the secret, while more advanced methods are required to release probabilistic information, which can amount to less than 1 bit at a time, see for example [18, 13]. It is even possible to make the choice of bits to release adaptive [17].

The resulting solutions would be very far from practical, however, and a solution that is both practical and provably secure does not seem to have appeared before.

Before looking at the basic problem with earlier practical solutions, we have to point out a fundamental fact: the demand that the released parts of the secret be correct makes sense only if the secret is itself determined by some public piece of information. Otherwise, not even the secret itself can be verified. Typically, something like $f(s)$ is public, where s is the secret and f is some one-way function.

Earlier practical release protocols assume a priori that the secret is given in some particular form, e.g. a discrete log in [4], a factorization in [6]. However, when trying to apply such protocols, we are likely to find that the application dictates the way in which the secret is given, i.e. the particular one-way function f involved is determined by the application. For example, if we want to release an RSA signature on a given message, f would be the function mapping a signature to the message it signs. Thus, if we wanted to use e.g. [4], we would have to make both $f(s)$ and g^s known, where g is chosen in some appropriate group. The problem is that this may release additional information about s, and so is very unlikely to lead to a provably secure scheme.

The release protocol of this paper solves the problem by using a new tool - an unconditionally hiding bit commitment scheme that allows commitment to a string of any length (but such that the commitment has constant length) and can be opened bit by bit. In addition, we present efficient protocols for checking that the contents of such a commitment has a particular form, for example that it is the Rabin, RSA or El Gamal signature on a given message. This leads to provably secure release protocols for such signatures.

Since digital signatures are obvious candidates for representing value or commitments in practical applications, methods for releasing or exchanging such signatures seem to be of very strong practical relevance.

1.3 Fair Exchange and Contract Signing

Intuitively, a fair exchange of secrets protocol is one that avoids a situation where A can obtain B's secret, while B cannot obtain that of A. If there is no assumption made that third party intervention is possible, the best we can do is to guarantee that if one party stops the protocol early, both parties are left with roughly the same computational task in order to find the other party's secret. This is the model used in this paper. In Section 2.2, we will discuss to what extent such a fair exchange follows from a gradual and verifiable release.

But in any case, it is clear that if for example A has much more powerful hardware than B the *actual time* A would need to find the secret in case of early stopping would be much smaller than for B. Depending on the application this may or may not be a problem. One example where this comes up is if we use exchange of secrets to implement fair contract signing. This is straightforward: A and B both sign the contract, and then exchange gradually their signatures. However, if the contract involves time related issues, such as a commitment taking effect at a certain date, the above "real-time" problem could be serious in case one party cheats.

In [2] Ben-Or, Goldreich, Micali and Rivest show how to avoid this problem, if we assume that a judge is available to settle disputes. Moreover, their protocol can make use of any signature scheme. The difference to our work is first the assumption about the judge, and secondly that the protocol of [2] is a dedicated protocol for solving the contract signing problem: it does not implement an exchange or a relase of secrets.

The protocol in [2] involves a certain computational overhead: the signature scheme must be employed a large humber of times by both parties. In Section 7 we show that this overhead can be avoided.

2 Basic Definitions

This section gives some basic definitions and some connections between them. Although the model is certainly not the most general possible, it does describe appropriately the protocols we present in the following. In subsection 2.3 we argue that the model is in fact useful in many practical situations.

2.1 Release Protocols

In this section we give a formal definition of a secure release protocol. Intuitively, we are modelling the situation, where party A has a secret s, which he will release bit by bit to B who knows t, where (if A is honest) t, s satisfy some predicate P. Typically, P will be satisfied if t is the image under some one-way function of s. At any point in the protocol, B should be able to compute some of the bits of s correctly, but not more, i.e. he should be in the same situation as if he had been given t and the bits of s by an oracle.

We will think of the pair (A, B) as interactive Turing machines as defined in [11]. In particular, both A and B will be polynomial time bounded in the input length, and are equipped with knowledge tapes containing their private inputs. In the following X will mean A or B. Following [11], we let $\bar{X}$ denote a machine following the protocol specified for party X, while $\tilde{X}$ denotes an arbitrary cheating participant playing the role of X. X will represent $\bar{X}$ or $\tilde{X}$.

The properties of (A, B) will be defined with respect to a fixed polynomial time computable predicate P. P takes as input a k-bit string t and a bit string s of length at most $f(k)$, where f is a polynomial.

A receives as private input on its knowledge tape a string s of length at most $f(k)$, while the k-bit string t is common input to A and B. B receives the string k_B on its knowledge tape. $s|_i$ denotes the first i bits of s ($s|_0$ is the empty string). The interesting case is of course when $P(t, s) = 1$.

The event that one party sends a message to the other is called a *pass*. Passes are numbered ordinarily, starting from 1. The protocol is required to define a series of increasing functions $\{p_k\}_{k=1}^{\infty}$, where

$$p_k : \{0...f(k)\} \rightarrow \mathbf{N},$$

and where $p_k(f(k))$ is polynomially bounded. The meaning of p_k is that, for input length k, $p_k(i)$ is the index of the first pass after which B is able to compute $s|_i$. After each pass, the participant receiving a message may output "reject" and stop, indicating that cheating has been detected. We say that (A, B) *completes pass* i if no party outputs reject after pass i.

As usual, the view of a participant is defined to be the ordered concatenation of the messages sent in the protocol, followed by the random bits read by the participant. This is denoted by $View_X(t, s, r_A, k_B, r_B)$, where r_X is the contents of the random tape of party X. $View_X^i(t, s, r_A, k_B, r_B)$ denotes party X's view of the truncated protocol where we only consider passes number 1 through i (note that this view may be shorter than i passes if the protocol stops earlier). In the following, $View_{\tilde{B}}(...)$ will always refer to a conversation with A, while $View_B(...)$ will refer to a conversation with $\bar{A}$.

Finally, the protocol must define a set of polynomial time computable functions $\{h_k^i | \; k = 1..\infty, i = 1..f(k)\}$, such that h_k^i takes as input a sample of $View_B^{p_k(i)}(t, s, r_A, k_B, r_B)$. As output, it produces an i-bit string. These functions should be used by B to compute the first i bits of the secret after pass $p_k(i)$ is completed. An h_k^i-value is said to be *correct* if there is a z of length at most $f(k)$ bits such that $P(t,z) = 1$ and $z|_i = h_k^i(View_B^{p_k(i)}(t,s,r_A,k_B,r_B))$. Otherwise it is incorrect.

We can now give the following

Definition 1 The pair (A, B) is called a *secure release protocol with respect to* $P, \{p_k\}$ *and* $\{h_k^i\}$ if the following three properties are satisfied:

- h_k^i-values computed on views of conversations between $\bar{A}$ and $\bar{B}$ are always correct.
- $\forall A \forall c \exists k_0 \forall |t| > k_0 \forall s, r_A, k_B \forall i = 1..f(k)$:

  $$Prob(h_k^i(View_{\tilde{B}}^{p_k(i)}(t,s,r_A,k_B,r_B)) \text{ incorr. and } (A,\bar{B}) \text{ completes pass } p_k(i)) \leq k^{-c}$$

  The probability is taken over the choice of r_B.
- For each B, and for each $i = 0..f(k)$, there exists an expected polynomial time machine M_B^i, which on input bit strings x, t, k_B and with random tape r_M simulates B's view of the first $p_k(i)$ passes of the conversation with $\bar{A}$. Let $M_B^i(x,t,k_B,r_M)$ denote M_B^i's output, considered as a random variable with distribution taken over r_M.
  We then require that for $i = 0...f(k)$, whenever $x = s|_i$ for some s with $P(t,s) = 1$, then the distribution of $M_B^i(x,t,k_B,r_M)$ is statistically indistinguishable from that of $View_B^{p_k(i)}(t,s',r_A,k_B,r_B)$, where the distribution is taken over r_A and r_B, and where s' is any string such that $s'|_i = s|_i$ and $P(t,s') = 1$.

Remark

- For simplicity, we only consider a bit by bit release in the above definition. The definition could trivially be generalized to talk about a release of a block of bits per pass.
- We need a simulator for each $i = 0..k$, because we want A to be protected, even if $\tilde{B}$ stops before all bits are released. The best we can do in such a case is to require that $\tilde{B}$ can compute only what he can get from the information he is entitled to know at the given time.

2.2 Exchange Protocols

In this section, we shall discuss to what extent release protocols can be used to build fair exchange protocols. It is clear that if parties X and Y possess secrets s_X, s_Y, resp., defined by (possibly) different predicates P, P', then if we have release protocols for these predicates as in Definition 1, it is natural to try to exchange the secrets by interleaving the release of s_X with that of s_Y.

Consider now the question whether this exchange protocol is fair, where we think of fairness as defined by Yao [24]: even a cheating Y (or, symmetrically, a cheating X) cannot force a situation where it is feasible for him to find s_X, but infeasible for X to find s_Y.

It is clear from Definition 1 that the interleaving approach forces the parties to send correct bits of their secrets, and also that each party knows at each point only a prescribed number of bits of the opponent's secret. Nevertheless, the exchange will not necessarily be a fair one *in general*: it is possible that for example s_X is uniquely determined already from the first half of its bits. If this is not the case for s_Y, then Y could gain an unfair advantage by quitting half way through the protocol, perhaps leaving X with only useless information about s_Y.

The point is of course that the problem Y has to solve to find s_X may be of a totally different nature than the one X is facing to find s_Y. We therefore have to restrict to a set of "nicer" cases, where it is possible to connect the two problems. A first step in this direction is to require that s_X, s_Y are defined by the same predicate (i.e. $P = P'$), and that the corresponding public strings t_X, t_Y are drawn independently from the same distribution. This leads to the following definition of the exchange protocol *induced* by a release protocol:

Definition 2 Let (A, B) be a release protocol secure with respect to P, $\{p_k\}$ and $\{h_k^i\}$ (see Definition 1). The following two-party protocol (X, Y) is called *the exchange protocol induced by* (A, B):

X and Y receive two common inputs t_X, t_Y, both of length k bits and drawn independently from the same probability distribution π_k. X and Y get as private input s_X resp. s_Y, such that $P(t_X, s_X) = P(t_Y, s_X) = 1$.

X simulates copies A_X, B_X of A and B, giving s_X, t_X as input to A_X and t_Y as input to B_X. Correspondingly Y runs copies A_Y, B_Y on inputs s_Y, t_Y and t_X. X, Y will now for $i = 1, 2, ...$ execute pass i of (A_X, B_Y) followed by pass i of (A_Y, B_X), until both protocols halt.

Even an induced exchange protocol is not guaranteed to be fair if we do not know anything about the predicate P: it is possible that for a non-negligible fraction of the t's, finding s, such that $P(t, s) = 1$ is much easier than for other t-values. If t_X happens to be such an easy case, Y is clearly in a better situation than X. What we need to avoid this is that the problem of finding s such that $P(t, s) = 1$ based on t and some bits of s is of about the same difficulty for nearly all choices of t under π_k. One way of stating such "uniform hardness" of a problem a little more precisely is to say that any algorithm that solves a non-negligble fraction of the instances of the problem can be turned into an algorithm that uses not much more time, and solves nearly all instances.

With this assumption on P, we can say the following about the induced exchange: assume that some $\tilde{Y}$ has a strategy for aborting the protocol at some stage and subsequently finding s_X with some non-negligible probability. When the protocol is aborted, this leaves $\tilde{Y}$ with t_X and, say, i bits of s_X. X is left with t_Y and i or $i-1$ bits of s_Y. It follows from Definition 1 that from this information only, the views of X, resp. Y can be simulatewd. So except for perhaps 1 bit X

has to guess, this means that both parties are faced with samples of the same problem, drawn from the same distribution. By the above assumption on P, this implies that whatever method $\tilde{Y}$ uses to find s_X will also work for X to find s_Y, and therefore the protocol is fair.

Using the simulators guaranteed by Definition 1, one can formalize this reasoning. In this paper, however, where we focus on practical protocols, we leave this to the reader. In stead we concentrate on the question whether the types of secrets one might want to exchange in practice are likely to have a uniform hardness property as the one we have discussed.

We have already discussed that digital signatures are interesting in this context. So as an example, assume that t specifies a message and an RSA public key, and that $P(t, s) = 1$ precisely if s is a valid RSA signature on the message. With our current knowledge, we can only conjecture that this predicate has an appropriate uniform hardness property. Some evidence is known in favor of this conjecture, however: from the multiplicative property of RSA, it follows easily that if you can sign in poynomial time a polynomial fraction of the messages for some modulus, then you can sign all messages using that modulus in expected polynomial time. Moreover the results of [1] give strong indications that something similar holds when some number of bits of s are given.

Since El Gamal signatures are known to satisfy a similar property, we conjecture that at least the signature schemes we consider in this paper have uniform hardness sufficient to make induced exchange protocols fair when using these signatures.

At this point one could perhaps complain that the assumptions made in the definition of induced exchange protocols are too demanding in practice, in particular the assumption that t_X and t_Y are identically distributed. What if Y could somehow manipulate the distribution of t_X and/or t_Y, presumably to make life easier for himself? However, if messages are hashed before they are signed - as is nearly always the case in practice - he is not likely to benefit from this: if the hash function used is strong, he will not be able to control the hash result and for example force t_X to be an easily signed hash value (of which there are only very few). This is the same kind of reasoning that underlies the Fiat-Shamir signature scheme.

In summary, we have argued that exchange protocols induced from release protocols are useful in many cases that are important in practice. As a side remark, it is also worth noting that more complicated exchange protocols that can deal with seemingly incompatible types of secrets typically work by choosing some auxiliary secret w, make public some information connecting w and the actucal secrets, and then release w bit by bit, see e.g. [24]. Thus a bit-by-bit release as defined here can also be useful as a building block in other protocols.

3 A Bit Commitment Scheme

In this section, we define the bit commitment scheme we will use, and prove its basic properties. To set up the commitment scheme, B must generate and make public a k-bit Blum integer N, and g, a random quadratic residue modulo N. Also, B must in zero-knowledge prove that he knows the two prime factors of N [21], prove that they are both congruent to 3 modulo 4 [1] [15], and that g is a quadratic residue [14]. The methods for doing this are well known and quite efficient, and in any case this step will only be necessary once, at system start-up time.

In this phase, either A acts as the verifier, or this role is played by a trusted third party. The latter case is the most likely one in practice, as setting up a large scale public key system nearly always requires a certification authority that registers users and certifies the relation between identities and public keys (moduli). Such a center might as well act as the verifier in the above, and certify by a digital signature that N and g have been verified successfully.

Let $SQ(N)$ denote the subgroup of quadratic residues modulo N. Having established N and g, the parties agree on a natural number l. A can now commit to any integer s satisfying $-2^{l-1} < s < 2^{l-1}$ by choosing R uniformly at random in $SQ(N)$ and computing the commitment

$$BC_g(R, s) := R^{2^l} g^s.$$

This is called a base-g commitment. A commitment is opened by revealing R and s, which allows B to verify the above equation.

The commitment scheme is based on the hash functions from [8]. In fact, $BC_g(R, s)$ is precisely the hash value of s computed with starting point R, using the factoring based hash function from [8]. The same type of function was used in [23] for the purpose of fail-stop signatures.

The basic properties of this commitment scheme are established in the following lemma:

Lemma 1 $BC_g(R, s)$ has distribution independent of s, when R is a uniformly chosen square mod N.
If A can open the same commitment using values R, s, resp. R', s', where $s \neq s'$, then A can compute a square root modulo N of g.

Proof The first statement is clear from the fact that squaring modulo a Blum-integer is a permutation, and that therefore a commitment is always a uniformly chosen element in $SQ(N)$. For the second statement, assume without loss of generality that $s' > s$ and write $s' - s = (2h+1)2^j$. Clearly $j < l$. Then the equation

$$R^{2^l} g^s = R'^{2^l} g^{s'} \bmod N$$

implies that

$$(R/R')^{2^{l-j}} = g^{2h+1} \bmod N$$

[1] Actually, [15] only proves that $N = p^r q^s$, where r, s are odd and p, q are 3 modulo 4. But even for such numbers, squaring is a permutation of the quadratic residues, and this is the property we need.

and therefore $(R/R')^{2^{l-j-1}} g^{-h}$ is a square root of g □

These properties of the function BC_g were also used in [8]. The crucial property in this context, however, is that these commitments can be opened *gradually*: given a commitment $BC_g(R,s)$ to a positive number s, A can reveal the least significant bit b of s by revealing X such that

$$X^2 \bmod N = BC_g(R,s) \text{ if } b = 0$$

and

$$g \cdot X^2 \bmod N = BC_g(R,s) \text{ if } b = 1.$$

After this, X can be regarded as a commitment to $s/2$ (with l replaced by $l-1$) and more bits of s can be opened.

By essentially the same argument as in Lemma 1, it is easy to see that if A knows how to open in one step the entire value of $s > 0$, he cannot open single bits of s with values that are inconsistent, unless he can compute a square root of g.

It is also clear that the procedure for opening 1 bit can be easily generalized to allow opening in one step of any number of the least significant bits of s.

Since computing square roots of random numbers mod N is equivalent to factoring N, we will need the following assumption on hardness of factoring:

Factoring Assumption There exists a probabilistic polynomial time algorithm Δ which on input 1^k outputs a k-bit Blum integer N, such that for any probabilistic polynomial size circuit families C, and any constant c, the probability that C factors N is at most k^{-c}, for all sufficiently large k. This probability is taken over the random choices of Δ and C.

It should be noted that from a practical point of view, this assumption is actually stronger than necessary for our protocol. What *is* needed is that the sender A cannot factor N before the protocol halts. Even a polynomial time factoring algorithm may not help him to do this.

4 Checking the Contents of Commitments

When A sends a commitment as above, there is no reason a priori to believe that this represents anything useful: A may not even know how to open the commitment he sends. We will therefore need the following protocol, which is based on the proof system from [4], and allows us to check that A knows how to open a commitment, and furthermore that the opening will reveal a number in a given interval.

We let the interval be $I =]a...b]$ and put $e = b - a$. We define $I \pm e =]a-e...b+e]$. These parameters must be chosen such that $I \pm e$ is contained in the legal range for openings of commitments $]-2^{l-1}...2^{l-1}[$. The protocol will be secure for A - in fact statistical zero-knowledge - if he knows how to open a given commitment $c = BC_g(R, s)$ to reveal $s \in I$. Moreover, it will convince B that $s \in I \pm e$.

PROTOCOL CHECK COMMITMENT

Execute the following k times in parallel:

1. A chooses t_1 uniformly in $]0..e]$, and puts $t_2 = t_1 - e$. He sends the unordered pair of commitments $T_1 = BC_g(S_1, t_1), T_2 = BC_g(S_2, t_2)$ to B.
2. B requests to see one of the following
   (a) opening of both T_1 and T_2

   (b) opening of $c \cdot T_i \bmod N$, where A chooses i such that $s + t_i \in I$.
3. In the first case of step 2, B checks that both numbers opened are in $]-e..e]$, and that their difference is e. In the second case, B checks that the number opened is in I.

B outputs reject and stops if any of the openings are not correctly done, or if any of the checks required are not satisfied.

The properties of this protocol are summarized in the following two lemmas:

Lemma 2 Given correct answers to both a) and b) in one instance of steps 1-3 above, one can efficiently compute a pair R, s such that $s \in I \pm e$ and $c = BC_g(R, s)$.

Proof By assumption, we are given X, x, Y, y such that

$$T_i = X^{2^l} g^x \text{ and } c \cdot T_i \bmod N = Y^{2^l} g^y$$

where $x \in]-e..e]$, $y \in I$. These two equations imply that we can write c in the form $c = BC_g(Y/X \bmod N, y - x)$ so that the result follows from putting $R = Y/X \bmod N$ and $s = y - x$ □

Lemma 3 Given the factorization of N, any B's view of CHECK COMMITMENT when talking to $\bar{A}$ can be simulated perfectly, provided $\bar{A}$ is given an s in I.

Proof With the factorization of N, modular square roots are easy to compute, and so given a square Q, for any s, we can compute R, such that $Q = BC_g(R, s)$. Armed with this observation, the simulation is quite trivial: we simply generate all the unordered pairs T_1, T_2 as random squares and send them to B. If for a given pair, we get request a) from B, we choose t_1, t_2 as A would have done, and open T_1, T_2 accordingly. If we get request b), we choose i at random to be 1 or 2, choose a random $x \in I$, and open $c \cdot T_i$ to reveal x. The simulation of case b) works since, in the real conversation, $s+t_i$ is always a uniformly chosen number in I, independently of s (provided $s \in I$)□

A slight variant allows us to show that two commitments c, c' contain the same number, even if the commitments use different bases, say g and h:

PROTOCOL COMPARE COMMITMENTS
Execute the following k times in parallel:

1. A chooses t_1 uniformly in $]0..e]$, and puts $t_2 = t_1 - e$. He sends to B the unordered pair $((T_1, T_1'), (T_2, T_2'))$, where each component of the pair is ordered and is defined by $(T_i, T_i') = (BC_g(S_i, t_i), BC_h(S_i', t_i))$.
2. B requests to see one of the following
   (a) opening of (T_i, T_i') for both $i = 1$ and 2.
   (b) opening of $c \cdot T_i \bmod N$ and $c' \cdot T_i'$, where A chooses i such that $s + t_i \in I$.
3. In the first case of step 2, B checks that opening T_i and T_i' has resulted in the same number, that both numbers opened are in $]-e..e]$, and that their difference is e. In the second case, B checks that opening $c \cdot T_i \bmod N$ and $c' \cdot T_i'$ reveals the same number, and that this number is in I.

B outputs reject and stops if any of the openings are not correctly done, or if any of the checks required are not satisfied.

The following two lemmas give the basic properties of this protocol:

Lemma 4 Given correct answers to both a) and b) in one instance of steps 1-3 above, one can efficiently compute R, R', s such that $s \in I \pm e$ and $c = BC_g(R, s)$, $c' = BC_h(R', s)$.

Proof Trivial from the proof of Lemma 2□

Lemma 5 Given the factorization of N, any B's view of COMPARE COMMITMENTS when talking to $\bar{A}$ can be simulated perfectly, provided $\bar{A}$ is given an s in I.

Proof Trivial from the proof of Lemma 3□

5 Release of Rabin and RSA Signatures

We are now ready to present a complete protocol for release of a Rabin signature. The common input to the parties will be a modulus n and a message $m \in]0...n[$, while A's private input will be a number s in $]n..2n]$ such that $s^2 \bmod n = m$. Thus k will be $2|n|$, where $|n|$ is the bit length of n. For all commitments in the following, we will use $l = 2|n| + 3$. The protocol has the following steps:

PROTOCOL RELEASE RABIN SIGNATURE

1. B chooses the parameters of the bit commitment scheme N, g. These are verified interactively as explained in section 3.
2. A sends to B the commitment $h = BC_g(R, s)$.
   A sends $v = BC_h(R', s)$ and $w = BC_g(R'', d)$, where d is defined by $s^2 = m + dn$.
   A opens (as a base-g commitment) the product $g^m w^n v^{-1} \bmod N$ to reveal a 0. Note that if h, v are constructed correctly, then $v = BC_g(R'R^s, s^2)$.
3. A uses the CHECK COMMITMENT protocol with $I =]n-1...4n-1]$ to prove that he knows how to open w to reveal a value in $]-2n-1..7n-1]$.
   A uses the COMPARE COMMITMENTS protocol with $I =]n...2n]$ to prove that he knows how to open h and v to reveal the same value, and that this value is in $]0..3n]$
4. A releases s bit by bit by opening h gradually as explained in Section 3. B checks each opening he receives and rejects if the check fails.

Note that in practice step 1 is only necessary once, and does not have to be repeated for every release. Nevertheless, we have included it here to make the formal proof easier.

Note also, that all the actions in Step 2-3 can be parallelized, so that they take only 3 passes. The definition of p_k below will be done with respect to this organization of the messages.

For this protocol, we define $P_k(m, n, s) = 1$ if and only if $s \in]0..3n]$ and $s^2 \bmod n = m$. Note that this predicate allows more than one possible s given m, n. This is no problem, however, because there are only 3 possible solutions for s given n, m, and from the first i bits of one solution, it is easy to compute the first i bits of any other solution.

Assume step 1 takes $a(k)$ passes. Then we define $p_k(i) = a(k) + 3 + i$.

The h_k^i functions are defined as follows: if the input view is shorter than $p_k(i)$ passes, then output i 0's. Else output the i bits opened by A in the final i passes of the input view.

We then have

Theorem 1 Under the factoring assumption, (A, B) is a secure release protocol with respect to the P, p_k and h_k^i functions defined above.

Proof The first property is trivial by inspection of the protocol.

The proof of the second property is by contradiction. So assume that there exists an A, a constant c and t's of infinitely many lengths, such that there are inputs s, r_A, k_B that make the probability of the definition be larger than k^{-c} for some $i = 1..k$.

Let k be any input length for which the above holds, and assume that we are given a k-bit Blum-integer N chosen with the same distribution B would have used. We now describe a poly-time non-uniform algorithm which factors N with probability at least a polynomial fraction, thus establishing a contradiction with the factoring assumption.

We first choose a random element x modulo N, square it and call the result g. We start up A with the inputs given by the assumption, and generate a random view of Steps 1 and 2.

To this end, we send N, g to A and simulate the proof of knowledge of the factorization of N and the proof that N is a Blum integer with A acting as the verifier. Since the proofs are almost perfect zero-knowledge, A's behavior in the sequel will have the same distribution as in "real life", except for a negligible amount of probability mass. The proof that g is a square we can do according to the protocol as we know a square root x. Note that since this proof is perfect zero-knowledge, it is in particular witness-indistinguishable, so since we will not use x in the sequel, any root of g that can later be derived from messages sent by A is independent of x, and so leads to factorization of N with probability 1/2.

A view of Step 1-2 is called *good*, if it can be completed up to pass $p_k(i)$ with probability at least $k^{-c}/2$, and the h_k^i-value that can be computed from the completed view is incorrect. The assumption implies that the probability of $(A, \bar{B})$ completing pass $p_k(i)$ with an incorrect h_k^i-value is at least k^{-c}. This means that a random view of Step 1-2 is good with probability at least $k^{-c}/2$.

Below we show how to factor N with probability at least 1/2 minus a super-polynomially small fraction, assuming that the view of Step 1 we just created is good. By the above, this will be sufficient.

By rewinding A to the start of Step 3 and issuing randomly chosen requests, we try to find correct answers to both requests in the same instance of the CHECK COMMITMENT, resp. the COMPARE COMMITMENTS protocol. Since the probability of acceptance is at least $k^{-c}/2$, we can do this in polynomial time and succeed with probability essentially 1. By Lemmas 2 and 4, this tells us how to open h and v with the same value s, and how to open w with some value d, where $s \in]0...3n]$ and $d \in]-2n-1...7n-1]$. This means that we can write v as a base-g commitment to s^2, which is a legal way of opening v, since $s^2 < 2^{l-1}$. This in turn implies that we know how to legally open the number $g^m w^n v^{-1} \bmod N$ as a base-g commitment, namely as $m + dn - s^2$. Since A has just told us how to open the same number as a 0, we get a factorization of N by Lemma 1 with probability 1/2 unless $s^2 = m + dn$, in other words, unless s is indeed a Rabin signature on m.

We now use rewinding A to its state at the start of Step 3, to generate random views of the conversation, until we find one where pass $p_k(i)$ is completed, and the bits released by A are incorrect. Once again, by the assumption, this can be done in polynomial time to succeed with probability essentially 1. But this means that we have two different ways of opening part of the contents of h, which by Lemma 1 gives us a factorization of N with probability 1/2.

The third condition in Definition 1 is proved by first observing that Step 1 contains a proof of knowledge of the factorization of N. Thus if B completes

Step 1 with probability more than a polynomial fraction, we can always find the factorization. Moreover, B cannot get a non-square accepted as g with probability more than 2^{-k}. Thus we see that except for negligibly few cases, we can simulate the conversation perfectly by sending random squares in place of all commitments, and opening them as needed using our knowledge of the factorization of N. In particular, Step 3 is simulated using Lemma 3 and 5, and Step 4 is simulated using the input we are given, which tells us what the least significant i bits of s are□

It is easy to see that this protocol can be modified to release for example an RSA signature with public exponent 3 by introducing a new commitment u, such that $h = BC_g(R, s)$, $v = BC_h(R', s)$, and $u = BC_v(R'', s)$, which will make u a base g commitment to s^3. It is also clear, however, that this quickly becomes impractical with increasing public exponents.

6 Release of El Gamal Signatures

In this section we sketch how to release El Gamal signatures. We first recall the usual setup of the El Gamal signature scheme: a k-bit prime p is chosen, together with a generator α of Z_p^*. A private key x is a number in $[0..p-1[$, while the corresponding public key is $y = \alpha^x \bmod p$. Messages are numbers in $[0..p-1[$, and a signature on message m is a pair (r, s) such that

$$\alpha^m \equiv y^r \cdot r^s \bmod p.$$

For the owner of x, a signature is easy to compute by choosing a random k relatively prime to $p-1$, putting $r = \alpha^k \bmod p$ and solving the equation $m = xr + ks \bmod p-1$ for s. It is conjectured that computing signatures from scratch is a as hard as finding x from y. In the following, we assume that α really is a generator of Z_p^*. In practice, this may be justified because p, α was generated by a trusted party, or because the factorization of $p-1$ is made public, which makes it easy to test α.

The following is based on the observation that gradual release of a discrete log mod p is sufficient for relase of an El Gamal signature. The idea is that we first reveal r and then release bit by bit s, which will be the discrete log base r of $\beta = \alpha^m y^{-r} \bmod p$. This reduces the problem to that of proving that the discrete log base r of β equals the contents of a base-g commitment $h = BC_g(R, s)$ computed as in Section 3.

We will assume that the prover (sender) A knows such a discrete log s in the interval $I =](p-1)..2(p-1)]$. Such an s can always be obtained from an El Gamal signature by adding $p-1$ to the last component.

Using a technique similar to that of COMPARE COMMITMENTS, we get the following protocol, which will be a proof that A knows a suitable s in $]0..3(p-1)]$. If A uses an s in I, the protocol will be zero-knowledge.

PROTOCOL TRANSFER DISCRETE LOG

Execute the following k times in parallel:

1. A chooses t_1 uniformly in $]0..p-1]$, and puts $t_2 = t_1 - (p-1)$. He sends to B the unordered pair $((T_1, T_1'), (T_2, T_2'))$, where each component of the pair is ordered and is defined by $(T_i, T_i') = (BC_g(S_i, t_i), BC_g(S_i', r^{t_i} \bmod p))$.
2. B requests to see one of the following
   (a) opening of (T_i, T_i') for both $i = 1$ and 2.
   (b) opening of $h \cdot T_i \bmod N$ and T_i', where A chooses i such that $s + t_i \in I$.
3. In the first case of step 2, B checks that the number contained in T_i is the discrete log base r of the number contained in T_i', and that this discrete log is in $]-(p-1)..p-1]$. In the second case, B checks that the number contained in $h \cdot T_i \bmod N$ is the discrete log base r of $\beta z \bmod p$ where z is the number contained in T_i', and that the discrete log revealed is in I.

B outputs reject and stops if any of the openings are not correctly done, or if any of the checks required are not satisfied. The following two lemmas give the basic properties of this protocol:

Lemma 6 Given correct answers to both a) and b) in one instance of steps 1-3 above, one can efficiently compute either R, s such that $s \in]0...3(p-1)]$ and $h = BC_g(R, s)$, $\beta = r^s \bmod p$; or a square root of g modulo N.

Proof Note that A must open T_i' in both case a) and b). If these openings are not consistent, we get a square root of g by Lemma 1. Otherwise, what we have from the correct answers is numbers u, U, v, V such that

$$T_i = BC_g(U, u) \quad h \cdot T_i = BC_g(V, v)$$

and z, Z such that

$$T_i' = BC_g(Z, z) \text{ and } z = r^u \bmod p, \quad \beta z = r^v \bmod p.$$

Furthermore, $v \in I$ and $u \in]-(p-1)..(p-1)]$. From this follows trivially that $h = BC_g(V/U, v-u)$ and that $\beta = r^{v-u} \bmod p$□

Lemma 7 Given the factorization of N, any B's view of TRANSFER DISCRETE LOG when talking to $\bar{A}$ can be simulated perfectly, provided $s \in I$.

Proof Follows by trivial modifications of the proof of Lemma 3□

To define the parameters of the release protocol, we put $k = |p| + 1$, the bit length of p plus 1, and we define the shared input to A and B to be p, α, y, r and m, all of length $|p|$. The private input to A is a k bit string s. The predicate

P for this situation is defined such that $P(p, \alpha, y, r, m, s) = 1$ if and only if $\alpha^m = y^r r^s \bmod p$ and $s \in]0..3(p-1)]$.

Note that with this definition of the shared input, we have implicitely assumed that the sender will make r known immediately at the start of the protocol. This does not lead to a security problem, because the receiver could easily by himself simulate such an r by first finding a k such that $(k, p-1) = 1$ and putting $r = \alpha^k$. For any such r there is an $s \in I$ such that (r, s) signs m. In other words, seeing r in the beginning does not help B to compute the signature ahead of time.

In the complete release protocol, B will set up the bit commitment scheme, A will commit to s by sending h as computed above, use TRANSFER DISCRETE LOG to show that the commitment really contains the discrete log base r of β, and will finally release s bit by bit as explained in Section 3.

The h_k^i and the p_k functions are defined similarly to what was done in the previous section.

Theorem 2 Under the factoring assumption, the protocol outlined above is a secure release protocol with respect to the P, h_k^i and p_k functions defined in this section.

Proof Sketch The first property is trivial. The second one is proved in essentially the same way as for Theorem 1: since the TRANSFER DISCRETE LOG protocol is a proof of knowledge, we can use rewinding of A to compute an s that both opens h and satisfies $r^s \bmod p = \beta$. Thus s is by definition the correct secret. Therefore a view of the protocol that leads to an incorrect value must give us a way of opening h that is inconsistent with s, and therefore enables us to compute a root of g, and factor N with large probability. The third property follows easily from Lemma 7 and the fact that B is required to give a proof of knowledge of the factorization of N□

We remark that the same basic idea can also be used for release of signatures in other discrete log based schemes such as NIST DSA and Schnorr's signature scheme. This is because these signatures, like El Gamal signatures, include a discrete logarithm that is hard to compute without knowledge of the secret key. Thus the sender can reveal all components of the signature except this discrete log, and release this gradually using the above methods.

7 Efficient Contract Signing

As explained in the introduction, one possible application of exchange of signatures is to fair contract signing.

However, under the assumption that intervention by a third party is possible, a different solution to contract signing was proposed in [2]. As pointed out there, that solution will sometimes be superior to simply exchanging signatures because it will work, even if one party has much more computing power than the other, and because any signature scheme can be used.

Very briefly, the solution works by having the parties first sign a message stating that they intend to use the protocol below to sign the contract C. We call this message $M(C)$. Signing $M(C)$ commits neither party to C, but prevents them from claiming a different contents of C later.

They then exchange signatures on messages of the form "this signature on contract C should be considered valid with probability p". This exchange is repeated with increasing values of p. When p reaches 1, the attached signature can be considered an ordinary signature on C. But if for example A stops early, B can appeal to a jugde, showing him the last signed message he received from A. Let p_A be the p-value used in this message. The judge then takes a biased random decision: if the signature is valid, then with probability p_A, he decides that the contract is binding for party A. This introduces some computational overhead, compared to simply signing the contract: a new signature is necessary each time we increase the p-value. Moreover, the number of signatures needed increases with the "granularity" with which the contract signing takes place.

The purpose of this section is to point out that all but one of these signatures can be replaced by simple computations of a one-way function. With the techniques known in practice today, this will much more efficient.

Our method makes use of an arbitrary one-way function f. Such a function always exists, if digital signatures do. Moreover, in practice, we have good candidates for one-way functions based on conventional cryptography that are much more efficient to compute than for example an RSA signature.

The idea now is to let A and B initially each choose a list of f-inputs, $a_1, ..., a_t$, resp. $b_1, ..., b_t$. They then exchange the f-values $f(a_i), f(b_i)$ for $i = 1..t$, and all these values are included in $M(C)$, which is signed initially by both parties.

In stead of exchanging signed messages with increasing p-values, the parties now exchange the f-preimages they have chosen, i.e. A starts by sending a_1, waits to receive b_1, if b_1 is valid he then sends a_2, etc. The only other change needed in [2] is in the procedure of the judge: we fix a rule, stating what the bias of his decision should be, as a function of how many valid preimages the complaining party can present to him.

Since as mentioned, almost all known digital signature schemes are much slower in practice than computing a conventional one-way function (such as MD4 for example), this protocol requires very little extra computational effort compared to simply signing the contract without being concerned about fairness.

References

1. Alexi,W., Chor, B., Goldreich, O. and Schnorr, C.P.: "RSA and Rabin Functions: Certain Parts Are as Hard as the Whole". Proc. of the 25th FOCS, 1984, pp. 449-457.
2. Ben-Or, Goldreich, Micali and Rivest: *A Fair Protocol for Signing Contracts*, IEEE Trans. Info. Theory, Vol.36, 1990, pp.40-46.

3. G.Brassard, D.Chaum and C.Crépeau: *Minimum Disclosure Proofs of Knowledge*, JCSS.
4. Brickell, Chaum, Damgård and van de Graaf: *Gradual and Verifiable Release of a Secret*, Proc. of Crypto 87, Lecture Notes in Computer Science, Springer Verlag.
5. Blum: *Three Applications of the Oblivious Transfer*, Dept. of EECS, University of California, Berkely, 1981.
6. Blum: *How to Exchange (Secret) Keys*, ACM Transactions on Computer Systems, vol.1, 1983, pp.175-193.
7. Cleve: *Controlled Gradual Disclosure Schemes for Random Bits and Their Applications*, Proc. of Crypto 89, Lecture Notes in Computer Science, Springer Verlag.
8. I.Damgård: *Collision Free Hash Functions and Public Key Signature Schemes*, Proc. of EuroCrypt 87, Lecture Notes in Computer Science, Springer Verlag.
9. Even, Goldreich and Lempel: *A Randomized Protocol for Signing Contracts*, Proceedings of Crypto 82, Plenum Press.
10. Even and Jacobi: *Relations Among Public Key Signature Systems*, Comp. Sci. Dept., Technion, Haifa Israel, March 1980.
11. U.Feige, A.Fiat and A.Shamir: *Zero-Knowledge Proofs of Identity*, J.Crypt. Voll, no.2, 1988.
12. O.Goldreich, S.Micali and A.Wigderson: *Proof that Yield Nothing but their Validity and a Methodology of Cryptographic Protocol Design*, Proc. of FOCS 86.
13. S.Goldwasser and L.Levin: *Fair Computation of General Functions in Presence of Immoral Majority*, Proc. of Crypto 90, Spinger Verlag LNCS series.
14. S.Goldwasser, S.Micali and C.Rackoff: *The Knowledge Complexity of Interactive Proof Systems*, SIAM J.Computing, Vol.18, pp.186-208, 1989.
15. J. van de Graaf and R.Peralta: *A simple and Secure Way to Show the Validity of your Public Key*, Proc. of Crypto 87, Lecture Notes in Computer Science, Springer Verlag.
16. Håstad and Shamir: *The Cryptographic Security of Truncated Linearly Related Variables*, Proc. of the ACM Symposion on the Theory of Computing, 1983, pp.356-362.
17. Impagliazzo and Yung: *Direct Minimum Knowledge Computations*, Proc. of Crypto 87, Lecture Notes in Computer Science, Springer Verlag.
18. Luby, Micali and Rackoff: *How to Simultaneously Exchange a Secret Bit by Flipping a Symmetrically-Biased Coin*, Proc. of the IEEE conference of the Foundations Of Computer Science 1983.
19. Rabin: *How to Exchange Secrets by Oblivious Transfer*, Tech. Memo, TR-81, Aiken Comp. Lab., Harward University, 1981.
20. Tedrick: *Fair Exchange of Secrets*, Proc. of Crypto 84, pp.434-438, Lecture Notes in Computer Science, Springer Verlag.
21. M.Tompa and H.Woll: *Random Self-Reducibility and Zero-Knowledge Proofs of Information Possession*, Proc. of FOCS 87.
22. Vazirani and Vazirani: *Trapdoor Pseudorandom Number Generators With Applications to Cryptographic Protocol Design*, Proc. of the IEEE conference on the Foundations Of Computer Science 1983, pp.23-30.
23. M.Waidner, B.Pfitzmann: *The Dining Cryptographers at the Disco: Unconditional Sender and Recipient Untraceability with Computational Secure Servicability*, Proc. of EuroCrypt 89, Lecture Notes in Computer Science, Springer Verlag.
24. Yao: *How to Generate and Exchange Secrets*, Proc. of the IEEE conference on the Foundations Of Computer Science 1986.

Subliminal Communication is ~~Possible~~ Easy Using the DSA

Gustavus J. Simmons
Sandia Park, NM 87047, USA

Abstract

In 1985, Simmons showed how to embed a subliminal channel in digital signatures created using the El Gamal signature scheme. This channel, though, had several shortcomings. In order for the subliminal receiver to be able to recover the subliminal message, it was necessary for him to know the transmitter's secret key. This meant that the subliminal receiver had the capability to utter undetectable forgeries of the transmitter's signature. Also, only a fraction of the number of messages that the channel could accommodate in principal could actually be communicated subliminally ($\phi(p-1)$ messages instead of p-1) and some of those that could be transmitted were computationally infeasible for the subliminal receiver to recover.

In August 1991, the U.S. National Institute of Standards and Technology proposed as a standard a digital signature algorithm (DSA) derived from the El Gamal scheme. The DSA accommodates a number of subliminal channels that avoid all of the shortcomings encountered in the El Gamal scheme. In fairness, it should be mentioned that not all are avoided at the same time. The channel in the DSA analogous to the one Simmons demonstrated in the El Gamal scheme can use all of the bits contained in the signature that are not used to provide for the security of the signature against forgery, alteration or transplantation, and is hence said to be broadband. All messages can be easily encoded for communication through this channel and are easily decoded by the subliminal receiver. However, this broadband channel still requires that the subliminal receiver know the transmitter's secret key. There are two narrowband subliminal channels in the DSA, though, that do not give the subliminal receiver any better chance of forging the transmitter's signature than an outsider has. The price one pays to secure this integrity for the transmitter's signature is a greatly reduced bandwidth for the subliminal channel and a large, but feasible—dependent on the bandwidth actually used—amount of computation needed to use the channel. In one realization of a narrowband subliminal channel, the computational burden is almost entirely on the transmitter while in the other it is almost entirely on the subliminal receiver.

In this paper we discuss only the broadband channel. The narrowband channels have been described by Simmons in a paper presented at the 3rd Symposium on State and Progress of Research in Cryptography, Rome, Italy, February 15-16, 1993. Space does not permit them to be described here. The reader who wishes to see just how easy it is to communicate subliminally using the DSA is referred to that paper as well. The inescapable conclusion, though, is that the DSA provides the most hospitable setting for subliminal communications discovered to date.

Prologue

One of the inevitable consequences of the development of digital signature standards such as the digital signature algorithm (DSA) proposed by the U.S. National Institute of Standards of Technology [2,3] will be their application to the authentication of documents, licenses, personal ID's, etc. Because a widely —even universally—accepted standard can be used, it will be possible for anyone, using only public information and the information contained in the signed instrument, to verify that that information could only have been signed by the specified issuing authority. The verifier can therefore be confident that the issuer is vouching for the integrity of the information he signed; i.e., that the issuer has independently established the identity of the bearer of the document and that he has verified the license, authority, credit or other information vouchsafed for in the document. As a result, any customs checkpoint in the world will be able to quickly verify the integrity of the information contained in a passport—including the photograph of the person to whom it was issued—and hence to determine whether it is being presented by its legitimate owner. Similarly, a teller at a bank, a clerk in a store, a guard at a restricted access facility, etc., will be able to validate identifying and authorizing information in personal ID's, licences, etc. There is no doubt that this is both possible and probable as a result of the development of widely accepted and easy to implement digital signature standards.

What isn't obvious, though, is that some digital signature schemes, and the DSA in particular, make it easy to conceal information in the signature which can be recovered by insiders, i.e., by persons possessing information about the scheme that was not made public, but which can't be recovered or whose presence can't even be detected by persons possessing only the public information. Information communicated in this way—indetectably concealed in a public communication—is said to be subliminal, and the mechanism for its communication is called a subliminal channel. The net result is that the digital signature on a passport which makes it possible for anyone who wishes to verify the authenticity of the document can also tell customs agents of the issuing nation that the bearer is a known terrorist, drug dealer, smuggler or felon. The digital signature on the driver's license that verifies the bearer's identity to the merchant or teller can tell law enforcement officers the bearer's driving while intoxicated (DWI) record, or his accident or traffic violations history, etc. Even commercial entities might have an interest in exploiting such a concealed channel. A financial institution issuing credit cards that are also used as personal ID's might well wish to conceal in the digital signature information about the customer's credit rating, payment history, etc., that could only be read by their agents.

The fact that subliminal messages can be concealed in digital signatures is not a new observation. What is new is that the shortcomings which limited the practical feasibility of earlier schemes can all be avoided in subliminal channels in the DSA.

Introduction

In 1985 Simmons [4] showed that in any digital signature scheme in which α bits are used to communicate a signature that provides only β bits of security against forgery, alteration or transplantation of a legitimate signature, where $\beta < \alpha$, the remaining α-β bits are potentially available for subliminal communication. However, the subliminal channels that have been found in most such digital signature schemes suffer from a number of deficiencies. If the subliminal channel uses all, or nearly all, of the α-β bits, it is said to be broadband, while if it can communicate only a very small fraction of the α-β bits, it is said to be narrowband. All broadband channels devised thus far suffer from a common, and serious, shortcoming: the subliminal receiver must know the transmitter's secret signing key in order to be able to recover the subliminal message(s). This means that the subliminal receiver has the capability to utter undetectable forgeries of the transmitter's signature. While there are situations in which a transmitter is willing to unconditionally trust the subliminal receiver, there are many more in which he isn't. Also in most cases, not all of the α-β bits available can actually be used for subliminal communications. Put another way, in a natural sense not all messages can be encoded for communication through the subliminal channel—and much more significantly not all of those that can be encoded can be recovered by the subliminal receiver. This is due to the fact that the decoding algorithm in most cases requires an extremely variable amount of computation by the subliminal receiver —ranging from the computationally trivial to the infeasible—depending on which message(s) are being decoded.

Subliminal channels that do not require the transmitter to entrust his secret key to the subliminal receiver are generally narrowband. The reason for this is that the security for the transmitter's signature against forgery by the subliminal receiver is bought at the expense of requiring at least one of the two to do computations that grow exponentially with the number of bits being communicated subliminally. Simmons has described two of the narrowband channels in the DSA at the 3rd Symposium on State and Progress of Research in Cryptography, Rome, Italy, February 15-16, 1993 [6]. In this paper, however, we will concentrate on the broadband channel that can be set up between a transmitter and subliminal receivers whom he trusts unconditionally since this is the most efficient (in an information theoretic sense) of the several channels permitted by the DSA.

A Subliminal Channel in the El Gamal DSS

The subliminal channel that Simmons discovered in the El Gamal digital signature scheme[1] —from which the DSA is derived—clearly illustrates all of the deficiencies mentioned above [5], so we start by discussing this channel in greater detail than has been done before in order to set the stage for a discussion of subliminal communications using the DSA. The likelihood of the DSA being adopted as a digital signature standard (DSS) by the U.S. makes such a discussion particularly timely and important; especially given the truly remarkable coincidence that the subliminal channels in the DSA avoid every one of the deficiencies mentioned above. In fairness, it should be said that no single channel (known to the author) avoids all of them simultaneously, however, several channels avoid all but a single one of them.

Simmons observed that if the transmitter[2], T_X, shares his secret signing key, x, with another person—the subliminal receiver, S_X—this person will under

1. A terse description of the El Gamal scheme is given in Appendix A for the reader's convenience.

2. In order to streamline the discussion, the signer or user who originates and/or transmits the digitally signed message will be denoted by T_X, the verifier or legitimate receiver(s) by R_X and the subliminal receiver(s) by S_X.

some circumstances be able to recover the secret session key, k, used by the T_x to generate the digital signature. When this is possible, the T_x can encode the subliminal message in the choice of k—which as we will show can only be recovered by someone in possession of the T_x's secret key, x. This, of course, requires that the T_x trust the S_x unconditionally since the S_x will have the capability to utter undetectable forgeries of the T_x's signature. A common reaction at this point is to say that no one in their right mind would accept such a condition, i.e., no one would unconditionally trust someone else with what amounts to a "power of attorney" for their signature. It should be pointed out, however, that all participants in single-key cryptosystems have always had to accept precisely this condition since any party can do anything the others can. In other words, it has only been since the discovery of public key (two-key) cryptography that participants have even had the option of having some degree of security for their communications without having to unconditionally trust the integrity of the other party(ies) to the communication.

Briefly, the subliminal channel in the El Gamal digital signature scheme works as follows. There is an overt message m which the T_x signs. Since m may be quite lengthy, it is first hashed using a publicly known hashing function $H(\cdot)$ whose range is contained in GF(p); the field over which the El Gamal signatures are computed. The signature is actually for $h = H(m)$, but since $H(\cdot)$ is chosen to be a strong one-way hashing function it is infeasible (impossible?) for anyone given a message m to find another message m' such that $H(m) = H(m')$, or given an h to find an m such that $H(m) = h$. Consequently, signing h is practically the same as signing m. The signed message is the triple (m;r,s) where

$$r = \alpha^k \mod p \tag{1}$$

and

$$s = k^{-1}(h - xr) \mod (p-1) \tag{2}$$

x is the T_x's secret signing key and k is a session key—supposedly randomly chosen from the interval $0 < k < p$ such that $(k,p-1) = 1$—to be used only in the generation of the signature for m.

The verification procedure for the signature is of no concern here. Suffice it to say that it is insensitive to the particular choice of k, so long as the conditions $0 < k < p$ and $(k,p-1) = 1$ are satisfied.

Now consider what the S_x can do by virtue of knowing x, that the public receivers, R_x, cannot do. Everyone knows p, α (a primitive element in GF(p)), the T_x's public key $y = \alpha^x \mod p$, the hashing function $H(\cdot)$ and the signed message (m;r,s). Also, they can easily compute $h = H(m)$. The S_x, though, can compute

$$w = h - xr \mod(p-1) \quad . \qquad (3)$$

It is easy to see, and to show, that for the R_X, w is unconditionally securely encrypted by the T_X's secret key, x. In other words knowing both h and r, the R_X's uncertainty about the value of w is precisely what it would be if he knew neither h nor r. The S_X knows from (2) that

$$s = k^{-1}w \mod(p-1) \quad . \qquad (4)$$

$(k^{-1},p-1) = 1$ since $(k,p-1) = 1$, so that

$$f = (s,p-1) = (w,p-1) \quad .$$

If the S_X is lucky, $f = 1$ and (4) can be easily solved to recover k directly,

$$k = s^{-1}w \mod (p-1) \qquad (5)$$

With probability $1 - \varphi(p-1)/(p-1) > 1/2$, though, $f > 1$ so that s does not have a multiplicative inverse with respect to p-1. The communications channel we have just described is a rather strange one. There are p-1 possible values that k could take, i.e., messages that conceivably might be communicated subliminally, but since $(h,p-1) = 1$, only $\varphi(p-1)$ of these can actually be encoded. Given that one of the permissable k is encoded, we have just seen that due to the random mapping effect of the T_X's secret key—which was chosen in advance and is fixed thereafter—that the probability that k can be recovered using the simple decoding scheme in (5) is $\varphi(p-1)/(p-1) < 1/2$. This is the subliminal channel in the El Gamal digital signature scheme as Simmons described it. As we shall see there is a great deal more to be said about this simple scheme.

If $(s,p-1) = f > 1$ so that s does not have a multiplicative inverse, we can still solve for k in almost all cases using the forward search cryptoanalytic technique devised by Holdridge and Simmons [7] as a means to attack a public key secrecy channel in those cases in which the source equivocation is small enough to make an exhaustive search feasible. In such situations, even if the encryption operation is impossible to invert, since the encryption key is public it is possible for the cryptanalyst to preencrypt plaintexts to generate a table of plaintext/ciphertext pairs, and then use table lookup to find the plaintext corresponding to a ciphertext—even though direct decryption is impossible. In the application being discussed here, inverting $r = \alpha^k$ to recover k directly is the hard problem of taking discrete logarithms (with respect to α). We can, however, in almost all instances, establish that k is an element in a set $\mathbb{K}$ whose cardinality is small enough that it is feasible to calculate the modular exponentiation, α^{k_i}, for all $k_i \in \mathbb{K}$. It is then an easy matter to identify k itself.

To carry out a forward search for the session key, k, when $(s,p-1) > 1$, we reduce (2) to form

$$u = \left(\frac{s}{f}\right)^{-1} \left(\frac{h-xr}{f}\right) \bmod \left(\frac{p-1}{f}\right) \tag{6}$$

where

$$\left(\frac{s}{f}\right)\left(\frac{s}{f}\right)^{-1} = 1 \bmod \left(\frac{p-1}{f}\right) .$$

We now know that k is a unique member of the set

$$\mathbf{K} = \left\{u + i\left(\frac{p-1}{f}\right)\right\} \qquad 0 \le i < f \quad . \tag{7}$$

k itself can be found by calculating $r_i = \alpha^{z_i} \bmod p$, $k_i \in Z$, to identify the element of $\mathbf{K}$ that reproduces r.

If f is small, the forward search is computationally easy. For large values of f, however, forward search may be infeasible. A result from elementary number theory states that if d is a divisor of N, the number of positive integers, $j < N$, for which $(j,N) = d$ is $\varphi(N/d)$. In the present case, $N = p-1$, so that for $\varphi(p-1)$ values of r, k can be recovered by the S_x without having to resort to forward search. If $d|N$, $d > 1$, then on average $(d+1)/2 - 1/d$ encryptions must be made before k is identified. We will later examine this result more closely, but for the moment, consider what it says about the difficulty of recovering the session key, k, from a knowledge of the secret key, x, if the prime p is of the form $p = 2p' + 1$. $\varphi(p) = 2p'$ in this case so that in $p'-1$ cases k could be recovered using (5). In $p'-1$ cases $f = 2$, so that precisely one modular exponentiation would need to be done to identify k, and in the one case in which $f = p'$, it would be computationally infeasible to use forward search to recover k. The statement that k can be recovered in almost all cases from a knowledge of x using forward search techniques is thus justified for p of the form $p = 2p' + 1$. The generalization to other p is obvious—but tedious.

To be precise, what we have called forward search—although closely related to the technique devised by Simmons and Holdridge—isn't truly forward search. Furthermore, the computational effort required to decode subliminal messages can be considerably less than indicated in the naive approach described. We defer a discussion of these points for the moment to exhibit a modest sized example illustrating all aspects of the subliminal channel.

Let $p = 4294969663$ so that $p-1 = 2 \cdot 3 \cdot 715828277$. 17 is a primitive element in GF(p), so we set $\alpha = 17$. p was chosen to be a ten-digit prime to make it possible to send four-letter words as subliminal messages. There is no secure hashing function over such a small range so we use a simple check sum as a message digest algorithm to compress arbitrarily long texts into 32-bit digests. The message digest H(m) will be the exclusive or of successive blocks of 32-bits of the binary encoded text. This is especially convenient if the text is encoded into 8-bit ASCII, since then the blocks each contain four characters, padded with blocks of 8 zeroes, Ø, if need be to make the text be a multiple of 32 bits. The hash of the text "NOT SEEING IS BELIEVING" then becomes.

$$(NOTsp) \oplus (SEEI) \oplus (NGspI) \oplus (SspBE) \oplus (LIEV) \oplus (ING\emptyset)$$

or

$$h = H(m) = (1011110101011110001001110011)_2 = 99283251 \quad .$$

Since $\varphi(p-1)/(p-1) \approx 1/3$ in this case, assuming that the ASCII encoding of four-letter words is a random mapping with respect to relative primality to p-1, only one out of three four-letter words has an ASCII code that is suitable for subliminal communication. For example $(FOUR)_{ASCII} = 3335509458 = 2 \cdot 3^2 \cdot 127^2 \cdot 11489$ so that ("FOUR," p-1) = 6 which says that the message "FOUR" can't be communicated subliminally, while $(FORE)_{ASCII} = 3335508677$ (a prime) so that "FORE" could be.

Assume also that the T_X's secret key, x, is 2045235856 and is known to the S_X. The T_X's public key, $y = \alpha^x \mod p$, isn't needed for the subliminal communication. One signature for the message NOT SEEING IS BELIEVING would be

(NOT SEEING IS BELIEVING; 870994817, 959793649) .

The S_X on receiving this signed message would first calculate (s,p-1) = 1 which tells him that he can solve for k directly using (5). He then calculates $s^{-1} = 1794768229$ and

$$w = h - xr = 3268808305 \mod (p-1) \quad .$$

Finally

$$k = ws^{-1} = 3335308741 \mod (p-1)$$

$$= (11000110\ 11001100\ 11000101\ 11000101)_2$$

which is the ASCII encoding of the subliminal message "FLEE."

As we've mentioned earlier, the verification of the signature—by either the R_X or the S_X—is unaffected by the presence of the subliminal communication. If the messages, both overt and covert, remain the same, and the same prime and primitive element are used—which means, of course, that $r = \alpha^k \mod p$ also remains the same, but the T_X's secret key, x, is smaller by only 1, i.e., x = 2045235855 instead of 2045235856 the result is quite different. $s = 2985295794 = 2 \cdot 3 \cdot 13 \cdot 38273023$ so that f = (s,p-1) = 6. The S_X calculates w = 4139803122 and knows that he must use (6) to solve for u.

$$u = (497549299)^{-1} \cdot 689967187 \mod \left(\frac{p-1}{6}\right)$$

$$u = 471995633 \quad .$$

He therefore knows that the subliminal message is the unique element, k_i, in the set

$$K = \{471995633 + i \cdot 715828277 \mod (p-1)\} \qquad 0 \le i \le 5$$

for which $\alpha^{k_i} = r \mod p$.

One way to find k would be to calculate the modular exponentiations $\alpha^{k_i} \mod p$ until r is recovered and k identified. This would require a large amount of computation, almost all of which would be wasted. A better approach that actually uses the forward search cryptanalytic technique is to note that for some i,

$$r = \alpha^{k_i} = \alpha^{u+i(p-1)/f} \mod p \quad . \tag{8}$$

u is determined by the subliminal communication, while (p-1)/f is simply a divisor of p-1. We rewrite (8) in the form

$$r\alpha^{-u} \in \left\{ \alpha^{i(p-1)/f} \mod p; \quad 0 \le i \le 5 \right\} = D_6 \quad . \tag{9}$$

Let $d = \alpha^{(p-1)/f} \mod p$, then we can simply form the powers of d as opposed to computing full modular exponentiations.

$$r\alpha^{-u} \in \{d^i \mod p; \ 0 \le i \le 5\} = D_6 \tag{10}$$

D_f can now be precomputed as soon as p is chosen. There are a number of time saving computational tricks to computing the various D_f since sets will have elements in common whenever the divisors, f, have factors in common. The important point is that D_f is now a precomputed table $\{i, d^i \mod p\}$ which is a true application of forward search. When the signed message (m;r,s) is received f is calculated from (s,p-1) using the Euclidean algorithm—which is only of order O(log(p)) difficulty—and u solved for using formula (5). $r\alpha^{-u}$ is then calculated and looked up in the table to identify the correct value of i. Finally

$$k = u + i\,\frac{p-1}{f} \tag{11}$$

is calculated to recover the subliminal message.

In the present example D_6 is given in Table I.

Table I.

i	d_i
0	1
1	983925568
2	983925567
3	4294969662
4	3311044095
5	3311044096

We next calculate

$$\alpha^{-u} = 3040091830$$

and

$$\alpha^{-u} r = 3311044095 \mod p$$

which is the fifth entry in the table; i = 4. We can, therefore, finally decode the subliminal message

$$k = 471995633 + 4 \cdot 715828277 = 3335308741$$

which we recognize from before as the ASCII encoding of the message "FLEE." This is an efficient decoding algorithm for the subliminal channel. When a message is received, a single modular exponentiation is required to compute α^{-u} mod p followed by a modular product and a table lookup. These steps are independent of the value of f which determines whether it is feasible to carry out the forward search calculations needed to precompute the table $\mathbf{D}_f$ in the first place.

We have seen that if the user's secret key is x = 2045235856 the subliminal message is directly recoverable without forward search, while if x = x-1 there are six possible values for k—but that the equivocation can be resolved by a forward search to identify the one that generates the known cipher, r. If x = x-2, however, f = 715828277 so that the forward search would have to be carried out in a set of size 715828277. Although this is barely within the state of the art for such a small example, it illustrates the infeasibility of recovering the subliminal message when f is a large divisor of p-1 for realistic sized primes, p, even though the message could be easily concealed and communicated in the subliminal channel. The reader may have guessed that our initial "random" choice of x = 2045235856 was not entirely random. In fact, it was chosen to illustrate the different levels of computational difficulty that can be involved in carrying out a forward search for keys that are very nearly the same. Table II summarizes this dramatically. For the entries in this table, the message (NOT SEEING IS BELIEVING), the modulus (4294969663) and the primitive element, 17, are all fixed. The secret key, x, and parameters dependent on x are variable. Since r is a function of only fixed elements, it is also fixed at r = 870994817 in all five cases.

Table II.

x	s	f = (s,p-1)
2045235856	959793649	1
2045235855	2985295794	6
2045235854	715828277	715828277
2045235853	2741330422	2
2045235852	471862905	3

The broadband subliminal channel in the El Gamal DSS has several deficiencies: some serious and some of only academic concern. The signature (•;r,s) consists of 2ℓ bits, $\ell = \lceil \log_2 p \rceil$, the equivocation of ℓ of which is "used up" to provide a security of $2^{-\ell}$ against forgery, alteration or transplantation of the signature. The remaining ℓ bits are potentially available for subliminal communications, however as we've seen, only $\log_2(\varphi(p-1))$ of these can be used. This says that one bit will be wasted in all cases, so a natural question is how significant is this loss of channel capacity. It is easy to verify directly that p = 421 is the smallest prime for which two bits are unavailable. This problem, however, is of academic interest only, since a number theoretic calculation shows that no prime with fewer than 71 decimal digits can lose as many as three bits and that the prime would have to have 7777 decimal digits in order for four bits to be unavailable! The conclusion is that for realistic sized primes only a bit or two out of the ℓ bits total will be unavailable for subliminal communications.

A somewhat more serious (practical) problem is that there is no natural encoding/decoding between the space of binary messages and the elements of M_p. As we've seen, we can only encode k for which $(k,p-1) = 1$ for communication in the subliminal channel, but we must have a practical way to associate these values of k with the messages we wish to convey, which is easy for both the T_x and the S_x to use. For large primes, and especially for primes for which p-1 has several factors, this is so difficult to do that it limits the potential usefulness of the subliminal channel.

The next most important problem—in increasing order of significance—is the fact that some messages that can be easily encoded for communication in the subliminal channel are completely infeasible for the S_x to recover but which message is lost will depend on the value of H(m). Depending on the factorization of p-1, this may be a serious problem (because of the number of messages that can't be conveyed to the S_x) or a problem of only academic interest as is the case when $p = 2p'+1$ where there is only a single message that can't be recovered by the S_x. It should be pointed out, however, that this "lost message" isn't fixed by the choice of x so it can't be avoided by the coding scheme, rather for any choice of the message m, and hence of h, and of x, there will be one message whose encoding will be infeasible for the S_x to recover. For primes not of the special form $p = 2p'+1$, the preceding comments apply to a larger set of messages.

Finally, there is the most serious deficiency of all; namely, that the S_x must know the T_x's secret key, x, in order to recover the subliminal messages.

A Subliminal Channel in the DSA

Since both components of the signature $(\cdot;r,s)$ created using the DSA are elements in GF(q), 2ℓ bits must be communicated with a message, m, to convey the signature; where $\ell = \lceil \log_2 q \rceil$. If for any given message either r or s is fixed, the probability of a would-be cheater finding the unique value of s or r, respectively, that would cause the triple (m;r,s) to be accepted as having been signed by the specified user is precisely the same as the probability of drawing a unique but unknown, element from GF(q) in a random drawing, i.e., 1/q. This statement holds, even though r almost certainly cannot assume all of the values in GF(q). The concatenated modular reductions in forming r

$$r = (g^k \bmod p) \bmod q$$

map the residues of g^k mod p onto residues mod q. Since g was constructed to be an element of order q in GF(p), as k ranges over the elements of GF(q), g^k mod p ranges over a unique cycle of order q in GF(p). Note: the number of elements of order d, where $d|p-1$, in GF(p) is $\varphi(d)$. Since d is the prime q in this case there are q-1 elements of order q in GF(p). This says that no matter which element of order q is chosen as g, that as k varies over GF(q), the same set of q elements (in permuted order of course depending on the choice of g) will be generated by g^k mod p. The residues mod q in turn of these q residues will be random elements of GF(q).

For example, if we choose $p = 547$ and $q = 13$, the set of residues of g^k mod p, g an element of order 13 in GF(p), is,

$$(1, 46, 237, 261, 293, 350, 353, 375, 440, 475, 509, 517, 519) .$$

The set of residues of this set mod 13, is,

$$(1^2, 2^2, 3, 7^3, 10, 11^2, 12^2)$$

where the superscripts indicate the multiplicities with which the indicated r occurs,

$$46 \equiv 293 \equiv 475 \equiv 7 \mod 13 \quad ,$$

etc. Note that in this example, r cannot be any one of the elements in the set (0, 4, 5, 6, 8, 9). The mapping of the q residues of g^k mod p onto the residues mod q is modeled by the random drawing of q elements from a set of q distinct objects with replacement. For q large (2^{16} is large) the expected number of distinct elements in the drawing will be $q(1 - 1/e)$. In other words, the probability that a randomly chosen r, $r \in GF(q)$, can actually occur in a signature will be only $p \approx 0.63$. On the other hand, it is easy to construct an r that does occur. Choose any $k \in GF(q)$ and compute $r = (g^k \bmod p) \bmod q$. Given any permissible r, all elements of GF(q) are possible values for s (depending on the values of h, x and k. Conversely, given any value of s, every value of r that can occur does occur (again depending on the values of h, x and k). Since it is computationally infeasible to exhaustively calculate the values that r can assume, and impossible to show that a particular value cannot occur except by an exhaustive search, the statement made earlier about the uncertainty of r (given an s) or of s (given an r) being ℓ bits is true. In other words, ℓ of the 2ℓ bits in the signature are expended to provide security for the signature.

The remaining ℓ bits are available for subliminal communications. We next show that all ℓ of them can be used.

As was the case for the broadband subliminal channel in the El Gamal digital signature scheme, the T_x shares his secret key, x, with the subliminal receiver(s) to set up the subliminal channel in the DSA as well. The object is for the T_x to be able to communicate an arbitrary element of $GF^*(q)$ to the S_x in such a way that it will be impossible for anyone to either recover the subliminal message m' without knowing x, or to detect that the subliminal channel is being used, even if he knows the subliminal message(s), m', being communicated.

In this case, since $m' \epsilon GF^*(q)$ and $k \epsilon GF^*(q)$, no encoding algorithm is needed, we can simply take k to be the desired subliminal message, $k = m'$. The T_X sends the overt signed message (m;r,s), where

$$r = (g^k \bmod p) \bmod q. \tag{12}$$

and

$$s = k^{-1}(h-xr) \bmod q \quad . \tag{13}$$

Since the modulus is a prime, any nonzero element of GF(q) has a multiplicative inverse. For the moment, assume that $w = h\text{-}xr \neq 0$, so that $s \neq 0$, then

$$k = s^{-1}w \bmod q \tag{14}$$

as before. It is important to note that because the modulus is a prime in this case (it was p-1 in the El Gamal scheme, which was necessarily composite) that $(s,q) = (w,q) = 1$, so that forward search is <u>never</u> needed to decode a subliminal message. Since $k \epsilon GF^*(q)$ $k \neq 0$ and $k^{-1} \neq 0$, however, w can be 0 so that $s = 0$ is possible, in which case k cannot be recovered. To see this, consider the same example used before: $p = 547$, $q = 13$, $g = 475$, $h = 6$ and $r \epsilon (1, 2, 3, 7, 10, 11, 12)$. Let the T_X's secret key $x = 3$. $w = 0$ if $r = 2$ in this case, i.e., if $k = 3$ or 4. What this says is that for this small example and for any message that hashes to produce a value of $h = 6$, neither $k = 3$ nor $k = 4$ can be communicated subliminally. The probability of s being 0, however, is 1/q, i.e., $\approx 2^{-160}$, and even if it is 0, the expected number of subliminal messages that cannot be communicated is only $e/(1-e) \approx 1.58$. To appreciate these numbers: the age of the universe measured in microseconds from the big bang to the present is less than 2^{79}. Thus, although there is a nonzero probability that an $m \epsilon GF(q)$ cannot be communicated subliminally, the channel capacity is $\ell\text{-}\epsilon$, where $\epsilon < 10^{-47}$.

To summarize, the broadband channel in the DSA is easy for the T_X to encode for and for the S_X to decode: forward search is never needed to resolve ambiguities. All messages can be encoded—and more importantly all messages can be decoded. To be precise, the latter statement is only $1\text{-}\epsilon$ true as discussed in the previous paragraph, but we will not continue to qualify every statement about the channel with this ϵ qualifier.

The channel just described does have a weakness, though. If anyone knows the subliminal message that the T_X wishes to communicate, he can easily recover the T_X's secret signing key, x. First, anyone can compute h using the public hashing function H(·) and the overt communication (m;r,s). From (13), he knows that

$$x = r^{-1}(ks-h) \quad \bmod q \tag{15}$$

if neither r nor ks-h is 0, so that with probability 1 - (1/q), i.e., with probability 1 for all practical purposes, knowledge of the intended subliminal message equates to knowledge of the signers secret key. As a matter of fact, the uncertainty about the signer's secret key, x, is reduced by precisely the amount known about the session key, k. Since x should pose ℓ bits of uncertainty to everyone except the T_X, this means that k must be ℓ bits uncertain as well—even when the intended subliminal message is known, i.e., even when there is no uncertainty at all about m'. Since k is an element in the multiplicative group GF*(q), it is easy to achieve this level of unconditional security. The T_X and the S_X secretly share in advance a one-time key consisting of a random sequence, V, of symbols drawn from GF*(q). They then use the next unused symbol, $v \in V$, from this sequence to Vernam encrypt/decrypt the subliminal message, m', using multiplication in GF(q). To recover the text m', S_X first recovers k as described above and then calculates $m' = kv^{-1}$ in GF(q). The concealment is perfect, irrespective of how many known plaintext's the public receiver may have, since the Vernam encryption always introduces at least as much equivocation (per encryption) as is removed by the exposure of a cipher/ plaintext pair. The bottom line is that the broadband subliminal channel in the DSA is perfectly concealed—even from an opponent who suspects its use and who knows exactly the messages being communicated through it.

Unconditional security for the subliminal messages and unconditional concealment of the fact that subliminal communications are occurring requires that the key sequence V be a truly random sequence. For practical purposes the T_X and S_X would simply share the key to any cryptographically secure key stream generator—the synchronized output of which would be used instead of the one-time key sequence described here.

The broadband channel still suffers from the necessity that the S_X know the T_X's secret signing key in order to be able to decode subliminal communication. Space doesn't permit a description of the narrowband subliminal channels allowed by the DSA that overcome this limitation. The interested reader is referred to a paper presented by Simmons at the 3rd Symposium on State and Progress of Research in Cryptography, Rome, Italy, February 15-16, 1993, for a detailed discussion of these channels.

Conclusion

The DSA provides the most hospitable setting for subliminal communications discovered to date.

Note to the Appendices

All digital signature schemes presuppose the existence of a secure hashing function H(•), which operates on messages to produce message digests. H(•) satisfies the condition that given a message m it is computationally infeasible (impossible?) to find another message m' such that H(m) = H(m'); i.e., H(•) is collision free. Equivalently, given an h, it is infeasible to find an m such that H(m) = h. H(•) is public information, and for all messages m, H(m) is easy to evaluate.

Appendix A.

The El Gamal digital signature scheme [1] involves four distinct steps.

Step 1 (performed by a trusted issuing authority).

- a large prime, p, is randomly selected, subject to the condition that $p-1$ has at least one large prime factor.
- a primitive element $\alpha \in GF(p)$ is chosen.

Note: p and α are made public and can be used by a community of users.

Step 2 (performed by each user who wishes to be able to sign messages).

- user chooses a random $x \in GF^{*}(p)$, which is his secret (signing) key.
- user publishes $y = \alpha^{x} \bmod p$ as his public (verification) key.

Note: each user's public key, y, is associated with his identity in a public, certified, directory.

Step 3 (performed by a user[3] wishing to sign a message m).

- user first calculates the message digest $h = H(m)$.
- user chooses a random $k \in GF^{*}(p)$ such that $(k,p-1) = 1$ (k is essentially a session key).
- user calculates $r = \alpha^{k} \bmod p$.
- user calculates $s = k^{-1}(h-xr) \bmod (p-1)$ where $k \cdot k^{-1} \equiv 1 \bmod (p-1)$.

Note: The triple $(m;r,s)$ is the signed message.

A receiver (verifier) on receiving a triple $(\underline{m};\underline{r},\underline{s})$ purporting to be a message signed by the user whose public key in the directory is y carries out Step 4.

Step 4 (performed by a receiver (verifier)).

- verifier first calculates the message digest $\underline{h} = H(\underline{m})$.
- verifier calculates $u = \alpha^{\underline{h}}$
- verifier calculates $v = y^{\underline{r}}\underline{r}^{\underline{s}} \bmod p$.
- The triple $(\underline{m};\underline{r},\underline{s})$ is accepted as having been signed by the user (associated with the public key y) if and only if $u = v$.

Appendix B

The U.S. digital signature algorithm (DSA) [2,3] involves four distinct steps.

Step 1 (performed by a trusted issuing authority).

- a large (512-1024 bits in 64-bit increments) prime, p, is randomly selected, subject to the condition that p is divisible by a 160-bit prime, q.
- an element, g, of order q in $GF(p)$ is constructed by choosing any element $h \in GF^{*}(p)$ for which $g = h^{(p-1)/q} > 1$.

Note: p, q and g are made public and can be used by a community of users.

Step 2 (performed by each user who wishes to be able to sign messages).

- user chooses a random $x \in GF^{*}(p)$, which is his secret (signing) key.
- user publishes $y = g^{x} \bmod p$ as his public (verification) key.

---

3. The user must have entered his public key into the public directory before receivers can verify his signature.

Note: each user's public key, y, is associated with his identity in a public, certified, directory.

Step 3 (performed by a user[4] wishing to sign a message m).

- user first calculates the message digest $h = H(m)$.
- user chooses a random $k \in GF^*(q)$.
  (k is essentially a session key.)
- user calculates $r = (g^k \bmod p) \bmod q$.
- user calculates $s = k^{-1}(h+xr) \bmod q$, where $kk^{-1} = 1 \bmod q$.

Note: The triple (m;r,s) is the signed message.

A receiver (verifier) on receiving a triple ($\underline{m}$;$\underline{r}$,$\underline{s}$) purporting to be a message signed by the user whose public key in the directory is y carries out Step 4.

Step 4 (performed by a receiver (verifier)).

- verifier first calculates the message digest $\underline{h} = H(\underline{m})$.
- verifier calculates $t = \underline{s}^{-1} \bmod q$.
- verifier calculates $u_1 = \underline{h}t \bmod q$.
- verifier calculates $u_2 = \underline{r}t \bmod q$.
- verifier calculates $v = (g^{u_1} y^{u_2} \bmod p) \bmod q$.
- The triple ($\underline{m}$;$\underline{r}$,$\underline{s}$) is accepted as having been signed by the user (associated with the public key y) if and only if $\underline{r} = v$.

References

1. El Gamal, T., "A Public Key Cryptosystem and a Signature Scheme Based on Discrete Logarithms," IEEE Trans. on Info. Theory, Vol. IT-31, No. 4, July 1985, pp. 469-72.

2. NIST, "A Proposed Federal Information Processing Standard for Digital Signal Standard (DSS)," Fed. Register, Vol. 56, No. 169, Aug., 1991, pp. 42980-2.

3. NIST, "Specifications for a Digital Signature Standard (DSS)," Federal Information Processing Standards Pub. xx (Draft), Aug. 19, 1991, 12 pps.

4. Simmons, G. J., "The Subliminal Channel and Digital Signatures," Eurocrypt'84, Paris, France, April 9-11, 1984, Advances in Cryptology, Ed. by T. Beth et al., Springer Verlag, Berlin, 1985, pp. 364-378.

5. Simmons, G. J., "A Secure Subliminal Channel (?)," Crypto'85, Santa Barbara, CA, August 18-22, 1985, Advances in Cryptology, Ed. by H. C. Williams, Springer-Verlag, Berlin, 1986, pp. 33-41.

6. Simmons, G. J., "The Subliminal Channels in the U.S. Digital Signature Algorithm (DSA)," presented at the 3rd Symposium on State and Progress of Research in Cryptography, Rome, Italy, February 15-16, 1993, to be published in the Proceedings of the SPRC'93.

7. Simmons, G. J., and D. Holdridge, "Forward Search as a Cryptanalytic Tool Against a Public Key Privacy Channel," Proc. of the IEEE Computer Soc. 1982 Symp. on Security and Privacy, Oakland, CA, April 26-28, 1982, pp. 117-128.

[4] The user must have entered his public key into the public directory before receivers can verify his signature.

Can O.S.S. be Repaired ?
- Proposal for a New Practical Signature Scheme -

David Naccache

Gemplus Card International
1 Place de Navarre, F-95200, Sarcelles, FRANCE.
E-mail : 100142.3240@compuserve.com

Abstract. This paper describes a family of new **Ong-Schnorr-Shamir-Fiat-Shamir-like** [1] identification and signature protocols designed to prevent forgers from using the **Pollard-Schnorr** attack [2].

Our first signature scheme (and its associated identification protocol) uses x, which is secret-free, as a commitment on which k will depend later. Therefore, the original quadratic equation is replaced by $x^2 - k(x)y^2 \equiv m \bmod n$ where k(x) is a non-polynomial function of x and since the **Pollard-Schnorr** algorithm takes as input value k (to output x and y), it becomes impossible to feed *à-priori* k(x) which is output-dependent.

The second signature method takes advantage of the fact that although an attacker can generate valid OSS signatures (solutions {x,y} of $x^2 - k y^2 \equiv m \bmod n$), he has no control over the internal structure of x and y and in particular, if we restrict the solution space by adding extra conditions on x and y, it becomes very difficult to produce forged solutions that satisfy the new requirements.

1. Introduction

In 1985, **Ong**, **Schnorr** and **Shamir** proposed a digital signature scheme which seemed to be very efficient [1]. In their system, the public-key consisted of a couple of integers k and n where n is an RSA [7] modulus of length N (bits) whose factorisation is kept secret.

A signature {x,y} of a message m (hashed value of a primitive file *M*) was considered valid if :

$$x^2 - k y^2 \equiv m \bmod n \qquad (1)$$

and it was shown by the authors that general solutions of this equation can be generated by the signer provided that he knows a secret u such that $u^2 k \equiv 1 \bmod n$.

Figure 1 : The original OSS scheme

In 1987, **Pollard** and **Schnorr** [2], exhibited a fast probabilistic algorithm for computing solutions of equation (1) and thereby broke OSS. This attack was later extended by **Adleman, Estes** and **McCurley** [3].

Figure 2 : Sketch of the attack introduced by Pollard and Schnorr

In this article, we describe a couple of signature methods (and an associated identification protocol which is zero-knowledge if the challenge size is bounded) intended to prevent forgery by this family of attacks.

The first (standard) signature protocol uses x (which is secret-free) as a commitment on which k will depend later. Therefore, the original quadratic equation is replaced by $x^2 - k(x)\, y^2 \equiv m \bmod n$ where k(x) is a non-polynomial function of x and since the **Pollard-Schnorr** attack takes as an input a value k (to output x and y), it is impossible to input in advance k(x) which is output-dependent and not yet known.

The second (interactive) signature method takes advantage of the fact that the attacker has no control over the solutions of the congruence $x^2 - k\, y^2 \equiv m \bmod n$. In particular, it is hard to produce an x such that the sub-equation $x = r + \frac{m}{r} \bmod n$ admits a solution r with a given internal redundancy.

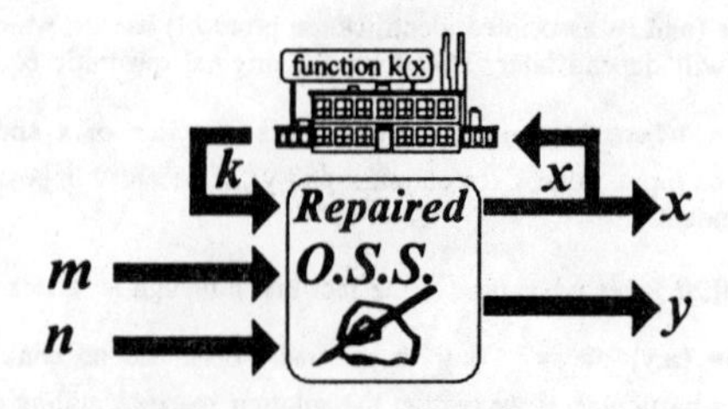

Figure 3 : Our repair strategy

2. Definitions

The system authorities select and publish a one-way function f hashing N-z bit strings into z bit words. Practically, we recommend $z \approx 160$ for $N = 512$ (for instance, SHA or a DES-based hashing).

As in the case of the **Fiat-Shamir** protocol [5], each user is provided with a set of c secret keys $u_1,\ldots,u_c$ and the corresponding public keys $k_1,\ldots,k_c$ (such that $u_i^2 k_i \equiv 1 \bmod n$ for all i) are communicated to the verifiers by any desired means (for instance, ID-based as suggested by **Shamir** in [4] but the key transfer protocol suggested in the original **Fiat-Shamir** [5], should be modified as described in [8]).

3. Protocol #1 (Standard Digital Signature)

① The signer picks t random numbers $\{r_i\}$, computes the set $\{x_i\}$ where : $x_i = r_i + \frac{m}{r_i} \bmod n$, hashes $f(\{x_i\})=\{e_i\}$ (where each e_i is of size c) and calculates $y_i = \prod_{e_{i,j}=1} u_j \left(r_i - \frac{m}{r_i} \right) \bmod n$ for i=1 to t.

② The signature $\{\{x_i\},\{y_i\}\}$ is checked by re-computing $f(\{x_i\}) = \{e_i\}$ and verifying that $x_i^2 - y_i^2 \prod_{e_{i,j}=1} k_j \equiv 4m \bmod n$ for $i = 1,\ldots,t$.

4. Security and Efficiency

The security level of this protocol is $\frac{1}{2^{tc}}$ (with typically $tc \approx 80$) but the size of the signature ("multiple evidence" that the signer affixed honestly his signature on m) is 2Nt.

This yields, (for instance : c=10, t=8), a system where 8192 bit signatures are generated in average $\frac{t(2+c)}{2}$ (=48) modular multiplications with an Nc (=5120) bit secret-key.

The key size and number of multiplications required to implement the new scheme are equivalent to those of the **Fiat-Shamir** but the computation of modular inverses (by using the extended Euclidean algorithm) is much slower than the squaring operation and the size of the resulting signatures is a bit less than the double of **Fiat-Shamir** ones (Things are much better for the identification protocol of section 5 which requires exactly the same amount of transmission as a **Fiat-Shamir**). However, the following variant offers exactly the same amount of transmission as a **Fiat-Shamir** but requires t additional modular squarings : $\{r_i\}$ and $\{x_i\}$ are as in section 5 but $\{e_i\}$ is the hashed value of $\{x_i^2 \bmod n\}$, the definition of the y_i remains unchanged (use the new $\{e_i\}$!) and the signature $\{\{e_i\},\{y_i\}\}$ is checked by comparing $\{e_i\}$ to $f(\{4m + y_i^2 \prod_{e_{i,j}=1} k_j \bmod n\})$.

Note that :

- The new scheme is in public domain (no patents) and can be freely used and implemented.
- We assume that the verifier checks out trivial weak instances (eg. x= ±2 and y= 0 in protocol 2 etc.).
- A simple and practical technique for delegating the computation of $\frac{m}{r}$ mod n to the verifier (which in many practical cases is much more powerful than the prover) is described in section 9.

5. Protocol #2 (Identification and Security Proof)

Repeat t times :

① The prover picks a random r, computes $x = r + \frac{1}{r} \bmod n$ and sends this value to the verifier.

② The verifier replies with a random binary string e, of length c.

③ The prover computes and sends $y = \prod_{e_i=1} u_i \left(r - \frac{1}{r}\right) \bmod n$.

④ And the verifier checks if : $x^2 - y^2 \prod_{e_i=1} k_i \equiv 4 \bmod n$.

The security of the identification scheme can be proved by transforming any algorithm breaking the protocol into a scheme for extracting roots modulo n (For simplicity, we consider the case c=1 since extension to bigger challenges is straightforward) :

Breaking the algorithm means being able to commit in advance a number x (no matter what its internal structure is) such that whatever c will be, both y_1 and y_2 (such that $x^2 - y_2^2 \equiv 4 \bmod n$ and $x^2 - y_1^2 k \equiv 4 \bmod n$) are efficiently computable.

Subtracting the two equations, we get $y_1^2 k \equiv y_2^2 \mod n$ which yields $\frac{y_2}{y_1} \equiv \sqrt{k} \mod n$.

6. Protocol #3 (Interactive Signature)

① The signer picks $t > 1$ random N-z bit numbers $h_1,...,h_t$, computes $r_i = 2^{N-z} f(h_i) + h_i$ and $x_i = r_i + \frac{m}{r_i} \mod n$ for i=1,...,t and sends $x_1,...,x_t$ to the receiver.

② The receiver replies with a randomly chosen index $1 \le j \le t$.

③ The signer reveals all the h_i except h_j.

④ The receiver checks that all the x_i (except x_j) are coherent with the above definition and if a false x_i is detected at this point, the signer is rejected and the protocol is aborted.

⑤ The receiver picks a random binary string e, of length c, and sends it to the signer.

⑥ The signer replies with $y_j = \left(r_j - \frac{m}{r_j}\right) \prod_{e_i = 1} u_i \mod n$.

⑦ The receiver checks that $x_j^2 - y_j^2 \prod_{e_i = 1} k_i \equiv 4m \mod n$ and if this test holds, he accepts the interactive signature $\{x_j, y_j, e\}$ of the message m.

7. Security

If the sender uses the **Pollard-Schnorr** attack his chances to remain undetected by the receiver are $\frac{1}{t2^c}$.

Therefore, the receiver can convince himself, with any desired probability, that the sender actually used a redundant random to generate x_j and y_j.

In case of dispute (the signer pretends that $\{x_j, y_j, e\}$ is a forgery), a judge (knowing the prime factors of n) can solve $X^2 - X x_j + m \equiv 0 \mod n$ and check that one of the solutions X presents the redundancy of step ①

(That is, $\exists$ h such that $X^2 - X x_j + m \equiv 0 \mod n$ where $X = 2^{N-z} f(h) + h$).

If yes, the signer is cheating and if not, the receiver used the **Pollard-Schnorr** method and is falsely accusing the signer.

Note that due to the interactive nature of the protocol an attacker is prevented from using pre-processing. The practical significance of this observation is a significant reduction in the size of the key (parameter c).

Also, it should be observed that although the receiver of the signature is convinced that the signature is valid, he cannot transmit this conviction to anybody else (except the judge).

8. Other Possible Research Directions

Except the extension of our scheme to higher degrees (for instance $x^3+ky^3+k^2z^3-3kxyz \equiv m \mod n$ with keys $\{u,w\}$ such that $u^3=k \mod n$, $w^3=1 \mod n$ and $1+w+w^2=0 \mod n$ as suggested by Ong, Schnorr and Shamir in [6]), other mono-key OSS-like variants are now being investigated.

These are based on a solution (for x and r) of the equation :

$$r+\frac{g(m,x)}{r}=x \mod n \qquad (2)$$

Where g is a public function.

Such a scheme should work as follows :

① **The signer solves the equation $r+\frac{g(m,x)}{r}=x \mod n$ for $\{x, r\}$.**

② **Then, he computes $y=u\left(r-\frac{g(m,x)}{r}\right) \mod n$ and sends the signature $\{x, y\}$ to the receiver.**

③ **The signature is checked by comparing that $x^2-ky^2 \equiv 4\,g(m,x) \mod n$.**

Although still incomplete, we demonstrate this idea with the concrete example $g(m,x)=m\oplus x$ (where "⊕" stands for a bitwise xor) and argument why we believe that efficient algorithms for solving $r+\frac{m\oplus x}{r}=x \mod n$ may exist.

Denoting : $x=\sum_{i=0}^{N-1} 2^i x_i$ and $m=\sum_{i=0}^{N-1} 2^i m_i$, a simple trick for getting rid of the "⊕" in the sub expression $x_i \oplus m_i$ is the observation that :

$$x_i \oplus m_i = x_i(1-2m_i)+m_i .$$

Replacing this into (2), we get : $$r+\sum_{i=0}^{N-1} 2^i\left(\frac{x_i(1-2m_i)+m_i}{r}-x_i\right)=0 \mod n$$

or (with $R=\frac{1}{r} \mod n$) : $$r+\sum_{i=0}^{N-1} 2^i\left(x_i(R-2Rm_i-1)+Rm_i\right)=0 \mod n$$

and finally : $$\sum_{i=0}^{N-1} x_i 2^i(2Rm_i-R+1)=r+Rm \mod n \qquad (3)$$

Defining $a_i=2^i\,(2Rm_i-R+1)$ and $b=r+Rm$, equation (3) clearly appears as a knapsack problem :

$$\text{Find } \bar{x}=\left\langle x_0,x_1,\ldots,x_{N-1}\right\rangle \in \{0,1\}^N \text{ such that } \sum_{i=0}^{N-1} x_i a_i = b \mod n$$

for which efficient algorithms may exist (under certain assumptions...) for small N (condition that can be softened by leaving some liberty to m).

The idea is that the forger should be unable to "mix" the number-theoretic operations of the **Pollard-Schnorr** algorithm with the knapsack solution but this is not a sufficient argument for proving security and a couple of open questions still persists :

1. Can the linearity in the proposed example be used by a forger in order to attack the system efficiently ?

2. Give a characterisation of the functions g such that repairing OSS by solving $r + \frac{g(m,x)}{r} = x \mod n$ ***is secure and still computable efficiently.***

9. Delegating the Extended Euclidean Algorithm

In many cases, the verifier (for instance, a smart-card reader, a terminal or a PC) is much more powerful than the prover (typically a smart-card) and therefore it seems attractive to delegate the computation of the term $\ell = \frac{m}{r} \mod n$ to the verifier :

① The signer picks a random d, computes and sends $s = r\,d \mod n$.

② The powerful verifier computes $v = \frac{m}{s} \mod n$ and sends v to the signer.

③ The signer retrieves $\ell = v\,d \mod n$ and checks that $\ell\, r \equiv m \mod n$.

Practically, this protocol presents the second advantage of not forcing the signer to keep in memory the whole message m : The signer can *secretly* and randomly select a group of 10 bytes in m and check in step ③ that these 10 bytes actually match with those of $\ell\, r \mod n$.

Note that :

☛ The size of d can be reduced to accelerate the multiplications.

☛ For $t>1$, the computation of the inverse of $\prod_{i=1}^{t} r_i$ allows to retrieve all the m/r_i by inter-multiplications.

If r is a (sufficiently big) prime, a second (theoretically) interesting delegation protocol is :

① The signer picks a big random prime d, computes and sends $s = r\,d$ (not modulo !).

② The verifier computes $v = \frac{m}{s} \mod n$ and sends v to the signer.

③ The signer retrieves $\ell = v\,d \mod n$ and checks that $\ell\, r \equiv m \mod n$.

10. Implementation

GemenOSS is one of the public key implementations realised by GEMPLUS CARD INTERNATIONAL.
The whole family includes **Guillou-Quisquater** signature and identification protocols implemented within the MIMOSA smart-card, a one-time identification scheme with low memory consumption based on the authentication tree concept, a superfast version of the DSS [9], trading the modular inverse computation against one modular multiplication and the transmission of a few bytes, and an RSA prototype due to be issued in the near future.

GemenOSS implements our identification scheme with tc = 20 and $|n| = 512$ bits. The verifier (a Compaq Deskpro 4/66i) and the prover (68HC05 clocked at 4 MHz) communicate via a 115,200 baud interface and calculations are done in parallel whenever possible.

Big numbers are manipulated in a redundant (proprietary) format wherein the number of 8-bit by 8-bit multiplications in each big modular multiplication grows as $\log\left(\frac{(\alpha+n)^2}{n-3.02\sqrt{n-\alpha}}\right)$ but transmission is polynomial in parameter α.

The prototype is expected to complete an identification session in about 1.7 seconds when operating at the best α communication / multiplication trade-off point.

11. Conclusion

We demonstrated a family of protocols that allow to reuse quadratic equations modulo n for digital signatures.

The cost of "repairing" the OSS is very acceptable and can be expressed differently (various trade-offs are possible) in terms of key size, number of modular multiplications and transmission overhead.

Due to progress made since the publication of the original OSS scheme, the author strongly encourages the cryptographic community to attack the proposed protocols (#1, #2 and #3) and try to degrade the basic security probabilities (respectively : 2^{-tc}, 2^{-tc} and $2^{-c}/t$) obtained by a brutal application of [2] and [3].

12. Acknowledgements

We thank **Claus Schnorr, Adi Shamir, David M'raïhi** and **Beni Arazi** for their useful remarks, **Jacques Stern** for outlining a subtle attack against a preliminary version of protocol 2 and **Serge Vaudenay** for warning against the choice of weak {x, y} values by a cheating prover (or signer).

References

[1] **H. ONG, C. SCHNORR & A. SHAMIR**, ***"An efficient signature scheme based on quadratic equations"*** in Proceedings of the 16th Symposium on the Theory of Computing, Washington, 1984, pp. 208-216.

[2] **J. POLLARD & C. SCHNORR**, ***"An efficient solution of the congruence $x^2 + k y^2 \equiv m \bmod n$"***, IEEE Transactions on Information Theory, vol. IT-33, no. 5., September 1987, pp 702-709.

[3] **L. ADLEMAN, D. ESTES & K. McCURLEY**, ***"Solving bivariate quadratic congruences in random polynomial time"***, Mathematics of Computation, vol. 48, no. 177, January 1987, pp 17-28.

[4] **A. SHAMIR**, ***"Identity-Based Cryptosystems and Signature Schemes"***, Proceedings of Crypto'84, Lecture Notes in Computer Science, no. 196, Springer-Verlag 1985.

[5] **A. FIAT & A. SHAMIR**, ***"How to Prove Yourself : Practical Solutions to Identification and Signature Problems"***, Proceedings of Crypto'86, Lecture Notes in Computer Science, no. 263, Springer-Verlag 1986.

[6] **H. ONG, C. SCHNORR & A. SHAMIR**, ***"Efficient Signature Schemes Based on Polynomial Equations"***, Proceedings of Crypto'84, Lecture Notes in Computer Science, no. 196, Springer-Verlag 1985.

[7] **R. RIVEST, A. SHAMIR & L. ADLEMAN**, ***"A Method for Obtaining Digital Signatures and Public-Key Cryptosystems"***, *Comm: ACM 21, 2 (Feb. 1978), pp 120-126.*

[8] **D. NACCACHE**, ***"Unless Modified Fiat-Shamir is Insecure"***, *Proceedings of the Third Symposium on State and Progress of Research in Cryptography : SPRC'93, Fondazione Ugo Bordoni (1993), pp 172-180.*

[9] **D. NACCACHE & D. M'RAÏHI**, ***"A strictly DSS-compatible scheme without 1/k mod q"***, *to appear.*

On a Limitation of BAN Logic*

Colin Boyd and Wenbo Mao

Communications Research Group
Electrical Engineering Laboratories
University of Manchester
Manchester M13 9PL
UK

Abstract. In the past few years a lot of attention has been paid to the use of special logics to analyse cryptographic protocols, foremost among these being the logic of Burrows, Abadi and Needham (the BAN logic). These logics have been successful in finding weaknesses in various examples. In this paper a limitation of the BAN logic is illustrated with two examples. These show that it is easy for the BAN logic to approve protocols that are in practice unsound.

1 Introduction

In recent years there has been great interest in the design and analysis of secure protocols. Various new techniques have been developed and used to find a great variety of different attacks on such protocols. One of the most important of these techniques is the Logic of Authentication of Burrows, Abadi and Needham [2], (the 'BAN logic') which transforms a protocol into a special form and then uses logical rules to analyse it. The BAN logic has been used to find new weaknesses in various cryptographic protocols. A number of variations and enhancements of the basic BAN logic have been developed [3, 4].

It has been recognised by the authors of the BAN logic, as well as others, that there are limitations to its power [7, 8]. These limitations can be attributed to its inability to express certain events. In this paper we investigate a different kind of difficulty that does not appear to have been discussed before. This is the lack of precision in moving from a protocol description to its expression in the logic itself - the process called *idealisation*.

Cryptographic protocols have not traditionally been expressed in a completely formal manner and so it is inevitable that there must be some conversion of an informal description to a formal description if formal analysis is to take place. However, the BAN logic does not correspond very well to usual formal descriptions of communications protocols such as process algebras [1]. Instead the protocol is reduced to a number of primitives which include the beliefs of principals. According to the BAN authors [2]:

* This work is funded by the UK Science and Engineering Research Council under research grant GR/G19787.

> Only knowledge of the entire protocol can determine the essential logical contents of (each) message.

This seems to us regrettable and to be the cause of difficulties. In our view, messages consist of items of information. Beliefs of principles are updated as a result of messages (information) received, but do not form parts of messages themselves. Thus formalisation should be possible one message at a time. Formal rules to allow this would lead the way to machine assisted analysis of protocols with all its benefits.

In this paper we present examples to show that the BAN analysis can be dangerous in that it allows protocols to be reasoned as secure that are in fact insecure. We do this by showing that certain variations of protocols cannot easily be distinguished in their BAN logic representations, but that these variations are critical in deciding whether or not the protocol is secure. These examples also serve to re-emphasise the difficulty in designing protocols correctly and the extreme sensitivity of protocols to subtle modifications.

We would like to emphasise that we are not suggesting that there are inconsistencies in the logical rules defining the BAN logic. What we are pointing out is a difficulty in the practical use of the logic which is equally a problem with other related logics. We cannot suggest practical measures to overcome these difficulties and instead would advocate a different approach altogether where it is desired to obtain a protocols of verified high security.

In the next section two detailed examples are given of failures in the analysis of protocols using the BAN logic, as analysed in the original paper [2]. The two attacks seem to exhibit a dilemma in practical use of the idealisation step of BAN logic analysis. If we regard the BAN logic idealisation technique as easy to apply (as [2] indicated) then straightforwardly following the guidance suggested by [2] for idealising protocols we find it possible to idealise a flawed protocol into a good one; otherwise we would have to say that the idealisation idea is vaguely specified and extremely difficult to apply correctly.

In the final section we briefly discuss possible alternative approaches which may overcome the problems highlighted.

2 The Problem

The two examples we will present both come from the same base protocol, that of Otway and Rees [9]. The BAN logic was used to analyse this protocol [2] and the conclusion reached was that the protocol is basically sound but that there are a number of redundancies.

2.1 Description of the Otway-Rees Protocol

This protocol involves two users A and B and a server S whose role is to pass a new session key, K_{AB}, to A and B. Initially S shares keys K_{AS} and K_{BS} with A and B respectively. The steps in a successful run of the protocol are as follows. Here, and throughout, the notation $\{X\}_K$ indicates the string X encrypted using the key K. In this protocol there is an implicit assumption that a symmetric encryption algorithm is used that provides both authentication and confidentiality of encrypted information.

1. A sends to B: $M, A, B, \{N_A, M, A, B\}_{K_{AS}}$
2. B sends to S: $M, A, B, \{N_A, M, A, B\}_{K_{AS}}, \{N_B, M, A, B\}_{K_{BS}}$
3. S sends to B: $M, \{N_A, K_{AB}\}_{K_{AS}}, \{N_B, K_{AB}\}_{K_{BS}}$
4. B sends to A: $M, \{N_A, K_{AB}\}_{K_{AS}}$

The values N_A and N_B are random *nonce* values chosen by A and B to ensure that their replies from S are new messages and not old ones replayed. The value M is another random nonce chosen by A. It can be seen that A relies on B to relay the messages between her and S. At the end of the protocol it is intended that A and B are both in possession of the shared key K_{AB} and believe it is good for communication with the other. The idealised version of the protocol, as presented by the BAN authors [2], is as follows.

1. A sends to B : $\{N_A, N_C\}_{K_{AS}}$
2. B sends to S : $\{N_A, N_C\}_{K_{AS}}, \{N_B, N_C\}_{K_{BS}}$
3. S sends to B : $\{N_A, (A \overset{K_{AB}}{\leftrightarrow} B), (B \mid\sim N_C)\}_{K_{AS}}$,
   $\{N_B, (A \overset{K_{AB}}{\leftrightarrow} B), (A \mid\sim N_C)\}_{K_{BS}}$
4. B sends to A: $\{N_B, (A \overset{K_{AB}}{\leftrightarrow} B), (A \mid\sim N_C)\}_{K_{BS}}$

In this description $P \mid\sim X$ and $P \overset{K}{\leftrightarrow} Q$ are parts of the BAN logic notation, the former meaning that the principal P sent the message X at some time in the past, and the latter meaning that K is a key good for communication between principals P and Q. For brevity, the three elements M, A, B have been amalgamated into the nonce N_C. It will be noted that there are various differences between the idealised version and the concrete protocol. Firstly, all cleartext information has disappeared. Secondly, message 3 from S has additional elements in the idealised version which cannot be mapped to any part of the concrete message. These have been added by understanding what the messages are meant to convey as well as the actual message elements. For example, A will associate the nonce N_A with user B and hence concludes that the key is meant for use with B. The analysis in the logic of the idealised protocol reveals a subtle difference between the final beliefs of A and B but concludes that both A and B will believe that the key is good for use with the other.

2.2 A Faulty Implementation of the Otway-Rees Protocol

Consider what is the point of sending the cleartext information M, A, B to S in message 2 of the concrete protocol. As to M, it is referred to in the original paper of Otway and Rees [9] as a *conversation identifier* and so can be used by A and B to identify that the messages they receive from S refer to this instance of the protocol. The reason for sending the names A and B is so that S is able to know what keys to use to decrypt the encrypted parts of the message. S then recovers the contents of these encrypted messages. It is an essential part of the protocol that S now checks the correctness of these messages. In [2] it is stated that S should check "whether the components M,A, and B match in the encrypted messages" and a similar (although more ambiguous) statement is contained in [9].

Thus we find it reasonable to make the following interpretation for the precise actions of S.

(a) S uses the cleartext identifiers to choose keys to decrypt the two messages received.
(b) S checks that the fields containing M, A, B are the same in both messages.
(c) S encrypts the new key and respective random numbers using the same keys used in step (a).

Somewhat surprisingly it turns out that if S does only this checking then the protocol is completely insecure. Any user C can masquerade as any other user B by choosing his own nonce N_C and sending the following message as the second one in the protocol.

2. C sends to S: $M, A, C, \{N_A, M, A, B\}_{K_{AS}}, \{N_C, M, A, B\}_{K_{CS}}$

By following the steps performed by S it can easily be seen that the message will be found correct by S and furthermore that C will get the session key K_{AB} encrypted with the key he shares with S, K_{CS}. Once this problem has been seen it is obvious that S must also check that the values M, A, B in the decrypted messages match with the cleartext versions, as well as with each other. It is rather curious to discover that the security of this protocol is critically dependant on cleartext information which can be considered as incidental to the protocol. Indeed, the BAN authors have said [2]:

> We have omitted cleartext information (in idealised protocols) simply because it can be forged, and so its contribution to an authentication protocol is mostly one of providing hints as to what might be placed in encrypted messages.

Our understanding of this guidance on how to perform idealisation leads us to the following dilemma for the user of BAN logic:

1. an unsound protocol, such as Otway-Rees protocol where the actions of S are interpreted by us as above, can be idealised into a "sound" one; the above example is delibrately made to demonstrate this point, or else
2. deleting cleartext should not be conducted as a simple, straightforward and universal treatment for all protocols (as suggested in [2]). Instead, for a flawed protocol like the one given above, the user should have to single it out for an *ad hoc* analysis (as suggested in [10])

We regard the above situation as a dilemma because the user will certainly not be happy with the first dangerous possibility, nor can s/he find it easy to become aware of the unsoundness of a protocol at the time of idealisation (which means to find a bad protocol without using BAN logical manipulation).

2.3 An Attack on a Simplified Version

As a result of their examination of the Otway-Rees protocol the BAN authors noticed that there would be no change in their logical analysis if a number of simplifications were made. In particular they say [2]:

> ...we may notice that there are various forms of redundancy in the protocol. Two nonces are generated by A; however the verification using N_A could just as well have been done using N_C. Therefore, N_A can be eliminated, so reducing the amount of encryption in the protocol. Moreover, it is clear from the analysis that N_B need not be encrypted in the second message.

We now demonstrate that this is definitely not the case. Such a simplification again results in an attack that completely defeats the protocol. According to BAN authors, the protocol attacked is the same as the Otway-Rees protocol except that messages 1 and 2 are now as follows.

1. A sends to B: $M, A, B, \{M, A, B\}_{K_{AS}}$,
2. B sends to S: $M, A, B, \{M, A, B\}_{K_{AS}}, N_B, \{M, A, B\}_{K_{BS}}$

In this attack, the attacker C masquerades as A in the protocol and is also assumed to be in control of communications between B and the server S. The essence of the attack is that C can change the names presented to S while using the nonce that B associates with A. In the version presented here it is assumed that C has possession of a message fragment $\{M, C, B\}_{K_{BS}}$ which was formed by B during a previous legitimate run of the protocol between C and B. (We are making the reasonable assumption that M is a random number selected by A for each run of the protocol. If M were a timestamp a similar real-time attack would be possible.) The attack proceeds as follows, with B and S acting exactly as in a normal run. Messages 2 and 3, which B and S intend for each other, respectively, are captured by C.

1. C sends to B: $M', A, B, \{M, C, B\}_{K_{CS}}$
2. B sends to C: $M', A, B, \{M, C, B\}_{K_{CS}}, N_B, \{M', A, B\}_{K_{BS}}$

2'. C sends to S: $M, C, B, \{M, C, B\}_{K_{CS}}, N_B, \{M, C, B\}_{K_{BS}}$

3. S sends to C: $M, \{M, K_{CB}\}_{K_{CS}}, \{N_B, K_{CB}\}_{K_{BS}}$

3'. C sends to B: $M', \{M, K_{CB}\}_{K_{CS}}, \{N_B, K_{CB}\}_{K_{BS}}$

4. B sends to C: $M', \{M, K_{CB}\}_{K_{CS}}$

At the end of this attack, B believes he shares the key K_{CB} with A whereas he in fact shares it with C. It may be noted that S needs to ignore the replay of

M if the attack is to succeed. Since M is not intended as a nonce for S, who is not supposed to record all of the clients nonces, this is the expected situation.

Looking at the idealised protocol we will see what is wrong this time. According to [2], the idealised protocol should be (note that N_C stands for M, A, B)

1. A sends to B : $\{N_C\}_{K_{AS}}$
2. B sends to S : $\{N_C\}_{K_{AS}}, \{N_C\}_{K_{BS}}$
3. S sends to B : $\{N_C, (A \stackrel{K_{AB}}{\leftrightarrow} B), (B \mid\sim N_C)\}_{K_{AS}}$,
   $\{N_B, (A \stackrel{K_{AB}}{\leftrightarrow} B), (A \mid\sim N_C)\}_{K_{BS}}$
4. B sends to A: $\{N_B, (A \stackrel{K_{AB}}{\leftrightarrow} B), (A \mid\sim N_C)\}_{K_{BS}}$

However, our attacking run shows that the server S has not told B anything like $A \mid\sim N_C$. If S really makes any statement to B for the attacking run, it should be $C \mid\sim N_C$ and $C \stackrel{K_{AB}}{\leftrightarrow} B$. In fact, S can never make such a statement because he actually does not read the syntactic specification of a protocol. What the server can read is messages in a *run*, i.e. an *instance*, of the protocol.

We believe that the idealisation scheme of BAN logic has a fundamental difficulty. This is apparent if we view a protocol specification as analogous to a program specification written in a programming language (e.g. Pascal). The specification only contains *identifiers*, such as principals, nonces, etc. These identifiers will be filled with *real values* during the time of run. Just as inputing a wrong value into a program can result in computational mistakes, running a protocol by filling it with wrong principals or wrong nonces can establish false statements as long as the protocol contains statements about identifiers, just like an idealised protocol under the scheme of BAN logic.

An extension of the BAN logic due to Gong, Needham and Yahalom (GNY) [3] resorted to the same idea of idealisation as BAN. The trivial difference is that in GNY a recipient of a message needs to "convey" a statement of formal parameters from the message.

Here the dilemma for the user of BAN logic still exists, i.e. either to risk the danger of "idealising" a bad protocol into a good one, or find a protocol to be bad without using BAN logical manipulation.

2.4 Implications for BAN logic

Both the above attacks can be easily avoided once they have been spotted. However the idealisation of these protocols is quite reaonsable within the rules defined for the BAN logic. The attacks show that there are instances where the translation depends on subtle factors and may be more difficult than it appears.

If we follow through the logical analysis given in [2] we see that S is able to deduce the origin of the two sub-messages sent in message 2. In the attacks, S can correctly deduce that the attacker C sent his part of message 2. However, in the idealised protocol the message that S sends back to the participants says that in fact it was A and B who said the nonce M and that the key sent is good for communication between A and B. We see that this interpretation is quite unjustified in either of the two attack scenarios.

Thus with hindsight it may be argued that the idealisation for the protocols in the two attacks should not in fact be the same as that of the original protocol where the server checks both plaintext and encrypted principal names. But this merely serves to illustrate how difficult the process is to get right since this idealisation was made by the BAN authors themselves. A formal analysis is not very helpful if the protocols it analyses have to be completely understood before analysis can begin.

3 Solutions

We regard the most effective and lasting solution to the problems identified by the above examples to be the development of a new approach to logical analysis of protocols which does not require an idealisation step in the same way as in the BAN approach. Such an approach is detailed in [5].

A second approach is to limit the kinds of protocols under consideration to include only a few message formats whose semantics are well understood. Such an approach is suggested by the lessons learned from formal analyis of computer programs. It is much easier to design a system to be correct in the first place rather than to take some given system and prove that it is correct *post hoc*. Our analysis has revealed that the Otway-Rees protocol is difficult to analyse because the actions of S are unnecessarily complex and the semantics of the messages from S are consequently difficult to pin down. We intend to explore this approach in more detail in future work.

Acknowledgements Anmar Alani carried out the formal specification of the Otway-Rees protocol which led to discovery of the first attack. We are grateful to Professor Roger Needham for helpful comments on an earlier draft of this paper and to Paul Van Oorschot for numerous interesting and lively discussions.

References

1. Colin Boyd, *A Formal Framework for Authentication*, Computer Security - ESORICS 92, pp.273-292, Springer-Verlag, 1992.
2. M.Burrows, M.Abadi, and R.Needham, *A Logic of Authentication*, Proceedings of the Royal Society, Vol A426,pp 233-271,1989.
3. L.Gong, R.Needham & R.Yahalom, *Reasoning about Belief in Cryptographic Protocols*, Proceedings of IEEE Symposium on Research in Security and Privacy, pp.234-248, 1990.
4. R.Kailar & V.D.Gligor, *On Belief Evolution in Authentication Protocols*, Proceedings of IEEE Symposium on Research in Security and Privacy, pp.103-116, 1991.
5. Wenbo Mao & Colin Boyd, *Towards Formal Analysis of Security Protocols*, Proceedings of Sixth IEEE Workshop on Foundations of Computer Security, pp 147-158, June, 1993.

6. Chris Mitchell & Andy Thomas, *Standardising Authentication Protocols based on Public Key Techniques*, circulated within BSI IST/33/-2, 1992.
7. R.M.Needham, *Reasoning about Cryptographic Protocols*, Presented at ESORICS 92.
8. D.M.Nessett, *A Critique of the Burrows, Abadi and Needham Logic*, ACM Operating Systems Review, 24,2,pp.35-38,1990.
9. Dave Otway & Owen Rees, *Efficient and Timely Mutual Authentication* ACM Operating Systems Review, 21,1,pp.8-10, 1987.
10. Paul van Oorschot, An Alternate Explanation of two BAN-logic "failures", Talk delivered at the Rump Session of Eurocrypt 93.

Efficient Anonymous Channel and All/Nothing Election Scheme

Choonsik PARK, Kazutomo ITOH and Kaoru KUROSAWA

Department of Electrical and Electronic Engineering,
Faculty of Engineering, Tokyo Institute of Technology
2-12-1 O-okayama, Meguro-ku, Tokyo 152 Japan
parkcs@ss.titech.ac.jp
kkurosaw@ss.titech.ac.jp

Abstract. The contribution of this paper are twofold. First, we present an efficient computationally secure anonymous channel which has no problem of ciphertext length expansion. The length is irrelevant to the number of MIXes (control centers). It improves the efficiency of Chaum's election scheme based on the MIX net automatically. Second, we show an election scheme which satisfies fairness. That is, if some vote is disrupted, no one obtains any information about all the other votes. Each voter sends $O(nk)$ bits so that the probability of the fairness is $1-2^{-k}$, where n is the bit length of the ciphertext.

1 Introduction

Chaum showed a computationally secure anonymous channel called a MIX net [1]. It hides even the traffic pattern, that is, who sends whom. The MIX net consists of a series of control centers called MIXes. However, the length of the ciphertext which each sender sends is very large. It grows proportionally to the number of MIXes.

Anonymous channels and election schemes are closely related to each other. An anonymous channel hides the correspondences between the senders and the receivers. An election scheme hides the correspondences between the voters and the content of each vote. From this point of view, Chaum proposed an election scheme based on the MIX net [1]. However, the election scheme based on the MIX net provides very low level of correctness. It doesn't satisfy even fairness. That is, suppose that only one vote is disrupted. Still, everyone can know all the other votes in his election scheme. Then, this information will influence the re-election greatly.

Chaum showed another anonymous channel called a DC net [3], and an election scheme based on the DC net [2]. While the DC net is unconditionally secure, the participants must share random numbers beforehand. It also has a problem of message collision. The election scheme based on the DC net has the same problems.

Benaloh showed a totally different yes/no election scheme which is based on zero knowledge interactive proof systems (ZKIP) and secret sharing schemes [4].

Benaloh's scheme provides very high level of correctness, that is, fault tolerancy. The total number of yes votes is successfully obtained even if less than a half of control centers (corresponding to MIXes) are dishonest. However, the disadvantage of Benaloh's scheme is efficiency. Let p_i be the cheating probability of voter i. To obtain that $p_i \leq 2^{-k}$, each voter has to send $O(nkN)$ bits, where n = the size of each ciphertext and N = the number of the control centers.

The contribution of this paper are twofold. First, we present an efficient computationally secure anonymous channel which has no problem of ciphertext length expansion. The length is irrelevant to the number of MIXes. It improves the efficiency of Chaum's election scheme based on the MIX net automatically. Second, we show an election scheme which satisfies the fairness. That is, if some vote is disrupted, no one obtains any information about all the other votes. Each voter sends $O(nk)$ bits so that the probability of the fairness is $1 - 2^{-k}$, where n is the bit length of the ciphertext.

2 Chaum's Work

2.1 Basic Usage of Public Key

Let E_A be a public key and E_A^{-1} be a secret key of Alice. We assume that, for any X,

$$E_A^{-1}E_A(X) = E_AE_A^{-1}(X) = X. \tag{1}$$

Let M_i be a plaintext and C_i be the ciphertext ($1 \leq i \leq n$). Suppose that M_i and C_i are made public. Also suppose that n is small enough. When we want to hide the correspondence between M_i and C_i, each M_i should be encrypted as follows.

$$C_i = E_A(M_i \circ R_i),$$

where R_i is a random number. If R_i is not attached, it is easy to find the correspondence between M_i and C_i.

The digital signature for a random number M can be given by

$$D = E_A^{-1}(M \circ 0^l).$$

Everyone can verify the validity of the signature by forming

$$E_A(D) = M \circ 0^l$$

and by checking 0^l, where l is a sufficiently large number.

2.2 Anonymous MIX Channel

Chaum showed a scheme which hides even the traffic pattern. The model is as follows. There are n senders, $A_1, \ldots, A_n$. Each A_i wants to send a message m_i to a receiver B_i in such a way that the correspondence between A_i and B_i is kept secret. It is assumed that there exists a shuffle machine agent S_1 (called a MIX). Let the public key of B_i be E_{B_i} and the public key of S_1 be E_1.

An anonymous channel is realized by the following protocol.

[Simple MIX Anonymous Channel]

Step 1. Each A_i chooses a random number R and writes

$$C_i = E_1(R \circ B_i \circ E_{B_i}(m_i)) \quad (2)$$

on the public board.

Step 2. S_1 decrypts it, throws away R, and writes $\{B_i \circ E_{B_i}(m_i)\}$ on the public board in a lexicographical order.

In this protocol, anyone except for S_1 cannot see the correspondence between $\{A_i\}$ and $\{B_i\}$. To hide the correspondence even from S_1, k MIXes $S_1, \ldots, S_k$ are used. The protocol is as follows. Let E_i be the public key of S_i.

[k MIXes Anonymous Channel]

Step 1. Each A_i chooses random numbers $R_1, \ldots, R_k$ and writes

$$E_1(R_1 \circ E_2(R_2 \cdots E_k(R_k \circ B_i \circ E_{B_i}(m_i)) \cdots))$$

on the public board. (We say that A_i sends $B_i \circ E_{B_i}(m_i)$ to the k MIXes anonymous channel.)

Step 2. S_1 writes

$$E_2(R_2 \cdots E_k(R_k \circ B_i \circ E_{B_i}(m_i)) \cdots)$$

on the public board in a lexicographical order.

Step 3. $S_2, S_3, \ldots,$ and S_{k-1} execute the same job as Step 2 in sequence.

Step 4. Finally, S_k writes $B_i \circ E_{B_i}(m_i)$ on the public board in a lexicographical order.

In this protocol, if at least one MIX is honest, the correspondence between $\{A_i\}$ and $\{B_i\}$ is kept secret even from the MIXes.

2.3 Election Scheme

Chaum proposed an election scheme based on the k MIXes anonymous channel. In the k MIXes anonymous channel, if S_k is dishonest, S_k may write something other than $B_i \circ E_{B_i}(m_i)$ on the public board. A_i can detect this error. However, if A_i claims, S_k can know the correspondence between A_i and B_i because S_k knows B_i. This is a serious problem if the anonymous channel is used for an election scheme. To overcome this problem, Chaum proposed the following election scheme.

Let P_i be a voter and V_i be his vote.

(Registration phase)

Step 1. Each P_i chooses (K_i, K_i^{-1}), where K_i is a public key and K_i^{-1} is the secret key. P_i writes

$$E_1(R_1 \circ E_2(R_2 \cdots E_k(R_k \circ K_i) \cdots))$$

on the public board with his digital signature. (P_i sends K_i to the k MIXes anonymous channel. In step 1 of the k MIXes anonymous channel, $B_i \circ E_{Bi}(m_i)$ is replaced by K_i.)

Step 2. The k MIXes anonymous channel shuffles $\{K_i\}$ in secret. (Step 2 and 3 of the k MIXes anonymous channel are executed.)

Step 3. S_k writes K_i on the public board in the lexicographical order.

Let the list be $(\hat{K}_1, \hat{K}_2, \ldots)$.
(Claiming phase)

Step 4. Each P_i checks that his K_i is in the list on the public board. If not, P_i claims and the election stops. If there are no claims in some period, goto the next phase.

(Voting phase)

Step 5. Each P_i writes

$$E_1(R_1 \circ E_2(R_2 \cdots E_k(R_k \circ (K_i \circ K_i^{-1}(V_i \circ 0^l))) \cdots))$$

on the public board with his digital signature. (P_i sends $K_i \circ K_i^{-1}(V_i \circ 0^l$) to the k MIXes anonymous channel.)

Step 6. After the deadline of the voting period, the k MIXes anonymous channel shuffles $K_i \circ K_i^{-1}(V_i \circ 0^l)$ in secret.

Step 7. S_k writes $K_i \circ K_i^{-1}(V_i \circ 0^l)$ on the public board in the lexicographical order. Let the list be $(u_1 \circ v_1), (u_2 \circ v_2), \ldots$.

Step 8. Everyone checks that $u_i = \hat{K}_i$, and $u_i(v_i) = * \cdots * 0^l$ for each i. If the check fails, stop.

Step 9. It is easy for everyone to obtain $\{V_1, \ldots, V_n\}$.

Remark. At Step 1 and Step 5, digital signatures are necessary to check the voters' identities.

3 Proposed Anonymous Channels

The problem of the k MIXes anonymous channel shown in 2.2 is that each sender A_i has to send a very long message at step 1. The length of $E_1(R_1 \circ E_2(R_2 \cdots E_k(R_k \circ B_i \circ E_{B_i}(m_i)) \cdots))$ is proportional to k, which is the number of the MIXes.

In this section, we will present an anonymous channel which has no problem of such ciphertext length expansion.

[Proposed Anonymous Channel]

The proposed scheme makes use of ElGamal cryptosystem. The authority publishes (q, g, c), where

- q is a large prime number.
- g is a primitive element of $GF(q)$.
- c is the factorization of $q-1$. (Everyone can check that g is a primitive element by using c.)

(Secret key of S_i) $X_i \in \{1, \ldots, q-1\}$
(Public key of S_i) Y_i $(= g^{X_i} \bmod q)$
(S_i chooses X_i and publicizes Y_i.)

Step 1. Each sender A_i chooses a random number R and computes

$$(C_{0i}, C_{1i}) \triangleq (g^R, (B_i \circ E_{B_i}(m_i)) \times (Y_1 \cdots Y_k)^R)$$

A_i writes (C_{0i}, C_{1i}) on the public board. Define $f_j(t, u, r)$ as

$$f_j(t,u,r) \triangleq \begin{cases} (t \times g^r, u \times (Y_{j+1} \cdots Y_k)^r / t^{X_j}) & \text{if } 1 \leq j \leq k-1 \\ u/t^{X_k} & \text{if j = k} \end{cases}$$

For $i = 1, \ldots, k$, do the following.

Step 2. Let the latest list on the public board be

$$(t_1, u_1), (t_2, u_2), \ldots, (t_n, u_n).$$

S_i chooses random numbers $r_1, \ldots, r_n$ and computes $f_i(t_j, u_j, r_j)$ for each j.

Step 3. S_i writes $\{f_i(t_j, u_j, r_j)\}$ $(j = 1, \ldots, n)$ on the public board in a lexicographical order.

Finally, we have a list of $\{B_i \circ E_{B_i}(m_i)\}$ in a lexicographical order on the public board.

In this protocol, (C_{0i}, C_{1i}) changes as follows for some random numbers $R_1, \ldots, R_{k-1}$.

$$\begin{aligned} (C_{0i}, C_{1i}) &= (g^R, (B_i \circ E_{B_i}(m_i)) \times (Y_1 \cdots Y_k)^R) \\ &\rightarrow (g^{R_1}, (B_i \circ E_{B_i}(m_i)) \times (Y_2 \cdots Y_k)^{R_1}) \\ &\vdots \\ &\rightarrow (g^{R_{k-1}}, (B_i \circ E_{B_i}(m_i)) \times Y_k^{R_{k-1}}) \\ &\rightarrow B_i \circ E_{B_i}(m_i) \end{aligned}$$

Note that $|(C_{0i}, C_{1i})| = 2 \times |q|$. Thus, the proposed anonymous channel has no problem of the ciphertext length expansion. It is also easy to see that, if there exists at least one honest S_i, the correspondence between A_i and B_i is kept secret from any adversary.

4 Proposed Election Scheme

The proposed anonymous channel of Sect. 3 can be directly applied to Chaum's election scheme in subsection 2.3. Then, the communication complexity is improved automatically.

However, the Chaum's election scheme has a problem of fairness as mentioned in the Introduction. That is, suppose that only V_1 is disturbed by S_k. Then, from the final list on the public board, everyone knows that some vote has been disrupted. However, at the same time, everyone knows $\{V_2, \ldots, V_n\}$. This information (for example, the number of yes votes and that of no votes) will affect the re-election greatly.

Let's study this problem more in detail. For simplicity, suppose that each voter P_i is honest. (It is clear that P_i cannot vote more than one vote in Chaum's scheme.) Consider the following two events.(We assume that there are some undisrupted votes.)

Event 1 : Some vote cannot be recovered.

Event 2 : Some undisrupted vote is made public.

Define P_d as follows.

$$P_d \triangleq P_r[\text{ Event } 2 \mid \text{Event } 1 \text{ }].$$

In the Chaum's scheme, if S_k behaves as above, then always $P_d = 1$.

This section will present an election scheme such that P_d is negligibly small. We gives a high level description of the proposed election scheme in this section. The details will be given in the next section. The proposed election scheme consists of three phases as Chaum's scheme of 2.3 does. Our registration phase and claiming phase are the same as those of Chaum's scheme. In what follows, we will show our voting phase protocol. In addition to $S_1, \ldots, S_k$, we use S_0 whose role is to flip a coin. (Instead of S_0, we can use a collective coin flipping protocol. Such S_0 or a coin flipping protocol is also necessary in Benaloh's election scheme [4].) In this protocol, we use a variation of the anonymous channel proposed in Sect. 3.

4.1 Proposed Voting Phase Protocol (1)

First, we will present our voting phase protocol which achieves that $P_d \leq 1/2$.

Step 1. Each P_i chooses two random numbers R_{i1} and R_{i2} such that

$$V_i = R_{i1} \oplus R_{i2}, \tag{3}$$

where $\oplus$ denotes bitwise exclusive OR.

Step 2. Each P_i sends the ciphertexts of $R_{i1} \circ 0^l$ and $R_{i2} \circ 0^l$ to $S_1 \sim S_k$. (A group public key cryptosystem given in 5.1 is used.)

Step 3. After the deadline of the voting period, $S_1 \sim S_k$ shuffles the ciphertexts of ($(R_{i1} \circ 0^l), (R_{i2} \circ 0^l)$) in secret.

Step 4. At this moment, we have a secretly shuffled list of ciphertexts of $((\hat{R_{11}} \circ 0^l), (\hat{R_{12}} \circ 0^l)), ((\hat{R_{21}} \circ 0^l), (\hat{R_{22}} \circ 0^l)), \ldots$.
For each i, one of $\hat{R}_{i1} \circ 0^l$ and $\hat{R}_{i2} \circ 0^l$ is randomly chosen and made open. More precisely, S_0 flips a coin for each i. If the coin is head, $S_1 \sim S_k$ decrypt the ciphertext of $\hat{R_{i1}} \circ 0^l$ and make it open. Otherwise, $\hat{R_{i2}} \circ 0^l$ is made open.

Step 5. Everyone checks the form of 0^l of the decrypted pieces (in the same way as step 8 of the protocol in 2.3). If some disruption is detected, the protocol stops.

Step 6. Otherwise, for each i, the remained pieces are made open. Then, the form of 0^l is checked. (The same check as step 8 in 2.3 is done.)

Step 7. For each i such that no disruption is detected for both pieces, V_i is obtained from $R_{i1} \circ 0^l$ and $R_{i2} \circ 0^l$ by using eq. (3).

Remark. Voter's identity checking is done in the same way as in Chaum's election scheme by using digital signatures.

Example 1. Let the number of voters be 3.
[Step 1 and 2.] (Voting)

voter 1 $(R_{11}, R_{12}) \Rightarrow$ *anonymous channel*
voter 2 $(R_{21}, R_{22}) \Rightarrow$ *anonymous channel*
voter 3 $(R_{31}, R_{32}) \Rightarrow$ *anonymous channel*

[Step 3.] (Shuffling)

$$(\boxed{R_{31}}, \boxed{R_{32}}), (\boxed{R_{11}}, \boxed{R_{12}}), (\boxed{R_{21}}, \boxed{R_{22}})$$

[Step 4.] (Cut and Choose)

$$(R_{31}, \boxed{R_{32}}), (\boxed{R_{11}}, R_{12}), (R_{21}, \boxed{R_{22}})$$

R_{31}, R_{12} and R_{21} are made open.

[Step 6.] (Opening)

$$(R_{31}, R_{32}), (R_{11}, R_{12}), (R_{21}, R_{22})$$

R_{32}, R_{11} and R_{22} are made open.

[Step 7.] (Reconstruction)

$$\hat{V}_1 = R_{31} \oplus R_{32}$$

$$\hat{V}_2 = R_{11} \oplus R_{12}$$

$$\hat{V}_3 = R_{21} \oplus R_{22}$$

Theorem 1. *In the above protocol, $P_{d1} \leq 1/2$.*

Proof. Note that
$P_d = P_r\{$ No disruption is detected at Step 5 | Event 1 $\}$.
Event 1 occurs if some dishonest S_j has rewritten at least one element of $\{R_{i1} \circ 0^l\} \cup \{R_{i2} \circ 0^l\}$. Suppose that one element of $\{R_{i1} \circ 0^l\} \cup \{R_{i2} \circ 0^l\}$ is disrupted. Then, this cheating is detected at Step 4 and Step 5 with probability 1/2. □

4.2 Proposed Voting Phase Protocol (2)

Next, we will show our voting phase protocol which achieves that $P_d \leq 1/2^h$, where h is a security parameter.

Step 1. Each P_i chooses h pairs of random numbers $(R_{11}, R_{21}), \ldots, (R_{1h}, R_{2h})$ such that

$$V_i = R_{11} \oplus R_{21} = \cdots = R_{1h} \oplus R_{2h}, \tag{4}$$

where $\oplus$ denotes bitwise exclusive OR.

Step 2. Each P_i sends the ciphertexts of

$$((R^i_{11} \circ 0^l, R^i_{21} \circ 0^l), \ldots, (R^i_{1h} \circ 0^l, R^i_{2h} \circ 0^l))$$

to $S_1 \sim S_k$.

Step 3. The anonymous channel shuffles

$$\{(R^i_{11} \circ 0^l, R^i_{21} \circ 0^l), \ldots, (R^i_{1h} \circ 0^l, R^i_{2h} \circ 0^l)\}$$

in secret.

Step 4. For each j, one of $R^i_{1j} \circ 0^l$ and $R^i_{2j} \circ 0^l$ is randomly chosen and made open (for $\forall i$).

Step 5. Check the form of 0^l of the opened pieces as Step 8 in 2.3. If some disruption is detected, stop.

Step 6. Open all of $R^i_{1j} \circ 0^l \cup R^i_{2j} \circ 0^l$. Check the form of 0^l.

Step 7. Let

$$G(i) \triangleq \{j \mid \text{No disruption is detected both for } R^i_{1j} \circ 0^l \text{ and } R^i_{2j} \circ 0^l\}.$$

$$J(i) \triangleq \mathbf{min} G(i) \text{ if } \mid G(i) \mid \geq 1.$$

V_i is reconstructed as $R^i_{1J(j)} \oplus R^i_{2J(i)}$.

Example 2. [Step 1 and 2.] (Voting)

voter 1 $(R^1_{11}, R^1_{21}), \ldots\ldots, (R^1_{1h}, R^1_{2h}) \Rightarrow$ *anonymous channel*
voter 2 $(R^2_{11}, R^2_{21}), \ldots\ldots, (R^2_{1h}, R^2_{2h}) \Rightarrow$ *anonymous channel*
voter 3 $(R^3_{11}, R^3_{21}), \ldots\ldots, (R^3_{1h}, R^3_{2h}) \Rightarrow$ *anonymous channel*

[Step 3.] (Shuffling)

$(\boxed{R^3_{11}}, \boxed{R^3_{21}}), \ldots\ldots\ldots, (\boxed{R^3_{1h}}, \boxed{R^3_{2h}})$
$(\boxed{R^1_{11}}, \boxed{R^1_{21}}), \ldots\ldots\ldots, (\boxed{R^1_{1h}}, \boxed{R^1_{2h}})$
$(\boxed{R^2_{11}}, \boxed{R^2_{21}}), \ldots\ldots\ldots, (\boxed{R^2_{1h}}, \boxed{R^2_{2h}})$

[Step 4.] (Cut and Choose)

$$(R^3_{11}, \boxed{R^3_{21}}), \ldots\ldots\ldots, (\boxed{R^3_{1h}}, R^3_{2h})$$
$$(\boxed{R^1_{11}}, R^1_{21}), \ldots\ldots\ldots, (R^1_{1h}, \boxed{R^1_{2h}})$$
$$(\boxed{R^2_{11}}, R^2_{21}), \ldots\ldots\ldots, (\boxed{R^2_{1h}}, R^2_{2h})$$

[Step 6.] (Opening)

$$(R^3_{11}, R^3_{21})$$
$$(\mathbf{Error}, R^1_{21}) \Rightarrow (R^1_{12}, \mathbf{Error}) \Rightarrow (R^1_{13}, R^1_{23})$$
$$(\mathbf{Error}, R^2_{21}) \Rightarrow (R^2_{12}, R^2_{22})$$

Error means that some disruption is detected.

[Step 7.] (Reconstruction)

$$\hat{V}_1 = R^3_{11} \oplus R^3_{21}$$

$$\hat{V}_2 = R^1_{13} \oplus R^1_{23}$$

$$\hat{V}_3 = R^2_{12} \oplus R^2_{22}$$

Theorem 2. *In the above protocol, $P_d \leq 1/2^h$.*

Proof. Note that

$P_d = P_r\{$ no disruption is detected at Step 5 | there exists V_a such that both or one of R^a_{1i} and R^a_{2i} is disrupted for $1 \leq \forall i \leq h\}$.

Suppose that there exists V_a such that one of R^a_{1i} and R^a_{2i} is disrupted for $1 \leq \forall i \leq h$. This disruption is detected at Step 5 with probability $1/2^h$. □

5 Full Description of the Proposed Election Scheme

The proposed election scheme uses a modification of the anonymous channel given in Sect.3. The modified anonymous channel makes use of a group public key cryptosystem [5].

5.1 Group Public Key Cryptosystem

Remember that we have used

(**Common public information**) p, g, c
(**Secret key of** S_i) $X_i \in \{1, \ldots, q-1\}$
(**Public key of** S_i) Y_i $(= g^{X_i} \bmod q)$

in Sect.3. This setting is the same as the group public key cryptosystem in [5]. The public key of the group is $Y_1 \cdots Y_k$. All S_i have to cooperate to decrypt ciphertexts.

Let m be a plaintext. The ciphertext of the group public key cryptosystem is given by

$$E(m, r) \triangleq (g^r, m_i \times (Y_1 \cdots Y_k)^r) \bmod q,$$

where r is a random number. The decryption protocol is given as follows. Let

$$a \triangleq g^r \bmod q,$$
$$b \triangleq m \times (Y_1 \cdots Y_k)^r \bmod q.$$

[Decryption Protocol]

Step 1. Each S_i computes $Z_i = a^{X_i} (= (g^r)^{X_i} = Y_i^r \bmod q)$ and makes Z_i open.
Step 2. Everyone computes
$b/(Z_1 \cdots Z_k) = m \times (Y_1 \cdots Y_k)^r/(Z_1 \cdots Z_k) = m.$

5.2 One more tool

Let

$$h(a, b, e) \triangleq (a \times g^e, b \times (Y_1 \cdots Y_k)^e) \bmod q.$$

Lemma 3. *If* $(a, b) = E(m, r)$, *then* $h(a, b, e) = E(m, r + e)$.

The proof is immediate.

From this Lemma 3, we see that applying h to $E(m, r)$ successively several times yields $E(m, x)$ for some x.

5.3 Modified Anonymous Channel

We show a modification of the anonymous channel shown in Sect. 3, which will be used in the next subsection.

Step 1. Each sender A_i writes $E(B_i \circ E_{B_i}(m_i), r_i)$ on the public board, where r_i is a random number.
Step 2. S_1 chooses random numbers $e_1, e_2, \ldots$, and computes

$$h(E(B_i \circ E_{B_i}(m_i), r_i), e_i) = E(B_i \circ E_{B_i}(m_i), r_i + e_i)$$

for each i. S_1 writes $\{E(B_i \circ E_{B_i}(m_i), r_i + e_i)\}$ on the public board in a lexicographical order.
Step 3. $S_2 \sim S_k$ do the same job in sequence. Then, we have a list of $\{E(B_i \circ E_{B_i}(m_i), x_i)\}$ in a lexicographical order on the public board, where x_i is a random number.
Step 4. $S_1 \sim S_k$ obtain $\{B_i \circ E_{B_i}(m_i)\}$ by executing the decryption algorithm in 5.1.
If at least one S_j is honest, nobody knows the correspondence between A_i and B_i.

5.4 Details of the Election Scheme in 4.1

We show the details of the election scheme shown in 4.1. The details of the protocol of 4.2 will be obtained similarly.

Step 1. Each voter P_i chooses two random numbers R_{i1} and R_{i2} such that

$$V_i = R_{i1} \oplus R_{i2}.$$

Step 2. Each P_i chooses r_{i1} and r_{i2} randomly. He computes

$$(a_{i1}, b_{i1}) = E(K_i \circ K_i^{-1}(R_{i1} \circ 0^l), r_{i1})$$

$$(a_{i2}, b_{i2}) = E(K_i \circ K_i^{-1}(R_{i2} \circ 0^l), r_{i2})$$

and writes them on the public board.
At this moment, there is a list on the public board such that

$$((a_{11}, b_{11}), (a_{12}, b_{12})), ((a_{21}, b_{21}), (a_{22}, b_{22})), \ldots.$$

Step 3. For $i = 1, \ldots, k$, do the following in sequence.
Let the latest list on the public board be

$$((\alpha_{11}, \beta_{11}), (\alpha_{12}, \beta_{12})), ((\alpha_{21}, \beta_{21}), (\alpha_{22}, \beta_{22})), \ldots.$$

S_i computes

$$(\hat{\alpha}_{j1}, \hat{\beta}_{j1}) = h(\alpha_{j1}, \beta_{j1}, e_{j1})$$

$$(\hat{\alpha}_{j2}, \hat{\beta}_{j2}) = h(\alpha_{j2}, \beta_{j2}, e_{j2})$$

for each j, where e_{j1} and e_{j2} are random numbers. S_i writes

$$\{((\hat{\alpha}_{j1}, \hat{\beta}_{j1}), (\hat{\alpha}_{j2}, \hat{\beta}_{j2}))\}$$

on the public board in a lexicographical order.
Step 4. Let the list on the public board at this moment be

$$((\hat{\alpha}_{11}, \hat{\beta}_{11}), (\hat{\alpha}_{12}, \hat{\beta}_{12})), ((\hat{\alpha}_{21}, \hat{\beta}_{21}), (\hat{\alpha}_{22}, \hat{\beta}_{22})), \ldots.$$

S_0 chooses a random bit d_i for each i. By using the decryption protocol given in 5.1,

$$S_1, \ldots, S_k \text{ decrypt } (\hat{\alpha}_{i1}, \hat{\beta}_{i1}) (= E(\hat{K}_i \circ \hat{K}_i^{-1}(\hat{R}_{i1} \circ 0^l), x_{i1})), \text{ if } d_i = 0$$

$$S_1, \ldots, S_k \text{ decrypt } (\hat{\alpha}_{i2}, \hat{\beta}_{i2})(= E(\hat{K}_i \circ \hat{K}_i^{-1}(\hat{R}_{i2} \circ 0^l), x_{i2}), \text{ if } d_i = 1.$$

Step 5. Everyone checks the form of 0^l of the decrypted pieces (in the same way as Step 8 of the protocol in 2.3). If some disruption is detected, the protocol stops.

Step 6. Otherwise, for each i, the remained pieces are made open. Then, the form of 0^l is checked. (The same check as Step 8 in 2.3 is done.)

Step 7. For each i such that no disruption is detected for both pieces, V_i is obtained from $R_{i1} \circ 0^l$ and $R_{i2} \circ 0^l$ by using eq. (3).

6 Conclusion

First, we have presented an efficient computationally secure anonymous channel which has no problem of ciphertext length expansion. The length is irrelevant to the number of MIXes. It improves the efficiency of Chaum's election scheme based on the MIX net automatically. Second, we have shown an election scheme which satisfies the fairness. That is, if some vote is disrupted, no one obtains any information about all the other votes. Each voter sends $O(nk)$ bits so that the probability of the fairness is $1 - 2^{-k}$, where n is the bit length of the ciphertext.

References

1. Chaum, D.L.: Untraceable Electronic Mail, Return Addresses, and Digital Pseudonyms. Communications of the ACM, Vol. 24, No.2, (1981), 84–88
2. Chaum, D.L.: Elections with Unconditionally-Secret Ballots and Disruption Equivalent to Breaking RSA. Advance in Cryptology — EUROCRYPT'89, (1989), 177–182
3. Chaum, D.L. : The Dining Cryptographers Problem: Unconditional sender and Recipient Untraceability. Journal of Cryptology, Vol.1, No.1, (1988), 65–75
4. Benaloh, J.C. : Secret Sharing Homomorphisms: Keeping Shares of a Secret Secret. Advance in Cryptology — CRYPTO'86, (1986), 251–260
5. Desmedt, Y., Frankel, Y. : Threshold cryptosystems. Advance in Cryptology — CRYPTO'89, (1990), 307–315

Untransferable Rights in a Client-Independent Server Environment

Josep Domingo-Ferrer

Informàtica de Recerca i Docència
Universitat de Barcelona
Travessera de les Corts, 131-159 (Pavelló Rosa)
08028 Barcelona

Abstract. A scheme for ensuring access rights untransferability in a client-server scenario with a central authority and where servers hold no access information about clients is presented in this paper; an extension to a multi-authority scenario is conceivable, since servers are also authority independent. Usurping a right with no information at all about other clients is for a client as hard as the discrete logarithm, and rights sharing between clients does not compromise their non-shared rights as long as RSA confidentiality holds. Transferring rights between clients without the authority's contribution cannot be done unless RSA confidentiality is broken; however, only control on *partial* rights transfers is addressed in this paper, which does not deal with *total* identity transfer or alienation.

1 Introduction

In a distributed computing system, the entities that require identification are hosts, users and processes —see [Woo 92][Linn 90] for a more detailed framework. When one of those entities requests a service from another entity, we will use the term *client* to denote the first entity, and the term *server* to denote the second entity. A server will provide the requested service only after checking that the would-be client possesses a *right* to obtain that service from him.

Consider a typical distributed scenario consisting of a large network with a central authority, a set of servers giving access to certain resources, and a community of clients. Clients are granted rights by the authority, and servers need only a certified list of available access rights in order to perform access control. *Servers store no access information about clients*, neither access lists nor capabilities, and thus the authority is able to perform client registration, rights granting and rights revocation independently of servers; in addition, the latter two are public operations. Finally, it is also thinkable that servers do not depend on the authority, *i. e.*, that they store no confidential information about the authority.

Keeping no access information in the servers is not common in conventional access control schemes such as [Harr 76] or [Grah 72], so that server control on the rights transfers between clients —like the one implemented with *copy flags* in the [Harr 76] version of the access matrix model— is not feasible. Now the question is: *How to achieve rights untransferability in a scenario where servers are client-independent?*

Remark 1. Note that if servers are trusted and client-dependent, *i.e.* if they hold some kind of access matrix exclusively updated by an authority, then the authority can trivially enforce untransferability by just having each client request checked by the servers against the client's rights in the access matrix. So, no right can be successfully transferred without the authority's contribution.o

The mechanism presented in this article fulfills all the requirements of the distributed scenario above, with the only additional constraint that servers be able to securely hold a private RSA key. The degree of security is such that

- Usurping a right with no information at all about other clients is for a client as hard as solving a discrete logarithm.
- As long as RSA confidentiality holds, rights sharing between clients does not compromise their non-shared rights.
- As long as RSA confidentiality holds, for a client to transfer some of her rights to another client, the transfer must be performed by the authority.

Remark 2. Our primary goal is to prevent a client c_k from unauthorizedly transferring *some* of her rights to another client c_l. Notice that it is always possible for c_k to completely reveal her identity to c_l, so that c_l could use all rights belonging to c_k, by impersonating her. The problem of *total* identity transfer or alienation will not be dealt with here; this possibility always exists because in our context c_k's *identity* consists of a secret number a_k owned by c_k and only shared with the authority.o

The initial assumptions for the scheme are listed in section 2. Section 3 contains the scheme itself along with a theorem on security when a client has no information at all about other clients. Section 4 assesses the risks of rights sharing. Untransferability is dealt with in section 5, where an algorithm to perform rights transfers with the authority's contribution is given as well. Finally, section 6 contains a functional summary and an extension of the system for a scenario with several authorities.

2 Initial Setting

Definition 1. A server is said to be client-independent if it does not store protected access information about its potential clients (neither access lists nor capabilities).

As it was pointed out above, client independence allows the authority to register clients, as well as granting and revoking rights to them without having to communicate secretly with every server.

Consider two public numbers p and α, with p a large prime and α a generator of $\mathcal{Z}/(p)^*$. Take $p \gg (mn)^2$, where m is the number of clients and n is the number of rights.

Let N be a public RSA modulus [Rive 78], *i. e.* $N = q_1 q_2$, where q_1 and q_2 are two secret large primes; choose N to be greater than p. Take an RSA key pair (e, d) with modulus N, so that e is *public* and d is *only known to the servers.*

Definition 2. Let $E_{e,N}()$ and $D_{d,N}()$ be the usual RSA encryption and decryption functions, such that $E_{e,N}(y) = y^e \bmod N$ and $D_{d,N}(y) = y^d \bmod N$.

3 The Identification Scheme

The mathematical structure used is a modification of the one described in [Domi 91]: thanks to the use of RSA, the scheme becomes much simpler, but also dependent on the difficulty of factoring and on the servers securely storing a private RSA key. The algorithm in [Domi 91] relies solely on the discrete logarithm problem, but does not address untransferability.

Let *Auth* be the central network authority. In the presence of several clients $c_0, \cdots, c_{m-1}$ a way must be found to be able to grant the same right to more than one client, while keeping a single numerical expression y for it (the rights are also client-independent). Algorithm 1 gives a solution to this problem. Let us make some definitions before giving the algorithm.

Define n_k as the number of rights to be granted to client c_k, for $k = 0$ to $m-1$. Also, assuming that rights are granted first to c_0, then to c_1 and so on, let t_k be the number of rights, among the n_k to be granted to c_k, that have already been granted to some client in $\{c_0, \cdots, c_{k-1}\}$. Now *Auth* runs

Algorithm 1 *For* $k = 0$ *to* $m - 1$

1. *Assume that the* t_k *rights having been already granted to someone else are* $y_{k_0}, \cdots, y_{k_{t_k-1}}$. *Choose* $n_k - t_k$ *random integers* $x_{k_i}, t_k \leq i \leq n_k - 1$ *over* $\mathcal{Z}/(p-1)$.
2. *Pick a random* a_k *over* $\mathcal{Z}/(p-1)$, *such that* a_k *is prime to* $p-1$.
3. *Generate* n_k *random integers* r_{k_i} *over* $\mathcal{Z}/(p-1)$, *for* $0 \leq i \leq n_k - 1$.
4. *Find* n_k *nonzero numbers* z_{k_i} *over* $\mathcal{Z}/(p-1)$, *for* $0 \leq i \leq n_k - 1$, *such that*

$$x_{k_0} + r_{k_0} = a_k z_{k_0} \bmod (p-1) \tag{1}$$

$$\vdots$$

$$x_{k_{n_k-1}} + r_{k_{n_k-1}} = a_k z_{k_{n_k-1}} \bmod (p-1)$$

   To compute z_{k_i} *solve the i-th equation for* z_{k_i} *using that* a_k *can be inverted over* $\mathcal{Z}/(p-1)$.
5. *Compute* $y_{k_i} := \alpha^{x_{k_i}} \bmod p, t_k \leq i \leq n_k - 1$ *and append them together with their meaning to the certified public rights list available to both servers and clients.*
6. *Give the numbers* $z_{k_i}, E_{e,N}(r_{k_i})$, $0 \leq i \leq n_k - 1$ *to* c_k *in a public way.*
7. *Give the number* a_k *to* c_k *in a confidential way.*

It is possible to *publicly* give a right $y_{k_{n_k}} = \alpha^{x_{k_{n_k}}} \bmod p$ to a client c_k having rights $y_{k_i} = \alpha^{x_{k_i}} \bmod p, 0 \leq i \leq n_k - 1$ and a secret number a_k. This is straightforward since, according to the previous algorithm, it is possible for the authority to pick a random $r_{k_{n_k}}$ over $\mathcal{Z}/(p-1)$ and compute an integer $z_{k_{n_k}} \in \mathcal{Z}/(p-1)$, such that $x_{k_{n_k}} + r_{k_{n_k}} = a_k z_{k_{n_k}} \bmod (p-1)$. After this, the resulting $z_{k_{n_k}}, E_{e,N}(r_{k_{n_k}})$ are given *in a public way* to the client and the procedure is finished.

A way to perform *rights revocation* is for the authority *Auth* to publish a new certified rights list; then *Auth* also publishes the new numbers $z_{k_i}, E_{e,N}(r_{k_i})$ corresponding to the rights y_{k_i} which are maintained for each client c_k.

Bearing the above in mind, the following result holds for each client c_k

Theorem 3. *If the authority Auth has completed algorithm 1 for a client c_k, then c_k is able to show possession of her rights $y_{k_0}, \cdots, y_{k_{n_k-1}}$ (or a subset of them) to any server in the network, that need not previously know about her. The proof can be zero-knowledge and, no matter the value of n_k, it consists of proving knowledge of one logarithm. Stealing a nongranted right is for a client with no information at all about other clients as hard as solving a discrete logarithm.*

Proof. Client c_k supplies the server with integers $A_k (\neq 1)$ and $z_{k_i}, E_{k_i} (\neq 0)$, for $i = 0, \cdots, n_k - 1$, satisfying the following set of equations

$$y_{k_0} \alpha^{D_{d,N}(E_{k_0})} = A_k^{z_{k_0}} \bmod p \qquad (2)$$

$$\vdots$$

$$y_{k_{n_k-1}} \alpha^{D_{d,N}(E_{k_{n_k-1}})} = A_k^{z_{k_{n_k-1}}} \bmod p$$

Now if c_k is able to prove her knowledge of $\log_\alpha A_k$ over $\mathcal{Z}/(p-1)$, then the server can check that, if c_k knew his key, she *could* express the y_{k_i}'s as powers of α, *i. e.* that c_k *could* obtain the logarithms of the y_{k_i}'s for $i = 0$ to $n_k - 1$. Notice that c_k has been given a_k in the last step of algorithm 1, and it is straightforward from equations 1 that $A_k := \alpha^{a_k} \bmod N$ satisfies equations 2 when for all E_{k_i}'s it holds that $E_{k_i} = E_{e,N}(r_{k_i})$ and the same z_{k_i}'s are used in both systems. Now Protocol 1 or 2 of [Chau 88] can be used to show possession of the logarithm of A_k in zero knowledge.

As for security, equations 2 are verifiable by the server since the y_i's are public and certified for $0 \leq i \leq n_k - 1$. Assume that c_k does not own a particular y_{k_i} but has invented or obtained the corresponding random number r_{k_i} (see equations 1); now if c_k is able to compute by herself a number z_{k_i} satisfying the i-th equation 2, then c_k is able to solve the discrete logarithm problem of finding x_{k_i}. On the other hand, if c_k invents z_{k_i} and manages to compute then a matching r_{k_i}, then c_k is also able to solve the discrete logarithm problem of finding x_{k_i}.

The proof is now complete and the server has needed no particular previous information about client c_k. Also, the construction can be applied to a subset of the y_{k_i}'s if the client does not wish to prove all of them. **QED**

4 How Dangerous Is Rights Sharing?

Theorem 4 Security of Rights Sharing. *If RSA confidentiality is not broken, then it is not feasible for a client c_k to derive the identity of another client c_l —and thus c_l's non-shared rights— by using the fact that c_k and c_l share a right —or a group of rights—.*

Proof. When clients c_k and c_l share a right y_i, they are not likely to share a left-hand side of any of equations 1, since each logarithm x_i has been added a random number. The probability of randomly picking different r_{k_i} over $\mathcal{Z}/(p-1)$ for all m clients and n rights is

$$\frac{(p-1)(p-2)\cdots(p-mn)}{(p-1)^{mn}}$$

which approaches unity if $p-1 \gg (mn)^2$.

So, the only equality in terms of the exponents that can be established when c_k and c_l share a right y_i results from equations 1 and is

$$a_k z_{k_i} - D_{d,N}(E_{e,N}(r_{k_i})) = a_l z_{l_i} - D_{d,N}(E_{e,N}(r_{l_i})) \bmod (p-1) \tag{3}$$

c_k knows $a_k, z_{k_i}, z_{l_i}, E_{e,N}(r_{k_i})$ and $E_{e,N}(r_{l_i})$ in equation 3. Now, if c_k can derive a_l from the above equation, then $D_{d,N}(E_{e,N}(r_{k_i})) - D_{d,N}(E_{e,N}(r_{l_i}))$ must be known to her. In general, this is only possible if c_k can get r_{k_i}, r_{l_i} from decryption under $D_{d,N}$ —notice that $r_{k_i} - r_{l_i} \neq 0$, according to the beginning of the proof.**QED**

5 Untransferability of Rights

Theorem 5 Untransferability. *If RSA confidentiality is not broken, then it is not feasible for a client c_k to transfer a right to another client c_l without the authority's contribution.*

Proof. Thanks to the use of randomization and subsequent encryption of the random numbers, neither of the integers on the left hand side of equations 1 is known to the client. For a client c_k to transfer a right y_i to another client c_l, it is necessary to find a pair z_{l_i}, E_{l_i}, such that

$$x_i + D_{d,N}(E_{l_i}) = a_l z_{l_i} \bmod (p-1) \tag{4}$$

But even if c_k knows a_l (collusion with c_l), c_k ignores x_i, because her own E_{k_i} is an encrypted random number. On the other hand, in order for a server to believe that c_l possesses y_i, α raised to the second term on the left-hand side of equation 4 times y_i must coincide with α raised to the right-hand side over $\mathcal{Z}/(p)$. So inventing a right-hand side of equation 4 and an E_{l_i} that decrypts into a coherent left-hand side second term is not feasible due to the ignorance of x_i by the clients.**QED**

If c_k wants to transfer y_i to c_l, then the only way is to have the job done (and monitored) by the authority. For example

Algorithm 2 (Authorized Transfer) *1. Client c_k shows possession of right y_i to the authority by following a procedure analogous to the one in the proof of theorem 3 (the logarithm being shown possession of is x_i). The procedure requires that c_k prove her knowledge of a_k, which allows the authority to authenticate the giving client.*

2. Client c_l shows possession of a_l to the authority in zero knowledge. In this way, the receiving client is authenticated by Auth.

3. The authority Auth gives y_i to client c_l using the procedure for granting new rights discussed in section 3 (the logarithm being granted is x_i).

6 Requirements and Conclusion

As it has been shown, the proposed scheme is very flexible, since client management can be done independently of servers and, thanks to the linear transformation 1, the secret piece held by the client is constant and does not depend on the rights she owns at a given moment. Actually, it suffices for the client c_k to prove her identity a_k in order to use any subset of her rights, because a_k is the only secret parameter she holds.

As for the storage required, we have

Authority 1. Secret storage for logarithms x_i.

2. Secret storage for all client numbers a_k.
3. Read-write access to α, p, N, e and the list of the y_i's and their meanings (public certified data).

Servers 1. Secret storage for the servers' secret exponent d.

2. Read access to α, p, N and the list of the y_i's and their meanings (public certified data).

Client c_k 1. Secret storage for her number a_k (if the client is a human user, a smart card protected ROM is a good place for a_k).

2. Normal storage for her numbers $z_{k_i}, E_{k_i}, 0 \leq i \leq n_k - 1$.
3. Read access to α, p and the list of the y_i's and their meanings (public certified data).

If we say that two elements A and B are mutually dependent when there is some secret information relating them, then we have shown that functional dependencies between the different element classes of the access control system are those in table 1. The only actual dependencies are between a community of clients and the authority that gave them their identity and their rights, and also between a set of rights and the authority that publishes and certifies them in a list. So we see that servers are also authority-independent, and thus we might think of extending the proposed scheme so that *several authorities each with its client community and rights list share the same set of servers* —compare to a network of teller machines shared by several credit card issuing corporations—.

Depends on	Authority	Client	Server	Right
Authority	-	Yes	No	Yes
Client	Yes	-	No	No
Server	No	No	-	No
Right	Yes	No	No	-

Table 1. Functional dependencies.

References

[Chau 88] Chaum, D., Evertse, J.-H., and Van de Graaf, J. 1988. An Improved Procotol for Demonstrating Possession of Discrete Logarithms and Some Generalizations. *Proceedings of Eurocrypt'87*, Springer-Verlag, pp. 127-141.

[Domi 91] Domingo-Ferrer, J. 1991. Distributed User Identification by Zero-Knowledge Access Rights Proving. *Information Processing Letters*, vol. 40, pp. 235-239.

[Grah 72] Graham, G. S., and Denning, P. J. 1972. Protection: Principles and Practices. *Proceedings of the AFIPS Spring Joint Computer Conference*, pp. 417-429.

[Harr 76] Harrison, M. A., Ruzzo, W. L., and Ullman, J. D. 1976. Protection in Operating Systems. *Communications of the ACM*, vol. 19, pp. 461-471.

[Linn 90] Linn, J. 1990. Practical Authentication for Distributed Computing. *Proc. IEEE Symposium on Research in Security and Privacy*, IEEE CS Press, pp. 31-40.

[Rive 78] Rivest, R. L., Shamir, A., and Adleman, L. 1978. A Method for Obtaining Digital Signatures and Public-Key Cryptosystems. *Communications of the ACM*, vol. 21, pp. 120-126.

[Woo 92] Woo, T. Y. C., and Lam, S. S. 1992. Authentication for Distributed Systems. *IEEE Computer*, vol. 25, pp. 39-52.

Interactive Hashing Simplifies Zero-Knowledge Protocol Design

Rafail Ostrovsky* Ramarathnam Venkatesan† Moti Yung‡

(Extended abstract)

Abstract

Often the core difficulty in designing zero-knowledge protocols arises from having to consider every possible *cheating* verifier trying to extract additional information. We here consider a compiler which transforms protocols proven secure only with respect to the *honest* verifier into protocols which are secure against any (even cheating) verifier. Such a compiler, which preserves the zero-knowledge property of a statistically or computationally secure protocol was first proposed in [BMO] based on Discrte Logarithm problem. In this paper, we show how such a compiler could be constructed based on any one-way permutation using our recent method of *interactive hashing* [OVY-90, NOVY]. This applies to both statistically and computationally secure protocols, preserving their respective security. Our result allows us to utilize DES-like permutations for such a compiler.

1 Introduction

An interactive proof involves two communicating parties, a prover and a verifier. The prover is computationally unbounded; alternatively, in applications, it is a polynomial-time machine possessing additional private knowledge. It tries to convince the probabilistic polynomial time verifier that a given theorem is true.

A zero-knowledge (ZK) proof is an interactive proof with an additional privacy constraint: the verifier does not learn why the theorem is true [GMR]. That is, whatever the polynomial-time verifier sees in a ZK-proof with the unbounded prover of a true theorem x, can be approximated by a probabilistic polynomial-time machine working solely on input x. A statistical zero-knowledge proof (SZK proof) is one for which this true view and approximate view are (information-theoretically) indistinguishable.

A methodology suggested in [BMO] is to design statistical or computational zero-knowledge protocols by assuming a canonical behavior of the verifier, and then translate such protocols to those where cheating is allowed. The mechanism proposed there, as well as the one in [GKa, NY] (for computational zero-knowledge proofs only) uses specific algebraic assumptions to achieve it.

The task of finding the necessary and sufficient complexity conditions needed for various primitives has attracted a lot of work, showing that many primitives, originally based on specific algebraic functions, need only one-way functions or permutations. For example, pseudo-random generators [BM-84], secure signature schemes [GoMiRi], computational ZK-

---

* University of California at Berkeley Computer Science Division, and International Computer Science Institute at Berkeley. E-mail: `rafail@melody.berkeley.edu`. Supported by NSF postdoctoral fellowship and ICSI. Part of this work was done at Bellcore and part at IBM T.J. Watson Research Center.

† Bellcore, Room 2M-344, 445 South St, Morristown, NJ 07960. E-mail: `venkie@bellcore.com`.

‡ IBM Research, T.J. Watson Research Center, Yorktown Heights, NY 10598. E-mail: `moti@watson.ibm.com`.

proofs [GMR] were shown to be equivalent to the existence of general one-way functions [ILL, Ha-90, NY, Ro, OW]. Such efforts, not only develop the theoretical foundations of cryptography, but also enable the primitive implementations to be based on a larger possible concrete choices of underlying functions, thus making them more plausible.

The recent method of *interactive hashing* [OVY-90, NOVY] has been applied to various cryptographic primitives, to information theoretically secure Oblivious Transfer protocols [OVY-90], and then to zero-knowledge arguments [NOVY] (as well as to commitments by/ to powerful non-polynomial parties [OVY-92]). Here we show an extended use of this method with zero-knowledge protocols to provide a ZK-protocol design tool along the line of [BMO], but based on the existence of any *one-way permutation.* In particular, assuming that one-way permutations exist, we show that if a language L has a honest-verifier statistical zero-knowledge proof, then L has a (general) statistical zero-knowledge proof. We remark that our method applies to computational zero-knowledge as well. Previously, specific algebraic assumptions were needed in order to implement such tools [BMO, GKa, NY].

1.1 Organization of the paper

In section 2, we give the model and definitions. In Section 3, we present the main result on compiling protocols zero-knowledge against a honest verifier to general zero-knowledge protocols, and we show some implications. Section 4 outlines the compiler and its proof.

2 Definitions

We use standard notions of Turing machines (TM) and probabilistic polynomial time TM's (PPT), and interactive Turing machines [GMR]. We adopt the standard definition of computational and statistical indistinguishability (see, for example, [ILL, GMR]). Let us recall definitions of interactive proofs and zero-knowledge proofs, introduced and formalized in [GMR].

We assume that prover P is a probabilistic, infinite power, interactive TM and verifier V is a probabilistic, poly-time interactive TM [GMR]. We consider interactions between P and V, where they share the same input and can communicate. We say P convinces V to accept on x if P and V have common input x, and after the interaction V accepts. Let *view* of V be the transcript of the conversation between P and V which consists of all the messages between P and V and the portion of the random tape used by V (i.e. random coin tosses of V).

P and V form an interactive protocol for language L with security parameter k (k is the length of the input string), if the following two conditions are satisfied:

- **Completeness**: For all $x \in L$, P convinces V to accept with probability greater than $1 - \frac{1}{2^k}$, where probability is taken over coin tosses of P and V.

- **Soundness**: For all P' and for all $x \notin L$ probability that P' convinces V to accept on x is less than $\frac{1}{2^k}$.

IP ($= PSPACE$) is the class of languages which can be accepted satisfying completeness and soundness conditions.

The zero-knowledge property:
For every PPT verifier V' let $M_{V'}$ be the probabilistic poly-time TM. The goal of $M_{V'}$ is to simulate the *view* of V', i.e. the conversation between P and V' on x. As such, it must produce a pair: <random tape used by V', conversation between P and V' >. We restrict

our simulators to be average-PPT TM. An interactive protocol is *Statistical Zero-Knowledge* if for all V' there exists $M_{V'} \in PPT$ such that for all $x \in L$, the distributions of the conversation between P and V' on x and $M_{V'}(x)$ is statistically close. If the two distributions are computationally indistinguishable, this corresponds to *Computational Zero-Knowledge.*

Zero-knowledge with respect to honest verifier:
Finally, we are ready to specify what does it mean to have a protocol which works for *honest verifier* only. An interactive protocol is *Statistical Zero-Knowledge for Honest Verifier* if for the honest V (i.e. the one specified in the description of P, V) there exists $M_V \in PPT$ such that for all $x \in L$, the distributions of the conversation between P and V on x and $M_V(x)$ are statistically close. Similar definition holds for *Computational Zero-Knowledge Protocols for Honest Verifier.*

Let f be a length preserving function $f: \{0,1\}^* \rightarrow \{0,1\}^*$ computable in polynomial time.

Definition 2.1 [One-way function.] *f is one-way if for every probabilistic polynomial time algorithm A, for all polynomials p and all sufficiently large n,*

$$Pr[f(x) = f(A(f(x))) \mid x \in_R \{0,1\}^n] < 1/p(n).$$

If addition, if f is a permutation on $\{0,1\}^n, n > 0$, then we say that f is a *one-way permutation.* The above definition is of a *strong one-way function.* Its existence is equivalent to the existence of the weak one-way function [Y82]; a stronger equivalence is possible in the case of permutations (see [GILVZ]). A weak one-way function has the same definition as above, except the probability of successful inversion above is $1 - 1/n^c, c > 0$.

3 Main Result

We show that if there is any one-way permutation, then "honest verifier zero knowledge" is in fact just as strong as zero-knowledge.

Theorem 3.1 Suppose a one-way permutation exists. If a language L has an honest verifier statistical (respectively computational) zero knowledge protocol, then L has a statistical (respectively computational) zero knowledge protocol.

We remark that our transformation is constructive and that error probabilities are preserved, as in [BMO], it also works for zero-knowledge proof of knowledge.

3.1 Implications

The theorem has a few implications on languages and their proof systems (beyond giving a design tool). We discuss those briefly.

- *Black-box simulation:*
  Oren [Or] formalized the black box notion by saying that the simulator is a PPT oracle machine M which when asked to simulate a particular verifier $\widehat{V}$ is given that verifier as an oracle. Thus the same simulator works for all verifiers. Using our method we show that assuming any one-way permutation, black box simulation is not a restriction on zero-knowledge, i.e.: Suppose L has a (honest verifier) SZK (ZK) protocol and one-way permutation exists. Then, L has a black box simulation SZK (ZK) protocol.

- *Error probability one-sidedness* :
  Goldreich, Mansour and Sipser [GMS] define a one-sided proof system to be one in which completeness holds with probability 1 (that is the prover can always convince the verifier). An implication of our protocol tool is: If L has a (honest verifier) SZK proof system and one-way permutation exists. Then, L has a SZK one-sided proof system.

4 The Protocol Compiler and its Proof

Given a zero-knowledge for honest verifier proof system $(\overline{P}, \overline{V})$, we have to construct another prover/verifier pair (P, V) such that (P, V) is still an interactive proof system for L and for *any* (possibly cheating) verifier $\hat{V}$ there exists a simulator $S_{\hat{V}}$. We specify the protocol below. For completeness sake, first we recall what is interactive hashing, and show the interactive hashing-based bit commitment protocol.

Remark: The bit commitment protocol parties are *efficient*, i.e. they need only perform polynomial time computations to execute the protocol.

<u>Commit to a bit a</u>

1. The verifier V selects $x \in_R \{0,1\}^n$ at random and computes $y \leftarrow f(x)$. V keeps both x and y secret from P.
2. The prover P selects $h_1, h_2, \ldots h_{n-1} \in \{0,1\}^n$ such that each h_i is a random vector over $GF[2]$

   such that $h_1, h_2, \ldots h_{n-1}$ are linearly independent over $GF[2]$
3. For j from 1 to $n-1$
   - P sends h_j to V.
   - V sends $r_j \leftarrow B(h_j, y)$ to P (where $B(u, v)$ is the bit resulting as the inner product of u and v).
4. At this point there are exactly two vectors $y_0, y_1 \in \{0,1\}^n$ such that for $i \in \{0,1\}$, $r_j = B(y_i, h_j)$ for all $1 \leq j \leq n-1$. y_0 is defined to be the lexicographically smaller of the two vectors. Both P and V compute y_0 and y_1. Let

$$d = \begin{cases} 0 & \text{if } y = y_a \\ 1 & \text{if } y = y_{1-a} \end{cases}$$

5. V computes d and sends it to R (d is "encrypting" the commitment bit a and given the inversion of one of y_0, y_1 and d, a is uniquely determined).

This committal reveals to P nothing about the committed bit (in the information-theoretic sense). On the other hand, V cannot later decommit to a value other than the one it committed without inverting a one-way permutation on a random challenge.

Next we present the compiler.

<u>Compiler Protocol</u>

1. V picks a sequence a_i, $1 \leq i \leq 2t$ of random bits, and commits to them using *Interactive Hashing.* The commitment can be done in parallel for all bits.

2. P chooses at random t-subset of $\{1,\ldots,2t\}$ and asks V to decommit bits a_j for j in the subset. Let $a'_i, i \leq t$ be the subsequence of unopened bits.
3. P picks t bits $b_1,\ldots,b_t$ at random and sends them to V.
4. V lets $c_i = b_i \oplus a'_i$ and $C = c_1c_2c_3\ldots c_t$ be its secret random (tape) string.
5. P, V execute an old $(\overline{P}, \overline{V})$ protocol with V, running an $\overline{V}$, but using C as its secret coinflips. Moreover, for every message sent from V to P is accompanied by a *zero-knowledge argument* that $\overline{V}$ would really have sent this message if its coinflips were C. (Remark: Such a proof is possible [NOVY] and users are engaged in *Interactive Hashing* based on one-way permutation as a subroutine).

   More specifically, V begins by sending the message α_1 that would have been the first message $\overline{V}$ sent on coins C, and proves that indeed it has done this. The prover checks this proof, and if it is incorrect it aborts. Otherwise it sends whatever response β_1 the old prover $\overline{P}$ would have sent. This continues till the proof ends. (The available strongly committed bits, and the specification of the original protocols are the witness to the proofs communicated).

4.1 Proof of correctness

We have to prove completeness, soundness and the zero-knowledge property.

Completeness: For all x in L, the prover can still convince the verifier, since the success probability of the new P is essentially equivalent to the old one (by the simple fact that it is following the protocol).

Soundness: Interactive hashing hides committed bits in the information-theoretic sense, and thus the prover does not get any information about the random tape of the verifier (other then what follows from the original protocol during the initialization stage). Since all the subsequent rounds use zero-knowledge arguments in addition to the messages of the old protocol, the soundness follows.

Zero-knowledge property: The simulator below proves this. We concentrate on statistical zero-knowledge here. The computational case is similar.

First, our new simulator runs the old simulator for honest verifier in order to obtain a pair $(\overline{C}, \alpha_1\beta_1\ldots\alpha_m\beta_m)$ consisting of coin tosses of the honest verifier $\overline{C} = \bar{c}_1\bar{c}_2\ldots\bar{c}_t$ and the transcript $\alpha_1\beta_1\ldots\alpha_m\beta_m$ of the conversation between the prover and the honest verifier. The new simulator, will now transform (with very high probability) this old transcript for honest verifier into one which is statistically close to the conversation between new prover/verifier pair as follows:

(1) It runs $\hat{V}$ for step 1 to get its commitment of $a_1,\ldots,a_{2t}$, using interactive hashing.
(2) At this point, the simulator uses the backtracking capability to run the protocol twice in order to learn what are the "unopened" bits. That is, it asks to reveal a random subset of t bits. Then it puts the verifier into the state it was in before the subset of t bits was requested to be revealed (but after the commitments) and now requests to open the complementary set of bits.
(3) Having the a_i, the simulator now picks $b_i = a_i \oplus \bar{c}_i$ for all $i = 1,\ldots,t$ as being the prover's response (modifying bits) of step 3, and has thus makes $\overline{C}$ be the secret random string for the new V.

Recall that the simulator has in its possession the old conversation with coins fixed to $\overline{C}$.

The zero-knowledge arguments executed at each round force cheating verifier to generate a conversation which is statistically close to the one we produced by using the honest verifier (with additional ZK arguments). The new simulator runs $\hat{V}$ and gets what is supposed to be $\overline{V}$'s first message if it had $\overline{C}$, together with a proof (i.e. a zero-knowledge argument based on interactive hashing and assuming one-way permutations exist) that this is indeed the case. It examines the proof and if it is found incorrect the simulator aborts as the prover would have. But if not, then with very high probability, the message $\hat{V}$ sent is *really* the message α_1 that the simulator expected at this stage. And to this message it can respond: it just has to send β_1. Continuing in this way the simulator soon has a transcript of the entire conversation, which (retracing through the argument) is statistically close to the real conversation. That is, the simulator generates exactly the correct conversation except if:

- $\hat{V}$ manages to break the commitment scheme (i.e. invert a one-way permutation), or
- if it is able to cheat the prover in a zero-knowledge argument (which as well implies it can invert a one-way permutation, given the underlying construction).

Thus, we are done.

Conclusions: To summarize, we have presented a uniform way to compile honest-verifier zero-knowledge protocols into general zero-knowledge ones. This gives a design method which seems to be easier than considering all possible verifiers as a starting design point. The proof has some implications to properties of languages and their proofs, and it further demonstrates a wider applicability of the recent notion of interactive hashing.

References

[BM-84] M. Blum, and S. Micali "How to Generate Cryptographically Strong Sequences Of Pseudo-Random Bits" *SIAM J. on Computing,* Vol 13, 1984, pp. 850-864.

[BMO] Bellare, M., S. Micali and R. Ostrovsky, "The (True) Complexity of Statistical Zero Knowledge" STOC 90.

[BCC] G. Brassard, D. Chaum and C. Crépeau, *Minimum Disclosure Proofs of Knowledge,* JCSS, v. 37, pp 156-189.

[BCY] Brassard, G., C. Crépeau, and M. Yung, "Everything in NP can be Argued in Perfect Zero Knowledge in a Bounded Number of Rounds," ICALP 89. (also in Theoretical Computer Science, special issue of ICALP 89).

[Dam] I. B. Damgaard, *Collision Free Hash Functions and Public Key Signature Schemes* , Eurocrypt, 1987.

[GKa] Goldreich, O. and A. Kahn, personal communication.

[GILVZ] O. Goldreich, R. Impagliazzo, L. Levin, R. Venkatesan, and D. Zuckerman, *Security Preserving Amplification of Hardness,* FOCS 90.

[GMS] Goldreich, O., Y. Mansour, and M. Sipser, "Interactive Proof Systems: Provers that never Fail and Random Selection," FOCS 87.

[GMW1] Goldreich, O., S. Micali, and A. Wigderson, "Proofs that Yield Nothing but their Validity", FOCS 86.

[GMR] Goldwasser, S., S. Micali, and C. Rackoff, "The Knowledge Complexity of Interactive Proofs," *SIAM J. Comput.*, **18**(1), 186-208 (February 1989).

[GoMiRi] Goldwasser, S., S. Micali, and R. Rivest, "A Digital Signature Scheme Secure Against Adaptive Chosen-Message Attacks," *SIAM J. Comput.*, **17**(2), 281-308 (April 1988).

[Ha-90] J. Hastad, "Pseudo-Random Generators under Uniform Assumptions" *STOC 90*

[ILL] I. Impagliazzo, L. Levin and M. Luby, *Pseudo-random generation from one-way functions*, Proc. 21st Symposium on Theory of Computing, 1989, pp. 12-24.

[NOVY] M. Naor, R. Ostrovsky, R. Venkatesan, and M. Yung. "Perfect Zero-Knowledge Arguments for NP Can Be Based on General Complexity Assumptions", *Advances in Cryptology – Crypto '92*, Lecture Notes in Computer Science, Springer, to appear.

[NY] Naor, M. and M. Yung, "Universal One-Way Hash Functions and their Cryptographic Applications," STOC 89.

[Or] Oren Y., "On The Cunning Power of Cheating Verifiers: Some Observations About Zero Knowledge Proofs", FOCS 87.

[OVY-90] R. Ostrovsky, R. Venkatesan, and M. Yung. "Fair Games Against an All-Powerful Adversary", *SEQUENCES '91*, Positano, June, 1991 (Proc. Springer Verlag), (also presented at Princeton Oct. 1990 Workshop on Complexity and Cryptography).

[OVY-92] R. Ostrovsky, R. Venkatesan, M. Yung, *Secure Commitment Against A Powerful Adversary, STACS 92*, Springer Verlag LNCS Vol. 577, p. 439-448, 1992.

[OW] R. Ostrovsky, A. Wigderson *One-Way Functions are Essential for Non-Trivial Zero-Knowledge*, The second Israel Symposium on Theory of Computing and Systems (ISTCS93) 1993.

[Ro] J. Rompel "One-way Functions are Necessary and Sufficient for Secure Signatures" STOC 90.

[Y82] A. C. Yao, *Theory and Applications of Trapdoor functions*, Proceedings of the 23th Symposium on the Foundation of Computer Science, 1982, pp 80-91.

One-Way Accumulators: A Decentralized Alternative to Digital Signatures (Extended Abstract)

Josh Benaloh[1] and Michael de Mare[2]

[1] Clarkson University
[2] Giordano Automation

Abstract. This paper describes a simple candidate one-way hash function which satisfies a *quasi-commutative* property that allows it to be used as an accumulator. This property allows protocols to be developed in which the need for a trusted central authority can be eliminated. Space-efficient distributed protocols are given for document time stamping and for membership testing, and many other applications are possible.

1 Introduction

One-way hash functions are generally defined as functions of a single argument which (in a "difficult to invert" fashion) reduce their arguments to a predetermined size. We view hash functions, somewhat differently here, as functions which take two arguments from comparably sized domains and produce a result of similar size. In other words, a hash function is a function h with the property that $h: A \times B \rightarrow C$ where $|A| \approx |B| \approx |C|$. There is, of course, no substantial difference between this view and the traditional view except that this view allows us to define a special *quasi-commutative* property which, as it turns out, has several applications.

The desired property is obtained by considering functions $h: X \times Y \rightarrow X$ and asserting that for all $x \in X$ and for all $y_1, y_2 \in Y$,

$$h(h(x, y_1), y_2) = h(h(x, y_2), y_1).$$

This property is not at all unusual. Addition and multiplication modulo n both have this property as does exponentiation modulo n when written as $e_n(x, y) = x^y \bmod n$. Of these, only exponentiation modulo n has the additional property that (under suitable conditions), the function is believed to be difficult to invert.

This paper will describe how to use the combination of these two properties (quasi-commutativity and one-wayness) to develop a *one-way accumulator* which (among other applications) can be used to provide space-efficient cryptographic protocols for time stamping and membership testing.

2 Definitions

We begin by formalizing the necessary definitions.

Definition 1. A family of *one-way hash functions* is an infinite set of functions $h_\ell: X_\ell \times Y_\ell \to Z_\ell$ having the following properties:

1. There exists a polynomial P such that for each integer ℓ, $h_\ell(x, y)$ is computable in time $P(\ell, |x|, |y|)$ for all $x \in X_\ell$ and all $y \in Y_\ell$.
2. There is no polynomial P such that there exists a probabilistic polynomial time algorithm which, for all sufficiently large ℓ, will when given ℓ, a pair $(x, y) \in X_\ell \times Y_\ell$, and a $y' \in Y_\ell$, find an $x' \in X_\ell$ such that $h_\ell(x, y) = h_\ell(x', y')$ with probability greater than $1/P(\ell)$ when (x, y) is chosen uniformly among all elements of $X_\ell \times Y_\ell$ and y' is chosen uniformly from Y_ℓ.

Note that the above definition only requires that "collisions" of the form $h(x, y) = h(x', y')$ for given x, y, and y' are hard to find. That is, given x, y, y', it is, in general, hard to find a *preimage* x' such that $h(x, y) = h(x', y')$. It may in fact be easy, given $(x, y) \in X \times Y$, to find a pair $(x', y') \in X \times Y$ such that $h(x, y) = h(x', y')$. It must, however, be the case that for a given (x, y) pair, there are relatively few $y' \in Y$ for which an $x' \in X$ can practically be found such that $h(x, y) = h(x', y')$.

Note also that this definition does not require that the "hash" value be smaller than its arguments. However, the hash functions considered here will have the property that $|X| \approx |Y| \approx |Z|$.

It follows from the above definition that a family of one-way hash functions is itself a family of one-way functions. Work by Naor and Yung ([NaYu89]) and by Rompel ([Romp90]) has shown that one-way hash functions exists if and only if one-way functions exist which, in turn, exist if and only if secure signature schemes exist. It has also been shown ([ILL89]) that the existence of one-way functions is equivalent to the existence of secure pseudo-random number-generators.

Definition 2. A function $f: X \times Y \to X$ is said to be *quasi-commutative* if for all $x \in X$ and for all $y_1, y_2 \in Y$,

$$f(f(x, y_1), y_2) = f(f(x, y_2), y_1).$$

By considering one-way hash functions for which the range is equal to the first argument of the domain, i.e. $h: X \times Y \to X$, we can exploit the properties of one-way hash functions which also have the quasi-commutative property.

Definition 3. A family of *one-way accumulators* is a family of one-way hash functions each of which is quasi-commutative.

3 Motivation

The quasi-commutative property of one-way accumulators h ensures that if one starts with an initial value $x \in X$, and a set of values $y_1, y_2, \ldots, y_m \in Y$, then the *accumulated hash*

$$z = h(h(h(\cdots h(h(h(x, y_1), y_2), y_3), \cdots, y_{m-2}), y_{m-1}), y_m)$$

would be unchanged if the order of the y_i were permuted.

In addition, the fact that h is a one-way hash function means that given $x \in X$ and $y \in Y$ it is difficult to, for a given $y' \in Y$, find an $x' \in X$ such that $h(x, y) = h(x', y')$.[3]

Thus, if the values $y_1, y_2, \ldots, y_m$ are associated with users of a cryptosystem, the accumulated hash z of all of the y_i can be computed. A user holding a particular y_j can compute a partial accumulated hash z_j of all y_i with $i \neq j$. The holder of y_j can then (presumably at a later time) demonstrate that y_j was a part of the original hash by presenting z_j and y_j such that $z = h(z_j, y_j)$. A user who wishes to *forge* a particular y' would be faced with the problem of constructing an x' with the property that $z = h(x', y')$.

This approach does *not* enable users to hide their individual y_j since all of the y_j are necessary to compute the accumulated hash z (although the y_j may themselves be encryptions of hidden information). However, using a one-way accumulator in this way keeps each user from having to *remember* all of the y_j.

A general application of this basic trick is as an alternative to digital signatures for credential authentication: if all parties retain the result z of the accumulated hash, then at a later time, any party can present its (y_j, z_j) pair to any other party who can then compute and verify $h(y_j, z_j) = z$ to authenticate y_j.

It might, of course, be possible for a dishonest user to construct a false pair (x', y') such that $h(x', y') = z$ by combining the various y_i in one way or another. It will, however, be seen in section 5.1 that this is not practical. Other methods of computing false (x', y') pairs may also be possible. However, by restricting the choice of y', constructing "useful" false pairs can be made impractical.

It should be emphasized that the advantage of this approach over the naive "save everything you see" approach is simply one of storage. In terms of storage, this protocol is comparable to that of retaining a public-key for a central authority and using it to verify that y_j has been signed by the central authority. However, using the one-way accumulator method can obviate the need for a central authority altogether.

Two applications of one-way accumulators will be presented in section 5. The first is a method to construct a time stamping protocol in which participants can archive and time stamp their documents in such a way as to allow the time stamped documents to be revealed to others at a later time. A second

---

[3] The assertion that the composition formed by applying h many times is one-way is not strictly the same as asserting that h itself is one-way. This will be addressed in section 4.

application shows how a membership testing system can be constructed without having to maintain membership lists. In both applications, storage requirements are minimized without having to rely upon a (potentially corruptible) central authority.

4 Modular Exponentiation

For any n, the function $e_n(x, y) = x^y \bmod n$ is clearly quasi-commutative. The commonly used RSA assumption ([RSA78]) is that for "appropriately chosen" n, computing x from $e_n(x, y)$, y, and, n cannot be done in time polynomial in $|n|$ except in an exponentially small number of cases. In [Sham81], Shamir gives a proof which, when applied in this context, shows that for these appropriately chosen n, if root finding modulo n is hard, then the family e_n constitutes a family of one-way hash functions. However, even this may not be enough if the e_n are to be used as one-way accumulators. The reason for this is that repeated application of an e_n may reduce the size of the image so much that finding collisions becomes feasible.

To alleviate this problem, we restrict our n even further than do most.

Definition 4. Define a prime p to be *safe* if $p = 2p' + 1$ where p' is an odd prime.

Definition 5. We define n to be a *rigid integer* if $n = pq$ where p and q are distinct safe primes such that $|p| = |q|$.

It is not hard to see that for $n = pq$ to be a rigid integer larger than 100, each of p, q, $\frac{(p-1)}{2}$ and $\frac{(q-1)}{2}$ must be primes congruent to 5 modulo 6.

4.1 Composition

The advantage of using a rigid integer $n = pq$ is that the group of squares (quadratic residues) modulo n that are relatively prime to n has the property that it has size $n' = \frac{(p-1)}{2}\frac{(q-1)}{2}$ and the function $e_{n,y}(x) = x^y \bmod n$ is a permutation of this group whenever y and n' are relatively prime. Thus, if the factorization of n is hidden, "random" exponentiations of an element of this group are extremely unlikely to produce elements of any proper subgroup. This means that repeated applications of $e_n(x, y)$ are extremely unlikely to reduce the size of the domain or produce random collisions.

Although constructing rigid integers is somewhat harder than constructing ordinary "difficult to factor" integers, it is still quite feasible. The process would be to select "random" p' congruent to 5 modulo 6 until one is found such that p' and $2p' + 1$ are both prime. Approximately one out of every $(\ln p')^2$ of the p' selected will have this property. Once a suitable p' has been found, a suitable q' is selected similarly. This allows $n = pq = (2p' + 1)(2q' + 1)$ to be formed within approximately $2(\ln p')^2$ trials. Thus, if the modulus n is to be approximately 200 digits in length, approximately 10,000 candidates for each of p' and q' would be expected to be examined before suitable choices are found. This would mean executing roughly 20,000 primality tests on 100 digit integers – an amount of work which is not terribly unreasonable.

In some sense, rigid integers may be the hardest of all integers to factor. Most cryptographic applications which depend upon the difficulty of factoring suggest that n be chosen as a product of two comparably sized primes p and q and further suggest that $p-1$ and $q-1$ each contain large prime factors. Such n are suitable for our purposes also. However, taking these parameters to the extreme case in which each of $p-1$ and $q-1$ have the largest of possible prime factors (namely $(p-1)/2$ and $(q-1)/2$) provides additional beneficial properties which can be exploited by our applications.

4.2 Collisions

The one-way property of one-way accumulators rests not on the difficulty of finding arbitrary collisions, but rather upon the difficulty of finding collisions (or alternate preimages) with specific constraints.

If an accumulated hash z, is formed from a given set of values taken modulo n, a new item y can be *forged* by finding an x such that $z = x^y \bmod n$. If y is itself the result of a one-way hash, a prospective forger must, for a y that it can change but not select, compute a yth root of z modulo n.

This, on the face of it, appears to be as hard as computing roots modulo a composite n which is believed to be computationally infeasible for large n of unknown factorization.

There are, however, other factors which may make the task easier for the prospective forger. First, together with z, the forger is provided with a number of roots of z modulo n. (These other roots are provided by the values used to form z.) Shamir, however, has shown ([Sham81]) that if basic root computation is difficult, then the roots $z^{1/r_1}, z^{1/r_2}, \ldots, z^{1/r_k}$ are insufficient to compute the value of $z^{1/\rho}$ unless ρ is a divisor of $R = \prod_{i=1}^{k} r_i$. Second, the forger may have had an opportunity to select some of the constituent y out of which the accumulated hash z was constructed. It is conceivable that a forger may weaken the system by choosing appropriate constituents which will facilitate a subsequent forgery.

We sketch below the result which says that even an actively participating (dynamic) forger cannot exist unless root finding is computationally feasible.

Theorem 6. *Suppose there exists a polynomial time algorithm $\mathcal{A}$ which when given x and n and a polynomial number of roots $y_1, y_2, \ldots, y_k$ and pre-selected indices $r_1, r_2, \ldots, r_k$ of x such that each $y_i^{r_i} \bmod n = x$ finds, for a given r, a y such that $y^r \bmod n = x$. Then there exists a polynomial time algorithm $\mathcal{B}$ which when given x, n, and $\rho = r/\gcd(r, r_1 r_2 \cdots r_k)$ will produce a y such that $y^\rho \bmod n = x$.* (In other words, the computation can be duplicated without the use of the roots $y_1, y_2, \ldots, y_k$.)

Proof. (sketch)

Algorithm $\mathcal{B}$ can be constructed from algorithm $\mathcal{A}$ as follows. Given x, n, and ρ, $\mathcal{B}$ computes $\hat{x} = x^{r_1 r_2 \cdots r_k} \bmod n$ and asks $\mathcal{A}$ for an rth root of $\hat{x}$ modulo n by providing $\mathcal{A}$ with the appropriate roots of $\hat{x}$ which can be easily computed from x and the r_i. $\mathcal{A}$ will return a $\hat{y}$ such that $\hat{y}^r \bmod n = \hat{x}$. Let $g = \gcd(r, r_1 r_2 \cdots r_k)$. The quotients $\frac{r}{g}$ and $\frac{r_1 r_2 \cdots r_k}{g}$ are now relatively prime, and the extended Euclidean algorithm can be used to construct cofactors a and b such that

$$a\left(\frac{r}{g}\right) + b\left(\frac{r_1 r_2 \cdots r_k}{g}\right) = 1.$$

The desired root $x^{1/\rho} \bmod n$ can now be constructed as $x^{1/\rho} \bmod n = x^a \hat{y}^b \bmod n$ since $(x^a \hat{y}^b)^\rho \bmod n = x^{(ar/g)} \hat{y}^{(br/g)} \bmod n = x^{(ar/g)} x^{(br_1 r_2 \cdots r_k/g)} \bmod n = x \bmod n$. □

In short, this theorem shows that (unless general root finding is feasible) an rth root of a given z modulo n can be computed only if one is given a set of known roots and indices $\{(x_i, r_i) : x_i^{r_i} \bmod n = z\}$ such that r is a divisor of $\prod r_i$.

It may, however, be possible for a forger to obtain a set of roots such that the product R of their indices *is* a multiple of the desired root index. But, it can be shown that the number of known roots which would have to be provided in order to have a non-negligible probability of their product being a multiple of a random number r selected later would be prohibitively large (see [KnTr76]). Asymptotically, for any polynomial P, it is the case $P(|n|)$ items can be combined into a single accumulated hash value with extremely high security. Numerically, even in a worst-case scenario in which an adversary is allowed to select all hash values (root indices) in advance, a 220 digit n would comfortably allow about 20 million items to be hashed with probability of forgery well below 10^{-30}. (See [Brui51], [Mitc68], and [LuWa69].)

A full asymptotic and numerical analysis will be included in the full version of this paper.

5 Applications

Two applications are described in this section.

5.1 Time Stamping

Haber and Stornetta ([HaSt90]) describe how documents can be *time stamped* by cryptographically chaining documents. By following the links in the chain, one can later determine where in the sequence a document occurred and thereby determine the relative positions of any two documents. This process, however, is somewhat cumbersome since it requires the active cooperation of other participants who have documents in the chain. Each link of the chain must be individually reconstructed to relocate the position of a document.

In the same work, Haber and Stornetta also describe a system by which documents are transmitted to a subset of the participants. The specific subset is determined by the document itself. With the appropriate cooperation of these participants, one can later substantiate to others that the document was sent at the claimed time.

Benaloh and de Mare ([BeMa91]) describe another method using a somewhat different model. They break time into rounds and assume the existence of broadcast channels (which can be simulated with any of a variety of consensus protocols — see, for example, [CGMA85], [Fisc83], [BenO83], and [Rabi83]). Benaloh and de Mare describe how time stamping can be accomplished without assumptions of cooperation. Within their model, they show how the amount of information which must be saved in each round of the protocol can be made proportional to the logarithm of the number of participants in the protocol. They pose as an open problem the question of whether the amount of information which must be saved can be made independent of the number of participants.

The time stamping protocol given here essentially solves the question posed by Benaloh and de Mare. Using modular exponentiation as a one-way accumulator, a simple protocol can be devised.

A Time Stamping Protocol. Before beginning, a rigid integer n is agreed to by all parties. This n can be supplied by a (trusted) outside source, constructed by a special purpose physical device, or (perhaps more likely) chosen by joint evaluation of a circuit for computing such an n which is supplied with random inputs by the participants (see [GMW86], [GMW87], [BGW88], [CCD88], [RaBe89], [Beav89], [BeGo89], [GoLe90], [MiRa90], and [Beav91] for work on secure multiparty computation). Since this n need be selected only once and may thereafter be used continuously, any extraordinary effort which may be required to construct such an n may be warranted.

Once n has been selected, a starting value x is agreed upon. This x may, for instance, be a representation of the current date. From this x, the value $x_0 = x^2 \bmod n$ is formed.

Each of the m participants takes any document(s) that it wishes to stamp in a given round and applies an agreed upon conventional one-way hash function to its document(s) to produce a y such that $y < n$. Let $y_1, y_2, \ldots, y_m$ denote the set of (conventionally hashed) documents to be stamped in a given round. Let $Y = \prod_{i=1}^{m} y_i$, and for each j let y'_j denote the product Y/y_j. The *time print* of the round z is computed as the accumulated hash

$$z = x_0^Y \bmod n = ((\cdots((x_0^{y_1} \bmod n)^{y_2} \bmod n)\cdots)^{y_m}) \bmod n.$$

The jth participant also computes and maintains the partial accumulated hash

$$z_j = x_0^{y'_j} \bmod n$$

which is also easily computed.

Now, for the jth participant to demonstrate at a later time that a given document (which presumably only it saved) has a claimed time stamp, the participant need only produce y_j and z_j. Anyone can check that $z_j^{y_j} \bmod n$ is equal to the time print z of the round and must therefore accept the time stamp of the document as legitimate. The claimant can then show that when the conventional hash function is applied to its document the value y_j is produced.

Is Forgery Possible? Before discussing whether or not forgery is possible, we must define precisely what forgery means within this context. A participant has the ability to time stamp many documents per round. These documents might contain contradictory information or promises. There is nothing, for instance, to stop a participant from time stamping a large number of predictions about the world series outcome and then (after the outcome is decided) revealing only the one time stamped document which correctly predicted the outcome.

Depending on the method of implementation, it might even be possible for a user who wishes to stamp (hashed) document y to, for example, submit (hashed) documents u and v for stamping where $y = uv$ and then later construct a time stamp for y out of the time stamps for u and v. Although this simple ploy can be remedied by requiring the submission of both pre-hash and post-hash documents (note that the documents may, of course, also be encrypted before any hashing to protect their contents), other similar ploys may be possible if the user knows the document for which a stamp is desired at the time of the stamp. This, however, does not pose a concern since we allow participants to stamp any and all documents within any round.

The only claim which we can make about forgery is that a user cannot produce a valid time stamp for a document that was not anticipated at the time indicated by the stamp. For example, an industrial spy who reads a patent application with a given date will not be able to change the name on the application and forge a time stamp to indicate an earlier date.

The results of theorem 6, however, show that forging unanticipated documents is infeasible.

5.2 Membership Testing

Suppose a large group of people (perhaps the attendees of a cryptography conference) want to develop a mechanism which will allow participants to recognize each other at a later time. Several solutions are possible.

The attendees could simply produce a membership list and distribute the list amongst themselves. However, this requires each member to maintain a large and bulky membership list. In addition, if the members do not want outsiders to know their identities, these membership lists would have to be carefully guarded by all members. Thus, it is never possible for a member to be identified to a non-member.

An alternative solution would be for the group to appoint a trusted secretary. The secretary can digitally sign "id cards" for each member and post its own public verification key. Each member need only remember its own signed information and the secretary's public key. At a later time, one member can be identified to another by providing its own signed id card. Additionally, it is possible to give the secretary's public key to outsiders so the members can identify themselves to non-members. The problem, of course, is that the secretary must be trusted to not produce additional "phony" id cards for non-members.

One-way accumulators offer a solution with the advantages of a single trusted secretary but without the need for such an authority. Each member selects a y_j consisting of its name and any other desired identifying characteristics. A base x is agreed upon, and the members exchange their information and compute the accumulated hash value

$$z = h(h(h(\cdots h(h(h(x, y_1), y_2), y_3), \cdots, y_{m-2}), y_{m-1}), y_m).$$

Each member saves the hash function h, its own y_j, and the value z_j which represents the accumulated hash of all y_i with $i \neq j$. For the holder of y_j to prove that it is a member of the group, it need only present the pair (y_j, z_j). By verifying that $h(y_j, z_j) = z$, any other participant can authenticate the membership of the holder of y_j. Note that it is not even necessary for each participant to retain the accumulated hash value z since each participant would hold its own (y_j, z_j) pair from which $z = h(y_j, z_j)$ can be easily generated.

Also, non-members can be given the hash function h and the value of the accumulated hash z. Thus, any member that wishes to can identify itself to a non-member without revealing the entire membership list.

In [Merk80], Merkle describes a similar application in which a directory of public keys is to be jointly maintained. He describes a "tree authentication" solution to the problem in which each user must retain its own key, a hash function h, and a number of additional partial hashes which is logarithmic in the number of participants. By using one-way accumulators, the same properties can be achieved while reducing to a constant the number of values which must be retained by each participant.

6 Other Applications, Generalizations, and Further Work

The idea of one-way accumulators can be applied to a variety of other problems. The special advantage offered by accumulators over signatures is that no one individual need know how to authenticate/sign/stamp a document or message. Thus, a class of applications of one-way accumulators is as a simple and effective method of forming collective signatures. There seem to be a variety of cryptographic problems which are closely related to membership testing, and it seems likely that such problems may be amenable to the approach of one-way accumulators. Many other applications may also be possible.

David Naccache has observed that the function $e_{n,c} = x^y c^{y-1} \bmod n$ is quasi-commutative for all constants c. This is a direct generalization of the function $e_n(x, y) = x^y \bmod n = e_{n,1}(x, y)$ used within this paper. A possible advantage of this more general form is that it facilitates the use of efficient Montgomery processors.[4] Naccache also observes that the Dixon polynomial generating function

[4] Montgomery processors can, for certain constants c which depend solely upon n, compute the modular product $abc \bmod n$ as quickly as ordinary processors can compute the integer product ab.

$$g(x,k) = \sum_{i=0}^{\lfloor k/2 \rfloor} \frac{k}{k-i} \binom{k-i}{i} x^{k-2i}$$

is quasi-commutative. It is not known whether these functions have appropriate one-way properties.

Clearly the existence of one-way accumulators implies the existence of one-way functions. The question of whether or not the existence of one-way accumulators is implied by the existence of arbitrary one-way functions is an area for future research. No relationship is known between the existence of one-way accumulators and that of one-way trap-door functions.

A related open question is that of whether a candidate one-way accumulator can be found which does *not* have a trap-door. There is no apparent reason why this should not be possible, and such a function could alleviate the need for the secure multiparty computation required to select an appropriate modulus n for the function $e_n(x,y) = x^y \bmod n$.

Acknowledgements

The authors would like to express their thanks to Narsim Banavara, Laurie Benaloh, Ernie Brickell, Joshua Glasser, David Greenberg, Kevin McCurley, David Naccache, Janice Searleman, Satish Thatte, Dwight Tuinstra, and anonymous reviewers for their helpful comments and suggestions regarding this work. The authors would also like to express their thanks to Paul Giordano.

References

[Beav91] **Beaver, D.** "Efficient Multiparty Protocols Using Circuit Randomization." *Advances in Cryptology — Crypto '91*, ed. by J. Feigenbaum in *Lecture Notes in Computer Science*, vol. 576, ed. by G. Goos and J. Hartmanis. Springer-Verlag, New York (1992), 420–432.

[Beav89] **Beaver, D.** "Multiparty Protocols Tolerating Half Faulty Processors." *Advances in Cryptology — Crypto '89*, ed. by G. Brassard in *Lecture Notes in Computer Science*, vol. 435, ed. by G. Goos and J. Hartmanis. Springer-Verlag, New York (1990), 560–572.

[BeGo89] **Beaver, D.** and **Goldwasser, S.** "Multiparty Computation with Faulty Majority." *Proc. 30th IEEE Symp. on Foundations of Computer Science*, Research Triangle Park, NC (Oct.-Nov. 1989), 468–473.

[BeMa91] **Benaloh, J.** and **de Mare, M.** "Efficient Broadcast Time-Stamping." *Clarkson University Department of Mathematics and Computer Science TR 91-1.* (Aug. 1991).

[BenO83] **Ben-Or, M.** "Another Advantage of Free Choice: Completely Asynchronous Agreement Protocols." *Proc. 2nd ACM Symp. on Principles of Distributed Computing*, Montreal, PQ (Aug. 1983), 27–30.

[BGW88] **Ben-Or, M., Goldwasser, S.,** and **Wigderson, A.** "Completeness Theorems for Non-Cryptographic Fault-Tolerant Distributed Computation." *Proc. 20st ACM Symp. on Theory of Computation*, Chicago, IL (May 1988), 1–10.

[Brui51] **de Bruijn, N.** "The Asymptotic Behaviour of a Function Occurring in the Theory of Primes." *Journal of the Indian Mathematical Society 15.* (1951), 25–32.

[CCD88] **Chaum, D., Crépeau, C.,** and **Damgård, I.** "Multiparty Unconditionally Secure Protocols." *Proc. 20st ACM Symp. on Theory of Computation*, Chicago, IL (May 1988), 11–19.

[CGMA85] **Chor, B., Goldwasser, S., Micali, S.,** and **Awerbuch, B.** "Verifiable Secret Sharing and Achieving Simultaneity in the Presence of Faults." *Proc. 26th IEEE Symp. on Foundations of Computer Science*, Portland, OR (Oct. 1985), 383–395.

[Denn82] **Denning, D.** *Cryptography and Data Security*, Addison-Wesley, Reading, Massachusetts (1982).

[Fisc83] **Fischer, M.** "The Consensus Problem in Unreliable Distributed Systems", *Proc. 1983 International FCT-Conference*, Borgholm, Sweeden (Aug. 1983), 127–140. Published as *Foundations of Computation Theory*, ed. by M. Karpinski in *Lecture Notes in Computer Science*, vol. 158, ed. by G. Goos and J. Hartmanis. Springer-Verlag, New York (1983).

[GMW86] **Goldreich, O., Micali, S.,** and **Wigderson, A** "Proofs that Yield Nothing but Their Validity and a Methodology of Cryptographic Protocol Design." *Proc. 27th IEEE Symp. on Foundations of Computer Science*, Toronto, ON (Oct. 1986), 174–187.

[GMW87] **Goldreich, O., Micali, S.,** and **Wigderson, A** "How to Play Any Mental Game or A Completeness Theorem for Protocols with Honest Majority." *Proc. 19st ACM Symp. on Theory of Computation*, New York, NY (May 1987), 218–229.

[GoLe90] **Goldwasser, S.** and **Levin, L.** "Fair Computation of General Functions in Presence of Immoral Majority." *Advances in Cryptology — Crypto '90*, ed. by A. Menezes and S. Vanstone in *Lecture Notes in Computer Science*, vol. 537, ed. by G. Goos and J. Hartmanis. Springer-Verlag, New York (1991), 77–93.

[HaSt90] **Haber, S.** and **Stornetta, W.** "How to Time-Stamp a Digital Document." *Jounal of Cryptology 3.* (1991), 99–112.

[KnTr76] **Knuth, D.** and **Trabb Pardo, L.** "Analysis of a Simple Factorization Algorithm." *Theoretical Computer Science 3.* (1976), 321–348.

[ILL89] **Impagliazzo, R., Levin, L.,** and **Luby, M.** "Pseudorandom Generation from One-Way Functions." *Proc. 21st ACM Symp. on Theory of Computation*, Seattle, WA (May 1989), 12–24.

[LuWa69] **van de Lune, J.** and **Wattel, E.** "On the Numerical Solution of a Differential-Difference Equation Arising in Analytic Number Theory." *Mathematics of Computation 23.* (1969), 417–421.

[Merk80] **Merkle, R.** "Protocols for Public Key Cryptosystems." *Proc. 1980 Symp. on Security and Privacy*, IEEE Computer Society (April 1980), 122–133.

[MiRa90] **Micali, T.** and **Rabin, T.** "Collective Coin Tossing Without Assumptions nor Broadcasting." *Advances in Cryptology — Crypto '90*, ed. by A. Menezes and S. Vanstone in *Lecture Notes in Computer Science*, vol. 537, ed. by G. Goos and J. Hartmanis. Springer-Verlag, New York (1991), 253–266.

[Mitc68] **Mitchell, W.** "An Evaluation of Golomb's Constant." *Mathematics of Computation 22.* (1968), 411–415.

[NaYu89] **Naor, M.** and **Yung, M.** "Universal One-Way Hash Functions and their Cryptographic Applications." *Proc. 21st ACM Symp. on Theory of Computation*, Seattle, WA (May 1989), 33–43.

[RaBe89] **Rabin, T.** and **Ben-Or, M.** "Verifiable Secret Sharing and Multiparty Protocols with Honest Majority." *Proc. 21st ACM Symp. on Theory of Computation*, Seattle, WA (May 1989), 73–85.

[Rabi83] **Rabin, M.** "Randomized Byzantine Generals." *Proc. 24th IEEE Symp. on Foundations of Computer Science*, Tucson, AZ (Nov. 1983), 403–409.

[Romp90] **Rompel, J.** "One-Way Functions are Necessary and Sufficient for Secure Signatures." *Proc. 22nd ACM Symp. on Theory of Computation*, Baltimore, MD (May 1990).

[RSA78] **Rivest, R., Shamir, A.**, and **Adleman, L.** "A Method for Obtaining Digital Signatures and Public-key Cryptosystems." *Comm. ACM 21*, 2 (Feb. 1978), 120–126.

[Sham81] **Shamir, A.** "On the Generation of Cryptographically Strong Pseudo-Random Sequences." *Proc. ICALP*, (1981).

The breaking of the AR Hash Function

Ivan B. Damgård and Lars R. Knudsen

Aarhus University, Denmark

Abstract. The AR hash function has been proposed by ***Algorithmic Research Ltd*** and is currently being used in practice in the German banking world. AR hash is based on DES and a variant of the CBC mode. It produces a 128 bit hash value.
In this paper, we present two attacks on AR hash. The first one constructs in one DES encryption two messages with the same hash value. The second one finds, given an arbitrary message M, an $M' \neq M$ with the same hash value as M. The attack is split into two parts, the first part needs about 2^{33} DES encryptions and succeeds with probability 63%, the second part needs at most about 2^{66} DES encryptions and succeeds with probability about 99% of the possible choices of keys in AR. Moreover, the 2^{33} respectively 2^{66} encryptions are necessary only in a one-time preprocessing phase, i.e. having done one of the attacks once with success, a new message can be attacked at the cost of no encryptions at all. Since the hash value is 128 bits long, the times for the attacks should be compared to 2^{64}, resp. 2^{128} DES encryptions for brute force attacks.
For the particular keys chosen in AR hash we implemented the first part of the second attack. In 2^{33} encryptions we found two messages that breaks AR hash.

1 The AR Hash Function

The AR hash function has been proposed by Algorithmic Research Ltd., it has been distributed in the ISO community [1] for informational purposes, but has not been considered a standard. It is currently in use in the German banking world.

In the following, $DES_k(y)$ will denote the DES-encryption of block y using key k.

The basic structure in AR-hash can be described as a variant of DES in CBC-mode, where the last 2 ciphertext blocks are added to the current input, and where the state consists of the last two "ciphertext" blocks computed. To do the entire function, the message is processed with two keys, yielding a result of 2 times 128 bits. This is then further compressed to get a result of 128 bits.

To define AR more precisely, we first divide the message m to be hashed into 8-byte blocks, denoted by $m_1, m_2, ..., m_n$ (0-padding is used on the last block if it is incomplete).

We then define a series of 64-bit blocks $o_{-1}, o_0, o_1, ..$by

$$o_{-1} = o_0 = 0$$

and

$$o_i = m_i \oplus DES_k(m_i \oplus o_{i-1} \oplus o_{i-2} \oplus \eta),$$

where k is an arbitrary DES key, and the constant η is defined by

$$\eta = 01\ 23\ 45\ 67\ 89\ AB\ CD\ EF$$

in hexadecimal notation. We now let $f_1(m,k), f_2(m,k)$ denote o_{n-1}, o_n, respectively.

In the actual hash function AR/DFP, two different keys k_1 and k_2 are used, specified as

$$k_1 = 00\ 00\ 00\ 00\ 00\ 00\ 00\ 00, \quad k_2 = 2A\ 41\ 52\ 2F\ 44\ 46\ 50\ 2A$$

One then first computes

$$c_1 = f_1(m,k_1),\ c_2 = f_2(m,k_1),\ c_3 = f_1(m,k_2),\ c_4 = f_2(m,k_2)$$

and the hash value is now the concatenation of the two 8 byte blocks

$$G(G(c_1,c_2,k_1), G(c_3,c_4,k_1), k_1) \text{ and } G(G(c_1,c_2,k_2), G(c_3,c_4,k_2), k_2),$$

where G is the function defined by

$$G(x,y,k) = DES_k(x \oplus y) \oplus DES_k(x) \oplus DES_k(y) \oplus y.$$

For convenience in the following, we will let $DFP(c_1,c_2,c_3,c_4,k)$ denote the final hash result.

2 Properties of f_1, f_2 and G

In the following, let A and B be messages of length a multiple of 8 bytes, and let $A|B$ be the concatenation of A and B. Choose a fixed, but arbitrary DES key k, and let $y = f_1(A,k)$, $z = f_2(A,k)$. Let m be an arbitrary 8-byte block. Let $C(A,m)$ be the three-block message

$$m \oplus \eta \oplus y \oplus z|\ DES_k(m) \oplus y|\ DES_k(m) \oplus z$$

Let $D(A,m)$ be the three-block message

$$m \oplus \eta \oplus y \oplus z|\ m \oplus y|\ m \oplus z$$

Let $E(A,m)$ be the three-block message

$$m \oplus \eta \oplus y \oplus z|\ m \oplus y|\ DES_k^2(m) \oplus z$$

Then we have the following result, showing that it is very easy to find collisions for the functions f_1, f_2:

Lemma 1 *For arbitrary A, B, k, m as above, we have that*

$$f_i(A|\ B, k) = f_i(A|\ C(A, m)|\ B, k), i = 1, 2$$

$$f_2(A, k) = f_2(A|\ E(A, m), k)$$

If k is a weak DES key, then we also have

$$f_i(A|\ B, k) = f_i(A|\ D(A, m)|\ B, k), i = 1, 2$$

Proof: By combining the definition of $C(A, m)$ and f_1, f_2 and by letting f_0 be the hash value produced just before f_1 we obtain

$$\begin{aligned}
f_0(A|\ C(A, m), k) &= m \oplus \eta \oplus y \oplus z \oplus DES_k(m \oplus \eta \oplus y \oplus z \oplus y \oplus z \oplus \eta) \\
f_1(A|\ C(A, m), k) &= DES_k(m) \oplus y \oplus DES_k(DES_k(m) \oplus y \oplus m \oplus \\
&\quad \eta \oplus y \oplus z \oplus DES_k(m) \oplus z \oplus \eta) \\
&= y \\
f_2(A|\ C(A, m), k) &= DES_k(m) \oplus z \oplus DES_k(DES_k(m) \oplus z \oplus y \oplus \\
&\quad m \oplus \eta \oplus y \oplus z \oplus DES_k(m) \oplus \eta) \\
&= z
\end{aligned}$$

This proves the first statement. The second and third are proved similarly, using for the third that if k is a weak key, then by definition we have that $DES_k(DES_k(m)) = m$ for all m. □

By inspection of the definition of G, it is trivial to show the following lemma:

Lemma 2 *The functions G, DFP have the following properties for arbitrary c_1, c_2, k:*

$$G(c_1, c_2, k) = G(c_1 \oplus c_2, c_2, k)$$

$$G(c_1, 0, k) = DES_k(0)$$

$$DFP(c_1, c_1, c_1, c_1, k) = (c_1, c_1), \quad DFP(c_1, 0, c_2, 0) = 0$$

Thus, it is also very easy to find collisions for G and DFP.

Although none of these properties imply directly a collision for the hash function itself, they will be useful in the following.

3 Attacks on AR Hash

3.1 Collision attack

A collision attack finds two messages m and m' that hash to the same value. This first attack on AR hash exploits the fact that for a weak key k it is easy to find fixpoints for DES, i.e. to find m s.t. $DES_k(m) = m$. There are exactly 2^{32} such fixpoints for a weak key [2] and each fixpoint can be found in half a DES encryption. Since all round keys for a weak key are equal, a necessary and sufficient condition for a fixpoint is that the halves of the encrypted value after 8 rounds of encryption are equal.

If A is the empty message in Lemma 1, then $y = z = 0$. Let $X(m)$ be the 3-block message $m \oplus \eta|\ DES_{k_2}(m)|\ DES_{k_2}(m)$. This means that by Lemma 1

$$f_1(X(m), k_2) = 0$$
$$f_2(X(m), k_2) = 0$$

for any m. Let m be a fixpoint for k_1, then

$$f_1(X(m), k_1) = DES_{k_2}(m) \oplus DES_{k_1}(DES_{k_2}(m))$$
$$f_2(X(m), k_1) = DES_{k_2}(m) \oplus DES_{k_1}(DES_{k_1}(DES_{k_2}(m))) = 0$$

since k_1 is a weak key. The above four values are also the c_i values produced by hashing $X(m)$. But by Lemma 2, a G-value is invariant in the first argument if the second is 0, so it is clear that for fixpoints (for k_1) $m \neq m'$, $X(m)$ and $X(m')$ will be hashed to the same value. Finding two fixpoints for k_1 takes in time one DES encryption, which leads to:

Theorem 1 *There exists an algorithm, which finds in time one DES encryption, two different messages with the same AR hash value.*

The above attack can be extended to attacks that in time $n/2$ encryptions find n messages that hash to the same value, where $n \leq 2^{32}$. By contrast, a brute force attack that finds two messages that hash to the same value would require computation of about 2^{64} hash values.

3.2 Preimage attack

A preimage attack takes a given message M as input and tries to find a new message with the same hash value.

AR hash uses two fixed keys. In the following we consider arbitrary keys, where one key, k_1, is a weak key[1].

The basic idea in this second attack on AR hash is to try to find a message which takes the initial state back to itself, i.e. leads to a set of all-zero c-values. If Z is such a message, then clearly $AR(M) = AR(Z|M) = AR(Z|Z|M) = \cdots$. It is also clear that once we have found such a Z, any message M can be attacked at no further cost.

In more detail, we try, inspired by Lemma 1, with Z of the form $Z = m_1 \oplus \eta|m_2|m_2$. It is now easy to write down the equations that m_1, m_2 must satisfy in order for $f_1(Z, k_i) = f_2(Z, k_i) = 0, i = 1, 2$. We get the following:

$$DES_{k_1}(m_1) \oplus m_1 = DES_{k_1}^{-1}(m_2) \oplus m_2 \quad (1)$$

$$DES_{k_2}(m_1) \oplus m_1 = DES_{k_2}^{-1}(m_2) \oplus m_2 \quad (2)$$

[1] The DES has 4 weak keys.

It is difficult in general to say anything about the number of solutions to these equations, or how hard it is to find them. There is a special case, however, that is easier:

Let m_1 be a fixpoint for k_1. Put $m_2 = DES_{k_2}(m_1)$. Then (2) is always satisfied and (1) is true if

$$DES_{k_1}(DES_{k_2}(m_1)) = DES_{k_2}(m_1) \tag{3}$$

which is true if also $DES_{k_2}(m_1)$ is a fixpoint for k_1. It is reasonable to assume that the mapping $DES_{k_2}(\cdot)$ distributes fixpoints for k_1 uniformly. Therefore the probability that $DES_{k_2}(m_1)$ is a fixpoint for k_1 is 2^{-32}. By running through all fixpoints for k_1 the probability that (3) is satisfied is

$$1-(1-2^{-32})^{2^{32}} \simeq 1-e^{-1} \simeq 0.63$$

Since checking whether a message is a fixpoint for a weak key takes half a DES encryption, the attack needs a total of $2 \times 2^{32} = 2^{33}$ DES encryptions. A similar attack appeared in [3].

To confirm the validity of the 0.63 probability, we did a computer simulation on a "scaled-down" version of DES, working with 32-bit blocks, thus making it easy to run through all fixpoints. The experiments confirmed the theory. The test ran through all 2^{16} fixpoints for 100 pairs of keys, where one key was a weak key in a 32 bit block version of DES. Out of 100 key pairs, the equation (3) had a solution for 62 pairs.

The above attack is quite feasible, and can be executed in at most a few days, even hours, using up to date hardware. Later in this section we give the results of an implementation of the attack on AR hash with the two keys given in [1].

The above probability can be improved to almost 1 on the cost of a squared complexity. In this case we proceed as follows (where m_1 is not necessarily a fixpoint for k_1):

If we put $m_2 = DES_{k_1}(m_1)$, then equation (1) is trivially satisfied, and (2) is satisfied as well, if

$$DES_{k_1}(m_1) = DES_{k_2}(m_1) \tag{4}$$

or

$$DES_{k_2}(m_1) \oplus m_1 = DES_{k_1}(m_1) \oplus DES_{k_2}^{-1}(DES_{k_1}(m_1)) \tag{5}$$

Symmetrically, we can put $m_2 = DES_{k_2}(m_1)$. This means that (2) is now always satisfied, and that (1) is true if either $DES_{k_1}(m_1) = DES_{k_2}(m_1)$ (same condition as (4)) or if

$$DES_{k_1}(m_1) \oplus m_1 = DES_{k_2}(m_1) \oplus DES_{k_1}^{-1}(DES_{k_2}(m_1)) \tag{6}$$

Finally, since k_1 is a weak key, there is another possibility, namely to put $m_1 = m_2$. Once again, this trivially satisfies (1), and (2) is in this case satisfied, if

$$DES_{k_2}^2(m_1) = m_1 \tag{7}$$

To summarize, if we can find a 64-bit block m_1 that satisfies (4), (5), (6) or (7) then we have a 3-block sequence Z that makes the attack successful. Checking if a block satisfies any of the equations requires at most 5 encryptions, so going through all possibilities for m_1 will require about $5 \cdot 2^{64} \simeq 2^{66}$ encryptions.

The remaining question is of course if there are any solutions to the equations at all. Simply doing the 2^{66} encryptions is not feasible today (although it probably will become feasible in the not too distant future). Therefore the best we can do is to see if we can estimate the probability that solutions exist, assuming that the two keys k_1, k_2 are randomly chosen, but where k_1 is a weak key.

Each of the 4 equations can be written in the form $h(m_1) = 0$, where h is some function that depends on the keys, and is built from a number of DES en- and decryptions. It is a generally accepted assumption that DES in a context like this one behaves like a random function. This means that the 3 equations (4),(5) and (7) each have solutions with an independent probability of

$$1 - (1 - 2^{-64})^{2^{64}} \simeq 1 - e^{-1} \simeq 0.63$$

However since (6) contains (3) as a special case this probability splits into two depending on whether fixpoints are examined or not, the probability that (6) has a solution therefore is

$$1 - ((1 - 2^{-64})^{2^{64}-2^{32}} \times (1 - 2^{-32})^{2^{32}}) \simeq 1 - e^{-2}$$

Thus we expect that the probability over the choice of k_1, k_2 with k_1 weak that solutions do exist is about $1 - e^{-5} \simeq 0.99$.
In summary we have the following:

Theorem 2 *There exists two attacks on AR hash that constructs from a given message M a new one $M' \neq M$ such that $AR(M) = AR(M')$. The attacks takes time at most about 2^{33} and about 2^{66} DES encryptions, respectively. Under reasonable heuristic assumptions, the attacks can be shown to be successful for respectively about 63% and 99% of the possible choices of keys in AR hash. Both attacks can be done in a preprocessing phase, after which each message can be attacked at no further cost.*

These attacks are much faster than a brute-force attack, which would require computation of about 2^{128} hash values.

For the keys chosen in AR hash we did an exhaustive search through all fixpoints for the weak key, $k_1 = 0$. We obtained

Theorem 3 *For AR hash there exists two 3-block messages Z_1 and Z_2, s.t. any message M can be prefixed with either Z_1 or Z_2 (or both) any number of times, yielding unchanged AR hash value, where*

Z_1 = 7a6199a238bb8643 | 8073d91a57ca1e2a | 8073d91a57ca1e2a
Z_2 = 02bb2604aafcbecf | 6421e999f02ddfd6 | 6421e999f02ddfd6

4 Conclusion

The weaknesses we have found in AR hash clearly make it very problematic to continue using the hash function as it is. The collision and preimage attacks can be thwarted by adding the message length to the message, however because of Theorem 3 collisions still can be obtained in constant time, because $Z_1|M$ and $Z_2|M$ would hash to the same value.

So the question arises whether one can repair the function so that our attacks are prevented.

We have of course exploited the fact, that there are 2^{32} fixpoints for a weak DES key and that they are easy to find. However, avoiding weak keys still would enable a preimage attack, since equations (4), (5) and (6) can be set up independently of the nature of the keys. The probability for success for this attack is expected to be $1 - e^{-3} \simeq 95\%$.

To confirm this we did another computer simulation on a "scaled-down" version of DES. The test used 16 bit blocks and ran through all 2^{16} possible messages for 100 pairs of random keys. Out of 100 key pairs, for only 3 key pairs none of the equations (4), (5) and (6) had solutions thus confirming the theory.

Furthermore we made essential use of the fact that the initial state is all-zero, in particular that it consists of 4 blocks that are equal. Trying to prevent attacks only by changing the initial values is extremely dangerous and it is shown in [3] how to find collisions even in this case.

Section 2 shows a number of problematic properties of f_1, f_2 and G that are independent on the initial state and on the chosen keys, we therefore believe that the basic design of f_1, f_2 and G should be reconsidered. One can perhaps guess that AR hash (or rather the f_1, f_2 functions) was designed starting from the standard MAC-mode for DES (which uses a secret key), obtaining a hash function by using a known, fixed key, and adding some extra elements (the feed-forward, etc.) to compensate for the weaknesses implied by the fact that the key is now known.

Our attacks can be seen as an illustration that constructing a hash function in this way from a MAC is not easy, and that it is perhaps a better strategy to build a hash function mode "from scratch".

5 Acknowledgements

We would like to thank Dr. Bart Preneel for helpful discussions and comments.

References

1. *AR fingerprint function.* ISO-IEC/JTC1/SC27/WG2 N179, working document, 1992.
2. D. Coppersmith. *The real reason for Rivest's phenomenon.* Proceedings of Crypto 85, Springer Verlag LNCS series.
3. B. Preneel. *Analysis and Design of Cryptgraphic Hash Functions.* Ph.D. Thesis, Katholieke Universiteit Leuven, January 1993.

Collisions for the compression function of MD5

Bert den Boer
Philips Crypto B.V.
P.O. Box 218
5600 MD Eindhoven
The Netherlands

Antoon Bosselaers
ESAT Laboratory, K.U. Leuven
Kard. Mercierlaan 94
B-3001 Heverlee, Belgium
antoon.bosselaers@esat.kuleuven.ac.be

Abstract. At Crypto '91 Ronald L. Rivest introduced the MD5 Message Digest Algorithm as a strengthened version of MD4, differing from it on six points. Four changes are due to the two existing attacks on the two round versions of MD4. The other two changes should additionally strengthen MD5. However both these changes cannot be described as well-considered. One of them results in an approximate relation between any four consecutive additive constants. The other allows to create collisions for the compression function of MD5. In this paper an algorithm is described that finds such collisions.
A C program implementing the algorithm establishes a work load of finding about 2^{16} collisions for the first two rounds of the MD5 compression function to find a collision for the entire four round function. On a 33MHz 80386 based PC the mean run time of this program is about 4 minutes.

1 Introduction

The MD5 Message Digest Algorithm [Rive91, Rive92b, Schn91] introduced by Ronald L. Rivest at Crypto '91 as a strengthened version of MD4 [Rive90, Rive92a] differs from MD4 on the following points:

- A fourth round has been added.
- The second round function has been changed from the majority function $XY \vee XZ \vee YZ$ to the multiplexer function $XZ \vee Y\overline{Z}$.
- The order in which input words are accessed in rounds 2 and 3 is changed.
- The shift amounts in each round have been changed. None are the same now.
- Each step now has a unique additive constant.
- Each step now adds in the result of the previous step.

The first four changes are clearly a consequence of the two existing attacks on the two round versions of MD4 [Merk90, dBBo91]. The last two changes should additionally strengthen MD5. However both these changes can hardly be described as well-considered.

The unique additive constant in step k contains the first 32 bits of the absolute value of $\sin(k)$. This together with the following relation between four consecutive sine values

$$(\sin(k) + \sin(k+2))\sin(k+2) = (\sin(k+1) + \sin(k+3))\sin(k+1)$$

establishes an approximate relation between any four consecutive additive constants. This could be easily avoided by choosing the next 32 bits in the binary expansion of the sine values.

The last change however has more serious implications: adding in the result of the previous step allows to create collisions for the compression function of MD5. In this paper an algorithm that finds such collisions is described. This means that one of the design principles behind MD5, namely to design a collision resistant hash function based on a collision resistant compression function, is not satisfied. The entire 640-bit input of the compression function is used to produce these collisions. Therefore they do not result in an attack on the MD5 hash function, having a single and fixed 128-bit initial value. This is why they are sometimes called pseudo-collisions.

In Section 2 the necessary notation and definitions are introduced. Section 3 describes and explains the actual collision search algorithm. Section 4 contains a discussion about the optimal value for a constant of the collision search algorithm. Finally, in Section 5 some details on the implementation as well as an example collision are given.

2 Notation and definitions

The following notation will be used:

XY, $X \vee Y$, $X \oplus Y$	respectively the bitwise AND, OR and XOR of X and Y
$\overline{X}$	the bitwise complement of X
$X \lll s$, $X \ggg s$	the rotation of X to respectively the left and right by s bit positions
$V \leftarrow E$	assign to variable V the value of the expression E
MSB, LSB	respectively most and least significant bit

A *word* is defined as an unsigned 32-bit quantity taking on only nonnegative values. The word wise application of the operation $\star$ on two 4-word buffers (A_1, B_1, C_1, D_1) and (A_2, B_2, C_2, D_2) is denoted by

$$(A_1, B_1, C_1, D_1) \star (A_2, B_2, C_2, D_2) = (A_1 \star A_2, B_1 \star B_2, C_1 \star C_2, D_1 \star D_2).$$

MD5 uses the following four functions (one for each round) to process the input. They all take a 3-word input and produce a single word of output.

$$f_1(X,Y,Z) = XY \vee \overline{X}Z \qquad f_3(X,Y,Z) = X \oplus Y \oplus Z$$
$$f_2(X,Y,Z) = XZ \vee Y\overline{Z} \qquad f_4(X,Y,Z) = (X \vee \overline{Z}) \oplus Y$$

Each round i $(1 \leq i \leq 4)$ consists of 16 steps, each of which contains a single application of the round function f_i. Hence each round function is used 16 times. In addition each step of round i uses one of the shift constants $si1$, $si2$, $si3$ or $si4$, each of which is used four times in each round (see Table 1). In total there are 48 steps in MD5, grouped in 4 rounds of 16 steps and numbered from 1

	Round i			
	1	2	3	4
$si1$	7	5	4	6
$si2$	12	9	11	10
$si3$	17	14	16	15
$si4$	22	20	23	21

Table 1. The 16 different shift constants of MD5

(step 1 of round 1) through 48 (step 16 of round 4). For a complete specification of MD5 the reader is referred to the original description [Rive92b, Schn91]. Note that in this original description of MD5 the designation f, g, h and i are used for respectively the round functions f_1, f_2, f_3 and f_4.

3 Description of the collision search algorithm

First the condition imposed on two inputs to produce the same image under the compression function of MD5 will be translated to a condition on the inputs to the round function of each step of MD5. Next we will show how these conditions can be (easily) met for the third and fourth round. Finally we will derive an algorithm that generates an input meeting the conditions for the first and second round.

3.1 Derivation of the round function input condition

The basis of the MD5 algorithm is a compression function G that takes as input a 4-word buffer (A, B, C, D) and a 16-word message block $(X[0], X[1], \ldots, X[15])$, and produces a 4-word output (AA, BB, CC, DD):

$$(AA, BB, CC, DD) = G((A, B, C, D), (X[0], X[1], \ldots, X[15])).$$

The idea of the collision search algorithm is to produce an input to the compression function such that complementing the MSB of each of the 4 words of the buffer (A, B, C, D) has no influence on the output of the compression function. In other words, finding an (A, B, C, D) and an $(X[0], X[1], \ldots, X[15])$ such that

$$G((A, B, C, D) \oplus (2^{31}, 2^{31}, 2^{31}, 2^{31}), (X[0], X[1], \ldots, X[15])) = G((A, B, C, D), (X[0], X[1], \ldots, X[15])). \quad (1)$$

The compression function G of MD5 consists of four 16-step rounds enclosed by a feedforward, that adds (modulo 2^{32}) to each of the 4 words A, B, C and D at the end of the fourth round the values they had at the beginning of the first round. Hence

$$G(A, B, C, D) = H(A, B, C, D) + (A, B, C, D), \quad (2)$$

where H consists of the four 16-step rounds. Substituting G in (1) by (2) together with the fact that $(A + 2^{31}) \bmod 2^{32} = A \oplus 2^{31}$ means that we are looking for an (A, B, C, D) and an $(X[0], X[1], \ldots, X[15])$ such that

$$H((A,B,C,D) \oplus (2^{31}, 2^{31}, 2^{31}, 2^{31}), (X[0], X[1], \ldots, X[15])) = \\ H((A,B,C,D), (X[0], X[1], \ldots, X[15])) \oplus (2^{31}, 2^{31}, 2^{31}, 2^{31}).$$

Consider a step of the MD5 algorithm

$$A \leftarrow B + ((A + f_i(B,C,D) + X[j] + t) \lll s),$$

where

- $1 \leq i \leq 4$,
- $X[j]$ is one of the 16 message words ($0 \leq j \leq 15$),
- t is the unique additive constant of the step,
- s is one of the 16 possible shift amounts $si1$, $si2$, $si3$ or $si4$ ($1 \leq i \leq 4$), and
- all additions are modulo 2^{32}.

The new value of the word A is obtained by adding to the result of the previous step the result of an addition rotated over s bits to the left. Complementing the MSB of each of the 4 words A, B, C and D in the right hand side of this assignment will result in a complementation of the MSB of the updated A, if the MSB of $f_i(B,C,D)$ is complemented when the MSBs of B, C and D are complemented. This observation leads to the following proposition.

Proposition 1. *Let T be a 20-word input to the compression function G and let X, Y and Z be the MSBs of the 3-word input to a round function f_i. If T produces in all steps inputs to the round functions f_i for which*

$$f_i(\overline{X}, \overline{Y}, \overline{Z}) = \overline{f_i(X,Y,Z)}$$

then the 20-word input T and the 20-word input in which the MSBs of the first four words of T are complemented have the same image under G.

Note that this is made possible by adding in, in each step, the result of the previous step. This is why this attack does not work for MD4. Note also that this collision has the property that the message part of the input is the same.

Proposition 2. *The condition $f_i(\overline{X}, \overline{Y}, \overline{Z}) = \overline{f_i(X,Y,Z)}$ is met by the following 3-tuples (X,Y,Z) for respectively f_1, f_2, f_3, and f_4.*

1. $(0,0,0)$, $(1,0,0)$ *and their complements* $(1,1,1)$ *and* $(0,1,1)$,
2. $(0,0,0)$, $(0,0,1)$ *and their complements* $(1,1,1)$ *and* $(1,1,0)$,
3. *all inputs,*
4. $(0,0,0)$, $(0,1,0)$ *and their complements* $(1,1,1)$ *and* $(1,0,1)$.

Proof.

1. $\overline{XY} \vee X\overline{Z} = \overline{XY \vee \overline{X}Z}$

   $\Leftrightarrow (\overline{\overline{XY} \vee X\overline{Z}}) \oplus (XY \vee \overline{X}Z) = 0$

   $\Leftrightarrow (X \vee Y)(\overline{X} \vee Z) \oplus (XY \vee \overline{X}Z) = 0$

   $\Leftrightarrow (\overline{X}Y \vee XZ) \oplus (XY \vee \overline{X}Z) = 0$

   $\Leftrightarrow Y \oplus Z = 0$

2. the same as above, but with X and Z interchanged.
3. $\overline{X \oplus Y \oplus Z} = \overline{X} \oplus Y \oplus Z = X \oplus \overline{Y} \oplus Z$.
4. $(\overline{X} \vee Z) \oplus \overline{Y} = \overline{(X \vee \overline{Z})} \oplus Y$

   $\Leftrightarrow (\overline{X} \vee Z) \oplus (X \vee \overline{Z}) = 0$

   $\Leftrightarrow X \oplus Z = 0$

□

3.2 Collisions for round 3 and 4

From Proposition 2 it follows that a random 20-word input to round 4 has a probability of 2^{-16} of fulfilling the condition $f_4(\overline{X}, \overline{Y}, \overline{Z}) = \overline{f_4(X, Y, Z)}$ for all 16 steps of the round. Round 3 imposes according to the same proposition no additional constraints. Due to the pseudo-random behaviour of round 3 it is save to assume that the input at the beginning of round 4 does not significantly deviate from a random one. A 20-word random input has therefore the same probability of 2^{-16} of meeting all conditions in both round 3 and 4. It remains to produce enough 20-word inputs fulfilling all conditions in the first two rounds in order to generate a collision for the compression function G.

3.3 Collisions for round 1 and 2

According to Proposition 2 the condition $f_1(\overline{X}, \overline{Y}, \overline{Z}) = \overline{f_1(X, Y, Z)}$ is met by both $(1,1,1)$ and $(1,0,0)$, and their complements. However in each step of the first round only one of A, B, C and D is updated. Therefore an appearance of $(1,0,0)$ in a particular step will lead to $(x,1,0)$ in the next step (where x is either 1 or 0). Since $f_1(\overline{x},0,1)$ is not equal to $\overline{f_1(x,1,0)}$, $(1,0,0)$ cannot appear as input to the function f_1 in the course of the first round. The same applies to its complement $(0,1,1)$ and to the inputs $(0,0,1)$ and $(1,1,0)$ to the second round function f_2. Hence only $(1,1,1)$ or its complement $(0,0,0)$ are allowed as inputs to the first and second round functions f_1 and f_2. This input condition is met by keeping the MSBs of A, B, C and D in the first two rounds equal to one, except for the value of A at the beginning of the first round and the value of B at the end of the second round, for which there are no constraints: they are not used as input to f_1 or f_2. The idea of the algorithm is therefore to choose the 16 words $X[0], X[1], \ldots, X[15]$ in precisely such a way that all the input words to the f_1 and f_2 function keep their MSB on one during the first two rounds. This is done in the following way.

We start halfway the first two rounds by generating random A, B, C and D values between the first and second round with MSBs equal to one. We walk through the second round making all the updated buffer words equal to a "magic value" N by specific choices for the 16 message words $X[0]$ through $X[15]$. This is called the forward walk. The best choice for N depends on the actual values of the shift constants in the first two rounds and will be discussed in Section 4. For the current values of the shift constants the best choice for N is F8000000 (hexadecimal notation).

Next we check whether the choices for the message words made in the second round are also good choices for the first round, i.e., whether they keep the MSB of the buffer words in the first round on one. We therefore start at the end of the first round and walk through the first round in the reverse direction. This is called the backward walk. When we find a buffer word with zero MSB, we adapt the most significant part of the message word used in that particular step in such a way that the buffer word now approximates the magic value N. We then once again start the forward walk at the second round step where this message word is used, and check whether this change has any influence on the MSBs of the remaining buffer words of the second round. If so, we make the necessary changes to the other (i.e., least significant) part of the message words in order that the buffer words approximate once again the magic value. These least significant parts of the message words become the most significant after the rotation in the forward walk steps. Next we start once again the backward walk. This way we go to and fro until we reach the beginning of the first round, at which point we found a message block keeping the MSBs of the buffer words in the first two rounds on one.

First a description of the initialization procedure is given, which consists of a forward walk and partial backward walk.

1. *Initialize* (A, B, C, D).
   Generate random A, B, C, D values between the first and second round with MSBs equal to one.
2. *Initialize* $(X[0], X[1], \ldots, X[15])$.
   2.1 Step forwards (i.e., into round 2) and make the updated buffer words in the first six steps of round 2 (step 17 through 22) equal to the magic value N by a specific choice of the message words used in the first six steps: respectively $X[1]$, $X[6]$, $X[11]$, $X[0]$, $X[5]$ and $X[10]$.
   2.2 Do the next step (step 23) forwards making the updated value of C equal to N by a specific choice for X[15].

$$X[15] = ((N - D) \ggg s23) - C - f_2(D, A, B) - 3634488961$$

   Do the last step of the first round (step 16) backwards making the value of B at the beginning of step 16 equal to N by another specific choice for X[15].

$$X[15] = ((B - C) \ggg s14) - N - f_1(C, D, A) - 1236535329$$

Of course we get different values for X[15] but we take the $s23$ MSBs of the backward step solution and the other $32 - s23$ bits of the forward step solution. This way both newly computed values of C (forward step) and B (backward step) are approximations of N:

$$C' = D + (C + f_2(D, A, B) + X[15] + 3634488961) \lll s23,$$

$$B' = ((B - C) \ggg s14) - X[15] - f_1(C, D, A) - 1236535329.$$

In the forward step the $s23$ bits of the backward solution become the LSBs after the rotation of the sum over $s23$ bits, in the backward step the bits of the forward solution are on the least significant positions as well.

2.3 Step forwards (steps 24 and 25) computing $X[4]$ and $X[9]$ as in step 2.1
2.4 Put $X[14]$ equal to the $s22$ MSBs of the backward solution of step 15 and the $32 - s22$ LSBs of the forward solution of step 26.
2.5 Step forwards (steps 27 and 28) computing $X[3]$ and $X[8]$ as in step 2.1
2.6 Put $X[13]$ equal to the $s21$ MSBs of the backward solution of step 14 and the $32 - s21$ LSBs of the forward solution of step 29.
2.7 Step forwards (steps 30 and 31) computing $X[2]$ and $X[7]$ as in step 2.1
2.8 Put $X[12]$ equal to the backward solution of step 12, as there are no constraints on the value of B at the end of the second round (step 32).

Next an informal and formal description of the actual algorithm is given. First we define three functions used in these descriptions. Let

- $s2[j]$ be the shift constant used in step j of the second round ($17 \le j \le 32$).
- $fw[i]$ be the step in the forward walk (i.e., the second round) using the input word $X[i-1]$ ($1 \le i \le 16$),
- $bw[j]$ be the step in the backward walk (i.e., the first round) using the message word that is used in the jth step of the forward walk ($17 \le j \le 32$). Hence the functions $fw[]$ and $bw[]$ are each others inverse: if $j = fw[i]$ is the step in the forward walk using $X[i-1]$, then $i = bw[j]$ is the step in the backward walk using the message word that is used in the jth step of the forward walk (i.e., $X[i-1]$).

After the initialization of both (A, B, C, D) and $(X[0], X[1], \ldots, X[15])$ as already described, we step backwards checking whether our choices for the $X[\cdot]$'s so far are also good choices for the backward walk i.e., whether *at the beginning* of each first round step the MSB of the buffer word being updated is equal to one. If that is not the case for step i the first $s2[fw[i]]$ (i.e., the shift constant of the step in the forward walk using $X[i-1]$) bits of $X[i-1]$ are adapted such that the value of the buffer word at the beginning of that step is, given these limitations, the best possible approximation of the magic value N. Alas, now all values in the forward walk from step $fw[i]$ onwards change. The first changes are mild, but soon they will accumulate. But as long as the MSBs of the buffer words A, B, C and D do not change, we keep the $X[\cdot]$ values as they are. However if in step j of the forward walk the MSB of a buffer word changes, we adapt all or part of the bits of the message word used in that step (i.e., $X[bw[j]-1]$) to let the updated value of the buffer word approximate once again the magic value N. For this purpose we can use all bits of $X[bw[j]-1]$ in case, up to this point, it has not been used yet in the backward walk (i.e., if $bw[j] < i$). Otherwise we combine the forward and backward solutions for $X[bw[j]-1]$. Having completed the entire forward walk in the same way, we once again start the backward walk at step k, where $X[k-1]$ is the message word with the highest index that was changed in the forward walk, and we check whether these new choices for the $X[\cdot]$'s are also good choices for the backward walk. This way we go to and fro, until we find a solution meeting all conditions in both rounds. Below the formal description of the algorithm is given together with a flowchart in Figure 1.

3. *The actual algorithm.*
   - 3.0 Set $i \leftarrow 12$.
   - 3.1 If $i = 1$, a solution has been found as there are no constraints on the value of A at the beginning of the first round.
   - 3.2 Do step i backwards. The value at the beginning of step i of the buffer word that is updated in this step, is calculated using the known value at the end of the step and the value of $X[i-1]$ from the forward walk.
   - 3.3 If the MSB of the new value is 1, decrement i and goto 3.1.
   - 3.4 Set $j \leftarrow fw[i]$, $k \leftarrow i$ (k keeps track of the highest first round step using a message word that has been adapted during the forward walk). Adapt the $s2[j]$ MSBs of $X[i-1]$ to let the value of the buffer word *at the beginning* of first round step i approximate the magic value N.
   - 3.5 If $j = 32$, set $i \leftarrow k$ and goto 3.1, as there are no constraints on the value of B at the end of the second round.
   - 3.6 Do step j forwards.
   - 3.7 If the MSB of the updated buffer word is 1, increment j and goto 3.5.
   - 3.8 If $bw[j] < i$, compute $X[bw[j]-1]$ as in step 2.1 (i.e., if the message word used in this step has not been used yet in the backward walk, then use all the bits of this message word to make the updated value of the buffer word equal to N). Increment j and goto 3.5.
   - 3.9 Adapt the $32 - s2[j]$ LSBs of $X[bw[j]-1]$ to let the updated value of the buffer word in step j approximate the magic value N (i.e., in case the message word used in this step has already been used in the backward walk).
   - 3.10 If $bw[j] > k$, set $k \leftarrow bw[j]$ (the highest first round step so far using a message word that has been changed during this forward walk, and hence the place to start a new backward walk).
   - 3.11 Increment j and goto 3.5.

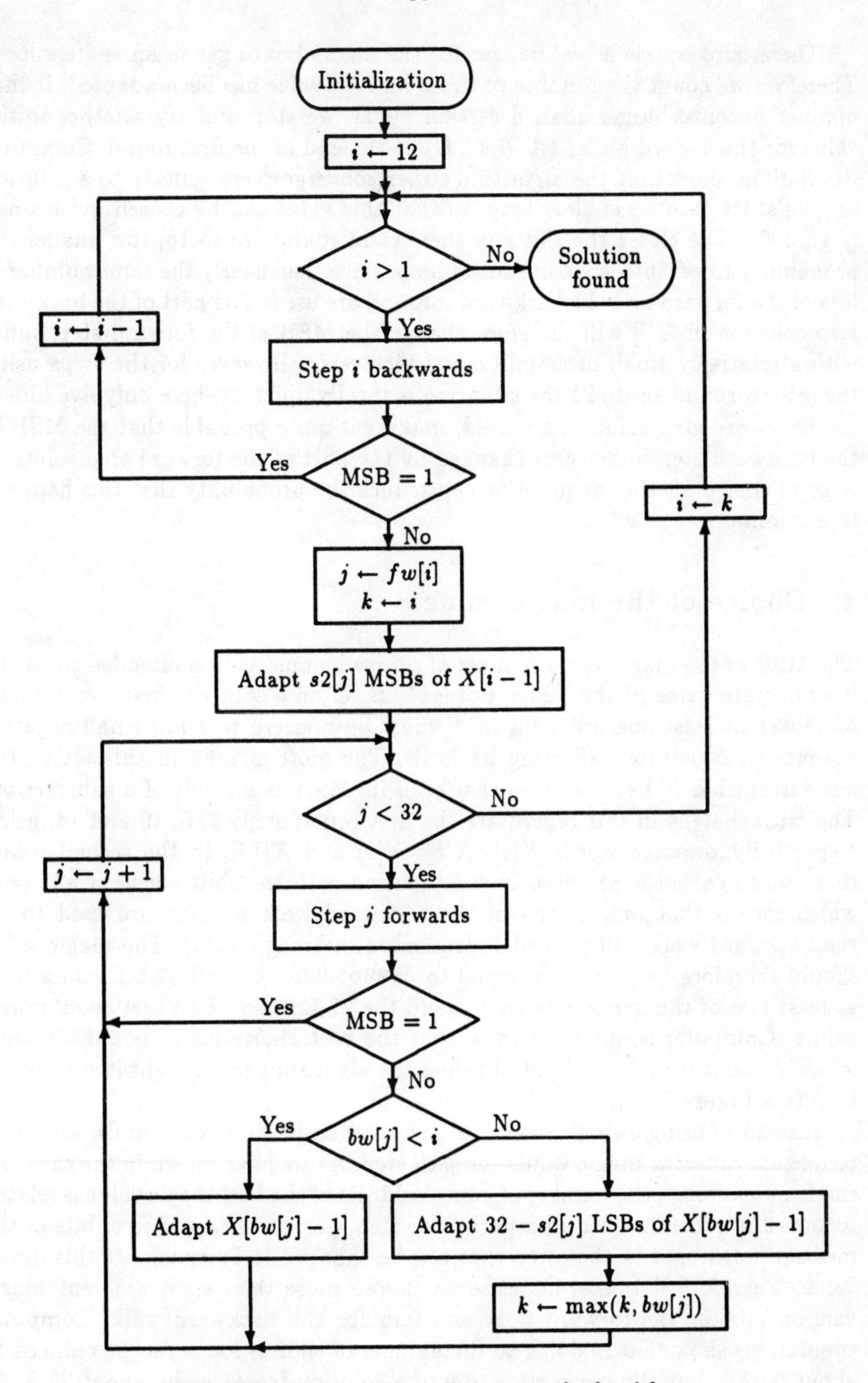

Fig. 1. Flowchart of the collision search algorithm

There is of course a real danger for the algorithm to get in an endless loop. Therefore we count the number of times an $X[\cdot]$ value has been adapted. If that number becomes larger than a certain value, we stop and try another initial value for the 4-word buffer (A, B, C, D) at the end of the first round. Computer simulations show that the algorithm either converges very quickly to a solution or gets stuck into an endless loop, so that this value can be chosen quite small (e.g., 300). The closer the shifts in the second round are to 16, the smaller the probability to get into such an endless loop, since then nearly the same number of bits of the forward and the backward solution are used. The part of the backward step solution of $X[\cdot]$ will therefore change the MSB of the forward step buffer with a relatively small probability, and vice versa. However for the steps using the second round shift $s21$ the situation is totally different: here only five bits of the backward step solution are used, making it quite probable that the MSB of the backward step buffer gets changed by the part of the forward step solution. A good choice for the magic value can reduce the probability that this happens to a minimum.

4 Choice of the magic value

The MSB of the magic value N must of course be one, as it is intended to be the intermediate value of the buffer words A, B, C and D in the first two rounds. Moreover at least one other bit of N must be nonzero to allow small negative changes to N without affecting its MSB. The more significant this bit is, the less susceptible N becomes to a change of its MSB as a result of a subtraction. The critical steps in this regard are the first round steps 2, 6, 10 and 14, using respectively message words $X[1]$, $X[5]$, $X[9]$ and $X[13]$. In the second round these message words are used in combination with the shift constant $s21 = 5$, which means that only 5 bits of the backward walk solution are used to let the backward walk buffer word approximate the magic value. The magic value should therefore be greater or equal to `0x88000000`, i.e., all 32-bit values with at least two of the five MSBs on one and the 27 LSBs on zero are 'good' magic values. Computer simulations show that the best choice for N is `0xF8000000`: for only about 0.15% of all initial values the algorithm gets caught in an endless loop (see Figure 2).

Instead of using a single magic value for the entire first two rounds, we can of course use different magic values for each step. As we have shown in the case of a single magic value, the number of nonzero MSBs of the best magic value is related to the shift constant used in a particular step, i.e., to the number of bits of the message word used in that step that can be changed to approximate this magic value. Therefore it makes no sense to choose more than eight different magic values: four for the forward walk and four for the backward walk. Computer simulations show that in doing so the number of endless loops can be reduced to about 0.02%, but the mean time to find a solution increases by about 25%. As the figure of 0.15% endless loops is very much acceptable, we decided to stick to a single magic value of `0xF8000000`.

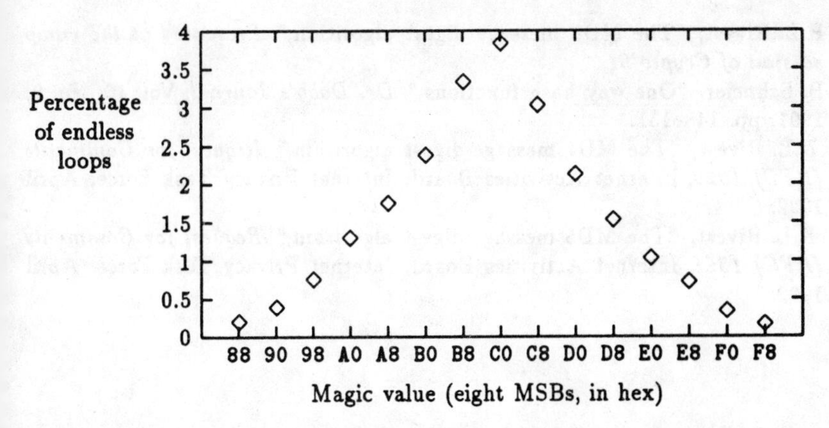

Fig. 2. Percentage of endless loops for the different 'good' magic values. Only the 8 MSBs of each magic value are indicated, the 24 LSBs are all zero.

5 Implementation

A C program has been written implementing the algorithm. It establishes a work load of finding about 2^{16} collisions for the first two rounds of the MD5 compression function to find a collision for the entire four round function. On a 33MHz 80386 based PC using a 32-bit compiler the mean time to find such a collision is about 4 minutes (2^{16} trials). However the variance is quite dramatic. Times have been observed ranging from about 1 second (317 trials) to more than 25 minutes (396324 trials). As an example, the following two 20-word inputs consisting of the common 16-word message part (hexadecimal notation)

```
5FFBB485 B73256D8 19DF08E4 11054A66 22C00E98 450A05C4 5F53A940 9DDC1CF8
DADAB3DB 8A43597A 4CA51993 E7DB12E5 1F1C0317 9A3BAAD6 B275B7BB 0F09CFD5
```

and respectively the 4-word input buffers I1 and I2

```
I1: 399E49D4 876C9442 F7DFE793 83D49001
I2: B99E49D4 076C9442 77DFE793 03D49001
```

are both compressed to the same 4-word output buffer

```
F80668D5 F8AB5C93 C93998F5 D007A636
```

References

[Rive90] R.L. Rivest, "The MD4 message digest algorithm," *Advances in Cryptology, Proc. Crypto'90, LNCS 537*, S. Vanstone, Ed., Springer-Verlag, 1991, pp. 303–311.

[Merk90] R.C. Merkle, Unpublished result, 1990.

[dBBo91] B. den Boer and A. Bosselaers, "An attack on the last two rounds of MD4," *Advances in Cryptology, Proc. Crypto'91, LNCS 576*, J. Feigenbaum, Ed., Springer-Verlag, 1992, pp. 194–203.

[Rive91] R.L. Rivest, "The MD5 message digest algorithm," *Presented at the rump session of Crypto'91.*

[Schn91] B. Schneier, "One-way hash functions," *Dr. Dobb's Journal*, Vol. 16, No. 9, 1991, pp. 148–151.

[Rive92a] R.L. Rivest, "The MD4 message-digest algorithm," *Request for Comments (RFC) 1320*, Internet Activities Board, Internet Privacy Task Force, April 1992.

[Rive92b] R.L. Rivest, "The MD5 message-digest algorithm," *Request for Comments (RFC) 1321*, Internet Activities Board, Internet Privacy Task Force, April 1992.

HOW TO FIND AND AVOID COLLISIONS FOR THE KNAPSACK HASH FUNCTION

Jacques PATARIN
Bull CP8, 68 route de Versailles - B.P.45 - 78430 Louveciennes - France

Abstract

Ivan Damgård [4] suggested at Crypto'89 concrete examples of hash functions including, among others, a knapsack scheme. In [3], P. Camion and myself have shown how to break this scheme with a number of computations in the region of 2^{32} and about 128 Gigabytes of memory. More precisely in [3] we showed how to find an x such that $h(x) = b$, for a fixed and average b. (1).

But in order to show that h is not collision free, we have just to find x and y, $x \neq y$ such that $h(x) = h(y)$. (2). This is a weaker condition than (1).

We will see in this paper how to find (2) with a number in the region of 2^{24} computations and about 512 Megabytes of memory. That is to say with about 256 times less computation and memory than [3]. Moreover, ways to extend our algorithm to other knapsacks than that (256, 128) suggested by Damgård are investigated.

Then we will see that for solving problems like (1) or (2) for various knapsacks it is also possible to use less memory if we are allowed to use a little more computing time. This is a usefull remark since the memory needed was the main problem of the algorithms of [3].

Finally, at the end of this paper, we will briefly study some ideas on how to avoid all these attacks by slightly modifying the knapsack Hash functions. However some different attacks could appear, and it is not so easy to find a colision free Hash function, both very quick and with very simple Mathematic expression.

The Proposed Knapsack

Let $a_1, \ldots, a_s$ be fixed integers of A binary digits, randomly selected. If T is a plaintext of s binary symbols, $T = x_1 \ldots x_s$, then $h(x) = \sum_{i=1}^{s} x_i a_i$ will be the proposed hashed value. In paragraph 1 and 2, values assigned are 256 for s and 120 for A, as suggested in [4]. Thus $h(x)$ has at most 120+8=128 binary digits.

1 The general scheme of our modified algorithm

Our algorithm for finding x and y such that $h(x) = h(y)$ will be mainly a variation of the algorithm described in [3] in order to find x such that $h(x) = b$, where b is a fixed and average value. But our modified algorithm will be in $0(2^{24})$ computations instead of $0(2^{32})$, and it will need about 512 Megabytes of memory instead of about 128 Gigabytes. Nowadays it is quite common to have 512 Megabytes but still quite unusual to have 128 Gigabytes of Memory. So our modified algorithm will appear more practical. Our modified algorithm will proceed in 16 steps, plus a step 0 at the beginning.

Step 0 : We choose integers m, m_1, m_2, m_3, m_4 and b such that :

a) $m = m_1 m_2 m_3 m_4 > 2^{80}$.

b) The m_i are pairwise coprime, and
• $m_1 \simeq 2^8$, • $m_2 \simeq 2^{24}$, • $m_3 \simeq 2^{24}$, • $m_4 \simeq 2^{24}$.
(So $m \simeq 2^{8+24+24+24} = 2^{80}$.).

c) Let b be a fixed integer, $b \simeq 2^{126}$, for example.
And $\forall i, 1 \leq i \leq 4$, we define $b_i = b \bmod m_i$.

In order to have a general view of our algorithm, the diagram given below shows the sequence of operations that will be carried out. Let us outline the meaning of the diagram before going into detail. Each black point represents a step of the algorithm. The number 2^{24} associated with the black point represents the evaluation of the number of partial solution that will be found for this step. Each step will study binary sequences. The length of those sequences is 32 for steps 1, 2, 3, 4, 5, 6, 7, 8. It is 64 for steps 9, 10, 11, 12. It is 128 for steps 13, 14. And then step 15 will produce about 2^{24} sequences of length 256 among which in step 16 we will find a collision with probability close to 1.

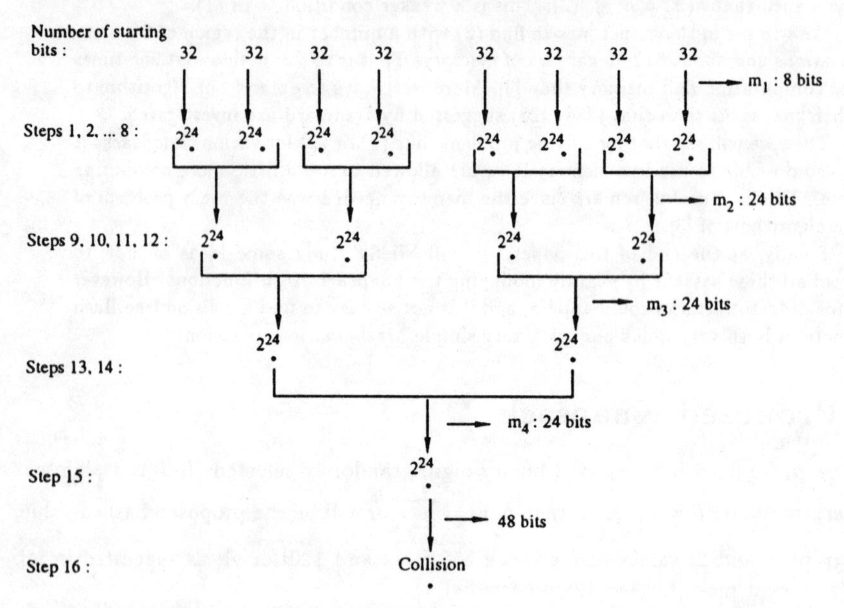

We will now go through each step in detail.

Step 1 : Let $b_1 = b \bmod m_1$.

We find all sequences $(x_i), 1 \leq i \leq 32, x_i = 0$ or 1, such that $\sum_{i=1}^{32} x_i a_i \equiv b_1 [m_1]$.

We will find about 2^{24} such sequences because there are 2^{32} sequences (x_i) of 32 bits, and m_1 is close to 2^8. In fact, we will see in Section 2 the number of solutions that we may expect to obtain when the algorithm is brought to completion.

It is important to notice that it is possible to do this step 1 with a number in the region of 2^{24} operations, with a memory of about 2^{24} words of 32 bits. Indeed, we just have to do the following :

a) Compute and store all values of $b_1 - \sum_{i=1}^{16} x_i a_i$ modulo m_1.

The "store" is done such that we will have easy access to all the sequences (x_i) such that $b_1 - \sum_{i=1}^{16} x_i a_i$ has a given value modulo m_1.

b) Compute, one by one, all the values of $\sum_{i=17}^{32} x_i a_i$ modulo m_1 and look if there are some sequences $(x_i), 1 \le i \le 16$, which gived the same value modulo m_1 in a). If it is so, keep all the pairs of sequences obtained $(x_i), 1 \le i \le 32$.

For a) we will need about one field of 2^{16} words of 16 bits. And for b) we will need about one file of 2^{24} words of 32 bits.

Step k, k=2 to 8 : In the same way, we find about 2^{24} sequences (x_i) such that $\sum_{i=32(k-1)+1}^{32k} x_i a_i \equiv 0[m_1]$.

Step 9 : We denote $\sum_{i=1}^{32} x_i a_i$ by s_1 and $\sum_{i=33}^{64} x_i a_i$ by s_2.

From the sequences (x_i) found at Steps 1 and 2, we find about 2^{24} sequences $(x_i), 1 \le i \le 64$ such that $s_1 + s_2 \equiv b_2[m_2]$. (where $b_2 = b \bmod m_2$). For there are about $2^{24} \times 2^{24} = 2^{48}$ sequences $(x_i), 1 \le i \le 64$ such that $(x_1, \ldots, x_{32})$ is a solution from Step 1 and $(x_{33}, \ldots, x_{64})$ is a solution from Step 2. So if the numbers $s_1 + s_2$ are about equally distributed modulo m_2, $m_2 \simeq 2^{24}$, we find about $\frac{2^{48}}{2^{24}} = 2^{24}$ among those sequences such that $s_1 + s_2 \equiv b_2[m_2]$. All sequences (x_i) to be found in Step 9 also have the following property :

$$s_1 + s_2 \equiv b_2[m_2] \quad \text{and} \quad s_1 + s_2 \equiv b_1[m_1].$$

This is because $s_1 \equiv b_1[m_1]$ and $s_2 \equiv 0[m_1]$. It is important to notice that it is possible to do this step 9 with a number in the region of 2^{24} operations, with a memory of about 2^{24} words of 64 bits. Indeed, we just have to do the following :
a) Compute and store all values of $b_2 - s_1$ modulo m_2, where s_1 has been found in step 1.
b) Compute and store all the values of s_2 modulo m_2, where s_2 has been found in step 2.
c) keep all pairs of sequences (x_i) which give the same value modulo m_2 in a) and b).

Step k, k = 10, 11 and 12 : The same way as step 9, we find about 2^{24} sequences (x_i) such that $\sum_{i=64(k-9)+1}^{64(k-8)} x_i a_i \equiv 0[m_i], i = 1, 2$.

Step 13 : Combining solutions of steps 9 and 10, we find about 2^{24} sequences (x_i) such that $\sum_{i=1}^{128} x_i a_i \equiv b_i[m_i], i = 1, 2, 3$. (This is done with about 2^{24} computations, the same way as step 9).

Step 14 : Combining solutions of steps 11 and 12, we find about 2^{24} sequences (x_i) such that $\sum_{i=129}^{256} x_i a_i \equiv 0[m_i], i = 1, 2, 3$.

Step 15 : Combining solutions of steps 13 and 14, we find about 2^{24} sequences (x_i) such that

$$\sum_{i=1}^{256} x_i a_i \equiv b_i[m_i], i = 1, 2, 3, 4. \tag{1}$$

The m_i are pairwise coprimes, so (1) means : we have about 2^{24} sequences (x_i) such that

$$\sum_{i=1}^{256} x_i a_i \equiv b[m], \quad \text{where } m > 2^{80}. \tag{2}$$

Step 16 : Among the 2^{24} sequences (x_i) found in Step 15, we will have with a "good" probability a collision, that is to say two sequences (x_i) and (y_i) such that :

$$\sum_{i=1}^{256} x_i a_i = \sum_{i=1}^{256} y_i a_i.$$

This is because all the sequences found in step 15 have the same value modulo m, where $m > 2^{80}$. So the probability that $h(x) = h(y)$, where x and y are found in step 15, $x \neq y$, is $\geq \frac{1}{2^{128-80}} = \frac{1}{2^{48}}$. But we have about $\frac{2^{24}.(2^{24}-1)}{2}$ couples (x, y) where x and y are found in step 15. So it is possible to prove (this is a classical "birthday paradox") that with a "good" probability we will obtain an x and an y such that $h(x) = h(y)$. And the collision will be found in about 2^{24} computations after step 15 : we just have to compute and store all the values $\sum_{i=1}^{256} x_i a_i$ where (x_i) has been found in step 15 (about 2^{24} such (x_i) have been found) and this will give us the collisions.

Differences between our algorithm and the original algorithm of [3]

There are three main differences in the design of the algorithm that we have described and the original algorithm of [3] :

1. The number of solutions after each step is about 2^{24} instead of 2^{32}, in order to require less memory and to do less computations.
2. We have one more stage where steps 1 to 8 are done. And for these steps we use reduction modulo m_1 where $m_1 \simeq 2^8$.
3. At the end we find a collision with a "Birthday Paradox" like attack.

We will now give more details about the "good" probability to find a collision with our algorithm, and the memory required.

2 More details and small improvements of our algorithm

What do we do if at the end of our algorithm no collision is found ? It is possible to use the algorithm again, but with new chosen values. For example we can replace b_1 by $s_1 \equiv b_1 - \lambda[m_1]$ at step 1, and 0 by $s_2 \equiv \lambda[m_1]$ at step 2, where λ is any fixed integer in $[0, m_1 - 1]$. Or we can permute the a_i's. We can also change the value of b or of the m_i's.

But it is much better to keep the same value for b and m : this is because the probability of success in Step 16 depends only on the number of (x_i) found such that $\sum_{i=1}^{256} x_i a_i \equiv b[m]$.

So all the solutions (x_i) found in step 15 with the first application of our algorithm will be useful with the second application of our algorithm. For this second application we can decide to find less solutions in step 15 than 2^{24}, because they will be combined with the "previous" solutions.

"Good" probability

But in fact, even one iteration of our algorithm has a probability of success near 1. This is because $h(x)$ is not equidistributed. If we denote by $P(b)$ the number of (x_i) such that $h(x) = b$, the function $P(b)$ will have a diagram as follows :

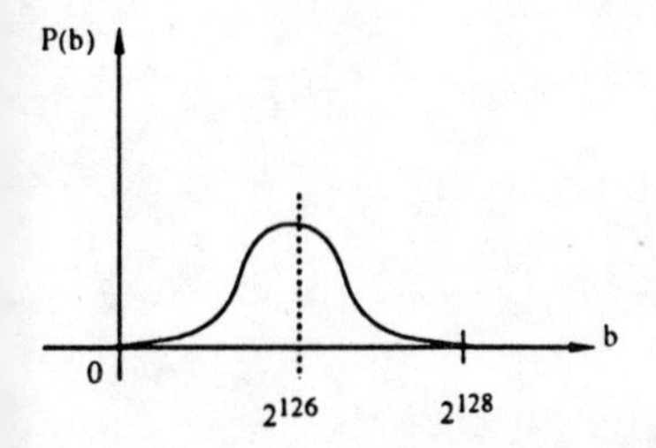

And it is possible to prove that, if the a_i's are random numbers of 120 bits, for about 99 % of the (x_i) we will have : $108.2^{119} \leq h(x) \leq 148.2^{119}$.
So for about 99 % of the (x_i), $h(x)$ will have less than 40.2^{119} values, thus less than 2^{125}. So the "collision" in step 16 of our algorithm is easier than expected, and then it is possible to prove that the probability of finding a collision after step 16 is near 1.

Memory

In order to use less memory, it is useful to begin with steps 1, 2 and 9, then steps 3, 4, 10 and 13. Then steps 5, 6 and 11, then steps 7, 8, 12 and 14. Then steps 15 and 16. Thus with a file of about 2^{24} words of 256 bits it will be possible to do all these steps. This is 512 Megabytes of memory. It is high but 256 less than what was needed in [3]. (In [3] it is explain that 64 Gigabytes are needed for one basic step. But it seams that at least about 128 Gigabytes are needed for all the steps). In paragraph 4 we will see some other ideas in order to use less memory but at the cost of a little more computations.

3 Generalization of the new algorithm for other sizes of Knapsacks

Values for complexity 2^{32}

Let $a_1, \ldots, a_s$ be fixed integers of A binary digits.
If x is a plaintext of s binary symbols, i.e. $x = x_1 \ldots x_s$, then $h(x) = \sum_{i=1}^{s} x_i a_i$ is the proposed hashed value. The hash value $h(x)$ has less than B binary digits, i.e. $B \simeq A + \log_2 s$.
In [3] some algorithms were given to find an x such that $h(x) = b$ (where b is a fixed and random value in $[1, 2^B]$). In complexity $0(2^{32})$ (in time and memory) these algorithms can find such x in the following cases :

Value of s (or more)	Value of B (or less)
128	96
256	128
512	160
1024	192
2048	224
	etc.

In paragraph 1, we have seen how to find a collision $h(x) = h(y)$ for $s = 256$ and $B = 128$ in $0(2^{24})$ complexity. By using the same ideas, we will now see that, in $0(2^{32})$ complexity our algorithms will find a collision $h(x) = h(y)$ in the following cases :

Value of s (or more)	Value of B (or less)
128	128
256	160
512	192
1024	224
2048	256
	etc.

Example 1 : let $s = 128$ and $B = 128$

In this case the function h is not a real "hash" function since from integers of 128 bits, it gives integers of about the same size. But the function is not collision free, even in this case. In the diagram below we will see how to find an x and y such that $h(x) = h(y)$ in $0(2^{32})$ complexity.
We will not give details because this algorithm is similar with algorithm of paragraph 1.

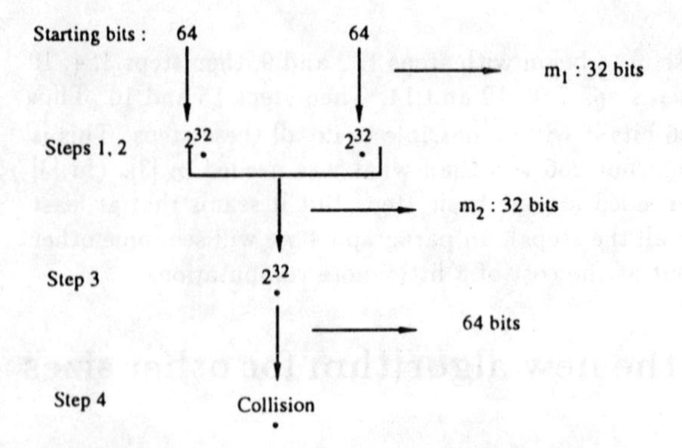

Example 2 : let $s = 256$ and $B = 160$

We will not give the details because it is just the same algorithm but with one more stage. The diagram is

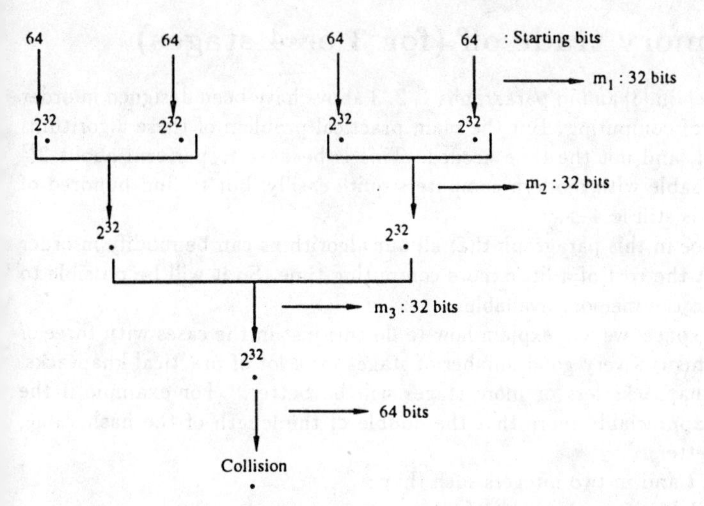

With the same technique, by using two, three, etc. more stages and still about 2^{32} operations we can solve the cases $s = 512$ and $B = 192$, or $s = 1024$ and $B = 224$, etc. And we obtained in this way the values given in this section before example 1. If we compare these values with the values given in [3], we see that, when s is given, we can obtain a collision on 32 extra bits. Or, when B is given, we can obtain a collision with a length of the Text which is twice as small.

Note. It is not a surprise that to find a collision we need a length of the text twice as small. If $x = y.z$ (that is to say if x is the concatenation of y and z) then
$h(x) = b \quad (1) \Leftrightarrow h(y) = b - h(z) \quad (2)$.
And (2) is similar to the problem of finding collision on texts y and z twice as small as x.

A general formula

With the same technique by using e stages, with at most about 2^m memories and in 2^m operations, and with eventually one extra-stage with a reduction modulo c bits (all the others stages perform reductions modulo m bits), we will obtain these general properties :

• Number of starting bits :	• $2^e(m+c)$ bits (or more).
• It invert h on :	• $(1+e)m+c$ bits (or less).
• It find collisions on :	• $(2+e)m+c$ bits (or less).

With $0(2^m)$ time and at most $0(2^m)$ memory.

Example a. With $m = 24, c = 8, e = 3$, we can find a collision for the (256,128) knapsack in $0(2^{24})$ complexity. This is exactly what we did in paragraph 1.

Example b. With $m = 32, c = 0, e = 3$, we can find a collision for the (256, 160) knapsack in $0(2^{32})$ complexity. This is what we did in example 2 above.

4 Time-Memory trade off (for 3 or 4 stages)

All the algorithms given in [3] and in paragraphs 1, 2, 3 above have been designed in order to minimize the time of computing. But the main practical problem of these algorithms is the memory needed, and not the time needed. This is because to perform about 2^{32} basic operations is faisable with modern computers quite easilly, but to find hundred of Gigabytes of memory is still less easy.

However, we will see in this paragraph that all our algorithms can be modify in order to use less memory, at the cost of a little more computing time. So it will be possible to adapt the algorithm to the memory available.

Due to the lack of space, we will explain how to do this just in the cases with three of four stages, because this is a very good number of stages for a lot of practical knapsacks. However for others knapsacks less or more stages will be better. (For example if the length of the text is appreciably more that the double of the length of the hash value, more stages will be better).

We will denote by t and m two integers such that :

- the memory available is on the order of 2^m.
- the time for computation available is on the order of 2^t.

We will assume that $\frac{t}{2} \leq m \leq t$. And we will denote by c a parameter such that : $t/2 + c/2 \leq m$ and $c \geq 0$. Before going into details, we give the diagram of the steps : The **general diagram** is :

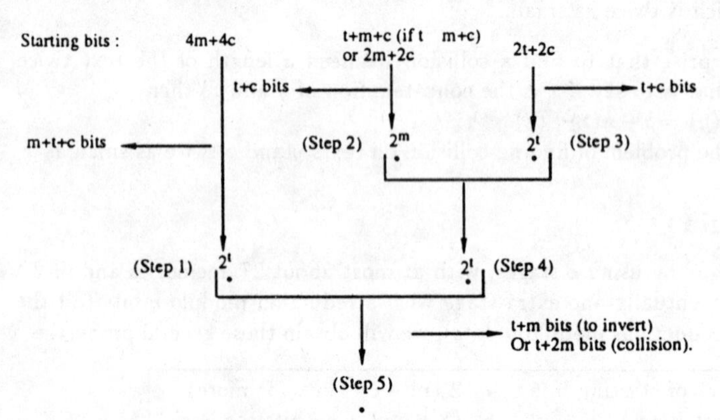

We now go through each step in more details.

Step 0 : We choose integers m_1, m_2, m_3, m_4 such that :

a) m_1, m_2, m_3, m_4 are pairwise coprime, and $m_1 \simeq 2^c$, $m_2 \simeq 2^m$, $m_3 \simeq 2^t$.

b) $m_4 \simeq 2^{t+2m}$ if we want a collision, or $m_4 \simeq 2^{t+m}$ if we want to invert (i.e. to find an x such that $h(x) = b$ where b is given).

Step 1 : The aim of step 1 is to find, and store about 2^m sequences of $4m + 4c$ bits such that $\sum_{i=1}^{4m+4c} x_i a_i \equiv 0[m_1 m_2 m_3]$ in $0(2^t)$ time and $0(2^m)$ memory.

Note. There are about $2^{4m+4c}/2^{t+m+c} \simeq 2^{3m-t+3c}$ such solutions. Among these we want only 2^m such solutions, and this is in fact less or equal because $m \leq 3m - t + 3c$ since $t \leq 2m$ and $c \geq 0$.

For step 1, we proceed in 2^{t-m} cases. Each of these 2^{t-m} cases follows a diagram like that :

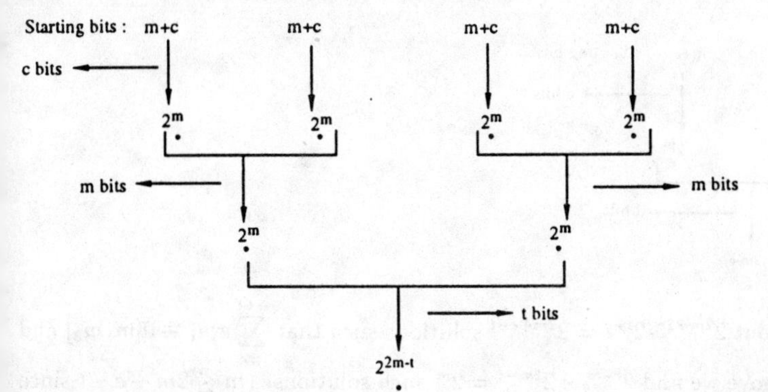

So for each case 2^{2m-t} solutions are found. But all the solution of one case are distincts from the solutions of another case because for example in each case we found solutions such that $\sum_{i=1}^{2m+2c} x_i a_i \equiv \lambda[m_1 m_2]$ where λ is a parameter distinct in each cases. So in this step 1 we will find $2^{t-m} \times 2^{2m-t} = 2^m$ solutions as claimed. And the total time for this step 1 in on the order of $2^{t-m} \times 2^m = 2^t$. (Note that the time for each reduction modulo c bits is about $2^{\frac{m}{2}+\frac{c}{2}} \leq 2^m$ since $c \leq m$ since $c/2 \leq m - \dfrac{t}{2} \leq \dfrac{m}{2}$). And the total memory is on the order of 2^m.

Step 2 : The aim of step 2 is to find, and store, about 2^m sequences of $\ell = t+m+c$ bits (if $t \geq m+c$) or of $\ell = 2m+2c$ bits (if $t \leq m+c$) such that $\sum_{i=\alpha}^{\ell+\alpha} x_i a_i \equiv 0[m_1 m_3]$ in $0(2^t)$ time and $0(2^m)$ memory, where $\alpha = 4m+4c+1$.

Case 1 : $t \geq m+c$
Then for step 2 we proceed in 2^c cases. Each of these 2^c cases follows a diagram like that :

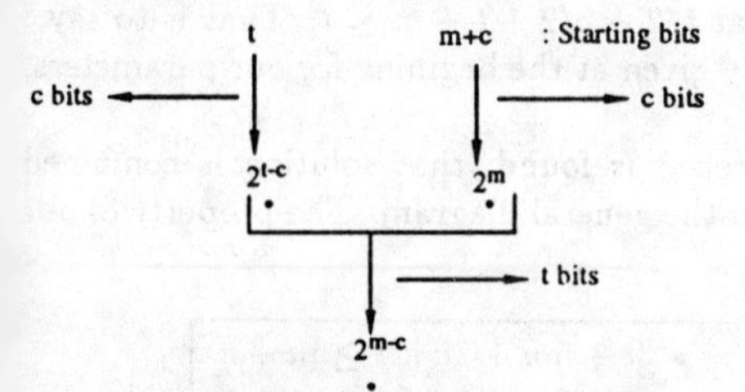

So we will found about $2^c \times 2^{m-c} \simeq 2^m$ solutions as wanted.

Note. Here we have 2^m solutions for step 2 and we found all of them. The time on the right side was $2^c \times 2^m = 2^{m+c}$ and this is $\leq 2^t$ because here $m+c \leq t$. So for step 2 the total time is in $0(2^t)$. And the value 2^{t-c} in the diagram is a number of sequences that are found one by one, and not stored. So the total memory is in $0(2^m)$.

Case 2 : $t \leq m + c$
Then for step 2 we proceed in 2^{t-m} cases. Each of these 2^{t-m} cases follows a diagram like that :

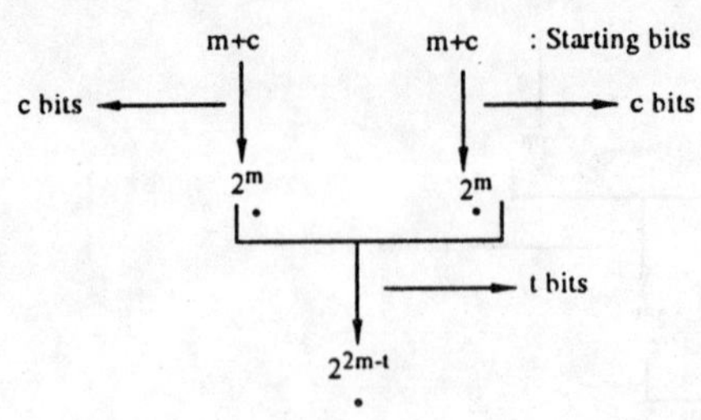

Here we have about $2^{2m+2c}/2^{t+c} = 2^{2m+c-t}$ solutions such that $\sum_{i=\alpha}^{\alpha+\ell} x_i a_i \equiv 0[m_1 m_3]$ and with the diagram above we find $2^{t-m} \times 2^{2m-t} = 2^m$ such solutions. ($m \leq 2m + c - t$ since $t \leq m + c$ here).

Step 3 : The aim of step 3 is to find, one by one, about 2^t sequences of $2t + 2c$ bits such that $\sum_{i=\beta}^{\beta+2t+2c} x_i a_i \equiv \xi[m_1 m_3]$ in $0(2^t)$ time and $0(2^m)$ memory, where $\beta = \ell + \alpha + 1$ (ℓ and α as in step 2) and where ξ is a fixed and given value. For step 3 we proceed in 2^{t-m} cases. Each of these 2^{t-m} cases follows a diagram like that :

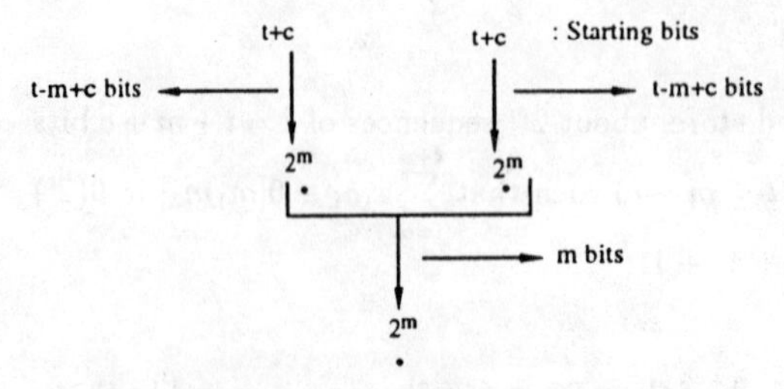

Note. Here, at the begining of this diagram, at least $2^{t/2+c/2}$ computations are done in each one of the 2^{t-m} cases. So we want that $t/2 + c/2 + t - m \leq t$. That is to say : $t/2 + c/2 \leq m$. And this was exactly a property given at the begining for our parameters.

Step 4 and 5 : Each time a solution of step 3 is found, that solution is combined with the solutions of steps 2 and 1, as shown in the general diagram. The property of our algorithm (with 3 or 4 stages) is finally :

Number of starting bits (or more) :	• $3t + 5m + 7c$ if $t \geq m + c$
	• $2t + 6m + 8c$ if $t \leq m + c$
It invert h on	• $2t + 2m + c$ bits (or less).
It find collision on	• $2t + 3m + c$ bits (or less).

With $0(2^t)$ time, $0(2^m)$ memory, and $t \geq m \geq t/2 + c/2$.

Examples.

Example 1. With $m = 32, t = 32, c = 0$ we can invert the (256, 128) knapsack. This was done in [3]. But with $m = 24, t = 39, c = 2$ we can solve the same problem with less memory (and a little more time).
Example 2. With $m = 48, t = 48, c = 16$, or with $m = 42, t = 56, c = 18$, we can find a collision for the (512, 256) knapsack. (However here the algorithm still need a huge amount of time and memory).

5 How to avoid these attacks

In [4], I. Damgård gave a great theorem which shows that if we can have a collision free hash function f from 256 to 128 bits (for example), then we can design a collision free hash function h from any size to 128 bits. And it will be possible to calculate h very quickly if we can calculate f very quickly. The Knapsack Hash function with $s = 256$ and $B = 128$ is quick to calculate. Moreover for example on a 32 bit computer it seems that it will be about 4 times slower than MD4. (MD4 is a concrete example of really used hash function, see [5] for details). Another problem of this Knapsack Hash function is that it requires an array of about 256 words of 120 bits for the numbers $a_1, 1 \leq i \leq 256$, (this is 3.75 Kbytes), and MD4 doesn't need this. However the main problem, of course, is that we have seen that this Knapsack Hash function is not collision free. But most of the hash functions that are used today (as MD4, MD5 or SHS) do not have a very simple mathematical description. So, is it possible to describe a candidate hash function which will be :

1. Very quick to calculate.
2. Collision free.
3. With a very simple mathematical description.
4. With about 128 bits in output.

In [6], G. Zémor suggested a candidate for points 1, 2, 3 above based on multiplication in the group $G = SL_2(\mathsf{F}_p)$ of 2×2 matrices of determinant 1 over F_p.

However, in order to avoid some potential attacks, G. Zémor suggested to take for p a prime of about 150 bits. So the hash value will be a hash of about 450 bits (instead of 128 for MD4 for example).

It is possible to suggest many different candidates (for example with modular multiplication, but then the function will be much slower than MD4). We will now give some example of functions obtained by modifying just a little the Knapsack Hash function. The hash values of these functions will be 128 bits long. However, in all the examples below our functions (designed in order to avoid the attacks of paragraphs 1-4) will not be collision free, due to other attacks. Nevertheless, we think that, from a theoretical point of view, it is very interesting to study simple mathematical hash functions, in order to gain a better understanding of what makes such a function "collision free" and what doesn't.

Example 1

Let x be a plaintext of 256 binary symbols. Let $h(x) = \sum_{i=1}^{256} x_i a_i$ be the (256,128) Knapsack Hash function that we have studied in paragraph 1. Let $x = (x_i), 1 \leq i \leq 256, x_i = 0$ or 1, and let $y = (x_i), 1 \leq i \leq 128$ and $z = (x_i), 129 \leq i \leq 256$. (So x is the concatenation of y and z). Then let $H(x) = h(x) + y + z$.

Is H collision free ? In fact, we will see that H is not collision free. If $1 \leq i \leq 128$, let $b_i = a_i + 2^{i-1}$. And if $129 \leq i \leq 256$, let $b_i = a_i + 2^{i-129}$. Then $H(x) = \sum_{i=1}^{256} x_i b_i$, so H is just another Knapsack like h. So H is not collision free as explained in [3] and in paragraph 1.

Example 2
Let $x, h(x), y$ and z be as in example 1. We now define an auxiliary function F that takes as input three 128-bit words and produces as output one 128-bit word : $F(X, Y, Z) = YX \vee (\rceil X)Z$.

(In this formula XY denote the bit-wise AND of X and Y, $X \vee Y$ denote the bit-wise OR of X and Y, and $\rceil X$ denote the bit wise complement of X).

Each bit position F acts as a conditional : if x then y else z. (In MD4 a similar function is used, but for 32 bit words). Finally, we define

$$H(x) = F(h(x), y, z).$$

Then H is just a slight variation of the Knapsack. The time needed to calculate H and h is about the same and all the attacks described for the Knapsack seem to be ineffective for H. However, this function is not collision free. In fact if $y = z = b$, then $H(x) = b$. So H is really easy to invert !

Example 3
Let $x, h(x), y, z$ and F be as in example 2.
And let $H(x) = F(h(x), y, z) + h(x)$.
This function H is defined in order to avoid the attacks of paragraphs 1-4 and the attack of example 2. (H computes only additions except one single use of F in order to have a simple mathematical description and be quick to calculate).
However, H is not collision free.
In fact, if $y = 0$ and $z = 1$, then $H(y, z) = 1$.
And more generally, if $y = 0$ then $H(0, z) = h \vee z$.
So by chosen a z with only one zero, the probability that $H(0, z) = 1 = H(0, 1)$ is about $1/2$. This property will give easily a collision.
As all our three examples show it can be very dangerous to add an extra composition of functions. It could be a good idea to re-use the input y and z with h in order to design an hash function H to avoid the attacks of paragraph 1-4, but this must be done very carefully.

6 Conclusion

We have seen how to modify the algorithms described in [3] in order to find collisions with less memory, or in order to find collisions for stronger Knapsack. For example, with about 512 Megabytes of memory (instead of about 128 Gigabytes) is it possible to find a collision for the (256,128) Knapsack. Or, in complexity $0(2^{32})$ it is possible to find a collision for the (128,128) Knapsack. The technique is very efficient for various values of the Knapsack. Moreover it is possible to adapt the algorithms by using less memory if a little more computing time is allowed. Finally, we have study some slight modifications of the Knapsack in order to avoid these attacks. Although these modifications did not avoid collisions we think that it is interesting to study simple mathematical candidate hash functions.

Acknoledgements

I want to thank you S. Vaudenay and J.S. Lair for usefull comments.

References

[1] P. Camion, *"Can a Fast Signature Scheme Without Secret Key be Secure ?"*, in AAECC-2, Lecture Notes in Computer Science n° 228, Springer Verlag.

[2] P. Camion and Ph. Godlewski, *"Manipulation and Errors, Localization and Detection"*, Proceedings of Eurocrypt'88, Lecture Notes in Computer Science n° 330, Springer Verlag.

[3] P. Camion and J. Patarin, *"The Knapsack Hash Function proposed at Crypto'89 can be broken"*, Proceedings of Eurocrypt'91, pp. 39-53, Springer Verlag.

[4] I. Damgård, *"Design Principles for Hash Functions"*, Proceedings of Crypto'89, Springer Verlag.

[5] R.L. Rivest, *"The MD4 Message Digest Algorithm"*, Crypto'90, Springer Verlag, pp. 303-311.

[6] G. Zémor, *"Hash Functions and Graphs with Large Girths"*, Proceedings of EUROCRYPT'91, pp 508-511, Springer Verlag.

Single Term Off-Line Coins

Niels Ferguson

CWI, P.O. Box 94079, 1090 GB Amsterdam, Netherlands.
e-mail: niels@cwi.nl

Abstract. We present a new construction for off-line electronic coins that is both far more efficient and much simpler than previous systems. Instead of using many terms, each for a single bit of the challenge, our system uses a single term for a large number of possible challenges. The withdrawal protocol does not use a cut-and-choose methodology as with earlier systems, but uses a direct construction.

1 Introduction

The main requirements for an electronic cash system are the following:

- Security. Every party in the electronic cash system should be protected from a collusion of all other parties (multi-party security).
- Off-line. There should be no need for communication with a central authority during payment.
- Fake Privacy. The bank and all the shops should together not be able to derive any knowledge from their protocol transcripts about where a user spends her money.
- Privacy (untraceable). The bank should not be able to determine whether two payments were made by the same payer, even if all shops cooperate.

The privacy requirement is obviously stronger then the fake privacy one. Under real-world circumstances, a payer is identified during some of the payments by means outside the electronic cash system. If two payments can be recognised as originating from the same payer, then knowing the identity of the payer during a single payment anywhere allows the tracing of *all* other payments made by that payer. Therefore, fake privacy provides no privacy at all in practice.

Electronic cash was first introduced by Chaum, Fiat and Noar [CFN90]. Their system (later improved in [CdBvH$^+$90]) meets all of the requirements, but is quite complex. The authors give a construction for electronic checks that allow a variable amount of money to be payed with a single signature. These checks can be thought of as a bunch of fixed-value coins sharing common overhead.

In [vA90] Hans van Antwerpen described a different scheme which is more efficient than [CFN90] and [CdBvH$^+$90], but has the same basic properties.

Okamoto and Ohta introduced the idea of divisibility of electronic cash [OO92]. This allows a piece of money to be divided into smaller pieces each of which can be spent separately. Their construction satisfies all of the above requirements except the privacy, and the fake privacy of the user is only computationally protected.

The most difficult fraud to counter in electronic cash systems is double-spending. A user can always spend a single coin twice in two different shops. This fraud cannot be detected at the time of spending as payments are off-line. The solution that all electronic cash systems use is to detect the double-spending after the fact. At each payment the user is required to release information in response to a challenge from the shop. One such release provides no clue to the user's identity, but two such releases are sufficient to identify the user uniquely.

In all earlier schemes, the identification of double-spenders requires multiple terms in each coin. Each term is used to answer one bit of challenge from the shop during payment. If both possible answers for any term is ever given, the user's identity is revealed. To achieve an acceptable probability of detection, a large number of terms is required. These schemes are therefore inefficient if a piece of money has a fixed value, so all of them provide some way to pay a variable amount of money with a single 'check'. The resulting refunds for partially spent checks pose further security and privacy problems [Hir93] as well as user-interfacing problems. Another major disadvantage of check systems is that they are very complex (the full but concise mathematical description of the protocols in [vA90] requires 40 pages), which makes them extremely difficult to understand, verify, implement or debug.

Recently, in independent work, Franklin and Yung gave a construction for a provably secure coin scheme in a slightly different setting where a trusted centre is available to produce 'blank' coins [FY93]. An alternative they discuss does not require the trusted centre, but is not provably secure. Like our construction, they use a secret sharing line instead of multiple challenge terms.

Up to this point all electronic cash schemes have used a cut-and-choose protocol for constructing the money. These protocols are by their very nature inefficient. To get a low enough probability of cheating, the cut-and-choose must consist of many terms, half of which are thrown away.

Our new system (described earlier in [Fer93]) is a coin scheme where each piece of money has a fixed value. Each coin consists of 3 numbers plus 2 RSA signatures and can be stored in about 250 bytes. (This assumes that we multiply the signatures of 4 different coins together to save storage space; recovering the original signatures is easy [Cha90].) During payment (of possibly several coins) the user has to perform about 30 modular multiplications plus two multiplications per coin being payed. The withdrawal protocol constructs the coins directly without resorting to cut-and-choose methods.

In [Bra93, Bra94] Stefan Brands recently showed a different construction for an efficient electronic coin scheme based on discrete logarithms. His construction does not use a cut-and-choose protocol or multiple challenge terms either.

I would like to thank David Chaum for all his support, and for helping to clean up and improve the withdrawal protocol. Stefan Brands and Ronald Cramer provided many helpful comments.

2 Efficient payments

Instead of using many challenge terms for double-spending detection we will use a polynomial secret sharing scheme [Sha79]. The user (Alice) gives a share of her identity to the shop at each payment in response to a random challenge. If Alice spends the same coin twice, she has to give two different shares which will allow the bank to recover her identity eventually. This solution gets rid of the large number of challenge terms in all previous systems.

The central idea (developed in cooperation with Hans van Antwerpen) is to represent a coin as 3 numbers, $C := f_c(c)$, $A := f_a(a)$, and $B := f_b(b)$ where the f functions are suitable oneway functions. Alice also gets two RSA-signatures from the bank: $(C^k A)^{1/v}$ and $(C^U B)^{1/v}$, where v is a prime, U is Alice's identity and k is a random number.

During payment, Alice sends the numbers c, a, and b to the shop. The shop replies with a randomly chosen (non-zero) challenge x. Finally Alice sends $r := kx + U \pmod{v}$ (which is the share of her identity) and the signature $(C^r A^x B)^{1/v}$ which she can easily compute from the two signatures she has. The shop, in turn, can verify the consistency of these two responses. When spending several coins at the same time, the same challenge is used for all the coins and Alice sends the product of all the signatures to the shop. (Coins of different values use a different v.) This reduces the computations done by Alice to two multiplications per coin plus one exponentiation to the power x. For a 20-bit challenge the exponentiation requires about 30 multiplications.

The payment protocol can obviously be converted to a one-move protocol by choosing x as a hash value on the coin(s) and the shops identity. This requires a larger challenge but eliminates the interaction between the payer and payee.

At the end of the day, the shop sends c, a, b, the challenge and the response to the bank. The bank can verify the correctness of the coin and credit the shop with the corresponding amount. If Alice spends the same coin twice, she must reveal two different points on the line $x \mapsto kx + U$ which immediately allows the bank to determine her identity U.

Given this idea for the payment protocol, the problem of withdrawing the coin from the bank remains. Alice must get the signatures on $C^k A$ and $C^U B$ while the bank learns nothing about c, a, b or k.

3 Randomized blind signatures

For our construction of an efficient withdrawal protocol we need a signature scheme with the following properties:

- Alice receives an RSA-signature on a number of a special form, which she cannot create herself.
- The bank is sure that the number it signs was randomly chosen.
- The bank receives no information regarding which signature Alice gets.

We call such a signature a *randomized blind signature.* The scheme that we use here is due to David Chaum [Cha92].

The protocol to get a randomized blind signature is shown in Fig. 1. All computations are done in an RSA system [RSA78] where the bank knows the factorization of the modulus n. The public exponent of the RSA system is v, a reasonably large prime (say 128 bits). Alice starts by choosing a random a_1, and two blinding factors σ and γ. She computes $\gamma^v a_1 g^\sigma$ where g is a (publicly known) element of large order in $\mathbb{Z}_n^*$ and sends the result to the bank. The bank chooses its own contribution a_2 and sends this back to Alice. Alice replies with $f(a_1 a_2) - \sigma$ where $f(\cdot)$ is a suitable oneway function mapping $\mathbb{Z}_n^*$ into $\mathbb{Z}_v$. The bank multiplies $\gamma^v a_1 g^\sigma$ by a_2 and $g^{f(a_1 a_2)-\sigma}$ to get $\gamma^v a_1 a_2 g^{f(a_1 a_2)}$, computes the v'th root of this number and sends it to Alice. Alice divides out the blinding factor γ to get the pair $(a, (ag^{f(a)})^{1/v})$. The number a is called the base number of the signature.

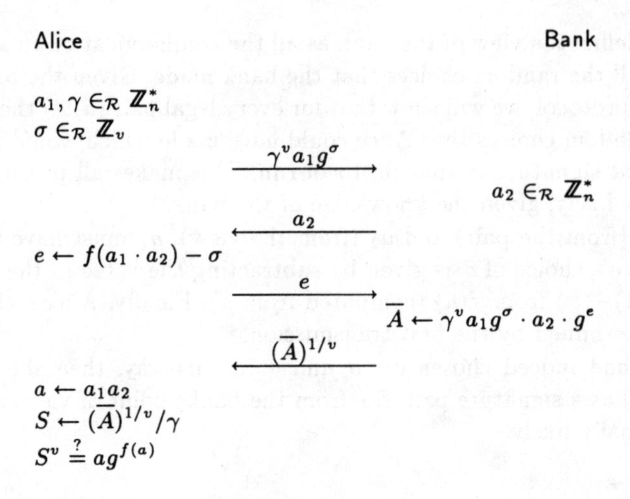

Fig. 1. Randomized blind signature scheme

Note: To make the blinding perfect, all computations involving exponents are done modulo v. For example, e is computed as $(f(a_1 a_2) - \sigma) \bmod v$. Alice can correct for the possible additional factor of g^v by multiplying the final signature by $g^{(f(a)-\sigma) \operatorname{div} v}$. In the rest of this paper we will assume implicitly that all computations involving exponents are done modulo v and that the necessary corrections are applied to the resulting signatures.

Assumption 1. *It is computationally infeasible to forge a signature pair of the form* $(a, (ag^{f(a)})^{1/v})$.

Reasoning. This assumption is a special case of the RSA signature assumption. Suppose Alice tries to forge a signature pair (a, A). If we define $t := f(a)$, then she must have solved the equation $t = f(A^v g^{-t})$. Two ways to solve this equation spring to mind. The first one is to choose A and then try different values of t until you get lucky (probability of success is $1/v$). The second one is to choose $A^v g^{-t}$, compute t, and then try to compute A. This requires the computation of a v'th root. Neither of these methods seems feasible.

Even if Alice has a large number of valid signature pairs, it is still difficult for her to find new ones. A result of Evertse and van Heyst [EvH92] implies (loosely stated) that the only new RSA signatures that can be computed from old ones are multiplicative combinations of the old signatures. If you multiply two signatures of the form $ag^{f(a)}$, you do not get another valid signature, unless $f(ab) = f(a) + f(b)$. One of the requirements for f is that it is infeasible to find relations of this kind.

Proposition 2. *The bank gets no information regarding* $(a, (ag^{f(a)})^{1/v})$ *from the protocol in Fig. 1.*

Proof. We define the view of the bank as all the communication to and from the bank, plus all the random choices that the bank made. Given the banks view of a run of the protocol, we will show that for every legal pair (a, A) there is exactly one set of random choices that Alice could have made which would result in her receiving that signature in that protocol run. This makes all possible signature pairs equally likely, given the knowledge of the bank.

Given a (from the pair) and a_2 (from the view), a_1 must have been chosen as a/a_2. Alice's choice of σ is given by subtracting the value in the third transmission $(f(a) - \sigma)$ from $f(a)$ (computed from a). Finally, Alice's choice of γ is uniquely determined by the first transmission.

If Alice had indeed chosen a_1, σ and γ in this way, then she would have gotten (a, A) as a signature pair. So, from the banks point of view, all signature pairs are equally likely. □

3.1 'Abuses' by Alice

There are several ways in which Alice can deviate from this protocol which we will investigate briefly. For this we rewrite the protocol as shown in Fig. 2. We assume that Alice is choosing the numbers A–E in some clever way. For practical reasons we have to restrict Alice's behaviour a bit. In choosing the numbers, she can use any construction, but we assume that the only computations that she does with the final signature are an exponentiation and a multiplication. (Other operations don't seem to make sense on an RSA signature.) Furthermore, Alice should end up with a v'th root on a number of the form $Kg^{f(K)}$ for some K that Alice can compute. Any other results are not of interest in the coin system.

First of all, observe that B only occurs as $C+B$ in the result. As C is chosen later then B, we can assume $B = 0$ without loss of generality.

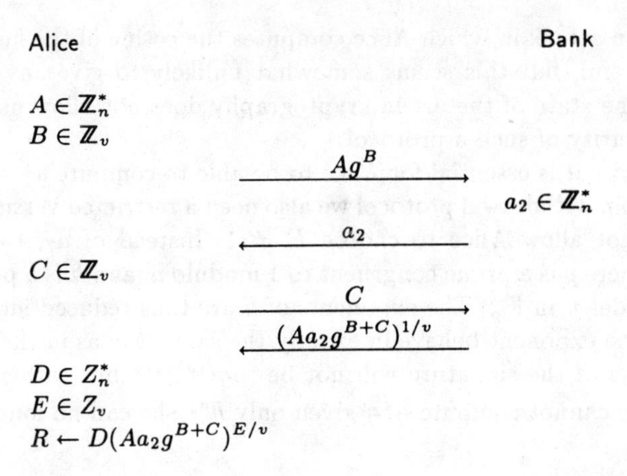

Fig. 2. Possible behaviours of Alice

To get a useful signature, Alice must have $R = (Kg^{f(K)})^{1/v}$, which is equivalent to $D^v A^E a_2^E g^{EC} = Kg^{f(K)}$. We can assumed that Alice doesn't use a factor g^x in A for some x. (Alice can get the same effect by adding x to C.) The exponents on g are modulo v, so D cannot contribute to them. As Alice doesn't know $g^{1/v}$, she cannot put any extra factors in herself. To get something useful she must therefore solve the following two simultaneous equations.

$$t + EC = f(K) \pmod{v}$$
$$D^v A^E a_2^E = Kg^t$$

for some $t \in \mathbb{Z}_v$. There seems only one way to do this, and that is to fix K by choosing D, A, E and t, and then choosing C to fit the first equation. (Any other way would involve inverting the oneway function, or computing a root.) But this means that D and E must have been fixed before sending C to the bank.

We conclude that Alice can 'abuse' this protocol by sending a slightly different reply in the third message. She can choose any D and E such that she gets a valid signature (on a number of the form $xg^{f(x)}$) after raising it to the E'th power and multiplying it by D. The factor D doesn't help Alice much in cheating. Alice can modify the base number after the bank has revealed a_2, but she can only multiply the base number by D^v. Because Alice cannot compute roots, any v'th power is essentially random to her, thus she cannot control the way in which she changes the base number. Alice could at most use the D factor to shift the uniform distribution of the base number slightly, but in the large set of possible base numbers this is hardly significant. The same basically holds for the E power.

Unfortunately, we cannot prove that there is no other way for Alice to cheat. The attacks allowed in Fig. 2 are only the most obvious ones. There might for example be an attack in which Alice computes the cosine of the last reply to get something useful, but this seems somewhat unlikely to give any useful result. At present, the state of the art in cryptography does not allow us in general to prove the security of such a protocol.

When $E \neq 1$ it is essential for Alice to be able to compute a_2^E after receiving a_2. For our coin withdrawal protocol we also need a restricted version (see Fig. 3) which does not allow Alice to choose $E \neq 1$. Instead of a_2, the bank sends $h^{a_2} \bmod p$ where p is a prime congruent to 1 modulo n, and h is a publicly known element of order n in $\mathbb{F}_p$. The exponents of h are thus reduced modulo n so the numbers in the exponent behave in exactly the same way as in the RSA system. The final form of the signature will not be $(ag^{f(a)})^{1/v}$ but rather $(ag^{f(h^a)})^{1/v}$. Because Alice cannot compute $h^{a_2^E}$ given only h^{a_2} she can no longer choose E.

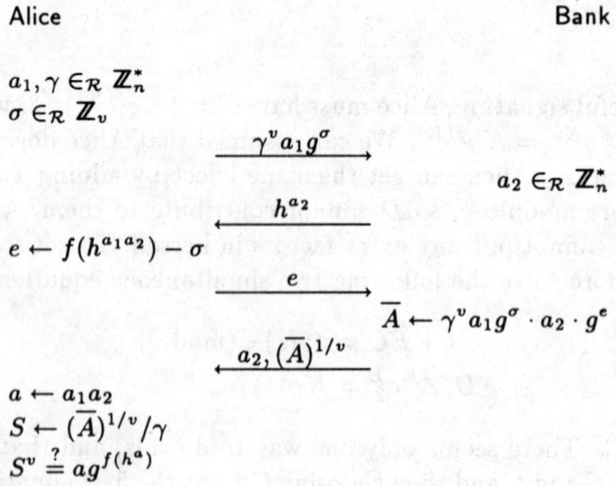

Alice		Bank
$a_1, \gamma \in_{\mathcal{R}} \mathbb{Z}_n^*$		
$\sigma \in_{\mathcal{R}} \mathbb{Z}_v$		
	$\xrightarrow{\gamma^v a_1 g^\sigma}$	
		$a_2 \in_{\mathcal{R}} \mathbb{Z}_n^*$
	$\xleftarrow{h^{a_2}}$	
$e \leftarrow f(h^{a_1 a_2}) - \sigma$		
	$\xrightarrow{e}$	
		$\overline{A} \leftarrow \gamma^v a_1 g^\sigma \cdot a_2 \cdot g^e$
	$\xleftarrow{a_2, (\overline{A})^{1/v}}$	
$a \leftarrow a_1 a_2$		
$S \leftarrow (\overline{A})^{1/v} / \gamma$		
$S^v \stackrel{?}{=} ag^{f(h^a)}$		

Fig. 3. Randomized blind signature scheme without exponential attack

4 Coin withdrawal protocol

For our system we need 3 numbers, C, A, and B. These will be of the form

$$C = cg_c^{f(h_c^c)}$$
$$A = ag_a^{f(a)}$$
$$B = bg_b^{f(h_b^b)}$$

where the numbers g_c, g_a and g_b are publicly known and of large order in the group Z_n^*. The numbers h_c and h_b are elements of order n from $\mathbb{F}_p$ where $p - 1$

is a multiple of n. The use of three different g values ensures that the numbers are distinct and do not mingle when multiplied together in a signature.

The coin withdrawal protocol (see Fig. 4) consists of three parallel runs of the randomized blind signature scheme. Two of the runs are the restricted version, and one is the unrestricted version. The E exponent is used to allow Alice to randomise the k parameter of the secret sharing line herself while the bank can ensure that the other parameter is Alice's identity U.

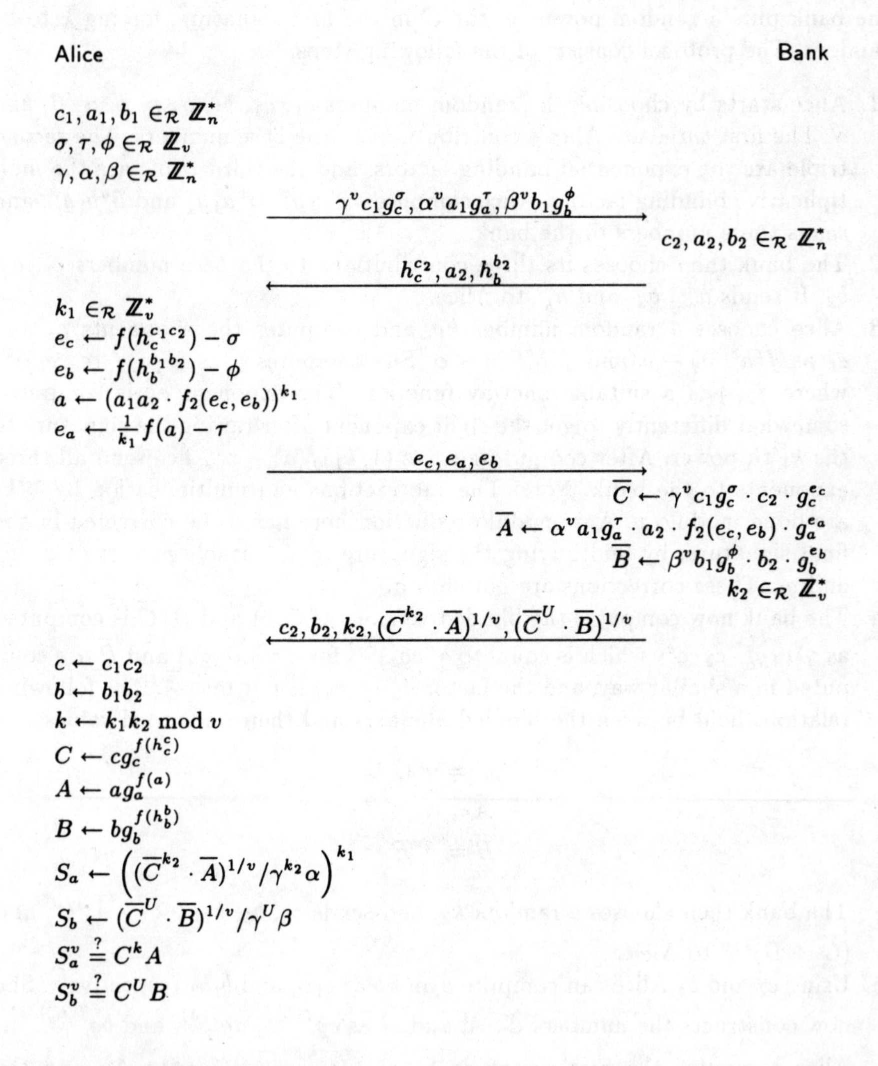

Fig. 4. Coin withdrawal protocol

There are 2 additions to this simple parallel-run view. One is that there is an extra one-way function that makes a depend on e_c and e_b. This prevents Alice form choosing e_c and e_b as a function of a. Were she able to do this, she could cancel some of the terms and get a signature on just C and B. Although this is not a threat as such against the payment scheme, it is undesirable that Alice has so much freedom.

The second modification is that the bank puts a random power on C in the first signature. Alice is going to end up with two signatures: $(C^k A)^{1/v}$ and $(C^U B)^{1/v}$. Here, U is the identity and k is a random number unknown to the bank. However, to prevent Alice from combining an old coin with the one currently being withdrawn the bank must ensure that k is indeed random. Therefore, the bank puts a random power on the C in the first signature, forcing k to be random. The protocol consists of the following steps:

1. Alice starts by choosing the random numbers c_1, a_1, b_1, σ, τ, ϕ, α, β, and γ. The first three are Alice's contributions to the base numbers. The second triple are the exponential blinding factors, and the third triple are the multiplicative blinding factors. Alice computes $\gamma^v c_1 g_c^\sigma$, $\alpha^v a_1 g_a^\tau$ and $\beta^v b_1 g_b^\phi$, and sends these numbers to the bank.
2. The bank then chooses its three contributions to the base numbers c_2, a_2, b_2. It sends $h_c^{c_2}$, a_2, and $h_b^{b_2}$ to Alice.
3. Alice chooses a random number k_1, and computes the exponents e_c and e_b as $f(h_c^{c_1 c_2}) - \sigma$ and $f(h_b^{b_1 b_2}) - \phi$. She computes a as $(a_1 a_2 f_2(e_c, e_b))^{k_1}$ where $f_2(\cdot)$ is a suitable oneway function. The exponent e_a is computed somewhat differently to get the right exponent after raising the signature to the k_1'th power. After computing e_a as $(1/k_1)f(a) - \tau$, Alice send all three exponents to the bank. Note: The subtractions and multiplication by $1/k_1$ are done modulo v. Any modulo reduction here has to be corrected in the final signature, by multiplying the signature by a suitable powers of g_c, g_a and g_b. These corrections are not shown.
4. The bank now computes the blinded versions of C, A and B. $\overline{C}$ is computed as $\gamma^v c_1 g_c^\sigma \cdot c_2 \cdot g_c^{e_c}$ which is equal to $\gamma^v c g_c^{f(h_c^c)}$ for $c = c_1 c_2$. $\overline{A}$ and $\overline{B}$ are computed in a similar way, and the factor $f_2(e_c, e_b)$ is put into $\overline{A}$. The following relations hold between the blinded numbers and their unblinded values:

$$\overline{C} = \gamma^v C$$
$$\overline{A} = \alpha^v A^{1/k_1}$$
$$\overline{B} = \beta^v B$$

   The bank then chooses a random k_2, and sends c_2, b_2, k_2, $(\overline{C}^{k_2} \cdot \overline{A})^{1/v}$, and $(\overline{C}^U \cdot \overline{B})^{1/v}$ to Alice.
5. Using c_2 and b_2 Alice can compute c and b as $c_1 c_2$ and $b_1 b_2$ respectively. She now constructs the numbers C, A, and B as $c g_c^{f(h_c^c)}$, $a g_a^{f(a)}$, and $b g_b^{f(h_b^b)}$.

   Alice computes the first signature S_a as $((\overline{C}^{k_2} \cdot \overline{A})^{1/v} / \gamma^{k_2} \alpha)^{k_1}$, and the second signature S_b as $(\overline{C}^U \cdot \overline{B})^{1/v} / \gamma^U \beta$. The k_1'th power is needed because the base number a was chosen as $(a_1 a_2)^{k_1}$ instead of $a_1 a_2$. All the necessary adjustments in the exponent of g_a were already made by Alice. The total effect of this k_1'th power is to get a v'th root on the number $C^k A$ where $k = k_1 k_2$.

Finally Alice checks that the signatures she received are correct by verifying that $S_a^v = C^k A$ and $S_b^v = C^U B$.

Alice ends up with the following set of numbers: c, a, b, k, S_a, and S_b which are the 3 base numbers, the random parameter for the secret sharing line and the 2 signatures. These 6 numbers plus the identity U are used as input to the payment protocol.

We still need a few additions to this protocol to protect Alice against framing by the Bank. To this end we let U be the concatenation of Alice's identity and a unique coin number. This makes the U's of all the coins distinct. Secondly, in the third transmission Alice includes a digital signature on U and all the data in the first three transmissions. If the bank now claims that Alice spent a coin twice, it must show a transcript of the withdrawal protocol for that coin. This transcript must include the correct data in the last transmission. (This is verifiable by a third party.) The bank also shows a, b and c from the doubly spent coin.

If Alice didn't spend the coin with that identity twice, then the bank can have *no* knowledge regarding a, b, or c. So if the bank tries to frame Alice, the triple (a, b, c) will (with high probability) be different from the actual values used by Alice. If Alice can provide a different triple (a, b, c) plus the corresponding blinding factors that match the transcript, then the bank must be framing Alice, as she cannot generate a new triple which matches a given transcript.

5 Remarks

It would be nice to make a similar system where C, A, and B are images under oneway functions of a single base number c. If c can also be made smaller (say in $\mathbb{Z}_v$ instead of $\mathbb{Z}_n$) then we could store a coin in about 70 bytes. However, constructing an efficient withdrawal protocol for such a coin remains an open problem.

Work is currently under way to implement this scheme on workstations to provide e-mail money. An extension of this coin system to n-spendable coins (which can be spent n times but not $n+1$ times) and the incorporation of observers in the coin system are described in [Fer94].

References

[Bra93] Stefan Brands. An efficient off-line electronic cash system based on the representation problem. Technical Report CS-R9323, CWI (Centre for Mathematics and Computer Science), Amsterdam, 1993. Anonymous ftp: ftp.cwi.nl:/pub/CWIreports/AA/CS-R9323.ps.Z.

[Bra94] Stefan Brands. Electronic cash systems based on the representation problem in groups of prime order. In *Proceedings of CRYPTO '93*, 1994. To appear.

[CdBvH+90] David Chaum, Bert den Boer, Eugène van Heyst, Stig Mjølsnes, and Adri Steenbeek. Efficient off-line electronic checks. In J.-J. Quisquater and J. Vandewalle, editors, *Advances in Cryptology—EUROCRYPT '89*, Lecture Notes in Computer Science, pages 294–301. Springer-Verlag, 1990.

[CFN90] David Chaum, Amos Fiat, and Moni Naor. Untraceable electronic cash. In S. Goldwasser, editor, *Advances in Cryptology—CRYPTO '88*, Lecture Notes in Computer Science, pages 319–327. Springer-Verlag, 1990.

[Cha90] David Chaum. Online cash checks. In J.-J. Quisquater and J. Vandewalle, editors, *Advances in Cryptology—EUROCRYPT '89*, Lecture Notes in Computer Science, pages 288–293. Springer-Verlag, 1990.

[Cha92] David Chaum. Randomized blind signature. Personal communications, April 1992.

[EvH92] Jan-Hendrik Evertse and Eugène van Heyst. Which new RSA-signatures can be computed from certain given RSA-signatures? *J. Cryptology*, 5(1):41–52, 1992.

[Fer93] Niels Ferguson. Single term off-line coins. Technical Report CS-R9318, CWI (Centre for Mathematics and Computer Science), Amsterdam, 1993. Anonymous ftp: ftp.cwi.nl:/pub/CWIreports/AA/CS-R9318.ps.Z.

[Fer94] Niels Ferguson. Extensions to single term off-line coins. In *Proceedings of CRYPTO '93*, 1994. To appear.

[FY93] Matthew Franklin and Moty Yung. Secure and efficient off-line digital money. In A. Lingas, R. Karlsson, and S. Carlsson, editors, *Automata, Languages and Programming, 20th International Colloquium, ICALP 93, Lund, Sweden*, Lecture Notes in Computer Science 700, pages 265–276. Springer-Verlag, 1993.

[Hir93] Rafael Hirschfeld. Making electronic refunds safer. In *Advances in Cryptology—CRYPTO '92*, 1993. To appear.

[OO92] Tatsuaki Okamoto and Kazuo Ohta. Universal electronic cash. In J. Feigenbaum, editor, *Advances in Cryptology—CRYPTO '91*, Lecture Notes in Computer Science, pages 324–337. Springer-Verlag, 1992.

[RSA78] Ronald Rivest, Adi Shamir, and Leonard Adleman. A method for obtaining digital signatures and public-key cryptosystems. *Communications of the ACM*, 21:120–126, February 1978.

[Sha79] Adi Shamir. How to share a secret. *Communications of the ACM*, 22(11):612–613, 1979.

[vA90] Hans van Antwerpen. Off-line electronic cash. Master's thesis, Eindhoven University of Technology, department of Mathematics and Computer Science, 1990.

Improved Privacy in Wallets with Observers

(Extended Abstract)

R. J. F. Cramer[1] and T. P. Pedersen[2]

[1] CWI, The Netherlands
[2] Aarhus University, Denmark***

Abstract. Wallets with observers were suggested by David Chaum and have previously been described in [Ch92] and [CP92]. These papers argue that a particular combination of a tamper-resistant-unit and a small computer controlled by the user is very suitable as a personal device in consumer transaction systems. Using such devices, protocols are constructed that, simultaneously, achieve high levels of security for organizations and anonymity for individual users. The protocols from [CP92] offer anonymity to users, under the assumption that the information stored by observers is never revealed to the outside world.
This paper extends [CP92] by defining additional requirements for the protocols which make it impossible to trace the behaviour of individuals in the system if one is also allowed to analyze a posteriori the information observers can collect. We propose two protocols satisfying our requirements, thus achieving a higher degree of privacy for individuals. This extra level of privacy is obtained at essentially no cost as the new protocols have the same complexity as those previously proposed.

1 Introduction

In [Cha83], the notion of blind signatures was proposed and used to construct an on-line electronic payment scheme for anonymous payments. Later, Even and Goldreich (see [EG84]) suggested an off-line electronic payment scheme in which the security relied on a tamper-resistant unit. However, such units inherently cannot offer proper protection of the individual's privacy as the unit in principle can send a lot of personal information to the counterpart during a transaction. An off-line payment system not depending on tamper-resistance and offering privacy was presented in [CFN90]. In this model (off-line and no tamper-resistant units) the same coin cannot be prevented from being spent several times, but [CFN90] solved this problem by guaranteeing that any person who spends a coin more than once would be identified.

Thus, as noted by Chaum in [Ch92], in off-line transactions a solution based on tamper-resistance can give good security but no satisfactory solution to the privacy aspect, whereas cryptographic solutions without tamper-resistant units although offering good privacy cannot prevent "double-spending". Therefore it

*** Research partly done while visiting CWI

was suggested in [Ch92] to obtain the best from these two worlds by combining the two approaches. This is done in an electronic wallet consisting of a small computer, trusted by the individual, and a tamper-resistant unit, called an *observer*. These observers are issued by a special issuing authority (IA) and are trusted by this authority. As the observer may contain (sensitive) information about the user, it is important for privacy reasons that the observer cannot communicate with the outside world (neither during the transactions nor when the observer is not used). During transactions all communication between an observer and the outside world (e.g. the organizations individuals do business with) therefore passes through the user's computer. As part of the protocols, this computer must ensure that the data transmitted to and from the observer contain no (Shannon) information (except for agreed upon messages). This is called "prevention of inflow and outflow". In [CP92], protocols are presented which satisfy this requirement.

In order to be able to recognize observers, these are equipped with a digital signature set-up. This so-called *native signature scheme* is used to prove that the user has a genuine observer. However, the native scheme cannot be used directly, as it would identify the observer (and thereby the user) in each transaction. Instead the native scheme is used to obtain a *validator* at a Validator Issuing Center (VIC). A validator consists of a secret key and a public key, together with a signature by VIC on the public key. One of the central ideas behind validators is that VIC doesn't know which validator it issues (VIC makes a blind signature), so that the wallet can later use it to sign messages that cannot be traced back to the user. Furthermore, the user (more precisely the user's computer) and the observer must cooperate in order to sign messages with respect to the public key of the validator. Such a signature therefore signals to the receiver that the observer (which the receiver is assumed to trust) has acknowledged the validity of the message.

Each validator can be used to sign multiple messages. Any other party can be sure that two messages signed with the same validator originated from the same observer. By using different validators at different times, the user can prevent linking.

As the protocols of getting and using a validator are the central components of most applications of wallets with observers, we concentrate on these two protocols in this paper. Such a pair of protocols will be called a *validator scheme*.

In [CP92] a validator scheme is proposed where unlinkability of a users transactions is based on the assumption that the observer's contents are never revealed. However, assuming an environment of mutual distrust and cheating, observers could be designed in such a way that their contents can be accessed by IA (using a secret trapdoor) in case it has complete physical control over the observer (either by means of stealing, or because of maintenance procedures). In Section 3 we present protocols where the executions of the signing protocol (unconditionally) cannot be linked to the user even if the contents of the observer is known to all organizations (including IA and VIC). First, however, Section 2 describes the set-up in more details and gives the necessary definitions.

The results in this paper are also presented in [BCCFP92]. Furthermore, to avoid misunderstandings it must be mentioned that the protocols were already presented by David Chaum as part of the presentation of [CP92] at Crypto'92. However, the protocols are not included in that paper. The contribution of this paper is to introduce the notion of *shared information* and to present the criteria according to which the protocols were constructed.

2 Security of Protocols Involving Observers

This section discuss the basic principles of wallets with observers, and the security of the basic protocols is defined.

2.1 Tracing an Observer if its Contents Are Known

When considering the privacy of the user we must assume that the observer stores whatever information can be collected from the transactions. In the validator issuing protocol of [CP92], observers know exactly which particular validator is being issued. When the validator is later used in a validator-based signature scheme to sign messages, the observer is able to link the execution of the validator-based signature scheme to that of the issuing protocol. Furthermore, the observer's native signatures, used in the issuing protocol, uniquely link VIC's views of the executions of the issuing protocol to the observer's views of that protocol. As the receiver of a validator-based signature knows the public key of the validator (and VIC's signature on it), the receiver's view of the execution of the validator-based signature scheme, can be linked to the observer's view of the same protocol. Consequently, if the contents of an observer are ever known to the outside world, its behaviour can be completely recovered (see Figure 1). Linking observers with individuals is just a practical matter. So in a scenario in which the contents of (some of) the observers may become available to the outside world, the anonymity of individuals in the system faces serious threats.

Moreover, this attack, which was based on the fact that the observer knows which particular validator is involved in each transaction, can be generalized:

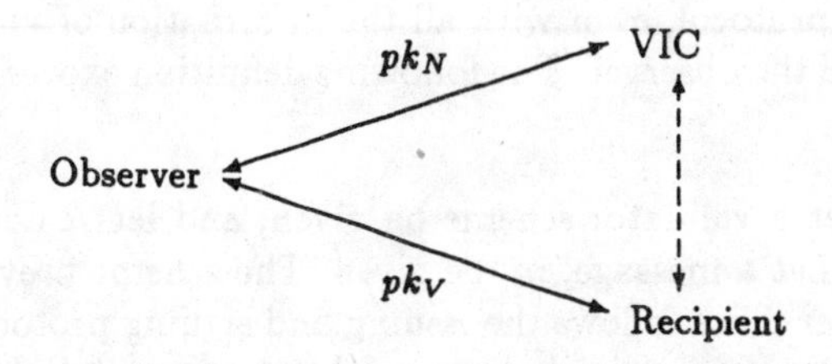

Fig. 1. Linking a validator based signature to the issuing of the validator using the public key of the native scheme, pk_N, and the public key of the validator, pk_V.

any transaction data that arises from the interaction between organizations and the wallet and is known by both observer and recipient can be used to establish the described links. Hence, when designing the protocols such data (called *shared information*) must be avoided.

2.2 Definitions

In view of the above attack we now describe additional criteria for the protocols, which, roughly speaking, require that the public key of validators be unconditionally hidden from observers and that the validator-based signature scheme prevents shared information (except for the message, m, to be signed). As the observer and the receiver in general must know the message, it should be chosen from a sparse set such that each message is likely to be signed by many different wallets. This prevents that m can be used to link transactions.[4]

Recall that a validator scheme consists of a pair of protocols. One, the validator issuing protocol, results in the user and her observer getting a validator, and the other, the validator signing protocol, can be used to make signatures with respect to the public key of the validator. We trust that this informal description is sufficient here.

First note that the validator issuing as well as the signing protocol are three-party protocols involving observer and user plus either VIC or the receiver of the signature. In this extended abstract we leave out a precise definition of our model. Briefly, each participant is modeled by an interactive probabilistic polynomial time Turing machine as in [GMR89]. In some places dishonest participants may have unlimited computing power, but this should be clear from the context. The user will be called Alice and her observer will be denoted by OA.

In order to capture what a participant learns from such a protocol we use the notion of views introduced in [GMR89].

Definition 1. For any execution of a protocol and for any participating party, X, which has input i_x, the view of X, $\text{VIEW}_X(i_x)$, is the set of all the messages that X sees plus all the random choices that X makes.

A validator scheme will be said to protect the privacy of the user, if whenever the user follows her protocol it is impossible for an unlimited powerful entity to find any information regarding which signature corresponds to which execution of the validator issuing protocol, even with all the information of VIC, the recipient of the signature, and the observer. The following definition expresses this informally in terms of views.

Definition 2. Let a validator scheme be given, and let R denote the recipient of the signature. Let a message, m, be given. The scheme prevents *shared information* if whenever Alice follows the issuing and signing protocols (when signing m) then no matter what unlimited powerful VIC, OA and R do in the protocols,

[4] Replay attacks can be prevented by incorporating random bits chosen by the recipient into the signature, see Section 3.5.

the (distribution of the) view of R (when receiving a signature on m) is independent of the combined views of VIC and OA (the views of the observer in both protocols).

This definition basically says that the receiver of the signature may not be able to obtain any information which can be linked to the observer. Namely, if such information exists the receivers view of the signature scheme can be linked to the observer, and the observer can easily link its view of the protocol to the corresponding validator issuing protocol (see also Section 2.1).

2.3 Requirements for the Validator Scheme

In light of the above definitions we now list the properties that the validator issuing and the signature protocol must have in order to be secure as well as protect the privacy of the user. The validator issuing protocol must have the following properties:

1. It must not be feasible for a cheating user to obtain a validator for which she knows the secret key herself (i.e., VIC must be ascertained that a valid observer is taking part).
2. VIC's signature scheme and the application of it in the validator issuing protocol must be secure, such that Alice (even if she breaks the observer) is unable to forge signatures.
3. No (Shannon) information flows between the observer and VIC (if Alice follows the protocol). This is true even if OA and VIC both have unlimited computing power (i.e., no inflow/outflow).
4. Only Alice knows the public key of the validator and VIC's signature on it (both should be unconditionally hidden from VIC and OA, see also requirement 5 to the validator signing scheme below).
5. OA and Alice each get a part of the secret key of the validator from VIC, such that they can only use the secret key by cooperating.

It is in properties 4 and 5 that our issuing scheme differs from that given in [CP92]. These properties ensure that the observer and Alice share the control over the validator, while the observer doesn't know the public key of the validator. The first requirement ensures that Alice is not able to acquire a validator for herself. If she were able to do so, then she could dispense with the observer and run all the other protocols herself without the help of an observer (see also point 3 below).

The signature scheme based on validators must have the following properties (we often refer to the recipient as the verifier):

1. If Alice and OA cooperate they can make a signature on a given message with respect to the public key of their validator.
2. If Alice follows the protocols, then it is not feasible for OA, VIC and the verifier to execute the validator issuing protocol and the signing protocol (several times) in such a way that they are able to learn the secret key of Alice.

3. If OA and VIC both follow the validator issuing protocol, then it is not feasible for Alice to execute this protocol and subsequent signing protocols in such a way that she can sign a message which the observer is not willing to sign.
4. In order to avoid replay, the verifier will be interactive in the protocol.
5. The validator scheme must prevent shared information (Definition 2).

The first four requirements are related to the security of the validator signing protocol, while the last refers to the privacy obtained by the validator scheme.

3 The Protocols

This section presents a validator scheme, which satisfies the requirements outlined above. The protocols work for every group of prime order q. For the sake of concreteness we take a prime order subgroup of $\mathbb{Z}_p^*$, where p is a prime. All protocols are based on the certified discrete logarithm problem. More precisely, it is assumed to be hard to compute discrete logarithms modulo a prime, p, even if the factorization of $p-1$ is given.

3.1 The Basic Signatures

The protocols in this paper use the signature scheme presented in [CP92]. We now briefly describe this scheme. Let two primes p and q be given such that q divides $p-1$, and let $g \in \mathbb{Z}_p^*$ be an element of order q. The group generated by g is denoted by G_q.

The secret key is $x \in \mathbb{Z}_q^*$ and the public key is

$$(p, q, g, h),$$

where $h = g^x$. Let $m \in G_q$ be a message. The signature on m consists of $z = m^x$ plus a proof that

$$\log_g h = \log_m z.$$

This proof bears some resemblance to two parallel executions of instances of Schnorr's scheme [S91] and works as follows:

1. The prover chooses $w \in \mathbb{Z}_q$ at random and computes $(a, b) = (g^w, m^w)$. This pair is sent to the verifier.
2. The verifier chooses a random challenge $c \in \mathbb{Z}_q$ and sends it to the prover.
3. The prover sends back $r = w + cx \bmod q$.
4. The verifier accepts the proof if and only if $g^r = ah^c$ and $m^r = bz^c$.

We shall often refer to this protocol as the *basic proof* (of equality of discrete logarithms).

Now let $\mathcal{H}$ be a one-way hash function mapping arbitrary inputs into the set $\mathbb{Z}_q$ (as in the Fiat-Shamir scheme, see [FS86]). Given this function and the above protocol the signature on m is

$$\sigma(m) = (z, c, r).$$

It is correct if

$$a = g^r h^{-c} \qquad \text{and} \qquad b = m^r z^{-c}$$

satisfy $c = \mathcal{H}(m, z, a, b)$.

The reader is referred to [CP92] for a discussion of the security of this signature scheme. In that paper it is argued that it is hard to forge signatures, and that the signer does not give away any information about his secret key except m^x. This is made more precise in the following assumption:

Assumption 1. *When the signer signs a message, he does not make it computationally easier to compute any function of x, than if he just gives away m^x.*

Assumption 2. *If it is possible to make a signature (z, c, r) on a message m with respect to the public key $h = g^x$ then $z = m^x$.*

The second assumption can be justified as follows. If $z \neq m^x$, then for every pair (a, b) there is at most one $c \in \mathbb{Z}_q$ for which there exists an $r \in \mathbb{Z}_q$ such that the signature is valid. Hence, if $z \neq m^x$ the forger must hope that the image of $\mathcal{H}$ is exactly this c. However, if the output of $\mathcal{H}$ is "hard to control" this seems extremely unlikely.

3.2 Blind Signatures

To get a blind signature on the message $m \in G_q$ one chooses a random $t \in \mathbb{Z}_q^*$ and asks the signer to sign $m_0 = mg^t$. Let $z_0 = {m_0}^x$. The signer then proves that $\log_g h = \log_{m_0} z_0$ in such a way that the messages are blinded:

1. The signer chooses a random $w \in \mathbb{Z}_q$ and computes $(a_0, b_0) = (g^w, m_0^w)$. This pair is sent to the verifier.
2. The verifier chooses $u \in_{\mathcal{R}} \mathbb{Z}_q^*, v \in \mathbb{Z}_q$ at random and computes

$$a = a_0^u g^v \qquad \text{and} \qquad b = (b_0/a_0^t)^u m^v.$$

   (If both parties follow the protocol, $a = (g^{uw+v})$ and $b = (m^{uw+v})$.) The verifier computes $z = z_0/h^t$, the challenge $c = \mathcal{H}(m, z, a, b)$ and the blinded challenge $c_0 = c/u \bmod q$. The verifier sends c_0 to the signer.
3. The signer sends back $r_0 = w + c_0 x \bmod q$.
4. The verifier accepts if

$$g^{r_0} = a_0 h^{c_0} \qquad \text{and} \qquad {m_0}^{r_0} = b_0 z_0^{c_0}.$$

   The verifier computes $r = ur_0 + v \bmod q$ and

$$\sigma = (z, c, r).$$

The following two propositions from [CP92] show that this really is a blind signature:

Proposition 3. *If the verifier accepts in the above protocol, then σ is a correct signature on m.*

Proposition 4. *The signer gets no information about m and σ if the verifier follows the protocol.*

The following assumption makes precise which signatures the verifier can obtain when executing the blind signature protocol:

Assumption 3. *On input m_0 to the above blind signature protocol, the verifier can only obtain a blind signature on a message $m \in G_q$ if he knows $s, t \in \mathbb{Z}_q$ such that $m = m_0^s g^t$.*

This assumption can be justified as follows. Assumption 2 implies that

$$z = m^x \quad \text{and hence} \quad \log_m b = \log_g a.$$

Thus the verifier must be able to compute m^x from m_0^x. This is presumably difficult unless m is a product of elements for which the x'th power is known. Furthermore, under Assumption 2 it can be shown that no matter how the verifier computes b it satisfies

$$b = m^{uw+v},$$

where u and v are defined by

$$u = c/c_0 \bmod q \quad \text{and} \quad v = r - ur_0 \bmod q.$$

It seems to be hard to compute a pair (m, b) satisfying this before getting r_0 from the signer unless m is of the prescribed form.

3.3 Signatures by OA

In [CP92] it is shown how this signature scheme can be used by the observer such that it cannot send any information to or receive any information from the verifier (i.e., without causing inflow of outflow). This scheme can therefore be used as the native signature scheme.

3.4 Validator Issuing

We now turn to the problem of creating a validator. The public key is going to be a number $h \in G_q$ such that the corresponding secret key, $\log_g h$, is of the form $s + t \bmod q$, where only OA knows t and only Alice knows s. Furthermore, Alice will obtain a (blind) signature from VIC on h.

Before this protocol can start, the native public key and the certificate on it are sent to Alice and VIC. VIC has a private key x and a public key $H = G^x$ where $G \neq g$ and G is a generator of G_q such that Alice does not know $\log_g G$ (hence, (G, H) will replace (g, h) when the blind signature scheme is used to issue a validator).

VIC will use the blind signature scheme when signing the public part of the validator. The resulting protocol to acquire a validator is:

1. Alice chooses $j, s \in \mathbb{Z}_q$ at random, computes $\alpha := g^s G^j$ and sends this to OA (s is Alice's part of the secret key, and G^j blinds it).
2. OA chooses a random $t \in \mathbb{Z}_q$ and sends $\beta := g^t$ to Alice ($s + t$ is the secret key).
3. Alice computes the public key $h := g^s \beta$ and chooses $k \in \mathbb{Z}_q$ uniformly at random. She then computes the blinded public key $B := hG^{j+k}$ and sends k to OA.
4. OA computes $B := \alpha\beta G^k$ and a signature, $\sigma(B)$, on B using its native signature scheme. OA sends the signature to Alice.
5. Alice verifies the signature received from OA and sends B and $\sigma(B)$ to VIC.
6. VIC verifies the signature to ensure that B was created in cooperation with an observer. He then chooses $w \in \mathbb{Z}_q$ at random, constructs $(a_0, b_0) := (G^w, B^w)$, and computes $V_0 := B^x$. VIC sends a_0, b_0, and V_0 to Alice.
7. Alice chooses $u \in \mathbb{Z}_q^*$ and $v \in \mathbb{Z}_q$ at random and computes $a := a_0^u G^v$ and $b := \left(b_0/a_0^{j+k}\right)^u h^v$. She computes $V := V_0/H^{j+k}$ and the corresponding challenge $c := \mathcal{H}(h, V, a, b)$. Alice then computes $c_0 := c/u \bmod q$ and sends c_0 to VIC.
8. VIC computes $r_0 := w + c_0 x$ and sends it to Alice.
9. Alice verifies that

$$G^{r_0} \stackrel{?}{=} a_0 H^{c_0} \qquad \text{and} \qquad h^{r_0} \stackrel{?}{=} b_0 V_0^{c_0}.$$

She then computes $r := v + ur_0$. The tuple (V, c, r) is the signature on h.

Proposition 5. *If* OA, *Alice and* VIC *all follow the prescribed protocol then Alice gets a correct signature on h.*

Proof. By straightforward computations. □

This protocol can in a natural way be split in two parts. In the first part (corresponding to step 1–4) Alice and OA choose a public key $h = g^{s+t}$, where Alice knows s and OA knows t. In the second part VIC signs this public key.

The first part can very well be performed off-line before the user contacts VIC. Furthermore, the computation of expressions of the form $a^d b^e \bmod p$, where $a, b \in G_q$ and $d, e \in \mathbb{Z}_q$ requires only a little more work than a single exponentiation (less than $2|q|$ multiplications in G_q), see [BC90, B92]. Hence, the on-line computation needed by the user is just a little more than 5 exponentiations in G_q, whereas VIC requires 3 exponentiations and a signature verification (which again requires approximately 2 exponentiations, if the scheme from Section 3.3 is used as the native scheme).

The next proposition says that the key selection part is secure for Alice independently of the computing power of OA (so that requirement 4 of the validator issuing protocol is satisfied)

Proposition 6. *If Alice follows the protocol then no matter what an unlimited powerful cheating observer and center do, they cannot (even together) obtain any Shannon information about the validator.*

Proof. Given a view of OA and a view of VIC it is sufficient to show that for all $h \in G_g$ and for all signatures on h there is exactly one possible choice of s, j, k, u, v by Alice such that OA and VIC would get these views when issuing a validator on h.

Except for the random bits the view of OA consists of

$$\alpha,\ \beta,\ k,\ B, \sigma(B)$$

and the view of VIC of

$$B,\ \sigma(B),\ (a_0, b_0, V_0),\ c_0,\ r_0.$$

Let h and the signature (V, c, r) on h be given. Then k is determined by the observer's view and s and j are determined by

$$s = \log_g(h/\beta) \qquad \text{and} \qquad j = \log_G(B/h) - k.$$

Finally, u and v are determined by

$$u = c/c_0 \bmod q \qquad \text{and} \qquad v = r - ur_0.$$

It can be shown that these values of u and v satisfy

$$a_0^u g^v = g^r h^{-c} \qquad \text{and} \qquad \left(b_0/a_0^{j+k}\right)^u h^v = h^r V^{-c}$$

as they will in an execution of the protocol. □

Supposing the native signature scheme prevents inflow and outflow this proof also implies that observer and center cannot use the protocol as a subliminal channel of information, so that requirement 3 of the validator issuing protocol is satisfied.

Finally we have to consider the security for the organization. In other words we want to show that no polynomially bounded user $\tilde{A}$ can get a validator of a public key for which she knows the corresponding secret key. Assume therefore that OA and VIC both follow the prescribed protocol, and recall that B is the blinded public key, which OA signs.

First note that Assumption 3 implies that $\tilde{A}$ cannot get a signature on a number h for which $\log_g h$ is known unless she knows a pair (d, e) such that $B = g^d G^e$. The following proposition shows that $\tilde{A}$ cannot know such a pair unless it is easy to compute $\log_g G$.

Proposition 7. *Let π denote the probability that $\tilde{A}$ can find a pair (e, d) such that $B = g^d G^e$ (over the random coins of* OA, VIC *and $\tilde{A}$). If π is greater than the inverse of some polynomial for p and q sufficiently large, then there is a probabilistic polynomial time algorithm for finding* $\log_g G$.

Proof. The idea is only sketched. Recall, that none of the participants need to know $\log_g G$. Hence, they may all be cooperate in order to find this logarithm. Consider the following algorithm, which uses the methods of $\tilde{A}$, OA and VIC.

1. $\tilde{A}$ computes α.
2. OA chooses $t \in \mathbb{Z}_q$ at random and computes $\beta = g^t$.
3. $\tilde{A}$ computes k, gets a signature on $B = \alpha\beta G^k$ from OA and a blind signature on B from VIC.
4. $\tilde{A}$ tries to find (d, e) such that $B = g^d G^e$.
   If this fails: stop.
5. Rewind $\tilde{A}$ to after Step 1.
6. Choose $\beta' = g^a G^b$ (a and b chosen at random).
7. Given this, $\tilde{A}$ computes k', gets a signature on $B' = \alpha\beta' G^{k'}$ from OA and a blind signature on B' from VIC.
8. $\tilde{A}$ tries to find (d', e') such that $B' = g^{d'} G^{e'}$.
   If this fails: stop.
9. If $d - t = d' - a$ then stop.
   Output $f := ((d - t) - (d' - a))/((e' - b - k') - (e - k)) \bmod q$.

If Step 9 is reached the algorithm will output $f = \log_g G$ with probability at least $1 - 1/q$ (namely, if $d - t \neq d' - a$). Next it is shown that this step is reached with probability at least $\frac{1}{8}\pi^3$, thus completing the proof.

For a given $\alpha \in G_q$ let p_α denote the probability that $\tilde{A}$ finds the pair (d, e) given α is chosen in the first step (over the random coins of all three parties). Let $Prob[\alpha]$ denote the probability that $\tilde{A}$ chooses α. Then

$$\pi = \sum_{\alpha} p_\alpha Prob[\alpha].$$

Let E denote the set $\{\alpha \in G_q \,|\, p_\alpha \geq \pi/2\}$. Step 9 will be reached with probability

$$\begin{aligned}\sum_{\alpha} p_\alpha^2 Prob[\alpha] &\geq \sum_{\alpha \in E} p_\alpha^2 Prob[\alpha] \\ &\geq \frac{\pi^2}{4} Prob[p_\alpha \geq \pi/2].\end{aligned}$$

Finally observe that

$$Prob[p_\alpha \geq \pi/2] \geq \pi/2.$$

□

Note that this shows that requirements 1 and 6 for the validator issuing protocol are satisfied.

Regarding requirement 2 the reader is referred to [CP92] for a discussion of the basic proof and the security of the blind signature protocol. It is argued that it is difficult to forge a signature after a number of executions of the blind signature protocol.

3.5 Validator Signing

It is now shown how Alice and OA can sign a message using the validator obtained above.

It is assumed that Alice initially sends the public key h and the signature on it to the recipient, who then verifies the signature. A message $m \in G_q$ is now signed as follows

1. OA chooses a random $w \in \mathbb{Z}_q$, computes $a_0 := g^w$, $b_0 := m^w$ and $z_0 := m^t$ and sends a_0, b_0 and z_0 to Alice.
2. The verifier chooses $\rho \in \{0,1\}^{|q|}$ at random and sends it to Alice.
3. Alice chooses u and v at random and computes $a := a_0^u g^v$, $b := b_0^u m^v$ and $z := z_0 m^s$. She then computes $c := \mathcal{H}(m, z, a, b, \rho)$ and $c_0 := c/u$. She sends the challenge c_0 to OA.
4. OA computes the response $r_0 := w + c_0 t$ and sends this to Alice.
5. Alice checks that $g^{r_0} = a_0 \beta^{c_0}$ (she knows β from the validator issuing protocol) and $m^{r_0} = b_0 z_0^{c_0}$ (up to this point OA and Alice have used the basic proof. OA provides m^t and proves that $\log_g g^t = \log_m z_0$).
   Alice computes $r := ur_0 + v + cs$ and sends (z, c, r) to the verifier.
6. The verifier computes
   $$a = g^r h^{-c} \quad \text{and} \quad b = m^r z^{-c}.$$
   and checks that $c = \mathcal{H}(m, z, a, b, \rho)$

By using the same technique as in the blind signature protocol the user can actually obtain a signature on a message of the form $m^l g^n$ for some $l, n \in \mathbb{Z}_q$ while the observer believes it signs m. To prevent this a hash-value of m should be used when b_0 and z_0 are computed.

First note that if all three parties follow the protocol then the verifier will end up with a correct signature on m (requirement 1 for the validator based signature scheme).

The purpose of the random string, ρ, is to prevent Alice from just replaying an old signature on m (see footnote 1). Hence requirement 4 to the validator signing protocol is satisfied.

The following proposition shows that Alice does not compromise her privacy by executing this protocol (requirement 5 for the validator based signature scheme).

Proposition 8. *The validator scheme satisfies Definition 2.*

Proof. Assume Alice follows the protocols. OA, VIC and the verifier may deviate arbitrarily from the protocols, and they may have unlimited computing power.

Let $\text{VIEW}^1_{\text{OA}}$ and $\text{VIEW}^2_{\text{OA}}$ be the views of OA in the validator issuing and signing protocols, respectively. Let furthermore VIEW_{VIC} be the view of VIC in the validator issuing protocol and VIEW_V be the view of the verifier in a signing protocol.

It is sufficient to show that for every triple $(\text{VIEW}^1_{\text{OA}}, \text{VIEW}^2_{\text{OA}}, \text{VIEW}_{\text{VIC}})$ belonging together there is exactly one sequence (s, j, k, u, v) such that this triple could correspond to the given VIEW_V. This can be done by a straightforward extension of the proof of Proposition 6 and is omitted here. □

In [OO90] a general construction of blind signatures based on divertible proofs is presented. Our blind signature uses the same principles, but it has the further (necessary) property that the observer can see the message without compromising the constraints from Definition 2 (the construction of [OO90] does not seem to allow the observer to see m without introducing shared information).

We finally show that it is secure for OA as well as Alice to execute this protocol (requirements 2 and 3 for the validator signing protocol).

Proposition 9. *This protocol satisfies:*

1. *If the basic proof system is secure for the prover then* OA *reveals no more information about its secret key than* m^t.
2. *The verifier alone just gets a random signature on* m.
3. *If* OA *and the verifier share all their information, Alice tells them no more than if she gives them* m^s *and executes the basic proof that* $\log_m m^s$ *equals* $\log_g g^s$.

Proof. The main ideas will be sketched.

The first property is obvious as OA just sends m^t to Alice and executes the basic proof system with her. The second follows from the fact that if Alice follows her protocol, then all possible signatures on m are equally likely.

As for the third assume that OA and VIC get $Z = m^s$ and a chance to execute the basic proof with Alice that it is correct. It can be shown that from such an execution they can generate both of their views from the signature protocol with the correct distribution.

First, the view of the observer can be generated by simply running OA.

Next, observe that OA actually proves knowledge of t such that $\beta = g^t$ (recall that β is the first message which OA sends to Alice in the validator issuing protocol). Hence, it is possible to obtain t by a machine with access to OA, and we can therefore assume that t is known from now on.

As part of the view of OA we got $z_0 = m^t$. In order to generate the view of the verifier we get $Z = m^s$ from Alice and execute the basic protocol proving that $\log_g(h/g^t)$ equals $\log_m Z$. Given (a, b) from Alice the challenge in this proof system is calculated as $c = \mathcal{H}(m, Zz_0, a, b, \rho)$, where the verifier supplies ρ. Alice then responds with some R, and we can compute the correct r corresponding to the signature as $r = R + ct \bmod q$. It is easy to see that this r satisfies

$$g^r = ah^c \quad \text{and} \quad m^r = b(Zz_0)^c.$$

Hence, the combined view of OA and V can be generated with almost the same distribution (the generated view is statistically indistinguishable from the real view). □

4 Conclusion and Open Problems

This paper has defined privacy in wallets with observers in such a way that the user is unconditionally protected even if the contents of the observer is later revealed to the organizations.

Furthermore, we have presented protocols for issuing a validator and using a validator to sign messages satisfying this definition. These protocols are just as efficient as previously proposed protocols offering less privacy.

However, the suggested scheme relies on quite strong, but trustworthy, assumptions. It would be interesting to prove (some of) these assumptions or construct another validator scheme based on weaker assumptions.

Acknowledgements

It is a pleasure to thank David Chaum for supporting and stimulating research on wallets with observers and we are grateful for the invitations to CWI which made this work possible. We also kindly acknowledge Jan-Hendrik Evertse's support and many helpful comments during the preparation of earlier work [Cr92] on which this paper is based. Last but not least we thank Stefan Brands, Niels Ferguson and Birgit Pfitzmann for many interesting discussions about wallets with observers.

References

[B92] J. Bos: Practical Privacy, PhD. thesis, Eindhoven University of Technology, The Netherlands, March 1992.

[BC90] J. Bos and M. Coster: Addition Chain Heuristics, Proceedings of Crypto '89, Santa Barbara, August 1989, pp. 400–407.

[BCCFP92] S. Brands, D. Chaum, R. Cramer, N. Ferguson, T. Pedersen: Transaction Systems with Observers. Survey report about wallets with observers, to appear as CWI-technical report.

[Cha83] D. Chaum. Blind signatures for untraceable payments. In *Advances in Cryptology - proceedings of CRYPTO 82*, pp.199–203, 1983.

[Ch92] D. Chaum: Achieving Electronic Privacy, Scientific American, August 1992, pp. 96–101.

[CFN90] D. Chaum, A. Fiat, and M. Naor. Untraceable electronic cash. In *Advances in Cryptology - proceedings of CRYPTO 88*, Lecture Notes in Computer Science, pages 319–327. Springer-Verlag, 1990.

[CP92] D. Chaum and T.P. Pedersen: Wallet Databases with Observers, Proceedings of Crypto '92, Abstracts, Santa Barbara, August 1992, pp. 3.1–3.6.

[Cr92] R.J.F. Cramer: Shared Information in the Moderated Setting, Master's thesis, R.U. Leiden/CWI, August 1992.

[DH76] W. Diffie and M.E.Hellman: New Directions in Cryptography, IEEE Transactions on Information Theory, IT-22(6), November 1976, pp. 644–654.

[EG84] S. Even and O. Goldreich: Electronic Wallet, Proceedings of Crypto'83, Plenum Press 1984, pp. 383–386.

[FS86] A. Fiat and A. Shamir: How to Prove Yourself: Practical Solutions to Identification and Signature Problems, Proceedings of Crypto '86, Santa Barbara, August 1986, pp. 186–194.

[GMR89] S. Goldwasser, S. Micali and C. Rackoff: The Knowledge Complexity of Interactive Proof-Systems, SIAM Journal of Computation, 18(1): 186–208, 1989.

[OO90] T. Okamoto and K. Ohta: Divertible Zero Knowledge Interactive Proofs and Commutative Random Self-Reducibility, Proceedings of Eurocrypt '89, pp. 134–149.

[S91] C.P. Schnorr: Efficient Signature Generation by Smart Cards, J. of Cryptology (1991) 4, pp. 161–174.

Distance-Bounding Protocols

(Extended abstract)

Stefan Brands[1] and David Chaum[2]

[1] CWI, Amsterdam, email: brands@cwi.nl
[2] CWI & DigiCash, Amsterdam, email: david@digicash.nl

Abstract. It is often the case in applications of cryptographic protocols that one party would like to determine a practical upper-bound on the physical distance to the other party. For instance, when a person conducts a cryptographic identification protocol at an entrance to a building, the access control computer in the building would like to be ensured that the person giving the responses is no more than a few meters away.
The "distance bounding" technique we introduce solves this problem by timing the delay between sending out a challenge bit and receiving back the corresponding response bit. It can be integrated into common identification protocols. The technique can also be applied in the three-party setting of "wallets with observers" in such a way that the intermediary party can prevent the other two from exchanging information, or even developing common coinflips.

1 Introduction

A prover convincing a verifier of some assertion is a frequently recurring element in many applications of cryptography. One potentially useful such assertion is that the prover is within a certain distance. It seems that this problem has not been specifically adressed, let alone solved in the literature. We introduce a technique called "distance bounding" that enables the verifying party to determine a practical upper-bound on the physical distance to a proving party.

In the literature, so-called "mafia frauds" have been adressed in which a party identifies himself to a verifying party using the identity of a third party, without that third party being aware of it. With our distance-bounding technique we can prevent these frauds as a special case.

Our distance-bounding protocols can be integrated with known public-key identification schemes, such that the verifier cannot obtain information that he could not have computed himself.

In the recently proposed setting of "wallets with observers," distance bounding can be incorporated in such a way that the verifying party can determine a practical upper-bound to the observer, whereas the intermediary party can prevent the other two parties from exchanging or developing information which can be used to compromise privacy.

This paper is organized as follows: In Section 2 we introduce the distance-bounding principle. We introduce our solution in parts and then unify them.

In Section 3, we describe how distance bounding can be integrated into known public key identification schemes. In Section 4, we describe a problem in the setting of wallets with observers. We then show how to use the distance-bounding technique to solve it. A final section ends this paper with some open problems.

2 Distance-bounding protocols

In this section, we first present the basic distance-bounding principle. We then discuss mafia frauds and previously proposed countermeasures. We show how distance bounding can be used to prevent these frauds. We go on to show how distance bounding can prevent frauds in which a party having access to the secret keys convinces a verifying party that he is within a certain distance whereas he is not. Both protocols are then merged into one protocol that prevents both attacks.

2.1 The distance-bounding principle

The essential element of a distance-bounding protocol is quite simple. It consists of a single-bit challenge and rapid single-bit response. In practice, a series of these rapid bit exchanges is used, the number being indicated by a security parameter k. Each bit of the prover $\mathcal{P}$ is to be sent out immediately after receiving a bit from the verifier $\mathcal{V}$. The delay time for responses enables $\mathcal{V}$ to compute an upper-bound on the distance.

What makes this approach really practical is that today's electronics can easily handle timings of a few nanoseconds, and light can only travel about 30cm during one nanosecond. For instance, even the timing between two consecutive periods of a 50 Mghz clock allows light to travel only three meters and back. (Later on we introduce exclusive-or operations on the bits exchanged, but 10113 chips have several such gates each with a throughput of two nanoseconds.)

2.2 Mafia frauds

A *mafia fraud*, first described in [9], is a real-time fraud that can be applied in zero-knowledge or minimum disclosure identification schemes by fraudulent prover $\overline{\mathcal{P}}$ and verifier $\overline{\mathcal{V}}$, cooperating together. It enables $\overline{\mathcal{P}}$ to convince an honest verifier $\mathcal{V}$ of a statement related to the secret information of an honest prover $\mathcal{P}$, without actually needing to know anything about this secret information. To this end, when $\mathcal{P}$ is about to perform the protocol with $\overline{\mathcal{V}}$, the latter establishes, say, a radio link with $\overline{\mathcal{P}}$, and will send any information transmitted to him by $\mathcal{P}$ straight on to $\overline{\mathcal{P}}$, who in turn sends it on to $\mathcal{V}$. The same strategy is applied by $\overline{\mathcal{P}}$, who sends information received from $\mathcal{V}$ on to $\overline{\mathcal{V}}$, who in turn sends it on to $\mathcal{P}$. In effect, $\overline{\mathcal{V}}$ and $\overline{\mathcal{P}}$ act as a single transparent entity, sitting in the middle between $\mathcal{P}$ and $\mathcal{V}$. This enables $\overline{\mathcal{P}}$ to identify himself to $\mathcal{V}$ as being $\mathcal{P}$, without any of $\mathcal{P}$ and $\mathcal{V}$ noticing the fraud.

$\mathcal{P}$		$\overline{\mathcal{V}}$ (radio link) $\overline{\mathcal{P}}$		$\mathcal{V}$
		(Repeat k times)		
$R_i \in_{\mathcal{R}} \mathbb{Z}_n^*$	$\xrightarrow{R_i^2}$	$\xrightarrow{R_i^2}$	$\xrightarrow{R_i^2}$	
	$\xleftarrow{\alpha_i}$	$\xleftarrow{\alpha_i}$	$\xleftarrow{\alpha_i}$	$\alpha_i \in_{\mathcal{R}} \{0,1\}$
	$\xrightarrow{X^{\alpha_i} R_i}$	$\xrightarrow{X^{\alpha_i} R_i}$	$\xrightarrow{X^{\alpha_i} R_i}$	
		(End of Repeat)		
				verify responses

Fig. 1. A mafia fraud in the basic Fiat-Shamir identification scheme.

In Figure 1, a mafia fraud is shown as it would be applied in the basic Fiat-Shamir identification scheme (see [12]). In order to enhance readability of the figures, we define the subscript i to run over the set $\{1,\ldots,k\}$, and computations are modulo n. In the most basic form of the Fiat-Shamir scheme $\mathcal{P}$ identifies himself to $\mathcal{V}$ by proving knowledge of a square root X of $X^2 \bmod n$, where $X^2 \bmod n$ in some way is related to $\mathcal{P}$'s identity or has been published in a trusted directory. As usual, n is the product of two distinct primes.

In [9], Desmedt proposed a countermeasure to mafia frauds which requires $\mathcal{P}$ to sign a message that contains his physical location on earth, and then prove to $\mathcal{V}$ knowledge of the signature. Usually in an identification scheme, $\mathcal{P}$ will be represented by some user-module, so it will be impracticable to implement this solution without requiring position detection or cooperation of the user. Also, it cannot guarantee that the verifier in the long run does not learn anything about the secret key of the prover.

In [2], Beth and Desmedt propose that all transmission times be accurately measured. This seems to be useless owing to the significant possible variations in speed of computation.

In [1], Bengio *et al* suggested that $\mathcal{V}$ shield $\mathcal{P}$'s module from the outside world (e.g., in a Faraday cage) when the protocol takes place. This countermeasure requires trust by $\mathcal{P}$ that $\mathcal{V}$ does not secretly modify his module in some way while shielded. One would rather like to identify oneself in such a way that the module remains visible (an infrared channel would be even better, the user-module never needing to leave the hands of the user). Futhermore, it requires special hardware equipment.

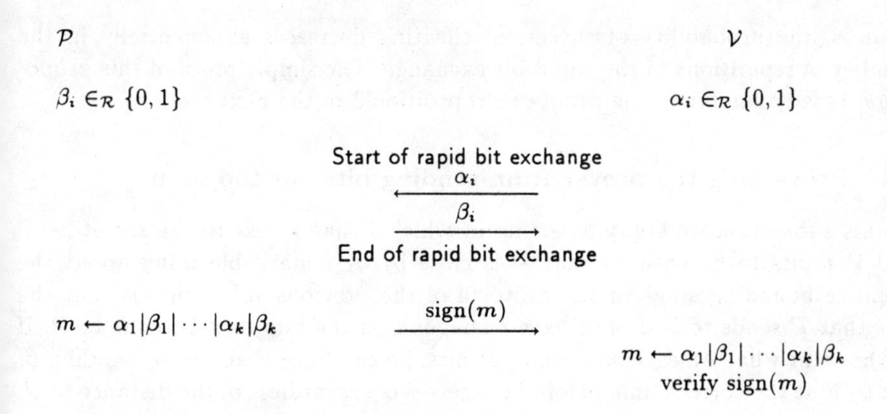

Fig. 2. Distance bounding to prevent mafia frauds.

2.3 Preventing mafia frauds using distance bounding

Consider how the distance-bounding principle can be used to prevent mafia frauds. We can assume that the distance between $\mathcal{P}$ and the fraudulent parties is not less than the accuracy that can be achieved with the apparatus being used, since otherwise obvious countermeasures can be taken. To ensure that the distance between $\mathcal{V}$ and the party $\mathcal{P}$ having access to the secret keys is measured, after the rapid bit exchanges have taken place the message formed by concatenating all the $2k$ bits sent back and forth in the distance-bounding stage is signed by $\mathcal{P}$, using his secret key (see Figure 2):

Step 1 $\mathcal{V}$ generates uniformly at random k bits α_i, and $\mathcal{P}$ generates uniformly at random k bits β_i. (Note: this can take place well beforehand.)

Step 2 Now the low-level distance-bounding exchanges can take place. The following two steps are repeated k times, for $i = 1, \ldots, k$.
- $\mathcal{V}$ sends bit α_i to $\mathcal{P}$.
- $\mathcal{P}$ sends bit β_i to $\mathcal{V}$ *immediately after* he receives α_i.

Step 3 $\mathcal{P}$ concatenates the $2k$ bits α_i and β_i, signs the resulting message m with his secret key, and sends the signature to $\mathcal{V}$. We denote concatenation by the symbol "|."

Now $\mathcal{V}$ can determine an upper-bound on the distance to $\mathcal{P}$ using the maximum of the delay times between sending out bit α_i and receiving bit β_i back, for $i = 1, \ldots, k$. $\mathcal{V}$ accepts if and only if $\mathcal{P}$ is close by, and the received signature is a correct signature of $\mathcal{P}$ on $m = \alpha_1|\beta_1|\cdots|\alpha_k|\beta_k$.

Proposition 1. *If the signature scheme is secure and $\mathcal{P}$ is not close by to $\mathcal{V}$, then a mafia fraud has probability of success at most $1/2^k$.*

That is, the probability of successful cheating decreases exponentially in the number of repetitions of the rapid bit exchange. The simple proof of this proposition is very similar to the proof of Proposition 3 in the next section.

2.4 Preventing the prover from sending bits out too soon

In this subsection we study a setting in which $\mathcal{P}$ has access to the secret keys, and $\mathcal{V}$ wants to be ensured that $\mathcal{P}$ is close by. A remarkable thing about the distance bounding stage in the protocol of the previous subsection is that the bits that $\mathcal{P}$ sends to $\mathcal{V}$ do not have to depend on the bits that $\mathcal{V}$ sends to $\mathcal{P}$. If $\mathcal{P}$ knows at what times $\mathcal{V}$ will send out bits, he can have $\mathcal{V}$ accept by sending β_i out to $\mathcal{V}$ at the correct time before he receives α_i, regardless of the distance to $\mathcal{V}$. Hence, the protocol we described for preventing mafia frauds does not prevent this fraud.

Two solutions suggest themselves. The first solution consists of $\mathcal{V}$ sending bits out with randomly chosen delay times. Since $\mathcal{P}$ cannot anticipate when $\mathcal{V}$ expects to have received back a bit, he cannot send out bits β_i before he has received bit α_i (since $\mathcal{V}$ will not accept if a response bit β_i arrives before he has sent out bit α_i). In fact, it is sufficient if $\mathcal{V}$ sends out bit α_i at random at one of two discrete times, say, at the rising edge of clock pulse $3i$ or $3i+1$, for $1 \leq i \leq k$. The probability of the strategy having success is at most $1/2^k$ if the choices are made independently.

The second solution consists of ensuring $\mathcal{V}$ that $\mathcal{P}$ must choose bits β_i depending on α_i. One way to do this involves creating a public bitstring $m_1|\cdots|m_k$ once (the choice of the bits m_i is irrelevant). The following protocol implements this (see Figure 3):

Step 1 $\mathcal{V}$ generates uniformly at random k bits α_i.
Step 2 Now the low-level distance-bounding exchanges can take place. The following steps are repeated k times, for $i = 1, \ldots, k$.
- $\mathcal{V}$ sends bit α_i to $\mathcal{P}$.
- $\mathcal{P}$ sends bit $\beta_i = \alpha_i \oplus m_i$ to $\mathcal{V}$ *immediately* after receiving bit α_i from $\mathcal{V}$.

$\mathcal{V}$ verifies whether the bit-string $(\alpha_1 \oplus \beta_1)|\cdots|(\alpha_k \oplus \beta_k)$ equals the public bitstring. If so, $\mathcal{V}$ computes an upper-bound on the distance to $\mathcal{P}$ using the maximum of the delay times between sending out bit α_i and receiving bit β_i back, for $i = 1, \ldots, k$. $\mathcal{V}$ accepts if and only if $\mathcal{P}$ is close by.

As before, it is easy to see that the probability that $\mathcal{V}$ accepts when $\mathcal{P}$ is not close by is at most $1/2^k$.

2.5 Preventing both types of fraud

By combining the two protocols, we can prevent both types of fraud. As before, it is assumed that a bit-string $m_1|\cdots|m_k$ is published. The following protocol can be used (see Figure 4):

Step 1 $\mathcal{V}$ generates uniformly at random k bits α_i.

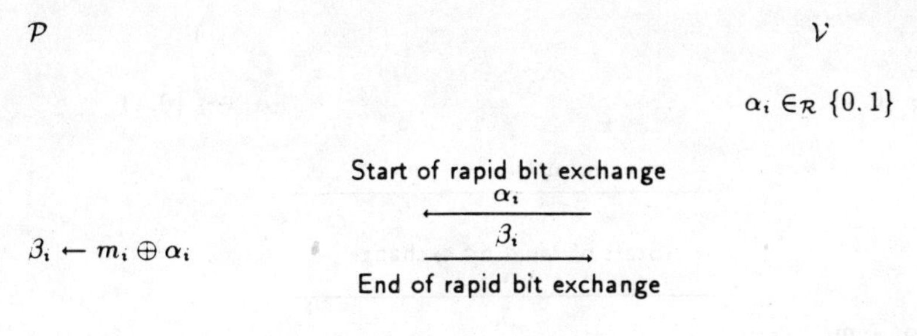

Fig. 3. Preventing the response bits from being sent out too soon.

Step 2 $\mathcal{P}$ generates uniformly at random k bits m_i. As before, both $\mathcal{P}$ and $\mathcal{V}$ can do so well beforehand. $\mathcal{P}$ commits to k bits m_i using a secure commitment scheme.

Step 3 Now the low-level distance-bounding exchanges can take place. The following steps are repeated k times, for $i = 1, \ldots, k$.
- $\mathcal{V}$ sends bit α_i to $\mathcal{P}$.
- $\mathcal{P}$ sends bit $\beta_i = \alpha_i \oplus m_i$ to $\mathcal{V}$ *immediately* after he receives α_i.

Step 4 $\mathcal{P}$ opens the commitment(s) on the bits β_i by sending the appropriate information to $\mathcal{V}$. $\mathcal{P}$ concatenates the $2k$ bits α_i and β_i, signs the resulting message m with his secret key and sends the resulting signature to $\mathcal{V}$.

With the information received in Step 4, $\mathcal{V}$ verifies whether the bits $\alpha_i \oplus \beta_i$ are indeed those commited to in Step 2. If this holds, then $\mathcal{V}$ computes m in the same way as $\mathcal{P}$ did and verifies whether the signature he received is indeed a correct signature of $\mathcal{P}$ on m. If so, he computes an upper-bound on the distance to $\mathcal{P}$ using the maximum of the delay times, and accepts if and only if $\mathcal{P}$ is close by.

3 Integration with public key identification schemes

The fact that a secure signature scheme must be used in the protocols of Subsection 2.3 and 2.5 can be a problem when the prover wishes only to identify himself by for example proving knowledge of a square root X of $X^2 \bmod n$ (as in the basic Fiat-Shamir identification scheme): it is not clear how he should sign the message by using his secret information X; also, since $\mathcal{V}$ receives information that he could not have computed himself, it is not clear whether he obtains useful information for computing the secret keys. In this section, we show how to integrate distance bounding with known public key identification schemes such that no useful information is transferred.

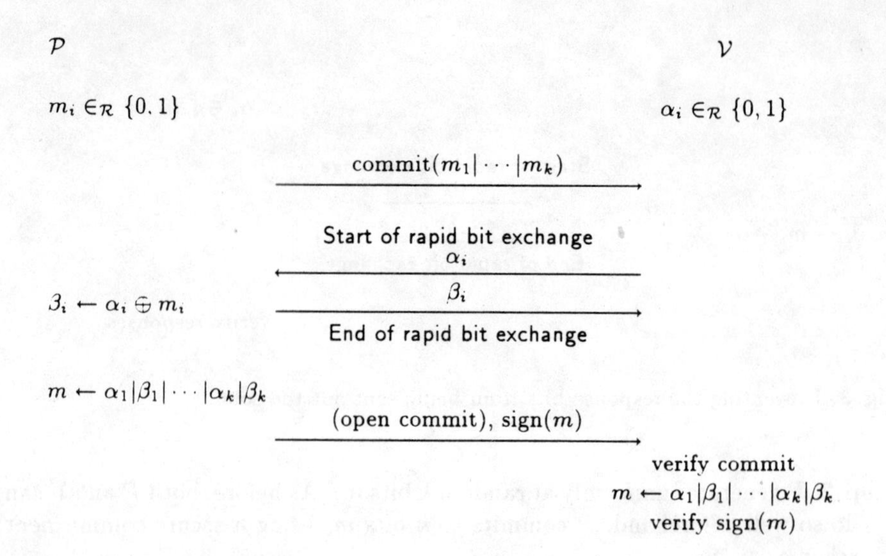

Fig. 4. Distance bounding to prevent both types of fraud.

3.1 Preventing mafia frauds

To prevent mafia frauds, we have the distance-bounding protocol dictate that $\mathcal{P}$ respond to challenges formed as the exclusive-or of the bits sent and received, instead of signing the concatenation. We illustrate this with the basic Fiat-Shamir scheme:

Step 1 $\mathcal{P}$ generates uniformly at random k numbers $R_i \in \mathbb{Z}_n^*$, and sends their squares R_i^2 mod n to $\mathcal{V}$. $\mathcal{P}$ also generates uniformly at random k bits β_i and commits to these bits (and their order) by sending a commitment on them to $\mathcal{V}$.

Step 2 $\mathcal{V}$ generates uniformly at random k bits α_i.

Step 3 Now the low-level distance-bounding exchanges can take place. Hereto, the following steps are repeated k times, for $i = 1, \ldots, k$.
- $\mathcal{V}$ sends bit α_i to $\mathcal{P}$.
- $\mathcal{P}$ sends bit β_i to $\mathcal{V}$ *immediately* after he receives α_i from $\mathcal{V}$.

Step 4 $\mathcal{P}$ opens the commitment on the bits β_i made in Step 1 by sending the appropriate information to $\mathcal{V}$. Furthermore, $\mathcal{P}$ determines the k responses $X^{c_i} R_i$ corresponding to challenges $c_i = \alpha_i \oplus \beta_i$, for $1 \le i \le k$, and sends them to $\mathcal{V}$.

$\mathcal{V}$ determines the k challenges c_i in the same way as $\mathcal{P}$ did, and verifies that the k responses are correct. Then $\mathcal{V}$ verifies whether the opening of the commitments by $\mathcal{P}$ is correct. If this holds, $\mathcal{V}$ computes an upper-bound on the distance to $\mathcal{P}$

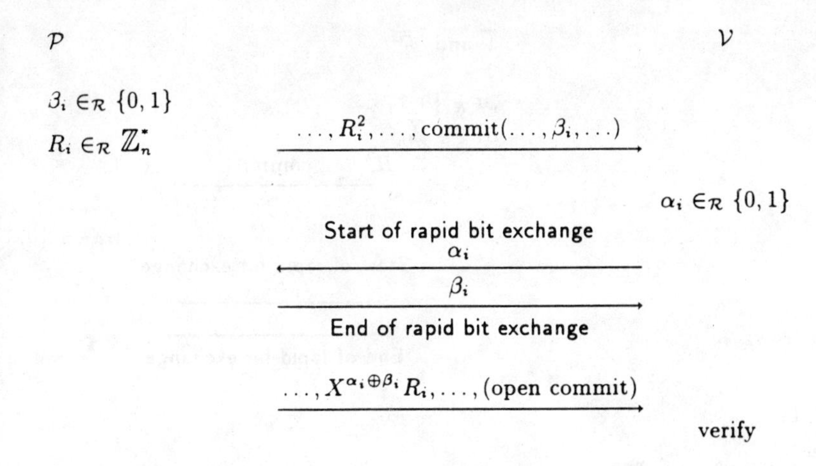

Fig. 5. Distance bounding in the Fiat-Shamir identification scheme.

using the maximum of the delay times between sending α_i and receiving β_i, for $i = 1, \ldots, k$. $\mathcal{V}$ accepts if and only if $\mathcal{P}$ is close by.

Proposition 2. *If the commitment scheme is secure, then this protocol is a proof of knowledge of a square root of* $X^2 \bmod n$ *that does not reveal any useful information for computing a root of* $X^2 \bmod n$.

Sketch of proof. In effect, this protocol is the parallel version of the basic Fiat-Shamir identification protocol. In [11] it is proven that this protocol reveals no useful information.

Since the binary challenges are chosen *mutually* random, the verifier cannot choose them as the outcome of a collision-free hash-function of the information known to him before Step 2. That is, the verifier does not receive information that he cannot compute himself. In particular, the transcript of an execution of the protocol cannot be used as a digital signature to convince others that the execution took place.

Proposition 3. *If the commitment scheme is secure,* $\mathcal{P}$ *is not close by to* $\mathcal{V}$ *and both follow the protocol, then the mafia fraud has probability of success at most* $1/2^k$.

Sketch of proof. In order to have any chance at all of having $\mathcal{V}$ accept, the fraudsters $\overline{\mathcal{P}}$ and $\overline{\mathcal{V}}$ must perform the rapid bit exchange first entirely with $\mathcal{V}$ and then with $\mathcal{P}$ (or vice versa), otherwise $\mathcal{V}$ will not accept because the computed upper-bound on the distance will not be tight enough (see Figure 6).

However, since a commitment was sent in Step 2, it is clear that with probability $1 - 1/2^k$ the fraudsters cannot prevent $\mathcal{P}$ and $\mathcal{V}$ from ending up with

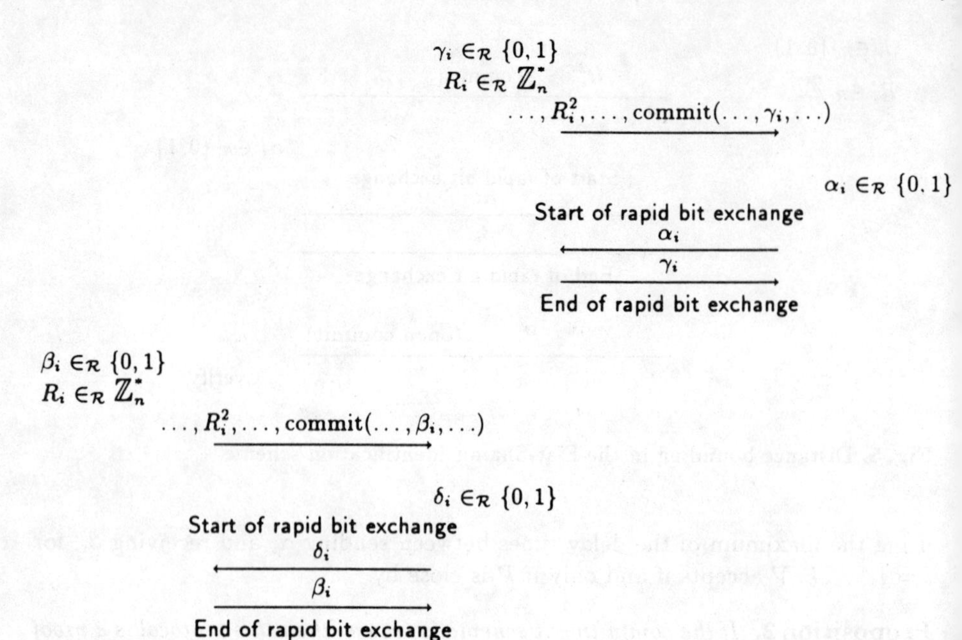

Fig. 6. Can $\overline{\mathcal{P}}$ and $\overline{\mathcal{V}}$ apply a mafia fraud?

at least one different challenge (i.e. $\beta_i \oplus \delta_i \neq \alpha_i \oplus \gamma_i$). Therefore, at least one response of $\mathcal{P}$ is correct with respect to a challenge that is complementary to the challenge $\mathcal{V}$ expects a response to. Clearly, one cannot convert $X^c R$ to $X^{c \oplus 1} R$ without knowing X.

Note that if at least one of $\mathcal{P}$ and $\mathcal{V}$ generates the challenge bits according to a distribution other than the uniform one (i.e., does not follow the protocol), this only increases the probability of successful cheating for $\overline{\mathcal{P}}$ and $\overline{\mathcal{V}}$.

Although we had the prover commit himself, it does not really matter whether the prover or the verifier commits. This holds for the protocol in the next section as well.

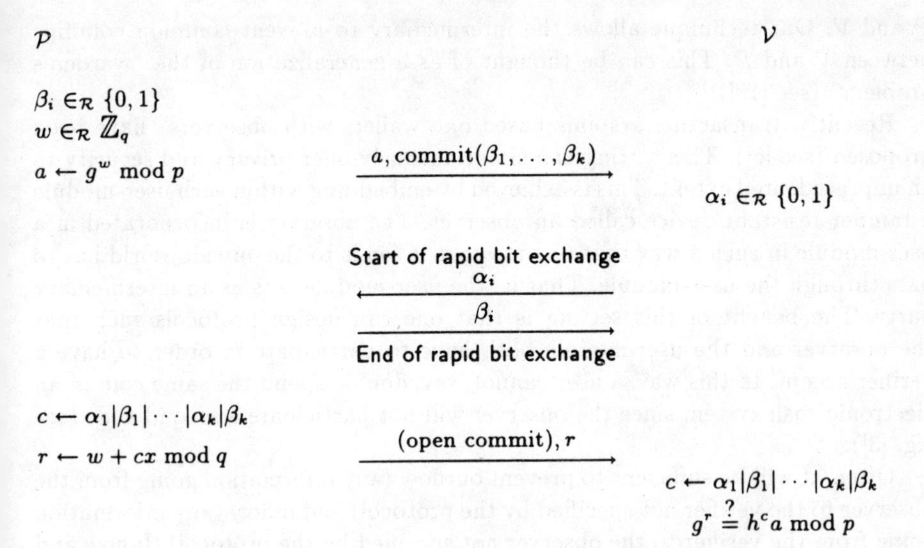

Fig. 7. Distance bounding in the Schnorr identification scheme.

3.2 Preventing both types of fraud

In order to prevent both types of frauds, as in Subsection 2.5, one can straightforwardly modify this protocol. To this end, in Step 1 $\mathcal{P}$ commits to k bits m_i, and in Step 3 $\mathcal{P}$ will reply with response bits $\beta_i = \alpha_i \oplus m_i$. Finally, the responses of $\mathcal{P}$ in Step 4 must be computed with respect to the multi-bit challenge $\alpha_1|\beta_1|\cdots|\alpha_k|\beta_k$ (using the concatenation of the xor-values does not prevent mafia frauds). This technique can be integrated in (minimum disclosure) identification schemes with multi-bit challenges, such as [5, 14, 15, 17], retaining the same security level for $\mathcal{P}$. Observe that the same propositions hold for this modified protocol; the only distinction is that the challenge bits *can* be chosen as the outcome of a collision-free hashfunction, and hence the transcript can serve as a digital signature.

Figure 7 shows how one might incorporate distance bounding into the Schnorr identification scheme. In this protocol, $(p, q, g, h = g^x \bmod p)$ is the public key of $\mathcal{P}$(as in [17]).

4 Distance bounding in wallets with observers

Up to now, we have considered distance-bounding protocols in a model with two legitimate parties. In this section, we will discuss distance bounding in a certain three-party setting. The goal of $\mathcal{V}$ is to determine an upper-bound to $\mathcal{P}$, and the task of the intermediary is to prevent undesired flow of information between

$\mathcal{P}$ and $\mathcal{V}$. Our technique allows the intermediary to prevent common coinflips between $\mathcal{V}$ and $\mathcal{P}$. This can be thought of as a generalization of the "warden's problem" (see [18]).

Recently, transaction systems based on "wallets with observers" have been proposed (see [6]). This setting can simultaneously offer privacy and security to an unprecedented extent. This is achieved by embedding within each user-module a tamper-resistant device called an observer. The observer is incorporated in a user-module in such a way that any message it sends to the outside world has to pass through the user-module. That is, the user-module acts as an intermediary party. The benefit of this setting is that one can design protocols such that the observer and the user-module both have to participate in order to have a verifier accept. In this way, a user cannot, say, double-spend the same coin in an electronic cash system since the observer will not participate a second time (see e.g. [3]).

Often, it will be sufficient to prevent outflow (any information going from the observer to the verifier not specified by the protocol) and inflow (any information going from the verifier to the observer not specified by the protocol). Inflow and in particular outflow can be a serious threat to the privacy of the user.

In [8] the privacy aspect of the wallet with observer setting has been investigated under an even more stringent requirement: even if an observer were to store all information it receives during the period it is embedded within a user-module, it still should be impossible (independent of computing resources) to link a payment to a user by examining afterwards the information inside the observer and all information gathered by the verifying parties. This possibility is not excluded by preventing inflow and inflow, since for example a single random number known to both an observer and a shop would enable linking: the fact that the user-module took part in generating it (so that no information could be encoded within it, thus preventing both inflow and outflow) is irrelevant in this matter. That is, one must also prevent "common coinflips." In [8], the term "shared information" is proposed, encompassing inflow, outflow, and common coinflips. The essential technique ("divertability") needed to prevent shared information in such a setting has been proposed earlier by Desmedt in [9], and was generalized in [16]. Prevention of shared information in some instances can be viewed as a slight generalization of divertability, in that the keys have to be shared together with the intermediary in a suitable way.

A fraud that can be applied in this three-party setting is one in which a user illegitimately uses an observer embedded within someone else's wallet. A possible motivation for doing so is that typically observers will gather (part of) negative credentials which can prevent the user from doing transactions he would like to do (see e.g. [7]). Also, another observer might have (part of) certain positive credentials the user would like to make use of. One can imagine a fraudulent organization specializing in lending, at a distance, observers with positive credentials (or without certain negative ones) to users who are willing to pay for this. In effect, when the user wants to do a transaction for which he needs certain positive credentials, he could use a radio link with the fraudulent organization and lets an appropriate observer authorize the transaction. We will call this the "observer fraud."

4.1 Preventing the observer fraud

Using our distance-bounding technique we show how the verifier in the three-party setting can determine an upper-bound on the distance to the observer, such that the user-module can prevent shared information. We only describe one protocol that meets the most stringent requirements: no shared information, no release of useful information for computing the secret key, and the verifier obtains no information that he could not have computed himself (transcripts cannot serve as proof that the protocol took place). For easy comparison with the distance-bounding protocol shown in Section 3, our discussion will be based on the three party version of the Fiat-Shamir protocol (see [9, 16]).

We need a new notion called a "xor-commitment scheme." This is a commitment scheme which enables one to commit to the exclusive-or $\alpha \oplus \beta$ of two bits α and β, whereas one only knows a commitment on β but not β itself. In addition, one should be able to open the xor-commit if and only if one knows how to open the commitment on β, and this opening information must leak no Shannon information on the bits α and β, and the random choices involved in the commitment on β.

An implementation of an xor-commitment scheme can be realized with RSA, based on the technique of probabilistic encryption (see [13]). Let n be a Blum integer. In order to encrypt a bit α, the commiter chooses $r \in \mathbb{Z}_n^*$ at random and computes $\text{commit}(\alpha) := (-1)^\alpha r^2 \bmod n$. According to the quadratic residuocity assumption (see [13]), it is infeasible to decide whether $\text{commit}(\alpha)$ is a quadratic residue or not (i.e., whether $\alpha = 1$ or 0), unless one knows the factorization of n. Given a commitment $\text{commit}(\alpha) = (-1)^\alpha r^2 \bmod n$ of a bit α and a commitment $\text{commit}(\beta) = (-1)^\beta s^2 \bmod n$ of a bit β, it follows that $\text{commit}(\alpha \oplus \beta) = (-1)^{\alpha \oplus \beta} r^2 s^2 \bmod n$ is an xor-commitment on $\alpha \oplus \beta$. When opening this commit, one reveals $rs \bmod n$, which does not contain any information on s.

We denote the observer by $\mathcal{O}$, the verifier by $\mathcal{V}$ and the user-module by $\mathcal{U}$. For clarity, we leave out the fact that to prevent shared information the secret and public keys must be shared between the observer and the user-module in a suitable way. It is not hard to see how to do this using some of the techniques suggested in [7].

In the protocol, $\mathcal{O}$ knows a square root X of $X^2 \bmod n$, and $\mathcal{O}$ wishes to convince $\mathcal{V}$ of this fact in such a way that $\mathcal{U}$ does not learn it, whereas $\mathcal{U}$ can be ensured that there is no shared information. $\mathcal{V}$ wants to be convinced not only of the fact that $\mathcal{O}$ knows a square root of $X^2 \bmod n$, but also that $\mathcal{O}$ is close by. In essence, this is the setting of the mafia fraud, with the intermediary party ($\overline{\mathcal{P}}$ and $\overline{\mathcal{V}}$ in mafia frauds, and $\mathcal{U}$ in this situation) also trying to prevent shared information.

The protocol is as follows (see Figure 8):.

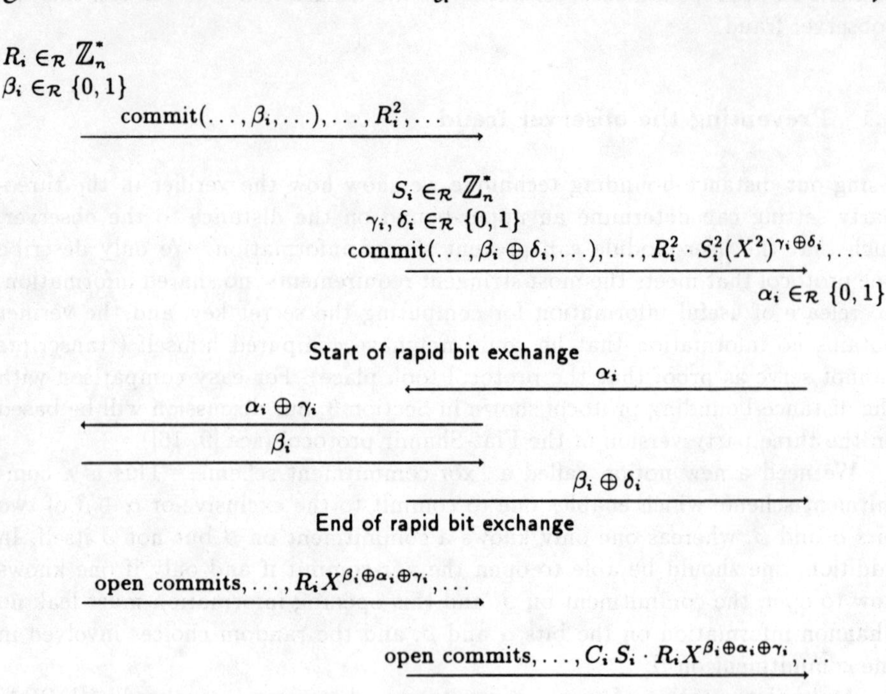

Fig. 8. Diverted Fiat-Shamir identification protocol with distance bounding.

Step 1 $\mathcal{O}$ generates k random numbers $R_i \in_{\mathcal{R}} \mathbb{Z}_n^*$ and sends the squares R_i^2 mod n of these numbers to $\mathcal{U}$. $\mathcal{O}$ also generates k bits β_i, and sends a xor-commitment on them to $\mathcal{U}$. (Clearly, if we use the specific xor-commitment just described, a commitment for each bit would be needed.)

Step 2 $\mathcal{U}$ first verifies that the numbers received from $\mathcal{O}$ all have Jacobi symbol 1. If this is the case, he generates at random k bits $\gamma_i \in_{\mathcal{R}} \{0,1\}$ as well as k bits $\delta_i \in_{\mathcal{R}} \{0,1\}$. $\mathcal{U}$ also generates k numbers $S_i \in_{\mathcal{R}} \mathbb{Z}_n^*$. He then computes the k products $R_i^2 \cdot S_i^2(X^2)^{\gamma_i \oplus \delta_i}$ mod n and sends them to $\mathcal{V}$. $\mathcal{U}$ also sends xor-commitment(s) on $\beta_i \oplus \delta_i$ to $\mathcal{V}$.

Step 3 $\mathcal{V}$ generates k challenge bits $\alpha_i \in_{\mathcal{R}} \{0,1\}$, which he will use for the rapid bit exchange.

Step 4 Now the rapid exchange of bits can take place. Hereto, the following four exchanges are repeated k times, for $i = 1,\ldots,k$:
- $\mathcal{V}$ sends bit α_i to $\mathcal{U}$.
- $\mathcal{U}$ sends $\alpha_i \oplus \gamma_i$ to $\mathcal{O}$ *immediately* after receiving α_i.
- $\mathcal{O}$ sends challenge bit β_i to $\mathcal{U}$ *immediately* after receiving $\alpha_i \oplus \gamma_i$.

– $\mathcal{U}$ sends $\beta_i \oplus \delta_i$ to $\mathcal{V}$ *immediately* after receiving β_i.

Step 5 $\mathcal{O}$ opens the k commits on the bits β_i to $\mathcal{U}$ $\mathcal{O}$ also computes the k responses $R_i X^{\beta_i \oplus \alpha_i \oplus \gamma_i}$ mod n and sends them to $\mathcal{U}$.

Step 6 $\mathcal{U}$ verifies whether the responses of $\mathcal{O}$ are correct with respect to the challenges $\alpha_i \oplus \gamma \oplus \beta_i$ and the squares received from $\mathcal{O}$ in Step 3. $\mathcal{V}$ verifies whether the bits that $\mathcal{O}$ sent to him in Step 4 are those he commited to. If all the verifications hold, then $\mathcal{U}$ computes the k responses $C_i\, S_i \cdot R_i X^{\alpha_i \oplus \gamma_i \oplus \beta_i}$ mod n by multiplying, for $1 \le i \le k$, the i-th response of $\mathcal{O}$ by S_i mod n and a correction-factor C_i. The correction-factor is equal to X^2 mod n if and only if $\gamma_i \oplus \delta_i = 1$ and $\alpha_i \oplus \beta_i \oplus \delta_i = 1$, otherwise it is equal to 1. $\mathcal{U}$ sends all these responses to $\mathcal{V}$. Furthermore, $\mathcal{U}$ opens the xor-commitments to the k values $\beta_i \oplus \delta_i$ to $\mathcal{V}$.

Afterwards, $\mathcal{V}$ verifies whether the responses of $\mathcal{U}$ are correct with respect to the challenges $\beta_i \oplus \delta_i \oplus \alpha_i$ and the squares received from $\mathcal{V}$ in Step 3. He also verifies whether the bits received from $\mathcal{U}$ in Step 4 are those $\mathcal{U}$ commited to. If all the verifications hold, then $\mathcal{V}$ derives an upper-bound on the physical distance to $\mathcal{O}$ by using the maximum of the delays between sending out α_i and receiving $\beta_i \oplus \delta_i$ from $\mathcal{U}$, for $1 \le i \le k$. $\mathcal{V}$ accepts if and only if $\mathcal{O}$ is close by.

Although we write commit$(\ldots, \beta_i, \ldots,)$ we do not mean to imply with this that a multi-bit commitment must necessarily be used: one might as well use k single-bit commitments.

It is straightforward to show that $\mathcal{V}$ accepts if all parties follow the protocol, and that Propositions 2 and 3 hold.

Since one can easily show that for each view of $\mathcal{V}$ and for each view of $\mathcal{O}$ in this protocol, there is exactly one set of random choices that could have been made by $\mathcal{U}$ such that the views are from the same execution of the protocol, there is no shared information. Clearly, for the protocol as we described it, this only holds for executions concerning proof of knowledge of the particular number X^2 mod n. However, as we noted before, if the knowledge of X^2 mod n is divided between $\mathcal{O}$ and $\mathcal{U}$ in a suitable way (as described in [8]), the property of absence of shared information holds for the set of *all* proofs of knowledge, regardless of the particular number X^2 mod n that the proof is concerned with.

Finally, as in Proposition 3 it is easy to see that the following must hold.

Proposition 4. *If $\mathcal{O}$ and $\mathcal{V}$ follow the protocol, then $\mathcal{U}$ cannot (with probability of success greater than $1/2^k$) trick $\mathcal{V}$ into believing that $\mathcal{O}$ is close by if this is not the case.*

As before, if at least one of $\mathcal{O}$ and $\mathcal{V}$ generates challenge bits according to a distribution other than the uniform one, then $\mathcal{U}$'s probability of successful cheating will only increase.

5 Open problems and further work

We would like to present two potentially fruitful areas for further investigation.

First is the physics and practical implementation of distance-bounding technology. We know little about the physical limits or precisely how best to use current technology. Some experimental work might also be interesting.

Second is dealing with a problem not adressed here. The techniques presented do not prevent frauds in which a distant party with access to the secret keys is *cooperating* with a party close by (without conveying the secret keys). The frauds were suggested informally under the name "the terrorist fraud" by Desmedt in [9]. We are currently working on some ideas preventing such frauds using distance bounding.

Acknowledgements

We are grateful to Niels Ferguson for some suggestions concerning distance bounding in wallets with observers, to Torben Pedersen for the particular realization we described of an xor-commitment, and to Corné Hoogendijk for discussions on the physical realizations of distance-bounding.

References

1. Bengio, S., Brassard, G., Desmedt, G., Goutier, C. and Quisquater, J., "Secure Implementation of identification schemes," Journal of Cryptology, 4 (1991), pages 175–183.
2. Beth, T. and Desmedt, Y., "Identification tokens–or: solving the chess grandmaster problem," Crypto '90, Lecture Notes in Computer Science, Springer-Verlag (1991), pages 169–176.
3. Brands, S., "An efficient off-line electronic cash system based on the representation problem," C.W.I. Technical Report CS - T9323, april 1993, The Netherlands.
4. Brassard, G., Chaum, D. and Crépeau, "Minimum Disclosure Proofs of Knowledge," Journal of Computer and System Sciences, Vol. 37 (1988), pages 156–189.
5. Brickell, E. and McCurley, K., "An interactive identification scheme based on discrete logarithms and factoring," Journal of Cryptology, Vol. 5, no. 1 (1992), pages 29–39.
6. Chaum, D., "Achieving electronic privacy," Scientific American, Aug. 1992, pages 96–101.
7. Chaum, D. and Pedersen, T., "Wallet Databases with Observers," Proceedings of Crypto '92, Abstracts, Santa Barbara, August 1992, pp. 3.1-3.6.
8. Cramer, R. and Pedersen, T., "Improved privacy in wallets with observers," in: these proceedings.
9. Desmedt, Y., "Major security problems with the 'unforgeable' (Feige)-Fiat-Shamir proofs of identity and how to overcome them," SecuriCom '88, SEDEP Paris, (1988), pages 15–17.
10. Desmedt, Y., Goutier, C., and Bengio, S., "Special uses and abuses of the Fiat-Shamir passport protocol," Crypto '87, LNCS 293, Springer-Verlag (1988), pages 16-20.
11. Feige, U., Fiat, A. and Shamir, A., "Zero-knowledge proofs of identity," Journal of Cryptology 1 (1988), pages 77–94.

12. Fiat, A. and Shamir, A., "How to prove yourself: practical solutions to identification and signature problems," Crypto '86, Springer-Verlag, (1987), pages 186–194.
13. Goldwasser, S. and Micali, S., "Probabilistic Encryption," Journal of Computer and System Sciences, Vol. 28 (1984), pages 270–299.
14. Guillou, L. and Quisquater, J.-J., "A 'paradoxical' identity-based signature scheme resulting from zero-knowledge," Crypto '88, Springer-Verlag, pages 216-231.
15. Okamoto, T., "Provably Secure and Practical Identification Schemes and Corresponding Signature Schemes," Proceeding of Crypto '92, pages (1-15) – (1-25).
16. Okamoto, T. and Ohta, K., "Divertible zero knowledge interactive proofs and commutative random self-reducibility," Eurocrypt '89, Springer-Verlag, pages 134-149.
17. Schnorr, C.P., "Efficient Signature Generation by Smart Cards," Journal of Cryptology, Vol. 4, No. 3, (1991), pages 161–174.
18. Simmons, G., "The prisoner's problem and the subliminal channel," Crypto '83, Santa Barbera (1983), Plenum, New York, pages 51–67.

On the Distribution of Characteristics in Bijective Mappings

Luke O'Connor [1]
University of Waterloo, Canada

Abstract

Differential cryptanalysis is a method of attacking iterated mappings which has been applied with varying success to a number of product ciphers and hash functions [1, 3]. The attack is based on predicting a series of differences $\Delta Y_1, \Delta Y_2, \ldots,$ ΔY_r known as a *characteristic* Ω. Partial information about the key can be derived when the differences are correctly predicted. The probability of a given characteristic Ω correctly predicting differences is derived from the XOR tables associated with the iterated mapping.

Even though differential cryptanalysis has been applied successfully to a number of specific iterated mappings such as DES, FEAL and LOKI, the effectiveness of the attack against an arbitrary iterated mapping has not been considered. In this paper we derive the exact distribution of characteristics in XOR tables, and determine an upper bound on the probability of the most likely characteristic Ω in a product cipher constructed from randomly selected S-boxes that are bijective mappings. From this upper bound we are then able to construct product ciphers for which all characteristics Ω occur with low probability.

Keywords: Differential cryptanalysis, iterated mapping, product cipher.

1 Introduction and Results

Differential cryptanalysis is a statistical attack popularized by Biham and Shamir in a series of well-known papers [1, 2, 3]. The attack has been applied to a wide range of iterated mappings including LUCIFER, DES, FEAL, REDOC, Kahfre [4, 5, 12, 13, 17, 19]. As explained below, the attack is based on a quantity Ω called a *characteristic*, which has some probability p^{Ω} of giving information about the secret key used in the mapping. The attack is universal in that characteristics Ω will always exist for any iterated mapping;

---

[1]The current employer of the author is the Distributed System Technology Center (DSTC), Brisbane, Australia. Correspondence should be sent to ISRC, QUT Gardens Point, 2 George Street, GPO Box 2434, Brisbane Q 4001, Australia; email oconnor@sleet.fit.qut.edu.au.

however p^{Ω} may be very small, and possibly less likely to furnish information concerning the key than the success of guessing the secret key at random. For this reason, differential cryptanalysis has had varying success against the iterated mappings it has been applied to, and little is known about how useful the attack is expected to be against an arbitrary iterated mapping.

In Figure 1 we present the basic substitution-transposition network (ST-network) [6]: each round consists of several small substitutions S (S-boxes) followed by a transposition T (anagram) of the current ciphertext. This model is generally acknowledged [6, 9, 18] as being a practical realization of product ciphers originally proposed by Shannon [16]. Most product ciphers such as LUCIFER, DES and IDEA are variations or extensions of the basic ST-network. The main result of this paper is to determine how well differential cryptanalysis is expected to perform against randomly generated instances of ST-networks.

We will give a brief description of differential cryptanalysis with reference to product ciphers, though any iterated mapping would suffice. For a product cipher E that consists of R rounds, let $E_r(X,K)$ be the encryption of the plaintext X under the key K for r rounds, $1 \leq r \leq R$. Note that $E_R(X,K) = E(X,K) = C$ is the ciphertext for X. Let $\Delta C(r) = E_r(X,K) + E_r(X',K)$ be the difference between the ciphertexts of two plaintexts X, X' after r rounds where $1 \leq r \leq R$. For our purposes the difference operator $+$ will refer to addition in the vector space Z_2^m. An r-round *characteristic* is defined as an $(r+1)$-tuple $\Omega_r(\Delta X, \Delta Y_1, \Delta Y_2, \ldots, \Delta Y_r)$ where ΔX is a plaintext difference, and the ΔY_i are ciphertext differences. A plaintext pair X, X' of difference $\Delta X = X + X'$ is called

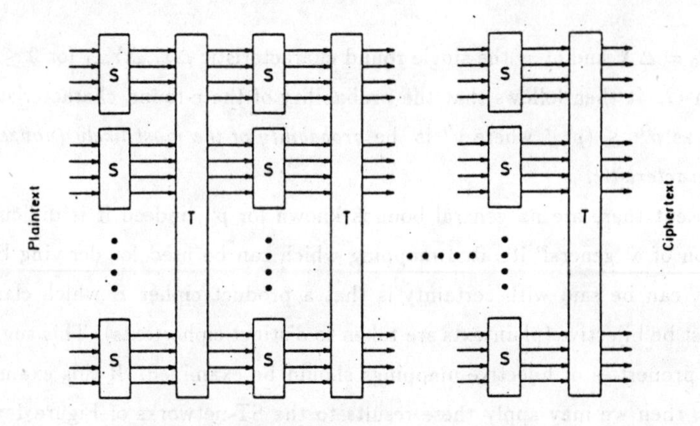

Figure 1: The ST-network product cipher

a *right pair* with respect to a key K and a characteristic $\Omega_r(\Delta X, \Delta Y_1, \Delta Y_2, \ldots, \Delta Y_r)$ if when X and X' are encrypted, $\Delta C(i) = \Delta Y_i$ for $1 \leq i \leq r$. That is, X, X' is a right pair if the characteristic correctly predicts the ciphertext differences at each round. The characteristic Ω_r has probability p^{Ω_r} if a fraction p^{Ω_r} of the plaintext pairs of difference

ΔX are right pairs. On the other hand, if X, X' such that $\Delta X = X + X'$ is not a right pair, then it is said to be a *wrong pair* (with respect to the characteristic and the key). A table which records the number of pairs of difference ΔX that give the output difference ΔY for a mapping π is called the *XOR table distribution* of π. A characteristic $\Delta X, \Delta Y$ is said to be *impossible* for π if its corresponding XOR table entry is zero. Also a characteristic will be called *nonzero* if $w(\Delta X), w(\Delta Y) > 0$, where $w(\cdot)$ is the Hamming weight function. Using a characteristic Ω of appropriate length it is then possible to devise a statistical experiment which when repeated a sufficient number of times will yield the subkey of the last round (see [1] for details).

Product ciphers such as LUCIFER, DES, FEAL and IDEA are iterated mappings that use a fixed mapping G at each round. For example, in DES the function G at round i, $1 \leq i \leq 16$ is defined as

$$L_i \circ R_i \; = \; G(L_{i-1} \circ R_{i-1}) \; = \; R_{i-1} \circ [\, L_{i-1} + P(S(E(R_{i-1}) + K_i))\,]$$

where $\circ$ denotes string concatenation, E is a 32-to-48-bit expansion, S is a substitution by 8×6-to-4-bit S-boxes, and P is a 32-element transposition. When the components of G are fixed, which for DES are the E, S and P mappings, we observe that an r-round characteristic is simply the concatenation of r 1-round, or single round, characteristics defined on the mapping G. For an r-round characteristic $\Omega_r(\Delta X, \Delta Y_1, \Delta Y_2, \ldots, \Delta Y_r)$ we have

$$p^{\Omega_r} \; = \; \Pr(\Delta C(i) = \Delta Y_i, \;\; 1 \leq i \leq r \mid X + X' = \Delta X) \;\; \leq \; \prod_{i=0}^{r-1} p^{\omega_i} \qquad (1)$$

where $\Delta Y_0 = \Delta X$ and ω_i is the single round characteristic $\Delta Y_i, \Delta Y_{i+1}$ for $0 \leq i \leq r-1$ defined on G. It then follows that the probability of the r-round characteristic Ω_r can be bound as $p^{\Omega_r} \leq (p^{\Omega})^r$ where p^{Ω} is the ***probability of the most likely (nonzero) single round characteristic***.

At present there are no general bounds known for p^{Ω}; indeed it is difficult to give a definition of a 'general' iterated mapping which can be used for deriving bounds on p^{Ω}. What can be said with certainty is that a product cipher E which claims to be useful must be bijective (plaintexts are taken to distinct ciphertexts). This suggests that the XOR properties of bijective mappings should be examined. If this examination is successful, then we may apply these results to the ST-networks of Figure 1 where the S-boxes themselves are bijective so as to ensure that the mapping itself is bijective.

Let $\pi : Z_2^m \rightarrow Z_2^m$ be a bijective mapping, referred to as an m-bit permutation. The set of all m-bit permutations is known as the symmetric group on 2^m objects and is denoted as S_{2^m}. Let $\Lambda_\pi(\Delta X, \Delta Y)$ be the value of the XOR table entry of the pair $\Delta X, \Delta Y \in Z_2^m$ for the permutation $\pi \in S_{2^m}$. Assuming the uniform distribution on the set S_{2^m} we prove (Theorem 2.1) that

$$\Pr(\Lambda_\pi(\Delta X, \Delta Y) = 0) \quad = \quad \frac{1}{2^m!} \cdot \sum_{k=0}^{2^{m-1}} (-1)^k \cdot \binom{2^{m-1}}{k}^2 \cdot 2^k \cdot k! \cdot (2^m - 2k)!.$$

We are then able to show (Corollary 3.1) that for large m, the expected probability of the most likely nonzero characteristic for an m-bit permutation is at most $\frac{m}{2^{m-1}}$. Equivalently, the expected maximum entry in the XOR table for nonzero characteristics is at most $2m$ for large m. The result of Corollary 3.1 can used to estimate the probability of the most likely 1-round characteristic p^Ω in an iterated mapping based on m-bit permutations. Consider a 16-round 64-bit product cipher E for which the round mapping consists of 8×8-bit permutations followed by a 64-bit transposition. Then to predict the input difference to the 16th round requires a 15-round characteristic Ω_{15} where the input difference to each of the first 15 rounds is nonzero. Let us assume that the permutations are selected uniformly from S_{2^8} and that at each round there is only one S-box which has a nonzero input difference. It then follows from Corollary 3.1 that

$$p^{\Omega_{15}} \;\leq\; \left(p^\Omega\right)^{15} \;\leq\; \left(\frac{8}{2^7}\right)^{15} \;=\; 0.86736 \times 10^{-18} \;\approx\; 2^{-59}.$$

Further, if Ω_{15} has nonzero input differences to two S-boxes at 7 out of the 15 rounds then

$$p^{\Omega_{15}} \;\leq\; \left(\frac{8}{2^7}\right)^{2 \cdot 7} \cdot \left(\frac{8}{2^7}\right)^{8} \;=\; 0.32311 \times 10^{-26} \;\approx\; 2^{-86}.$$

Corollary 3.1 indicates that the individual entries of an XOR table are expected to be distributed in the interval $[0, 2, \ldots, 2m]$. At this point we are not able to determine the exact distribution of entries within this interval, but we are able to prove that most XOR table entries are in fact zero. We prove (Theorem 3.2) that the expected fraction of the XOR table for nonzero characteristics that is zero approaches $e^{-\frac{1}{2}} = 0.60653$. In another way, approximately 60% of the entries for nonzero characteristics will be zero for a permutation selected uniformly.

The full proofs of the theorems to follow are ommitted since the final version of this paper has been accepted for publication in the *Journal of Cryptology.*

1.1 Notation

We will now formalize some of the notation given in the introduction. Let $[\cdot]$ be a boolean predicate that evaluates to 0 or 1 such as $[n \text{ is prime}]$. For a given $\pi \in S_{2^m}$, define $\Lambda_\pi(\Delta X, \Delta Y)$ as

$$\Lambda_\pi(\Delta X, \Delta Y) \; = \; \sum_{\substack{X,X' \in Z_2^m \\ \Delta X = X + X'}} [\pi(X) + \pi(X') = \Delta Y]. \qquad (2)$$

Thus $2^{-m} \cdot \Lambda_\pi(\Delta X, \Delta Y)$ is a random variable giving the probability that the difference in the output of the mapping π is ΔY when the difference of the input pair X, X' is ΔX. For all $\pi \in S_{2^m}$, observe that when $\Delta X = 0$ or $\Delta Y = 0$ it follows that $\Lambda_\pi(\Delta X, \Delta Y) = 0$, unless $\Delta X = \Delta Y = 0$ whereupon $\Lambda_\pi(\Delta X, \Delta Y) = 2^m$. The distribution of $\Lambda_\pi(\Delta X, \Delta Y)$ taken over all possible $\Delta X, \Delta Y \in Z_2^m$ is known as the *pairs XOR distribution table for* π, or simply the XOR table for π.

Example 1.1 For an m-bit permutation π, let XOR_π be the $2^m \times 2^m$ matrix where $\mathrm{XOR}_\pi(i,j) = \Lambda_\pi(i,j)$, $0 \leq i,j \leq 2^m - 1$, where i,j are treated as 3-bit binary vectors. Observe that $\mathrm{XOR}_\pi(0,0) = 8$, and all other entries in the first row or column of $\mathrm{XOR}(\pi)$ are zero. For $\pi = (7,2,4,1,5,6,3,0)$ the corresponding XOR table is given as:

$$\mathrm{XOR}_\pi = \begin{bmatrix} 8 & 0 & 0 & 0 & 0 & 0 & 0 & 0 \\ 0 & 0 & 0 & 4 & 0 & 4 & 0 & 0 \\ 0 & 0 & 0 & 4 & 0 & 0 & 4 & 0 \\ 0 & 0 & 0 & 0 & 0 & 4 & 4 & 0 \\ 0 & 2 & 2 & 0 & 2 & 0 & 0 & 2 \\ 0 & 2 & 2 & 0 & 2 & 0 & 0 & 2 \\ 0 & 2 & 2 & 0 & 2 & 0 & 0 & 2 \\ 0 & 2 & 2 & 0 & 2 & 0 & 0 & 2 \end{bmatrix}. \qquad (3)$$

Notice that if each entry in the XOR table is divided by 2^m then the resulting matrix will be doubly stochastic. □

The XOR table for an m-bit permutation π has the following general form:

$$\mathrm{XOR}_\pi = \begin{bmatrix} 2^m & 0 & 0 & \cdots & 0 \\ 0 & a_{1,1} & a_{1,2} & \cdots & a_{1,2^m-1} \\ 0 & a_{2,1} & a_{2,2} & \cdots & a_{2,2^m-1} \\ \vdots & \vdots & \vdots & \cdots & \vdots \\ 0 & a_{2^m-1,1} & a_{2^m-1,2} & \cdots & a_{2^m-1,2^m-1} \end{bmatrix} \stackrel{\text{def}}{=} \begin{bmatrix} 2^m & 0 \\ 0 & A_\pi \end{bmatrix}. \qquad (4)$$

We are interested in the properties of the $(2^m - 1) \times (2^m - 1)$ submatrix $A_\pi = [a_{i,j}]$, $1 \leq i,j \leq 2^m - 1$, which corresponds to that portion of the XOR table entries attributed to nonzero characteristics. In this paper we will show that for large m, approximately 60% of the entries in A_π are zero and largest entry in A_π is expected to be bounded by $2m$.

2 The Pairing Theorem

Observe that a characteristic $\Delta X, \Delta Y$ corresponds to a pairing of the inputs and outputs of a permutation π (namely the pairs X, X' and Y, Y' where $\Delta X = X + X'$ and $\Delta Y = Y + Y'$). For $\phi : A \to B$, let Π_A and Π_B be pairings on the sets A and B, respectively. Theorem 2.1 determines the number of functions ϕ which take no pair of Π_A to a pair in Π_B, and will be referred to as the Pairing Theorem.

Theorem 2.1 (Pairing Theorem) Let $A = \{a_1, a_2, \ldots, a_{2d}\}$ and $B = \{b_1, b_2, \ldots, b_{2d}\}$ be sets of distinct elements. Let $\Pi_A \subseteq A \times A$ and $\Pi_B \subseteq B \times B$ be unordered pairs, such that $a_i(b_i)$ occurs in one pair of $\Pi_A(\Pi_B)$ for $1 \le i \le 2d$. Then the number $\Phi(d)$ of bijective functions $\phi : A \to B$ such that for $\forall (a_i, a_j) \in \Pi_A$, $(\phi(a_i), \phi(a_j)) \notin \Pi_B$ is

$$\Phi(d) = \sum_{k=0}^{d} (-1)^k \cdot \binom{d}{k}^2 \cdot 2^k \cdot k! \cdot (2d-2k)!. \tag{5}$$

Proof. Order the elements of Π_B as (b'_i, b'_{d+i}), $1 \le i \le d$. For $1 \le i \le d$ define $P(i)$ as

$$P(i) = \{\, \phi \mid (\phi(a), \phi(a')) = (b'_i, b'_{d+i}),\ (a, a') \in \Pi_A \,\}$$

which is the number of functions ϕ that map some pair of Π_A to the pair $(b'_i, b'_{d+i}) \in \Pi_B$. It follows that

$$\Phi(d) = (2d)! - \left| \bigcup_{1 \le j \le d} P(j) \right| = (2d)! + \sum_{\substack{S \subseteq \{1,2,\ldots,d\} \\ S \neq \{\emptyset\}}} (-1)^{|S|} \cdot \left| \bigcap_{j \in S} P(j) \right| \tag{6}$$

using the inclusion-exclusion principle [8]. For $1 \le k \le d$ define the integers

$$P(i'_1, i'_2, \ldots, i'_k) = \left| \bigcap_{1 \le j \le k} P(i'_j) \right| \tag{7}$$

and by symmetry $P(1, 2, \ldots, k) = P(i'_1, i'_2, \ldots, i'_k) \stackrel{\text{def}}{=} P(d, k)$. From eq. (6) it then follows that

$$\Phi(d) = (2d)! + \sum_{k=1}^{d} (-1)^k \cdot \binom{d}{k} \cdot P(d, k). \tag{8}$$

It remains to determine $P(d, k)$ for $1 \le k \le d$. To this end, order the pairs within Π_A as (a'_i, a'_{d+i}) for $1 \le k \le d$. Then $P(d, k)$ is the number of functions ϕ for which there are k pairs (a''_i, a''_{d+i}) such that $\{\phi(a''_i), \phi(a''_{d+i})\} = \{(b'_i, b'_{d+i})\}$, $1 \le i \le k$. There are $\binom{d}{k}$ ways to choose the k pairs (a''_i, a''_{d+i}) from Π_A, $k!$ ways of assigning the (a''_i, a''_{d+i}) to the (b'_i, b'_{d+i}), and 2^k ways of assigning (a''_i, a''_{d+i}) to a particular pair in Π_B. It then follows that

$$P(d, k) = \binom{d}{k} \cdot 2^k \cdot k! \cdot (2d-2k)! \tag{9}$$

where $(2d-2k)!$ is the number of ways to assign the elements in $A - \{a''_i, a''_{d+i} \mid 1 \le i \le k\}$. We then have that

$$\Phi(d) = (2d)! + \sum_{k=1}^{d} (-1)^k \cdot \binom{d}{k} \cdot P(d,k) = \sum_{k=0}^{d} (-1)^k \cdot \binom{d}{k}^2 \cdot 2^k \cdot k! \cdot (2d-2k)!$$

which completes the proof of the theorem. □

It is a simple matter to observe that for a fixed mapping π the expected value of each entry in the XOR table is 1 since there are 2^m entries in each row which sum to 2^m. Using the Pairing Theorem we are now able to derive the exact distribution of the random variable $\Lambda_\pi(\Delta X, \Delta Y)$.

Corollary 2.1 For any fixed nonzero $\Delta X, \Delta Y \in Z_2^m$, assuming π is chosen uniformly from the set S_{2^m}, and $0 \le k \le 2^{m-1}$

$$\Pr(\Lambda_\pi(\Delta X, \Delta Y) = 2k) = \binom{2^{m-1}}{k}^2 \cdot \frac{k! \cdot 2^k \cdot \Phi(2^{m-1} - k)}{2^m!}. \tag{10}$$

□

3 Two properties of XOR tables

Recall that p^Ω was defined in the introduction as the probability of the most likely single round characteristic for an iterated mapping. In this section we will derive bounds on the expected value of p^Ω assuming that the round mapping G is based on m-bit permutations selected uniformly from S_{2^m}. Let G consist of s S-boxes implementing m-bit permutations $\pi_1, \pi_2, \ldots, \pi_s$ such that $G : Z_2^{m \cdot s} \to Z_2^{m \cdot s}$ where π_1 operates on the first block of s bits, π_2 operates on the second block of s bits, and so on, as in Figure 1. For example, G may operate on 48 bits of ciphertext which is used as the input to 8×6-bit permutations $\pi_1, \pi_2, \ldots, \pi_8$. Then define Λ_m^* as

$$2^m \cdot p^\Omega \le \Lambda_m^* \stackrel{\text{def}}{=} \max_{\substack{\pi \in \{\pi_1, \pi_2, \ldots, \pi_s\} \\ \Delta X, \Delta Y \in Z_2^m \\ w(\Delta X), w(\Delta Y) > 0}} \Lambda_\pi(\Delta X, \Delta Y)$$

from which it follows that $\frac{\Lambda_m^*}{2^m}$ is the probability of the most likely characteristic across all s permutations in G. Then for any nonzero r-round characteristic Ω_r it follows that

$$p^{\Omega_r} \le \left(\frac{\Lambda_m^*}{2^{m-1}}\right)^r. \tag{11}$$

At present there are no known general bounds on Λ_m^*. For a randomly selected set of

m-bit permutations $\{\pi_1, \pi_2, \ldots, \pi_s\}$ we may use the Pairing Theorem to determine an expected upper bound on Λ^*_m.

Theorem 3.1 Assuming that the S-boxes π_i are selected uniformaly from the set S_{2^m}

$$\lim_{m\to\infty} \frac{\mathbf{E}[\Lambda^*_m]}{2m} \leq 1. \tag{12}$$

□

Sketch of proof. For $1 \leq k \leq 2^{m-1}$, let $\Lambda_{m,2k}$ be the expected number of nonzero characteristics $\Delta X, \Delta Y$ for which $\Lambda_\pi(\Delta X, \Delta Y) = 2k$. Further let $\Pr(\Lambda_\pi = 2k)$ be the probability that an m-bit permutation has a nonzero characteristic $\Delta X, \Delta Y$ for which $\Lambda_\pi(\Delta X, \Delta Y) = 2k$. The proof rests on the following inequality:

$$\Pr(\Lambda^*_m = 2k) \;<\; \Pr(\Lambda_\pi = 2k) \;\leq\; \Lambda_{m,2k}.$$

Using the Pairing Theorem it can be shown that

$$\Lambda_{m,2k} \;=\; \frac{(2^m-1)^2}{2^m!} \cdot \binom{2^{m-1}}{k}^2 \cdot 2^k \cdot k! \cdot \Phi(2^{m-1}-k). \tag{13}$$

The theorem follows from proving that $\sum_{k>m} 2k \cdot \Lambda_{m,2k} = o(1)$. □

Corollary 3.1 For large m and assuming the uniform distribution on the set S_{2^m}, the expected probability of the most likely nonzero characteristic is bounded by $\frac{m}{2^{m-1}}$.

Proof. The expected probability of the most likely nonzero characteristic is $\frac{\Lambda^*_m}{2^m}$. □

Let $\overline{\Lambda_m} = \sum_{k=m+1}^{2^{m-1}} 2k \cdot \Lambda_{m,2k}$ be an upper bound on the last $2^{m-1} - m$ terms in the sum for $\mathbf{E}[\Lambda^*_m]$. Also let $\overline{\Lambda^*_m}$ be an empirical estimate of $\mathbf{E}[\Lambda^*_m]$ based on a sample of m_p random permutations. Further, let min (max) be the smallest (largest) maximum XOR entry found across the m_p permutations. Table 1 lists these quantities for $m = 4, 5, \ldots, 10$. We see that $2(m+1) \cdot \Lambda_{m,2(m+1)}$ is a good approximation of $\overline{\Lambda^*_m}$, and by $m = 6$ the tail of the summation for $\mathbf{E}[\Lambda^*_m]$ beginning at $k = 2(m+1)$ is negligibly small.

The presence of impossible characteristics assists in discarding certain plaintext pairs which cannot give any probabilistic information concerning the key. It has been observed that 20% –30% of the characteristics in the S-boxes of DES are impossible. Let $\Lambda_{m,0}$ be the expected number of nonzero characteristics $\Delta X, \Delta Y$ which have zero entries in the XOR table of a uniformly selected m-bit permutation. We are able to compute $\Lambda_{m,0}$ as a direct application of the Pairing Theorem.

Theorem 3.2 For any fixed nonzero $\Delta X, \Delta Y \in Z_2^m$ and assuming π is chosen uniformly from the set S_{2^m}

$$\lim_{m\to\infty} \Lambda_{m,0} \;=\; \frac{(2^m-1)^2}{e^{\frac{1}{2}}}. \tag{14}$$

m	$2(m+1)\cdot\Lambda_{m,2(m+1)}$	$\overline{\Lambda_m}$	Λ_m^*	min	max	m_p
4	.76863	.87258	3.114	2	6	10,000
5	.25973	.28436	3.839	3	6	10,000
6	.80244 $\times 10^{-1}$	.86489 $\times 10^{-1}$	4.495	3	7	10,000
7	.22027 $\times 10^{-1}$	.23498 $\times 10^{-1}$	5.126	4	8	10,000
8	.53856 $\times 10^{-2}$	.57019 $\times 10^{-2}$	5.606	5	8	1000
9	.11818 $\times 10^{-2}$	.12438 $\times 10^{-2}$	6.190	6	8	1000
10	.23470 $\times 10^{-3}$	.24584 $\times 10^{-3}$	6.700	6	9	1000

Table 1: The distribution of characteristics.

Sketch of proof. Recall from the Pairing Theorem that $\Phi(2^{m-1})$ will give the number of m-bit permutations that π for which a given characteristic $\Delta X, \Delta Y$ is impossible. It can be shown that the alternating sum in eq. (5) is dominated by its first term ($k = 0$), and that $\Phi(d) \sim (2d!)/e^{\frac{1}{2}}$. □

It now follows that approximately 60% of the entries of the A_π submatrix defined in eq. (4) are zero since $e^{-\frac{1}{2}} = 0.6065$.

4 Conclusion and Remarks

Our results then show that a relatively simple design can produce product ciphers for which all characteristics Ω are expected to (correctly) predict differences with low probability. We further note that random m-bit permutations can be generated efficiently [15], and that the fraction of permutations that are with linear [7] or degenerate [14] in any output bit is tending to zero rapidly as a function of m. On the other hand, Biham and Shamir [3] found that replacing the S-boxes of DES by random 4-bit permutations yielded systems that were far weaker than the original DES. The weakness of these S-boxes appears to be due to the dimension of the permutation rather than the use of permutations *per se*.

References

[1] E. Biham and A. Shamir. Differential cryptanalysis of DES-like cryptosystems. *Journal of Cryptology*, 4(1):3–72, 1991.

[2] E. Biham and A. Shamir. Differential cryptanalysis of the full 16-round DES. Technical Report 708, Technion, Israel Institute of Technology, Haifa, Israel, 1991.

[3] E. Biham and A. Shamir. Differential cryptanalysis of Snefru, Khafre, REDOC-II, LOKI and LUCIFER. *Advances in Cryptology, CRYPTO 91, Lecture Notes in Computer Science, vol. 576, J. Feigenbaum ed., Springer-Verlag*, pages 156–171, 1992.

[4] L. P. Brown, J. Pieprzyk, and J. Seberry. LOKI - a cryptographic primitive for authentication and secrecy applications. *Advances in Cryptology, AUSCRYPT 90, Lecture Notes in Computer Science, vol. 453, J. Seberry and J. Pieprzyk eds., Springer-Verlag*, pages 229–236, 1990.

[5] T. Cusick and M. Wood. The REDOC-II cryptosystem. *Advances in Cryptology, CRYPTO 90, Lecture Notes in Computer Science, vol. 537, A. J. Menezes and S. A. Vanstone ed., Springer-Verlag*, pages 545–563, 1991.

[6] H. Feistel. Cryptography and computer privacy. *Scientific American*, 228(5):15–23, 1973.

[7] J. Gordon and H. Retkin. Are big *S*-boxes best? In T. Beth, editor, *Cryptography, proceedings, Burg Feuerstein*, pages 257–262, 1982.

[8] M. Hall. *Combinatorial Theory*. Blaisdell Publishing Company, 1967.

[9] J. B. Kam and G. I. Davida. A structured design of substitution-permutation encryption networks. *IEEE Transactions on Computers*, 28(10):747–753, 1979.

[10] X. Lai. *On the design and security of block ciphers*. ETH Series in Information Processing, editor J. Massey, Hartung-Gorre Verlag Konstanz, 1992.

[11] X. Lai, J. Massey, and S. Murphy. Markov ciphers and differential analysis. In *Advances in Cryptology, EUROCRYPT 91, Lecture Notes in Computer Science, vol. 547, D. W. Davies ed., Springer-Verlag*, pages 17–38, 1991.

[12] X. Lai and J. L. Massey. A proposal for a new block encryption standard. In *Advances in Cryptology, EUROCRYPT 90, Lecture Notes in Computer Science, vol. 473, I. B. Damgård ed., Springer-Verlag*, pages 389–404, 1991.

[13] R. Merkle. Fast software encryption functions. *Advances in Cryptology, CRYPTO 90, Lecture Notes in Computer Science, vol. 537, A. J. Menezes and S. A. Vanstone ed., Springer-Verlag*, pages 476–501, 1991.

[14] L. J. O'Connor. Enumerating nondegenerate permutations. *Advances in Cryptology, EUROCRYPT 91, Lecture Notes in Computer Science, vol. 547, D. W. Davies ed., Springer-Verlag*, pages 368–377, 1991.

[15] E. M. Reingold, J. Nievergeld, and N. Deo. *Combinatorial Algorithms: Theory and Practice.* Prentice-Hall, 1976.

[16] C. E. Shannon. Communication theory of secrecy systems. *Bell System Technical Journal*, 28:656–175, 1949.

[17] A. Shimizu and S. Miyaguchi. Fast data encipherment algorithm FEAL. *Advances in Cryptology, EUROCRYPT 87, Lecture Notes in Computer Science, vol. 304, D. Chaum and W. L. Price eds., Springer-Verlag*, pages 267–278, 1988.

[18] N. J. A. Sloane. Error correcting codes and cryptography, part 1. *Cryptologia*, 6(2):128–153, 1982.

[19] A. Sorkin. LUCIFER: a cryptographic algorithm. *Cryptologia*, 8(1):22–35, 1984.

On the Security of the IDEA Block Cipher *

Willi Meier

HTL Brugg-Windisch, CH-5200 Windisch, Switzerland
email: meierw@htlulx.htl-bw.ch

Abstract. IDEA is an iterated block cipher proposed by Lai and Massey and is based on the design concept of "mixing operations from different algebraic groups". New arithmetic properties of the basic operations used in the round function are found and investigated with respect to the security of this block cipher. Evidence is given that these properties can be exploited in the cryptanalysis of the first 2 rounds of IDEA but that they are of no assistance in the cryptanalysis of the full IDEA block cipher containing 8 rounds.

1 Introduction

In [3] J. Massey and X. Lai introduced a new iterated block cipher, the Proposed Encryption Standard (PES). The differential cryptanalysis of PES carried out in [4] suggested a minor modification, called Improved PES (IPES). It was shown in [4] that this modification of PES improves the security against differential cryptanalysis. In recent work of Lai [5], the modified block cipher IPES is named IDEA (International Data Encryption Algorithm). The IDEA contains 8 computationally identical rounds plus an output transformation. The plaintext and the ciphertext are 64 bit blocks, while the secret key is 128 bit long. The cipher is based on the design concept of "mixing (arithmetic) operations from different algebraic groups".

Our aim is to contribute to a systematic investigation of arithmetic properties of both the basic operations and the round function of IDEA with respect to the security of this block cipher. The basic operations used in the design are multiplication modulo $2^{16}+1$ (where 0 is taken as 2^{16}), integer addition modulo 2^{16}, and bit-by-bit exclusive-OR of two 16 bit subblocks. In [3] the interaction of these operations is studied as it contributes to the "confusion" required for a secure cipher. In particular, it is stated in [3], that the 3 operations are incompatible in the sense that no pair out of them satisfies a distributive law. In Section 3 we shall show however, that the multiplication and the integer addition satisfy a "partial" distributive law, stemming from arithmetic modulo $2^{16}+1$. This fact made a detailed investigation necessary and may also be of interest for the construction of other cryptographic algorithms based on arithmetic operations. The interaction of the group operations is further studied in Section 4, where arithmetic properties in the context of a class of one-round differentials are in-

* This work is supported by Stiftung Hasler-Werke, Switzerland

vestigated. Our considerations extend a result in [5] and are useful in the cryptanalysis of few rounds of IDEA in Section 5. We give estimates for the computational complexity to break the first few rounds by combining results related to one-round differentials and the partial distributive law. We give evidence that the newly found arithmetic properties can be exploited in the cryptanalysis of the first 2 rounds of IDEA, but that they are of no assistance in the cryptyanalysis of a block cipher containing 3 or more rounds of IDEA. This estimate fits nicely with a conclusion drawn in [5] saying that IDEA will be secure against a differential cryptanalysis attack after only 4 of its 8 rounds.

2 Description of IDEA

For our analysis we recall the description of the IDEA-algorithm as given in [4] and [5]. In the block cipher IDEA (International Data Encryption Algorithm) plaintext and ciphertext are 64 bit blocks and the key is 128 bits long. The cipher is based on a novel design concept of mixing different arithmetic operations rather than using boolean functions (e.g., in terms of lookup tables). The cipher structure is chosen to provide confusion and diffusion and to facilitate both hardware and software implementation. For the latter aspect we refer to [2].

The IDEA-algorithm is an iterated cipher consisting of 8 computationally identical rounds followed by an output transformation. The complete first round as well as the output transformation are depicted explicitly in the computational graph shown in Figure 1.

2.1 Encryption

In the encryption process, three different (arithmetic) group operations on pairs of 16-bit subblocks are used, namely

- bit-by-bit exclusive-OR of two 16 bit subblocks, denoted as $\oplus$;
- addition of integers modulo 2^{16} where the 16 bit subblock is treated as the usual radix-two representation of an integer; the resulting operation is denoted as $\boxplus$;
- multiplication of integers modulo $2^{16}+1$ where the 16 bit subblock is treated as the usual radix-two representation of an integer except that the all-zero subblock is treated as representing 2^{16}; the resulting operation is denoted as $\odot$.

The 64 bit plaintext block X is partitioned into four 16 bit subblocks X_1, X_2, X_3, X_4, i.e., $X = (X_1, X_2, X_3, X_4)$. The four plaintext subblocks are transformed into the four 16 bit ciphertext subblocks Y_1, Y_2, Y_3, Y_4 under the control of 52 key subblocks of 16 bits that are formed from the 128 bit secret key to be described in the key schedule. For $r = 1, 2, ..., 8$, the six key subblocks used in the r-th round are denoted as $Z_1^{(r)}, Z_2^{(r)}, ..., Z_6^{(r)}$. Four 16 bit key subblocks are used in the output transformation; these subblocks are denoted as $Z_1^{(9)}, Z_2^{(9)}, Z_3^{(9)}, Z_4^{(9)}$.

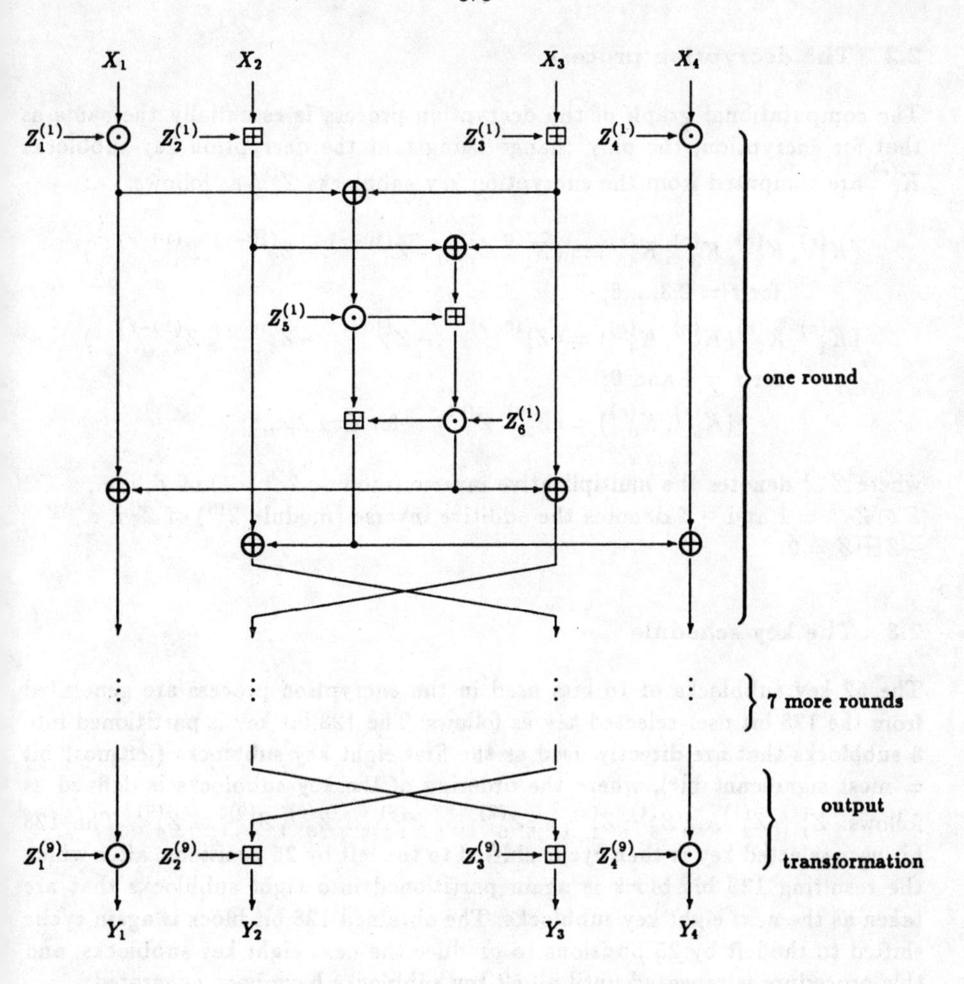

X_i : 16-bit plaintext subblock
Y_i : 16-bit ciphertext subblock
$Z_i^{(r)}$: 16-bit key subblock
$\oplus$: bit-by-bit exclusive-OR of 16-bit subblocks
$\boxplus$: addition modulo 2^{16} of 16-bit integers
$\odot$: multiplication modulo $2^{16}+1$ of 16-bit integers with the zero subblock corresponding to 2^{16}

Fig. 1. Computational graph for the encryption process of the IDEA cipher.

2.2 The decryption process

The computational graph of the decryption process is essentially the same as that for encryption, the only change being that the decryption key subblocks $K_i^{(r)}$ are computed from the encryption key subblocks $Z_i^{(r)}$ as follows:

$$(K_1^{(r)}, K_2^{(r)}, K_3^{(r)}, K_4^{(r)}) = (Z_1^{(10-r)^{-1}}, -Z_3^{(10-r)}, -Z_2^{(10-r)}, Z_4^{(10-r)^{-1}})$$
$$\text{for r} = 2,3,...,8;$$
$$(K_1^{(r)}, K_2^{(r)}, K_3^{(r)}, K_4^{(r)}) = (Z_1^{(10-r)^{-1}}, -Z_2^{(10-r)}, -Z_3^{(10-r)}, Z_4^{(10-r)^{-1}})$$
$$\text{for r} = 1 \text{ and } 9;$$
$$(K_5^{(r)}, K_6^{(r)}) = (Z_5^{(r)}, Z_6^{(r)}) \quad \text{for r=1,2,...,8;}$$

where Z^{-1} denotes the multiplicative inverse (modulo $2^{16}+1$) of Z, i. e., $Z \odot Z^{-1} = 1$ and $-Z$ denotes the additive inverse (modulo 2^{16}) of Z, i. e., $-Z \boxplus Z = 0$.

2.3 The key schedule

The 52 key subblocks of 16 bits used in the encryption process are generated from the 128 bit user-selected key as follows: The 128 bit key is partitioned into 8 subblocks that are directly used as the first eight key subblocks (leftmost bit = most significant bit), where the ordering of the key subblocks is defined as follows: $Z_1^{(1)}, Z_2^{(1)}, ..., Z_6^{(1)}, Z_1^{(2)}, ..., Z_6^{(2)}, ..., Z_1^{(8)}, ..., Z_6^{(8)}, Z_1^{(9)}, ..., Z_4^{(9)}$. The 128 bit user-selected key is then cyclic shifted to the left by 25 positions, after which the resulting 128 bit block is again partitioned into eight subblocks that are taken as the next eight key subblocks. The obtained 128 bit block is again cyclic shifted to the left by 25 positions to produce the next eight key subblocks, and this procedure is repeated until all 52 key subblocks have been generated.

3 A partial distributive law

The basic operations used in the design of IDEA are: $\odot$: multiplication modulo $2^{16}+1$ of 16 bit integers, with the zero number corresponding to 2^{16}, $\boxplus$: addition modulo 2^{16} of 16 bit integers and $\oplus$: bit-by-bit exclusive-OR of 16 bit integers. It is stated in [3] that the three operations are *incompatible* in the sense that no pair of these satisfies a distributive law. In this section we show however that the operations $\odot$ and $\boxplus$ satisfy a partial distributive law stemming from arithmetic modulo 2^n+1. This law also carries over to a partial arithmetic property of the MA structure in the round function (see Figure 2 in Section 4).
In the ring of integers modulo 2^n+1 one has a distributive law, i.e., for any integers a, b, $delta \in \{0, 1, ..., 2^n\}$

$$a \cdot (b + delta) = a \cdot b + a \cdot delta \pmod{2^n + 1} \tag{1}$$

where addition and multiplication are taken modulo $2^n + 1$. Obviously, this law carries over to the operations $\odot$ and $+$:

$$a \odot (b + delta) = a \odot b + a \odot delta \pmod{2^n + 1}. \tag{2}$$

We now ask whether this law even holds modulo 2^n for some fraction of integers $a, b, delta$. For this we compare $a \odot (b \boxplus delta)$ with $a \odot b \boxplus a \odot delta \pmod{2^n}$.

Proposition 1. *(1) If $a = 0$ the equation $a \odot (b \boxplus delta) = a \odot b \boxplus a \odot delta$ is satisfied for no b and delta;*
(2) If $a = 1$ the equation $a \odot (b \boxplus delta) = a \odot b \boxplus a \odot delta$ is satisfied for every b and delta;
(3) If $a \neq 0, 1$ the equation $a \odot (b \boxplus delta) = a \odot b \boxplus a \odot delta$ is satisfied for no triple of the form $(a, 0, delta)$ or $(a, b, 0)$;
(4) If $a \neq 0, 1$, $b \neq 0$ and $delta \neq 0$, the equation $a \odot (b \boxplus delta) = a \odot b \boxplus a \odot delta$ is satisfied if and only if the two conditions

$$b + delta \leq 2^n$$
$$a \odot b + a \odot delta \leq 2^n$$

are satisfied.

Proof. Statement (1) follows easily by using the fact that $0 \odot b = 1 - b \pmod{2^n}$ for any b. Staments (2) and (3) are trivial. For stament (4) observe that the equation $a \odot (b \boxplus delta) = a \odot b \boxplus a \odot delta \mod 2^n$ holds if $b + delta \leq 2^n$, and $a \odot b + a \odot delta \leq 2^n$. For $a > 1$ also the converse holds: If $b + delta > 2^n$, this leads to a calculation modulo 2^n, giving difference $a \pmod{2^n + 1}$ between $a \odot (b + delta)$ and $a \odot (b \boxplus delta)$, which cannot be equalized by computing $a \odot b + a \odot delta$ modulo 2^n rather than $2^n + 1$. This also shows that for the equation in (4) to hold, the condition $a \odot b \boxplus a \odot delta \leq 2^n$ is necessary.

If b and *delta* are supposed to be random, one has $b + delta \leq 2^n$ with probability 1/2 and similarly, $a \odot b + a \odot delta \leq 2^n$ with probability 1/2.
If a is random, we heuristically assume that these two events may be considered to be independent. This will roughly be justified. Hence, for random a, b and *delta* we may expect the equation in Proposition 1 to hold with probability about 1/4.

We give first an exact relationship in two cases, where *delta* is fixed (namely the opposite cases $delta = 1$ and $delta = 2^n - 1 = -1 \pmod{2^n}$).

Proposition 2. *Let $n \geq 2$ be an integer. Then for random integers $a, b \in \{0, 1, ..., 2^n - 1\}$ the distributive law*

$$a \odot (b \boxplus 1) = a \odot b \boxplus a \pmod{2^n} \tag{3}$$

holds with probability $1/2 - 2^{-n-1} + 2^{-2n} \approx 1/2$, and the distributive law

$$a \odot (b \boxplus (2^n - 1)) = a \odot b \boxplus a \odot (2^n - 1) \pmod{2^n} \tag{4}$$

holds with probability $2^{-n} + 2^{-n-1} - 2^{-2n} \approx 2^{-n}$.

Proof. According to Proposition 1 we can suppose $a \geq 1$, and for $a = 1$ the law always holds. As $delta = 1$ for the first part, one has $b + delta = b + 1 \leq 2^n$ for every $b \in \{0, ..., 2^n - 1\}$. Therefore, for every fixed $a > 1$ we count the number of cases, where $a \odot b + a \leq 2^n$, or $a \odot b \leq 2^n - a$. As $a \odot b \pmod{2^n + 1}$ for varying b can take every value between 1 and 2^n, we thus have to exclude a values for b. Hence counting the number of possibilities for $a = 1, 2, ..., 2^n - 1$ we obtain $2^n + (2^n - 2) + (2^n - 3) + ... + 2 + 1$ possibilities in which the law (3) holds. This number is $2^n + (2^n - 1)(2^n - 2)/2 = 2^{2n-1} - 2^{n-1} + 1$. Hence the probability for the law (3) to hold is $(2^{2n-1} - 2^{n-1} + 1) \cdot 2^{-2n} = 1/2 - 2^{-n-1} + 2^{-2n}$ as claimed. If $delta = 2^n - 1$ we again have the case $a = 1$ where the law (4) always holds. For $a > 1$ we have $b + delta = b + (2^n - 1) \leq 2^n$ only for $b = 0$ and $b = 1$. But for $a > 1$ and $b = 0$ the law doesn't hold. So let $b = 1$. Then (4) holds if and only if $a \odot (2^n - 1) = a \odot 2^n - a \pmod{2^n}$. This is true as long as $a \odot 2^n > a$. As $2 \cdot 2^n = 2^{n-1} \pmod{2^n + 1}$, $3 \cdot 2^n = 2^n - 2 \pmod{2^n + 1}$,..., one can see that this is the case for $a = 2, ..., a^{n-1}$. Hence we have $2^n + 2^{n-1} - 1$ possibilities where (4) holds. Therefore the probability is $2^{-n} + 2^{-n-1} - 2^{-2n}$.

Proposition 2 and the considerations made in this section show that, depending on *delta*, the probability for the partial distributive law to hold strongly varies between the values given in (3) and (4), and that this probability decreases for increasing *delta*. But even in the case $delta = 2^n - 1$ with lowest probability, this value is slightly higher than the probability 2^{-n} which one would expect if the validity of the distributive law (mod 2^n) would behave purely randomly. Experiments have shown that the decrease of probability is approximately linear in the increase of *delta*. In fact, for $n = 4$ and $n = 8$, the probability for the partial distributive law to hold for the "average" value $delta = 2^{n-1}$ is extremely near to the arithmetic mean of the two values given in (3) and (4). For $n = 4$ and $n = 8$ these values are 0.28125 and 0.25195, respectively. As these values (especially for $n = 8$) are only slightly higher than 1/4, this further confirms our heuristic considerations, that for random $a, b, delta$ the distributive law holds with probability about 1/4.

3.1 Applications of the partial distributive law

As an immediate consequence of the previous results we give a relationship between $a \odot (b \boxplus delta) - a \odot b$ and $a \odot delta$.
We have $a \odot (b \boxplus delta) = a \odot b \boxplus a \odot delta \pmod{2^n}$ with probability p depending on *delta* (as explained in Proposition 2). Thus we obtain

$$a \odot (b \boxplus delta) - a \odot b = a \odot delta \pmod{2^n} \tag{5}$$

with the same probability. In particular,

$$a \odot (b \boxplus 1) - a \odot b = a \pmod{2^n} \tag{6}$$

with probability $\approx 1/2$ if n is sufficiently large (e.g., $n = 8$ or $n = 16$).

In view of the IDEA-algorithm, the question will be whether the partial distributive law is of any assistance for cryptanalysis. As a first observation in this direction, we obtain that a key block Z which acts as multiplication $\odot$ can be determined with a certain probability, provided the input and output differences (or sums) to this block are supposed to be known. This probability depends on the input difference and on the magnitude of Z (see Proposition 2 and its proof). More importantly, consider the distributive law for differences ($a > 0, b > 0$, $c > 0$)

$$a \odot (b - c) = a \odot b - a \odot c \pmod{2^n + 1}. \tag{7}$$

This holds also modulo 2^n as long as $b > c$ and $a \odot b > a \odot c$. Let $delta = b - c > 0$. Then there are $2^n - delta$ pairs $(b, c) = (delta + 1, 1), ..., (2^n, 2^n - delta)$ for which "difference" modulo 2^n and modulo $2^n + 1$ has the same meaning, and ordinary multiplication agrees with $\odot$ modulo $2^n + 1$. If for $a > 0$ we take $a \odot b - a \odot c$ modulo $2^n + 1$ rather than modulo 2^n (and interpreting products $\odot$ as 2^n if they take the value 0), it is the correct difference.
Suppose $Z > 0$ and we know the outputs after multiplication with Z (not only their difference). Take their difference modulo $2^n + 1$. Then, knowing the input difference $delta \pmod{2^n}$ we can guess the key block Z with probability $(2^n - delta) \cdot 2^{-n} = 1 - delta \cdot 2^{-n}$. Using considerations as in Proposition 1 we can also detect a subblock $Z = 0$ from knowledge of input and output differences.

Partially arithmetic properties of the MA structure As the MA structure (see Figure (2)) is composed only of multiplications $\odot \pmod{2^n + 1}$ and additions $\boxplus \pmod{2^n}$, the partial distributive law carries over to certain relations holding between differences in the inputs p, q and the outputs u, t.

First suppose p is fixed and q_1, q_2 are two different input values with $q_2 = q_1 \boxplus delta$. Then $s_1 = q_1 \boxplus r$ and $s_2 = q_2 \boxplus r = q_1 \boxplus delta \boxplus r = s_1 \boxplus delta$. Hence $t_2 - t_1 = Z_6 \odot s_2 - Z_6 \odot s_1 = Z_6 \odot (s_1 \boxplus delta) - Z_6 \odot s_1$.
As $Z_6 \odot (s_1 \boxplus delta) = Z_6 \odot s_1 \boxplus Z_6 \odot delta$ with probability P depending on $delta$, we have

$$t_2 - t_1 = Z_6 \odot delta \text{ and } u_2 - u_1 = Z_6 \odot delta \tag{8}$$

with this probability P.
If q is fixed and p_1 and p_2 are two different input values, one gets a more complicated but weaker relationship for the outputs u, t, which we omit to formulate.
In the other direction, suppose the input difference $delta = p_2 - p_1$ and the output differences $t_2 - t_1$ and $u_2 - u_1$ are known. Then the difference $r_2 - r_1 = u_2 - t_2 - (u_1 - t_1)$ can be computed and thus Z_5 can be determined with a probability depending on $delta$. A similar conclusion also holds for Z_6 if both the input differences $p_2 - p_1$ and $q_2 - q_1$ are known.

4 A Class of High-Probability Differentials of IDEA

In [5] various considerations lead to three candidate classes of differentials of potential interest for differential cryptanalysis of the IDEA-algorithm. We give a new derivation of the class having highest probability (under the condition of key independence). Our approach allows for a quite general analytic treatment of these differentials. The arguments give new insight into the interaction of the 3 basic operations and will have consequences in the analysis of few rounds of IDEA in Section 5.

In our discussion we use the notation in [4], [5]. In particular the "difference" ΔX is given by $\Delta X = X \otimes X^{*-1}$, where the operation $\otimes$ is defined on 64 bit blocks by

$$X \otimes X^* = (X_1 \odot X_1^*, X_2 \boxplus X_2^*, X_3 \boxplus X_3^*, X_4 \odot X_4^*) \tag{9}$$

and where X^{*-1} denotes the inverse of X^* under the group operation $\otimes$.

The round function of IDEA is illustrated in Figure 2,where X_i, Y_i denote 16 bit subblocks of the 64 bit plaintext and ciphertext blocks respectively, and $Z_1^{(1)},...,Z_6^{(1)}$ denote 16 bit key subblocks of the first round according to the key scheduling as described in [4], [5]. One can also consider "mini ciphers" where the subblocks are $n = 2$, 4 or 8 bit integers.

For any two n-bit integers a and a^*, write $\delta a = a \odot (a^*)^{-1}$, and $\partial a = a - a^* = a \boxplus (-a^*)$. Then the differences ΔX and ΔY are expressed as $\Delta X = (\delta a, \partial b, \partial c, \delta d)$ and $\Delta Y = (\delta v, \partial w, \partial x, \delta y)$. From Figure 2 one has $(\delta a, \partial b, \partial c, \delta d) = (\delta e, \partial f, \partial g, \delta h)$.
The most probable one-round differentials (thus far known) which may be of use in differential cryptanalysis are of the form

$$(\alpha, \beta) = (1, o_a, 0, 0; 1, 0, o_b, 0) \text{ or } (0, 0, o_a, 1; 0, o_b, 0, 1) \tag{10}$$

Here o_a in the input difference α denotes a (fixed) odd integer between 1 and $2^n - 1$, i.e. $o_a \in \{1, 3, ..., 2^n - 1\}$ and o_b in the output difference β is a (fixed) element of a subset of the odd integers, where this subset is dependent on o_a and will be specified.

This class of differentials is referred to in ([5], Ch. 5) as "differentials based on the trivial transparency of the MA structure". The idea is to choose the input difference α such that the probability $P((\delta p, \partial q) = (1, 0))$ is maximized. This is achieved by fixing , e.g., X_1, X_3 and choosing the difference of the other input blocks appropriately.
In ([5], Ch. 5, Property 7) the values for α were determined by a direct computational search:
For $n = 2, 4, 8$ and 16 and for α of the form $(1, o_a, 0, 0)$ or $(0, 0, o_a, 1)$ where o_a is an odd integer between 1 and $2^n - 1$

$$P((\delta p, \partial q) = (1, 0) | \Delta X = \alpha) = \max_{\sigma} P((\delta p, \partial q) = (1, 0) | \Delta X = \sigma)$$
$$= 2^{-(n-1)}. \tag{11}$$

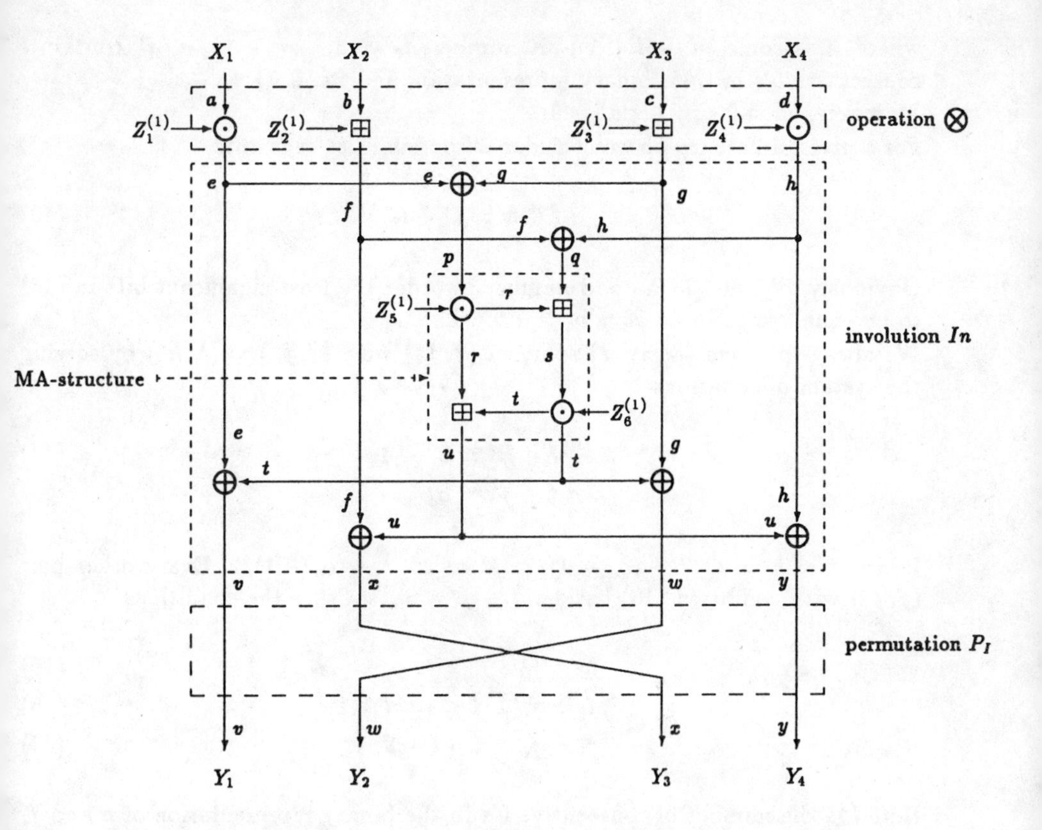

Fig. 2. The round function of IDEA and notation used for analysis.

If we already restrict to input differences with $\delta d = 0$, this computational fact can even be proved analytically for every integer n of interest.

Proposition 3. *Let $n \geq 2$ be an integer such that $2^n + 1$ is prime. Suppose the input difference ΔX is of the form $\Delta X = (1, \partial b, 0, 0)$. Then the probability*

$$P((\delta p, \partial q) = (1, 0) | \Delta X = \alpha) \tag{12}$$

is maximized by taking $\alpha = (1, o_a, 0, 0)$, where o_a is any (fixed) odd integer between 1 and $2^n - 1$, and this probability is $2^{-(n-1)}$.

Proof. Denote by $\bar{a}$ the bit-by-bit complement of the n-bit number a. We start with the fact that for any such a

$$\delta a = 0 \Longleftrightarrow \begin{cases} a & = & (A, 10...0, \theta) \\ a^* & = & (\bar{A}, 10...0, \bar{\theta}) \end{cases}$$

where A is some $[n-(l+1)]$-bit number, $l \in \{0,1,...,n-1\}$, a, a^* contain l consecutive 0's in their binary representation and $\theta \in \{0,1\}$.
Moreover, $\delta a = 0 \Longleftrightarrow a \boxplus a^* = 1$.
For a given difference ∂b we consider all possibilities such that

$$f \oplus h = f^* \oplus h^*. \tag{13}$$

Obviously, ∂b must be an odd number in order the least significant bits in (13) to be equal, so suppose $\partial b = o_a \geq 1$ fixed.
We always get one (positive) solution (f, f^*) with $(f, f^*) = (h, h^*)$ by solving the system of equations

$$\begin{aligned} h + h^* &= 2^n + 1 \\ f - f^* &= o_a \end{aligned} \tag{14}$$

Hence $f = h = 2^{n-1} + (o_a+1)/2$, $f^* = 2^{n-1} - (o_a - 1)/2$. Every other pair (f, f^*) with condition (13) besides $f - f^* = o_a$ satisfies the conditions

$$f_0 = \bar{f_0}^* \tag{15}$$
$$f_i = f_i^*, \quad 1 \leq i \leq l+1 \tag{16}$$
$$f_i = \bar{f_i}^*, \quad i > l+1 \tag{17}$$

Here l is the number of consecutive 0's in the binary representation of h and f_i denotes the i-th bit in the binary representation of f. We show that the number l is uniquely determined by the difference o_a. Suppose, on the contrary, that we have pairs (f, f^*), and (k, k^*) both having difference o_a and satisfying conditions (15),(16),(17) for integers l and l', respectively, with $l' < l$, say. Then already $k_i = \bar{k_i}^*$ for $l+1 \geq i > l'+1$. Compared to the bits in (f, f^*),these bits necessarily contribute to a change in the difference $k - k^*$ of the form $2^{l'+2} \cdot o$, where o is an odd integer. But this change cannot be compensated by simultaneously complementing (some or all) bits k_i, k_i^* for $i > l+1$ or $i = 0$, as such a change would be of the form $2^j \cdot o_1 + \epsilon, \epsilon = 0$ or ± 2, where o_1 is odd and where $j \geq l+2 > l'+2$. As a consequence we have

- The number l can be determined out of the system of equations (14)
- One gets all possibilities for (f, f^*) out of the solution of (14) by either simultaneously complementing some or all bits in f and f^* which agree, i.e., for $1 \leq i \leq l+1$, or the most significant bit.

As the number of equal bits in (f, f^*) is $l+1$, the number of possibilities for f is 2^{l+2}, or the probability to get an f of this form is 2^{-n+l+2}. The probability to get a h fitting this l is 2^{-l-1}. Hence, if we choose b and d (or f and h) randomly, the probability for $f \oplus h = f^* \oplus h^*$ to hold is $2^{-n+l+2} \cdot 2^{-l-1} = 2^{-n+1}$. Note that this probability is independent of the initially chosen o_a.

As a consequence of Proposition 3 we get

Proposition 4. *Let $n \geq 2$ be an integer such that $2^n + 1$ is prime. Suppose the input difference ΔX is of the form $\Delta X = (1, o_a, 0, 0)$ where o_a is a fixed odd integer between 1 and $2^n - 1$. Then*

$$P((\delta v, \partial w) = (1, 0)) = 2^{-n+1} \tag{18}$$

Proof. $\partial w = 0$ implies $\partial t = 0$, as $\partial g = 0$. Hence $\partial q = 0$ as $\partial s = 0$ and $\partial r = 0$. This means that $(\delta v, \partial w) = (1, 0) \Longleftrightarrow (\delta p, \partial q) = (1, 0)$. Therefore Proposition 3 applies.

In order to get also a statement for ∂x and δy, we compare $f \oplus u$ with $f^* \oplus u$ and similarly, $h \oplus u$ with $h^* \oplus u$.
As observed in the proof of Proposition 3 the chosen difference o_a determines the number l of consecutive 0's in h in order to satisfy the equation $f \oplus h = f^* \oplus h^*$. We have $\delta y = 0$ exactly if all these consecutive 0's together with the subsequent bit 1 remain unchanged after XORing with u. This means that $u_i = 0$ for $1 \leq i \leq l+1$. As u takes its values in $\{0, 1, 2, ..., 2^n - 1\}$ uniformly and independently of f and h, this event happens with probability 2^{-l-1}.
On the other hand, the difference o_a may change into a difference ∂x by XORing f with u. A change by ± 2 happens if the least significant bit is complemented. Moreover, a change necessarily also happens, if some of the bits f_i, $l+1 < i < n$, are complemented. Note that the change in difference is always an even number. Therefore the output difference is always an odd number o_b. The probability that $f \oplus u - f^* \oplus u = o_a$, i.e., that $u_i = 0$ for $i = 0$ and $l + 1 < i < n - 1$, is 2^{-n+l+1}. As the bits of u causing changes in ∂x and δy are disjoint *except* u_{n-1}, the differences ∂x and δy remain unchanged *simultaneously* iff $u = 0$ or $u = 2^{n-1}$. The probability for this event is 2^{-n+1}.
Depending on o_a (and therefore on l), output differences $\Delta Y = (1, 0, o_b, 0)$ for o_b in a restricted set of odd numbers are possible. Hereby all differences ΔY for a realizable o_b have the same probability. In particular, $o_b = o_a$ is always realizable.
Thus our considerations together with Proposition 4 prove the following result:

Theorem 5. *Let $n \geq 2$ be an integer such that $2^n + 1$ is a prime number and let o_a denote an arbitrary odd integer between 1 and $2^n - 1$. Then*

$$P(\Delta Y = (1, 0, o_a, 0) | \Delta X = (1, o_a, 0, 0)) = 2^{-2(n-1)}. \tag{19}$$

This extends a result in ([5], Ch. 5) from $o_a = 1$ to arbitrary odd integers.

Theorem 5 deserves some remarks:
Instead of differentials $(\alpha, \beta) = (1, o_a, 0, 0; 1, 0, o_b, 0)$ one can also consider differentials of the form $(\alpha, \beta) = (0, 0, o_a, 1; 0, o_b, 0, 1)$. Analytic considerations as well as experiments show that a result equivalent to Theorem 5 also holds for this class of differentials.

In Proposition 4 we have seen that probabilities for output differences restricted to the first two output subblocks are much higher than the probability given in Theorem 5 for differences considered over all output subblocks simultaneously. Our considerations leading to Theorem 5 show that depending on the integer l, and therefore on the input difference o_a, a similar statement is still true, if we restrict to differences in only three of the four output subblocks:

$$P((\delta v, \partial w, \partial x) = (1, 0, o_a) | \Delta X = (1, o_a, 0, 0)) = 2^{-n+1} \cdot 2^{-n+l+1} = 2^{-2(n-1)+l} \tag{20}$$

$$P((\delta v, \partial w, \delta y) = (1, 0, 0) | \Delta X = (1, o_a, 0, 0)) = 2^{-n+1} \cdot 2^{-l-1} = 2^{-n-l} \tag{21}$$

Here l is the number determined by equations (14) for given o_a, e.g., for $o_a = 1$ one has $l = n - 2$, and hence these two probabilities are 2^{-n} and $2^{-2(n-1)}$ respectively. The other extreme case is $l = 0$, which occurs, e.g., if $o_a = 3$. The probabilities in (20) and (21) are then $2^{-2(n-1)}$ and 2^{-n}, respectively.

4.1 Information on Subkeys for Known Input-Output Differences

The probability derived in Theorem 5 for the class of differentials $(\alpha, \beta) = (1, o_a, 0, 0; 1, 0, o_b, 0)$ is independent of the subkeys $Z_1^{(1)}, ..., Z_6^{(1)}$ of one round. Nevertheless, the known occurence of (part of) such a differential for a known plaintext pair (X, X^*) with difference α allows to derive considerable information on the subkeys $Z_1^{(1)}, Z_2^{(1)}, Z_3^{(1)}$ and $Z_4^{(1)}$.

Suppose a plaintext pair (X, X^*) with difference $\Delta X = (1, o_a, 0, 0)$ is submitted and produces the known (or anticipated) output difference $\delta v = 1$ (or equivalently $\partial w = 0$) after one round. Hence $t = t^*$. As $p = p^*$ this implies $u = u^*$ as $r = r^*$, and hence $q = q^*$. So we know that with this pair (X, X^*) of plaintexts the event $h \oplus f = h^* \oplus f^*$ has occured. (According to Proposition 4 for IDEA we have to make 2^{15} trials in the average until this event occurs). Suppose now $o_a = 1$. Then the equations (14) give $l = n - 2$, so there remain only four possibilities for h, namely $0 = (0, ..., 0)$, $1 = (0, ..., 1)$, $2^{15} = (1, 0, ..., 0)$ and $2^{15} + 1 = (1, 0, ...0, 1)$. Hence for known X there remain only four possibilities for $Z_4^{(1)}$. (In addition, the least significant bit of $Z_2^{(1)}$ is determined). Similar (but slightly weaker) conclusions can also be drawn on the subkey $Z_1^{(1)}$ for differentials of type $(\alpha, \beta) = (0, 0, o_a, 1; 0, o_b, 0, 1)$.

On the opposite side, suppose o_a is such that the equations (14) give $l = 0$. This is the case, e.g., for $o_a = 3$. Then there remain only four possibilities for f, namely $2^{n-1} + (o_a + 1)/2$, $2^{n-1} - (o_a - 1)/2$, $(o_a + 1)/2$ and $-(o_a - 1)/2$. Thus, if X is supposed to be known, there remain only four possibilities for $Z_2^{(1)}$.

5 Analysis of IDEA with a reduced number of rounds

The aim of this section is to give estimates for the computational complexity to break the first few rounds of IDEA by combining known results as well as the arithmetic properties we have found in previous sections. In our discussion we are unable to break more than 2 rounds of IDEA but a (rough) estimate indicates the number of rounds that are at least needed so that a complete exhaustive search will be necessary in order to find the secret key. We outline first a proposal how to break a 2-round IDEA.

2-round IDEA: The problem of analysing a 2-round IDEA is split up into determining the subkey blocks $Z_5^{(2)}$ and $Z_6^{(2)}$ (in the MA structure of the second round) under the assumption that the output $(Y_1^{(2)},...,Y_4^{(2)})$ after the second round (without permutation P_I and without "output transformation") is known, and then analysing a "one and a half" round IDEA (i.e., the first round with output transformation).

The idea is to make a search over $Z_5^{(2)}$, $Z_6^{(2)}$, by composing the 2-round IDEA block cipher, denoted by $F(\mathbf{x},\mathbf{k})$, with the involution In (see Figure 2) with a chosen pair $(Z_5^{(2)}, Z_6^{(2)})$ of key subblocks. Note that this composition $In \circ F$ agrees with the one and a half round IDEA provided we have found the correct pair $(Z_5^{(2)}, Z_6^{(2)})$. We further observe that the partial distributive law applied to the MA structure is of limited use to determine $Z_5^{(2)}$, $Z_6^{(2)}$, as this would need *simultaneous* knowledge of the differences $t^{(2)} - t^{(2)*}$ and $u^{(2)} - u^{(2)*}$ in the second round. But the final XOR's in the involution In leave many choices in general for these differences, even with some knowledge of input differences to In (e.g., using differentials in the first round). Therefore we make an exhaustive search over $(Z_5^{(2)}, Z_6^{(2)})$. A choice $(Z_5^{(2)}, Z_6^{(2)})$ could be tested for correctness by choosing plaintexts X, X^* with $\Delta X = (1,1,0,0)$, according to the differentials studied in Section 4. A faster method appears to be based on a consideration in [5] and essentially going back to [4], namely that for $n = 16$

$$P(\delta v = 1, \delta w = 0, \delta x = 0, \partial y = 2^{16} - 1 | \Delta X = (0,1,0,0)) \approx 2^{-9}. \qquad (22)$$

Thus we choose plaintexts X, X^* with $\Delta X = (0,1,0,0)$. Then by (22) and by the (refined) partial distributive law applied to the key subblock $Z_4^{(2)}$ and with negative input difference $-\partial y = 1$ we have, at the beginning of the second round (using notation similar as in Figure 2),

$$P(\delta e^{(2)} = 1, f^{(2)} \boxplus f^{(2)*} = 2Z_2^{(2)} \boxplus 1, g^{(2)} \boxplus g^{(2)*} = 2Z_3^{(2)} \boxplus 1, \\ h^{(2)*} - h^{(2)} = Z_4^{(2)}) \approx 2^{-9}. \qquad (23)$$

Although the (constant) key subblocks in (23) are unknown, and (22) is not a differential for IDEA in terms of difference as defined by (9), we may still use relation (23) as a test whether we have found the correct pair $(Z_5^{(2)},Z_6^{(2)})$ in an exhaustive search. This is based on the (unproven but plausible) hypothesis that for chosen plaintexts X, X^* with $\Delta X = (0,1,0,0)$ the outputs (v,x,w,y) and (v^*,x^*,w^*,y^*) of the cipher $In \circ F$ satisfy:

If the MA structure in the involution In has been loaded with the correct pair $(Z_5^{(2)}, Z_6^{(2)})$ of key subblocks there exist odd integer numbers o_x, o_w and an integer c_y (which in general are not unique) such that the probability

$$P(\delta v = 1, x \boxplus x^* = o_x, w \boxplus w^* = o_w, y^* - y = c_y) \qquad (24)$$

is significantly higher than the corresponding probability for most other (incorrect) pairs of key subblocks.

This hypothesis has been tested and verified experimentally in the case of the IDEA mini-cipher with $n = 4$. The ambiguity of the constants o_x, o_w and c_y is due to the (experimental) fact that there exist key-dependent one-round differentials of high probability. However, our experiments suggest, that for most keys the cipher $In \circ F$ has no such differentials.

Informally, a pair $(Z_5^{(2)},Z_6^{(2)})$ is accepted to be correct if for this pair the corresponding expressions in the output subblocks of the composition $In \circ F$ satisfy (24) for suitable 16 bit integer constants. According to (23) the computational complexity of this search is roughly of magnitude $2 \cdot 2^9 \cdot 2^{32} = 2^{42}$, which is on the verge of practical feasibility. We are thus reduced to find the other key blocks by breaking (part of) the one and a half rounds of IDEA. According to Section 4.1 the number of possible subkeys $Z_1^{(1)}, ..., Z_4^{(1)}$ can be reduced to 256 possibilities in less than 2^{20} trials. In order to determine the other key subblocks of the first round we make explicit use of the key scheduling. Recall that the 128 bit user-selected key is partitioned into 8 subblocks that are directly used as the first eight key subblocks, where the ordering of the key subblocks is defined as follows: $Z_1^{(1)}, Z_2^{(1)}, ..., Z_6^{(1)}, Z_1^{(2)}, ..., Z_6^{(2)}, ...$ The 128 bit user-selected key is then cyclic shifted to the left by 25 positions to give the next 8 key subblocks, and so on.

Suppose the key subblocks $Z_5^{(2)}, Z_6^{(2)}$ have been determined by the procedure as described above. Then according to the key scheduling $Z_5^{(2)}$ agrees with the last 7 bits of $Z_4^{(1)}$ (which may reduce the uncertainty in the previous estimate of $Z_4^{(1)}$) and the first 9 bits of $Z_5^{(1)}$. Similarly, $Z_6^{(2)}$ determines the last 7 bits of $Z_5^{(1)}$ and the first 9 bits of $Z_6^{(1)}$. We complete our knowledge of the remaining key subblocks entering the one and a half rounds of IDEA by a search over the 7 unknown bits of $Z_6^{(1)}$. For this we choose one of the remaining 256 (or less) possibilities of the quadruple $(Z_1^{(1)}, ..., Z_4^{(1)})$. Then every choice of the last 7 bits of $Z_6^{(1)}$ for given input determines (v, w, x, y) and thus $(Z_1^{(2)}, ..., Z_4^{(2)})$, as the output is supposed to be known. Hence the actual choice of the eight key subblocks can now be found by at most $2^7 \cdot 2^8 = 2^{15}$ trials. This shows that the previous search for the pair $(Z_5^{(2)},Z_6^{(2)})$ is more time consuming than breaking one and a half rounds of IDEA. Hence an optimistic estimate (from the point of view of a cryptanalyst!) predicts about 2^{42} trials to be necessary for breaking the first two rounds of IDEA.

r-round IDEA, $r \geq 3$: For the estimation of the computational complexity of more than 2 rounds we first note that to date no key-independent 2-round differentials with high probability have been found (see [5]). This has also been confirmed by experiments with a mini-IDEA for $n = 4$. Moreover we have found no arithmetic property that might facilitate breaking more than two rounds. Thus to find the subkeys $Z_1^{(3)}, ..., Z_4^{(3)}$ we have no better method than exhaustive search. Therefore, the computational amount to break a two and a half round IDEA is at least $2^{42} \cdot 2^{64} = 2^{106}$.

Proceeding further, breaking a 3-round IDEA needs a full exhaustive search. This suggests that the newly found arithmetic properties for random keys give no advantage in the cryptanalysis of the full IDEA block cipher containing 8 rounds. However these properties show the importance of the fact that in the design of IDEA *three different* group operations have been chosen.

Acknowledgements

I am grateful to X. Lai and J. L. Massey (ETH Zürich) and Th. Brüggemann, H. Bürk, K. Messerli, J.-M. Piveteau and D. Profos (Ascom Tech AG) for many helpful discussions and for their support of this work.

References

1. **E. Biham, A. Shamir, Differential Cryptanalysis of DES-like Cryptosystems, Journal of Cryptology, Vol.4, No. 1, 1991, pp. 3-72.**
2. **Th. Brüggemann, H. Bürk, Der Verschlüsselungsalgorithmus $IDEA^{TM}$, Elektronik, Heft 10, Francis-Verlag, 1993.**
3. **X. Lai and J. L. Massey, A Proposal for a New Block Encryption Standard, Advances in Cryptology-EUROCRYPT'90, Proceedings, Lecture Notes in Computer Science, Springer-Verlag, Berlin 1991, pp. 389-404.**
4. **X. Lai, J. L. Massey and S. Murphy, Markov Ciphers and Differential Cryptanalysis, Advances in Cryptology - EUROCRYPT'91, Proceedings, Lecture Notes in Computer Science, Springer-Verlag, Berlin 1991, pp. 17-38.**
5. **X. Lai, On the Design and Security of Block Ciphers, ETH Series in Information Processing, Editor: J. L. Massey, Vol. 1, 1992.**

Linear Cryptanalysis Method for DES Cipher

Mitsuru Matsui

Computer & Information Systems Laboratory
Mitsubishi Electric Corporation
5-1-1, Ofuna, Kamakura, Kanagawa 247, Japan
E-mail matsui@mmt.isl.melco.co.jp

Abstract

We introduce a new method for cryptanalysis of DES cipher, which is essentially a known-plaintext attack. As a result, it is possible to break 8-round DES cipher with 2^{21} known-plaintexts and 16-round DES cipher with 2^{47} known-plaintexts, respectively. Moreover, this method is applicable to an only-ciphertext attack in certain situations. For example, if plaintexts consist of natural English sentences represented by ASCII codes, 8-round DES cipher is breakable with 2^{29} ciphertexts only.

1 Introduction

Differential Cryptanalysis has been one of main topics in cryptology since the first paper by Biham and Shamir in 1990 [1]. They have broken FEAL cipher in the subsequent paper [2], and recently succeeded in breaking the full 16-round DES cipher by a chosen-plaintext attack [3].

Although Differential Cryptanalysis is a technique for a chosen-plaintext attack, it is more noteworthy that it can be applied to a known-plaintext attack on condition that sufficiently many plaintexts are available.

On the other hand, several new approaches to known-plaintext attacks have been also studied in special cases. As regards FEAL cipher, for example, Tardy-Corfdir and Gilbert have presented a statistical method to break FEAL-4 and FEAL-6 [4], and Matsui and Yamagishi have described a deterministic method to break FEAL-8 by a known-plaintext attack [5], respectively.

In this paper we introduce an essentially known-plaintext attack of DES cipher. The purpose of this method is to obtain a linear approximate expression of a given cipher algorithm. For this purpose, we begin by constructing a statistical linear path between input and output bits of each S-box. Then we extend this path to the entire algorithm, and finally reach a linear approximate expression without any intermediate value.

Our main results on the known-plaintext attack of DES cipher are as follows. The experiments were implemented with C language programs on HP9750 workstation computer (PA-RISC/66MHz).

- 8-round DES is breakable with 2^{21} known-plaintexts in 40 seconds;
- 12-round DES is breakable with 2^{33} known-plaintexts in 50 hours;
- 16-round DES is breakable with 2^{47} known-plaintexts faster than an exhaustive search for 56 key bits.

Generally speaking, there exist many linear approximate expressions for a given cipher algorithm. Moreover, if plaintexts are not random, we may even find an expression which has no plaintext bit in it. This suggests that our method finally leads to an only-ciphertext attack. As regards the only-ciphertext attack of DES cipher, we have obtained the following results.

- If plaintexts consist of natural English sentences represented by ASCII codes, 8-round DES is breakable with 2^{29} ciphertexts only;
- If plaintexts consist of random ASCII codes, 8-round DES is breakable with 2^{37} ciphertexts only.

We shall also illustrate a situation in which 16-round DES is still breakable faster than an exhaustive search for 56-bit keys by the only-ciphertext attack.

2 Preliminaries

Figure 1 shows a data randomization part of DES cipher. We omit the initial permutation IP and the final permutation IP^{-1} unless otherwise indicated. The following notations are used throughout this paper, where the right most bit is referred to as the zero-th bit.

P : The 64-bit plaintext.
C : The corresponding 64-bit ciphertext.
P_H : The left 32-bit of P.
P_L : The right 32-bit of P.
C_H : The left 32-bit of C.
C_L : The right 32-bit of C.
X_i : The 32-bit intermediate value in the i-th round.
K_i : The 48-bit subkey in the i-th round.
$F_i(X_i, K_i)$: The i-th round F-function.
$A[i]$: The i-th bit of A.
$A[i, j, ..., k]$: $A[i] \oplus A[j] \oplus, ..., \oplus A[k]$.

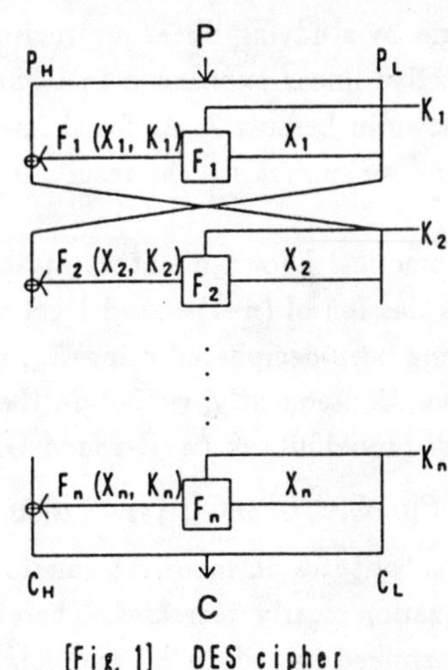

[Fig. 1] DES cipher

3 Principle of Linear Cryptanalysis

The purpose of Linear Cryptanalysis is to find the following "effective" linear expression for a given cipher algorithm:

$$P[i_1, i_2, .., i_a] \oplus C[j_1, j_2, .., j_b] = K[k_1, k_2, .., k_c], \tag{1}$$

where $i_1, i_2, .., i_a, j_1, j_2, .., j_b$ and $k_1, k_2, .., k_c$ denote fixed bit locations, and equation (1) holds with probability $p \neq 1/2$ for randomly given plaintext P and the corresponding ciphertext C. The magnitude of $|p - 1/2|$ represents the effectiveness of equation (1).

Once we succeed in reaching an effective linear expression, it is possible to determine one key bit $K[k_1, k_2, .., k_c]$ by the following algorithm based on the maximum likelihood method:

Algorithm 1

Step1 Let T be the number of plaintexts such that the left side of equation (1) is equal to zero.

Step2 If $T > N/2$ (N denotes the number of plaintexts),
then guess $K[k_1, k_2, .., k_c] = 0$ (when $p > 1/2$) or 1 (when $p < 1/2$),
else guess $K[k_1, k_2, .., k_c] = 1$ (when $p > 1/2$) or 0 (when $p < 1/2$).

The success rate of this method clearly increases when N or $|p - 1/2|$ does. We now refer to the most effective linear expression (i.e. $|p - 1/2|$ is maximal) as the best expression and the probability p as the best probability. Then our main concern is the following:

P1 How to find effective linear expressions.

P2 An explicit description of the success rate by N and p.

P3 A search for the best expression and a calculation of the best probability.

The first aim of this paper is to solve these problems for DES cipher. For this purpose, we begin by studying linear approximations of S-boxes in Chapter 4, and will reach an effective linear expression in Chapter 5. In this stage, the success rate will be also shown in Lemma 2. As for the search problem, which was solved by a computer program, we summarize the results in the annex.

For a practical known-plaintext attack of n-round DES cipher, we make use of the best expression of (n–1)-round DES cipher; that is to say, regarding the final round as having been deciphered using K_n, we accept a term of F-function in the linear expression. Consequently, we obtain the following type of expression which holds with the best probability of (n–1)-round DES cipher:

$$P[i_1, i_2, .., i_a] \oplus C[j_1, j_2, .., j_b] \oplus \ F_n(C_L, K_n)[l_1, l_2, .., l_d] = K[k_1, k_2, .., k_c]. \tag{2}$$

If one substitutes an incorrect candidate for K_n in equation (2), the effectiveness of this equation clearly decreases. Therefore, the following maximum likelihood method can be applied to deduce K_n and $K[k_1, k_2, .., k_c]$:

Algorithm 2

Step1 For each candidate $K_n^{(i)}$ $(i = 1, 2, ...)$ of K_n, let T_i be the number of plaintexts such that the left side of equation (2) is equal to zero.

Step2 Let T_{max} be the maximal value and T_{min} be the minimal value of all T_i's.

- If $|T_{max} - N/2| > |T_{min} - N/2|$, then adopt the key candidate corresponding to T_{max} and guess $K[k_1, k_2, .., k_c] = 0$ (when $p > 1/2$) or 1 (when $p < 1/2$).
- If $|T_{max} - N/2| < |T_{min} - N/2|$, then adopt the key candidate corresponding to T_{min} and guess $K[k_1, k_2, .., k_c] = 1$ (when $p > 1/2$) or 0 (when $p < 1/2$).

The success rate of this method will be discussed in Lemma 4 and Lemma 5.

The next aim of this paper is to consider the case where plaintexts are not random. Assume that, for example, the probability that $P[i_1, i_2, .., i_a] = 0$ is not equal to 1/2. Then even if we eliminate the term $P[i_1, i_2, .., i_a]$ from equation (2), the resultant equation may be still effective. This concludes that Algorithm 2 can be directly applied to an only-ciphertext attack of DES cipher.

We will study the known-plaintext attack of DES cipher in Chapter 6 and develop the only-ciphertext attack procedure in Chapter 7.

4 Linear Approximation of S-boxes

In this section we study linear approximation of S-boxes. Similar motivation can be found in articles of Shamir [6] and Rueppel [7]. Our first approach is to investigate the probability that a value on an input bit coincides with a value on an output bit. More generally, it is useful to deal with not only one bit position but also an XORed value of several bit positions. We now start with the following definition:

Definition 1 *For given S-box S_a $(a = 1, 2, .., 8)$, $1 \leq \alpha \leq 63$ and $1 \leq \beta \leq 15$, we define $NS_a(\alpha, \beta)$ as the number of times out of 64 input patterns of S_a, such that an XORed value of the input bits masked by α coincides with an XORed value of the output bits masked by β; that is to say,*

$$NS_a(\alpha, \beta) \stackrel{\text{def}}{=} \#\{x | 0 \leq x < 64, \ (\bigoplus_{s=0}^{5}(x[s] \bullet \alpha[s])) = \ (\bigoplus_{t=0}^{3}(S_a(x)[t] \bullet \beta[t]))\}, \quad (3)$$

where the symbol $\bullet$ denotes a bitwise AND operation.

Example 1

$$NS_5(16, 15) = 12. \quad (4)$$

When $NS_a(\alpha, \beta)$ is not equal to 32, we may say that there is a correlation between the input and the output bits of S_a. For example, equation (4) indicates that the fourth input bit of S_5 coincides with an XORed value of all output bits with probability $12/64 = 0.19$. Consequently, taking account of the E expansion and the P

permutation in F-function, we see the following equation which holds with probability 0.19 for fixed K and randomly given X:

$$X[15] \oplus F(X, K)[7, 18, 24, 29] = K[22]. \tag{5}$$

Table 1 describes part of distribution table of S-box S_5, where the vertical and the horizontal axes indicate α and β respectively, and each entry shows $NS_5(\alpha, \beta) - 32$. A complete table tells us that equation (4) is the most effective linear approximation in all S-boxes (i.e. $|NS_a(\alpha, \beta) - 32|$ is maximal); therefore, equation (5) is the best approximation of F-function.

The following Lemma is now trivial from the definition of S-boxes.

Lemma 1

(1) *$NS_a(\alpha, \beta)$ is even.*

(2) *If $\alpha = 1, 32$ or 33, then $NS_a(\alpha, \beta) = 32$ for all S_a and β.*

	1	2	3	4	5	6	7	8	9	10	11	12	13	14	15
1	0	0	0	0	0	0	0	0	0	0	0	0	0	0	0
2	4	-2	2	-2	2	-4	0	4	0	2	-2	2	-2	0	-4
3	0	-2	6	-2	-2	4	-4	0	0	-2	6	-2	-2	4	-4
4	2	-2	0	0	2	-2	0	0	2	2	4	-4	-2	-2	0
5	2	2	-4	0	10	-6	-4	0	2	-10	0	4	-2	2	4
6	-2	-4	-6	-2	-4	2	0	0	-2	0	-2	-6	-8	2	0
7	2	0	2	-2	8	6	0	-4	6	0	-6	-2	0	-6	-4
8	0	2	6	0	0	-2	-6	-2	2	4	-12	2	6	-4	4
9	-4	6	-2	0	-4	-6	-6	6	-2	0	-4	2	-6	-8	-4
10	4	0	0	-2	-6	2	2	2	2	-2	2	4	-4	-4	0
11	4	4	4	6	2	-2	-2	-2	-2	-2	2	0	-8	-4	0
12	2	0	-2	0	2	4	10	-2	4	-2	-8	-2	4	-6	-4
13	6	0	2	0	-2	4	-10	-2	0	-2	4	-2	8	-6	0
14	-2	-2	0	-2	4	0	2	-2	0	4	2	-4	6	-2	-4
15	-2	-2	8	6	4	0	2	2	4	8	-2	8	-6	2	0
16	2	-2	0	0	-2	-6	-8	0	-2	-2	-4	0	2	10	-20
17	2	-2	0	4	2	-2	-4	4	2	2	0	-8	-6	2	4
18	-2	0	-2	2	-4	-2	-8	4	6	4	6	-2	4	-6	0
19	-6	0	2	-2	4	2	0	4	-6	4	2	-6	4	-2	0
20	4	-4	0	0	0	0	0	-4	-4	4	4	0	4	-4	0
21	4	0	-4	-4	4	-8	-8	0	0	-4	4	8	4	0	4
22	0	6	6	2	-2	4	0	4	0	6	2	2	2	0	0
23	4	-6	-2	6	-2	-4	4	4	-4	-6	2	-2	2	0	4
24	6	0	2	4	-10	-4	2	2	0	-2	0	2	4	-2	-4
25	2	4	-6	0	-2	4	-2	6	8	6	4	10	0	2	-4
26	2	2	-8	-2	4	0	2	-2	0	4	2	0	-2	-2	0
27	2	6	-4	-6	0	0	2	6	8	0	-2	-4	-6	-2	0
28	0	-2	2	4	0	-6	2	-2	6	-4	0	2	-2	0	0
29	4	-2	6	-8	0	-2	2	10	-2	-8	-8	2	2	0	4
30	-4	-8	0	-2	-2	-2	2	-2	2	-2	6	4	4	4	0
31	-4	8	-8	2	-6	-6	-2	-2	2	-2	-2	-8	0	0	-4
32	0	0	0	0	0	0	0	0	0	0	0	0	0	0	0

Table 1. A distribution table of S5 (part).

5 Linear Approximation of DES Cipher

In this section we extend linear approximations of F-function to the entire algorithm. The first example is 3-round DES cipher (Figure 2). By applying equation (5) to the first round, we see the following equation which holds with probability 12/64:

$$X_2[7,18,24,29] \oplus P_H[7,18,24,29] \oplus P_L[15] = K_1[22]. \tag{6}$$

The same is true of the final round:

$$X_2[7,18,24,29] \oplus C_H[7,18,24,29] \oplus C_L[15] = K_3[22]. \tag{7}$$

Consequently, we obtain the following linear approximate expression of 3-round DES cipher by canceling common terms:

$$P_H[7,18,24,29] \oplus C_H[7,18,24,29] \oplus P_L[15] \oplus C_L[15] = K_1[22] \oplus K_3[22]. \tag{8}$$

The probability that equation (8) holds for given random plaintext P and the corresponding ciphertext C is $(12/64)^2+(1\text{-}12/64)^2=0.70$. Since equation (5) is the best linear approximation of F-function, equation (8) is the best expression of 3-round DES cipher. We can now solve equation (8) to deduce $K_1[22] \oplus K_3[22]$ using Algorithm 1. The following lemma describes the success rate of this method:

Lemma 2 *Let N be the number of given random plaintexts and p be the probability that equation (1) holds, and assume $|p-1/2|$ is sufficiently small. Then the success rate of Algorithm 1 is*

$$\int_{-2\sqrt{N}|p-1/2|}^{\infty} \frac{1}{\sqrt{2\pi}} e^{-x^2/2}\,dx. \tag{9}$$

Table 2 shows a numerical calculation of expression (9).

N	$\frac{1}{4}\lvert p-1/2\rvert^{-2}$	$\frac{1}{2}\lvert p-1/2\rvert^{-2}$	$\lvert p-1/2\rvert^{-2}$	$2\lvert p-1/2\rvert^{-2}$
Success Rate	84.1%	92.1%	97.7%	99.8%

Table 2. The success rate of Algorithm 1.

Next, we show an example of 5-round DES cipher (Figure 3). In this case, we apply equation (5) to the second and fourth rounds, and the following linear equation (which is deduced from $NS_1(27,4)=22$) to the first and final rounds:

$$X[27,28,30,31] \oplus F(X,K)[15] = K[42,43,45,46]. \tag{10}$$

Then an easy calculation leads to a linear approximate expression of 5-round DES cipher:

$$\begin{aligned} &P_H[15] \oplus P_L[7,18,24,27,28,29,30,31] \oplus C_H[15] \oplus C_L[7,18,24,27,28,29,30,31] \\ &= K_1[42,43,45,46] \oplus K_2[22] \oplus K_4[22] \oplus K_5[42,43,45,46]. \end{aligned} \tag{11}$$

The next lemma gives a simple method to calculate the probability that this type of equation holds:

Lemma 3 *(Piling-up Lemma) Let X_i $(1 \leq i \leq n)$ be independent random variables whose values are 0 with probability p_i or 1 with probability $1-p_i$. Then the probability that $X_1 \oplus X_2 \oplus \ldots \oplus X_n = 0$ is*

$$1/2 + 2^{n-1} \prod_{i=1}^{n} (p_i - 1/2). \tag{12}$$

This indicates that equation (11) holds with probability $1/2+2^3(-10/64)^2(-20/64)^2 = 0.519$. Therefore, according to Lemma 2, if $|0.519 - 1/2|^{-2} = 2800$ known-plaintexts are given, one can guess the right side of the equation (11) with success rate 97.7%.

The annex shows a table of the best expression and the best probability of DES cipher up to 20 rounds. Each entry describes from left to right, the number of round, the best expression, the best probability and the linear approximation of F-function used in each round. The sign '-' represents that no approximation is needed in the round. Moreover, it should be noted that there are two best expressions in some cases, which are indicated by sign '*' in the table, because DES cipher has the round symmetry; that is, the other best expression is easily obtained by exchanging P and K_i with C and K_{N+1-i}, respectively.

It follows from this table that two key bits of 16-round DES can be deduced with high success rate using $|1.49 \times 2^{-24}|^{-2} \simeq 2^{47}$ known-plaintexts. In next chapter, we will describe a method to derive more key bits at a time.

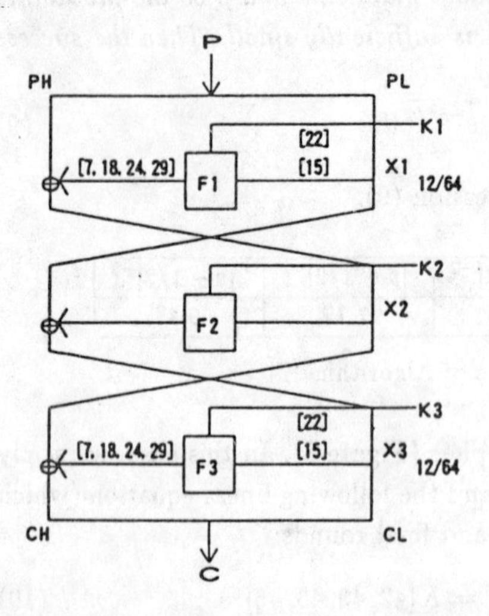

[Fig. 2] 3-round DES cipher

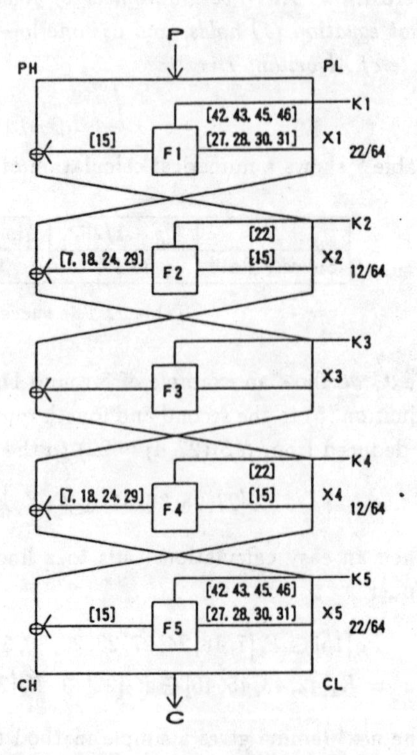

[Fig. 3] 5-round DES cipher

6 Known-Plaintext Attack of DES Cipher

We are now ready for a known-plaintext attack of DES cipher. Our first example is 8-round DES cipher. As mentioned in Chapter 3, we obtain the following 8-round expression which holds with the 7-round best probability $0.5 + 1.95 \times 2^{-10}$:

$$P_H[7,18,24] \oplus P_L[12,16] \oplus C_H[15] \oplus C_L[7,18,24,29] \oplus F_8(C_L, K_8)[15] \\ = K_1[19,23] \oplus K_3[22] \oplus K_4[44] \oplus K_5[22] \oplus K_7[22]. \tag{13}$$

Although this equation contains 48-bit subkey K_8, the number of subkey bits which essentially influences $F_8(C_L, K_8)[15]$ is only six, namely, $K_8[42]$–$K_8[47]$. Therefore, we need 64 counters to carry out Algorithm 2. As regards the success rate of this method, we can prove the following lemma, which generalizes Lemma 2.

Lemma 4 *Let N be the number of given random plaintexts and p be the probability that equation (2) holds, and assume $|p - 1/2|$ is sufficiently small. Then the success rate of Algorithm 2 depends on $l_1, l_2, ..., l_d$, and $\sqrt{N}|p - 1/2|$ only.*

Generally speaking, it is not easy to calculate numerically the accurate probability above. However, under a condition it can be possible as follows.

Lemma 5 *With the same hypotheses as Lemma 4, let $q^{(i)}$ be the probability that the following equation holds for a subkey candidate $K_n^{(i)}$ and a random variable X:*

$$F_n(X, K_n)[l_1, l_2, .., l_d] = F_n(X, K_n^{(i)})[l_1, l_2, .., l_d]. \tag{14}$$

Then if $q^{(i)}$'s are independent, the success rate of Algorithm 2 is

$$\int_{x=-2\sqrt{N}|p-1/2|}^{\infty} \Big(\prod_{K_n^{(i)} \neq K_n} \int_{-x-4\sqrt{N}(p-1/2)q^{(i)}}^{x+4\sqrt{N}(p-1/2)(1-q^{(i)})} \frac{1}{\sqrt{2\pi}} e^{-y^2/2} dy \Big) \frac{1}{\sqrt{2\pi}} e^{-x^2/2} dx, \tag{15}$$

where the product is taken over all subkey candidates except K_n.

Although $q^{(i)}$'s are not independent in our situation, our experiments have shown that Lemma 5 gives a practically good approximation of the success rate, as can be seen in the following.

Now let $d = 1$ and $l_1 = 15$ in equation (2). Then a numerical calculation of expression (15) is as follows.

N	$2\vert p-1/2\vert^{-2}$	$4\vert p-1/2\vert^{-2}$	$8\vert p-1/2\vert^{-2}$	$16\vert p-1/2\vert^{-2}$
Success Rate	48.6%	78.5%	96.7%	99.9%

Table 3. The success rate of Algorithm 2 by Lemma 5 ($d = 1$, $l_1 = 15$).

Since this method can be also applied to deduction of the subkey bits of the first round, we finally obtain 14 subkey bits by carrying out Algorithm 2 twice with negligible memory. It is easy to deduce the remaining key bits, and we omit the detail. Our computer experiments indicate results better than Table 3: The program completes deriving the whole key bits in 20 seconds using $4|1.95\times 2^{-10}|^{-2} \simeq 2^{20}$ known-plaintexts and in 40 seconds using $8|1.95\times 2^{-10}|^{-2} \simeq 2^{21}$ known-plaintexts. The success rate of each attack is 88% and 99%, respectively.

The method to break 12-round DES cipher is almost same as 8-round DES cipher. We have succeeded in deriving the key completely in 50 hours using $8|1.91\times 2^{-16}|^{-2} \simeq 2^{33}$ known-plaintexts. Similarly, according to Lemma 4, it is possible to break 16-round DES using $8|1.19\times 2^{-22}|^{-2} \simeq 2^{47}$ known-plaintexts by solving the following expression:

$$\begin{aligned} &P_H[7,18,24] \oplus P_L[12,16] \oplus C_H[15] \oplus \ C_L[7,18,24,29] \oplus F_{16}(C_L,K_{16})[15] \\ &= K_1[19,23] \oplus K_3[22] \oplus K_4[44] \oplus \ K_5[22] \oplus K_7[22] \oplus K_8[44] \oplus \\ &\quad K_9[22] \oplus K_{11}[22] \oplus K_{12}[44] \oplus \ K_{13}[22] \oplus K_{15}[22]. \end{aligned} \tag{16}$$

Once finding 14 key bits, the remaining 42 key bits should be deduced exhaustively. Then one can break 16-round DES cipher with negligibly small memory faster than an exhaustive search for 56 key bits.

7 Only-Ciphertext Attack of DES Cipher

Now we apply Algorithm 2 to an only-ciphertext attack of DES cipher. We start with an example of 8-round DES cipher again, which has a linear approximation illustrated in Figure 4. Then we easily obtain the following expression which holds with probability $1/2+2^4(-2/64)(4/64)^2(-4/64)^2 = 1/2-2^{-17}$:

$$\begin{aligned} &P_L[27] \oplus C_H[27] \oplus C_L[0] \oplus F_8(C_L,K_8)[27] \\ &= K_2[1] \oplus K_3[8] \oplus K_4[1] \oplus K_6[1] \oplus K_7[8]. \end{aligned} \tag{17}$$

We note that $P_L[27]$ corresponds to the 39-th bit of the "real" plaintext before the initial permutation IP. Therefore, assuming that the plaintexts consist of ASCII codes, this bit must be equal to zero; that is, equation (17) has no plaintext bit. In fact, under this assumption, a similar discussion to the previous chapter tells us that seven key bits can be derived from equation (17) with high success rate using $8|2^{-17}|^{-2} = 2^{37}$ ciphertexts only.

Moreover, assuming that the plaintexts consist of natural English sentences represented by ASCII codes, we can also make use of a linear approximation illustrated in Figure 5. Then we easily see the following expression which holds with probability $1/2+2^5(-2/64)(-6/64)(10/64)(-20/64)^3 = 1/2-1.83\times 2^{-12}$:

$$\begin{aligned} &P_L[7,18,24] \oplus C_H[7,18,24,29,30] \oplus C_L[15] \oplus \ F_8(C_L,K_8)[30] \\ &= K_2[22] \oplus K_3[44] \oplus K_4[22] \oplus K_6[22] \oplus K_7[45] \oplus \ K_8[22]. \end{aligned} \tag{18}$$

We note that $P_L[7]$,$P_L[18]$ and $P_L[24]$ correspond to the first, 45-th and 63-rd bit of the "real" plaintext, respectively. As far as we know, when the plaintexts consist of natural English sentences represented by ASCII codes, the probability of $P_L[7,18,24]=0$ is at most 0.35. Therefore, under this assumption, the linear expression which is obtained by eliminating $P_L[7, 18, 24]$ from equation (18) holds with probability $1/2 - 2 \times (0.35 - 0.5) \times 1.83 \times 2^{-12} = 1/2 + 1.10 \times 2^{-13}$. This indicates that seven key bits can be deduced with high success rate using $8|1.10 \times 2^{-13}|^{-2} \simeq 2^{29}$ ciphertexts only.

Finally, we show a situation in which 16-round DES cipher is still breakable faster than an exhaustive search for 56 bits key. We now return to equation (16), which contains five plaintext bits, and suppose that these bits are independently equal to zero with probability 80% and all other plaintext bits are random. Then the linear equation which is obtained by eliminating these five bits from equation (16) holds with probability $1/2 + 2^5(0.8 - 0.5)^5 \times 1.19 \times 2^{-22} = 1/2 + 1.48 \times 2^{-26}$. This concludes that seven key bits can be obtained with high success rate using $8|1.48 \times 2^{-26}|^{-2} = 1.82 \times 2^{53}$ ciphertexts only.

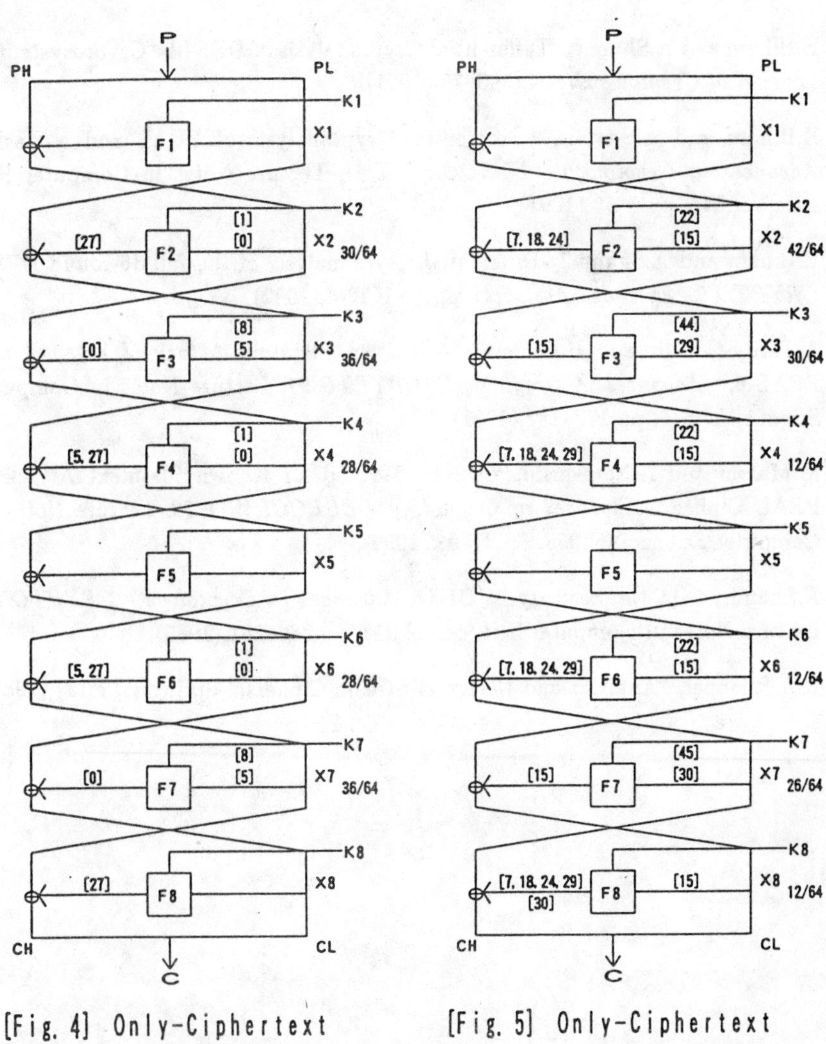

[Fig. 4] Only-Ciphertext Attack of 8-round DES (1)

[Fig. 5] Only-Ciphertext Attack of 8-round DES (2)

8 Concluding Remarks

We have introduced a new method for cryptanalysis of DES cipher. This method has enabled us the first known-plaintext attack of the full 16-round DES cipher and the initial step toward an only-ciphertext attack. To go more deeply into the only-ciphertext attack, however, we have to deal with several problems resulting from non-randomness of plaintexts. The detail discussion of this type of attack including complete tables and proofs, which we have omitted for lack of space, will appear in the full paper.

References

[1] E.Biham and A.Shamir, "Differential Cryptanalysis of DES-like Cryptosystems," *Journal of Cryptology*, Vol.4,pp.3–72,(1991).

[2] E.Biham and A.Shamir, "Differential Cryptanalysis of FEAL and N-Hash," *Advances in Cryptology - EUROCRYPT'91*, Lecture Notes in Computer Science, Vol.547, pp.1–16,(1991).

[3] E.Biham and A.Shamir, "Differential Cryptanalysis of the full 16-round DES," *CRYPTO'92 Extended Abstracts*, pp.12-1–12-5,(1992).

[4] A.Tardy-Corfdir and H.Gilbert, "A Known Plaintext Attack of FEAL-4 and FEAL-6," *Advances in Cryptology - CRYPTO'91*, Lecture Notes in Computer Science, Vol.576, pp.172–182,(1991).

[5] M.Matsui and A.Yamagishi, "A New Method for Known Plaintext Attack of FEAL Cipher," *Advances in Cryptology - EUROCRYPT'92*, Lecture Notes in Computer Science, Vol.658, pp.81–91,(1992).

[6] A.Shamir, "On the Security of DES," *Advances in Cryptology - CRYPTO'85*, Lecture Notes in Computer Science, Vol.218, pp.280–281,(1985).

[7] R.A.Rueppel, "Analysis and Design of Stream Ciphers," Springer Verlag,(1986).

3	$P_H[\alpha] \oplus P_L[15] \oplus C_H[\alpha] \oplus C_L[15]$ $= K_1[22] \oplus K_3[22]$	$1/2 + 1.56 \times 2^{-3}$	A-A
*4	$P_H[\alpha] \oplus P_L[15] \oplus C_H[15] \oplus C_L[\alpha,\beta]$ $= K_1[22] \oplus K_3[22] \oplus K_4[\gamma]$	$1/2 - 1.95 \times 2^{-5}$	A-AB
5	$P_H[15] \oplus P_L[\alpha,\beta] \oplus C_H[15] \oplus C_L[\alpha,\beta]$ $= K_1[\gamma] \oplus K_2[22] \oplus K_4[22] \oplus K_5[\gamma]$	$1/2 + 1.22 \times 2^{-6}$	BA-AB
*6	$P_L[\delta] \oplus C_H[\alpha] \oplus C_L[15]$ $= L_2 \oplus K_6[22]$	$1/2 - 1.95 \times 2^{-9}$	-DCA-A
*7	$P_H[\delta] \oplus P_L[12,16] \oplus C_H[\alpha] \oplus C_L[15]$ $= K_1[19,23] \oplus L_3 \oplus K_7[22]$	$1/2 + 1.95 \times 2^{-10}$	E-DCA-A
*8	$P_H[\delta] \oplus P_L[12,16] \oplus C_H[15] \oplus C_L[\alpha,\beta]$ $= K_1[19,23] \oplus L_3 \oplus K_7[22] \oplus K_8[\gamma]$	$1/2 - 1.22 \times 2^{-11}$	E-DCA-AB
*9	$P_H[15] \oplus P_L[\beta,\delta] \oplus C_H[15] \oplus C_L[\alpha,\beta]$ $= K_1[\gamma] \oplus K_2[22] \oplus L_4 \oplus K_8[22] \oplus K_9[\gamma]$	$1/2 - 1.91 \times 2^{-14}$	BD-DCA-AB
*10	$P_L[\alpha] \oplus C_H[\alpha] \oplus C_L[15]$ $= L_2 \oplus L_6 \oplus K_{10}[22]$	$1/2 - 1.53 \times 2^{-15}$	-ACD-DCA-A
11	$P_H[\alpha] \oplus P_L[15] \oplus C_H[\alpha] \oplus C_L[15]$ $= K_1[22] \oplus L_3 \oplus L_7 \oplus K_{11}[22]$	$1/2 + 1.91 \times 2^{-16}$	A-ACD-DCA- A
*12	$P_H[\alpha] \oplus P_L[15] \oplus C_H[15] \oplus C_L[\alpha,\beta]$ $= K_1[22] \oplus L_3 \oplus L_7 \oplus K_{11}[22] \oplus K_{12}[\gamma]$	$1/2 - 1.19 \times 2^{-17}$	A-ACD-DCA- AB
13	$P_H[15] \oplus P_L[\alpha,\beta] \oplus C_H[15] \oplus C_L[\alpha,\beta]$ $= K_1[\gamma] \oplus K_2[22] \oplus L_4 \oplus L_8 \oplus K_{12}[22] \oplus K_{13}[\gamma]$	$1/2 + 1.49 \times 2^{-19}$	BA-ACD-DCA -AB
*14	$P_L[\delta] \oplus C_H[\alpha] \oplus C_L[15]$ $= L_2 \oplus L_6 \oplus L_{10} \oplus K_{14}[22]$	$1/2 - 1.19 \times 2^{-21}$	-DCA-ACD-D CA-A
*15	$P_H[\delta] \oplus P_L[12,16] \oplus C_H[\alpha] \oplus C_L[15]$ $= K_1[19,23] \oplus L_3 \oplus L_7 \oplus L_{11} \oplus K_{15}[22]$	$1/2 + 1.19 \times 2^{-22}$	E-DCA-ACD- DCA-A
*16	$P_H[\delta] \oplus P_L[12,16] \oplus C_H[15] \oplus C_L[\alpha,\beta]$ $= K_1[19,23] \oplus L_3 \oplus L_7 \oplus L_{11} \oplus K_{15}[22] \oplus K_{16}[\gamma]$	$1/2 - 1.49 \times 2^{-24}$	E-DCA-ACD- DCA-AB
*17	$P_H[15] \oplus P_L[\beta,\delta] \oplus C_H[15] \oplus C_L[\alpha,\beta]$ $= K_1[\gamma] \oplus K_2[22] \oplus L_4 \oplus L_8 \oplus L_{12} \oplus K_{16}[22] \oplus K_{17}[\gamma]$	$1/2 - 1.16 \times 2^{-26}$	BD-DCA-ACD -DCA-AB
*18	$P_L[\alpha] \oplus C_H[\alpha] \oplus C_L[15]$ $= L_2 \oplus L_6 \oplus L_{10} \oplus L_{14} \oplus K_{18}[22]$	$1/2 - 1.86 \times 2^{-28}$	-ACD-DCA-A CD-DCA-A
19	$P_H[\alpha] \oplus P_L[15] \oplus C_H[\alpha] \oplus C_L[15]$ $= K_1[22] \oplus L_3 \oplus L_7 \oplus L_{11} \oplus L_{15} \oplus K_{19}[22]$	$1/2 + 1.16 \times 2^{-28}$	A-ACD-DCA- ACD-DCA-A
*20	$P_H[\alpha] \oplus P_L[15] \oplus C_H[15] \oplus C_L[\alpha,\beta]$ $= K_1[22] \oplus L_3 \oplus L_7 \oplus L_{11} \oplus L_{15} \oplus K_{19}[22] \oplus K_{20}[\gamma]$	$1/2 - 1.46 \times 2^{-30}$	A-ACD-DCA- ACD-DCA-AB

A:	$X[15] \oplus F(X,K)[7,18,24,29] = K[22]$	$p = \frac{12}{64}$	α:	7,18,24,29
B:	$X[27,28,30,31] \oplus F(X,K)[15] = K[42,43,45,46]$	$p = \frac{22}{64}$	β:	27,28,30,31
C:	$X[29] \oplus F(X,K)[15] = K[44]$	$p = \frac{30}{64}$	γ:	42,43,45,46
D:	$X[15] \oplus F(X,K)[7,18,24] = K[22]$	$p = \frac{42}{64}$	δ:	7,18,24
E:	$X[12,16] \oplus F(X,K)[7,18,24] = K[19,23]$	$p = \frac{16}{64}$	L_i:	$K_i[22] \oplus K_{i+1}[44] \oplus K_{i+2}[22]$

Annex. The best expression and the best probability of DES cipher.

New Types of Cryptanalytic Attacks Using Related Keys

Eli Biham[1]
Computer Science Department
Technion - Israel Institute of Technology
Haifa 32000, Israel

Abstract

In this paper we study the influence of key scheduling algorithms on the strength of blockciphers. We show that the key scheduling algorithms of many blockciphers inherit obvious relationships between keys, and use these key relations to attack the blockciphers. Two new types of attacks are described: New chosen plaintext reductions of the complexity of exhaustive search attacks (and the faster variants based on complementation properties), and new low-complexity chosen key attacks. These attacks are independent of the number of rounds of the cryptosystems and of the details of the F-function and may have very small complexities. These attacks show that the key scheduling algorithm should be carefully designed and that its structure should not be too simple. These attacks are applicable to both variants of LOKI and to Lucifer. DES is not vulnerable to the related keys attacks since the shift pattern in the key scheduling algorithm is not the same in all the rounds.

1 Introduction

In this paper we describe new types of attacks on blockciphers: chosen plaintext reductions of the complexity of exhaustive search and chosen key attacks in which only the relations between pairs of related keys are chosen by the attacker, who does not know the keys themselves. The chosen plaintext attacks reduce the complexity of exhaustive search and the complexity of the faster chosen plaintext attacks based on complementation properties by a factor of three. The chosen key attacks have very low complexities, however, they can be used only whenever the attacker can choose the relationships between unknown keys and wish to know the keys themselves.

These attacks are based on the observation that in many blockciphers we can view the key scheduling algorithm as a set of algorithms, each of which extracts one particular subkey from the subkeys of the previous few rounds. If all the algorithms of extracting the subkeys of the various rounds are the same, then given a key we can shift all the subkeys one round backwards and get a new set of valid subkeys which can be derived from some other key. We call these keys *related keys*.

An interesting feature of the attacks based on related keys is that they are independent of the number of rounds of the attacked cryptosystem. These attacks are applicable to both variants of LOKI[5,4], and to Lucifer[16]. Nevertheless, they are not applicable to DES[14]

---

[1]Acknowledgment: This research was supported by the fund for the promotion of research at the Technion.

due to the observation that the number of shifts of the key registers (C and D) in the key scheduling algorithm is not the same in all the rounds. However, if the shifts by one bit in the key scheduling of DES would be replaced by shifts by two bits, DES would become vulnerable to this kind of attack as well.

Another potential application of related keys is to analyze hash functions (either hash functions based on blockciphers or general hash functions). It may be possible in such functions to choose the message in a way that the related keys property suggest an additional message with the same hash value. Currently, we are not aware of a particular such application to hash functions, but designers of hash functions should be careful to design their functions immune to this weakness.

The results of the attacks are as follows: The complexity of a chosen plaintext attack on LOKI89 is about $1.5 \cdot 2^{54}$, which is almost three times faster than previously reported chosen plaintext attacks. The chosen key chosen plaintext attack takes a few seconds on a personal computer and its complexity is about 2^{17}, and the complexity of the chosen key known plaintext attack is about 2^{32}. The corresponding complexities of the attacks on the newer LOKI91 are $1.375 \cdot 2^{61}$, 2^{32}, and 2^{48} respectively. The complexity of the chosen key chosen plaintext attack on Lucifer is about 2^{33}. The DES, the IDEA cipher[11,12] and the FEAL cipher[15,13] are not vulnerable to these attacks.

Recently, Lars Ramkilde Knudsen found independently[9] the basic concept of the chosen plaintext related keys attacks and applied it to LOKI91. However, his attack (whose complexity is $1.07 \cdot 2^{62}$) is still 50% slower than the corresponding attack we present in this paper.

2 Description of LOKI89 and LOKI91

LOKI is a family of blockciphers with two variants: The original LOKI cipher, which was renamed to LOKI89[5], and the newer variant LOKI91[4]. Both variants have a structure similar to DES[14], with replaced F function and initial and final permutations and a replaced key scheduling algorithm. The new F function XORs the right half of the data with the subkey and expands the result to 48 bits, which enter into four 12-bit to 8-bit S boxes. The output of the S boxes is concatenated and permuted to form the output of the F function. In LOKI89 (see Figure 1), the initial and the final permutations are replaced by transformations which exclusive-or the data with the key. The key scheduling algorithm takes a 64-bit key, declares its left half as the value of K1 and its right half as the value of K2. Each other subkey Ki (out of K3,...,K16) is defined by rotating the subkey Kj of round $j = i - 2$ by 12 bits to the left (Ki = ROL12(Kj)). Thus, all the subkeys of the odd rounds share the same bits and all the subkeys of the even rounds share the same bits.

LOKI91 (see Figure 2) differs from LOKI89 by the choice of the S boxes, which are chosen to hold better against differential cryptanalysis. The initial and the final permutations are eliminated. The new key scheduling algorithm declares the value of the left half of the key to be K1 and the same value rotated 12 bits to the left is declared to be K2. The value of the right half of the key is declared to be K3 and the same value rotated 12 bits to the left is declared as K4. Each other subkey Ki (out of K5,...,K16) is defined by rotating the subkey Kj of round $j = i - 4$ by 25 bits to the left (Ki = ROL25(Kj)). Still, the subkeys share bits with a very structured order.

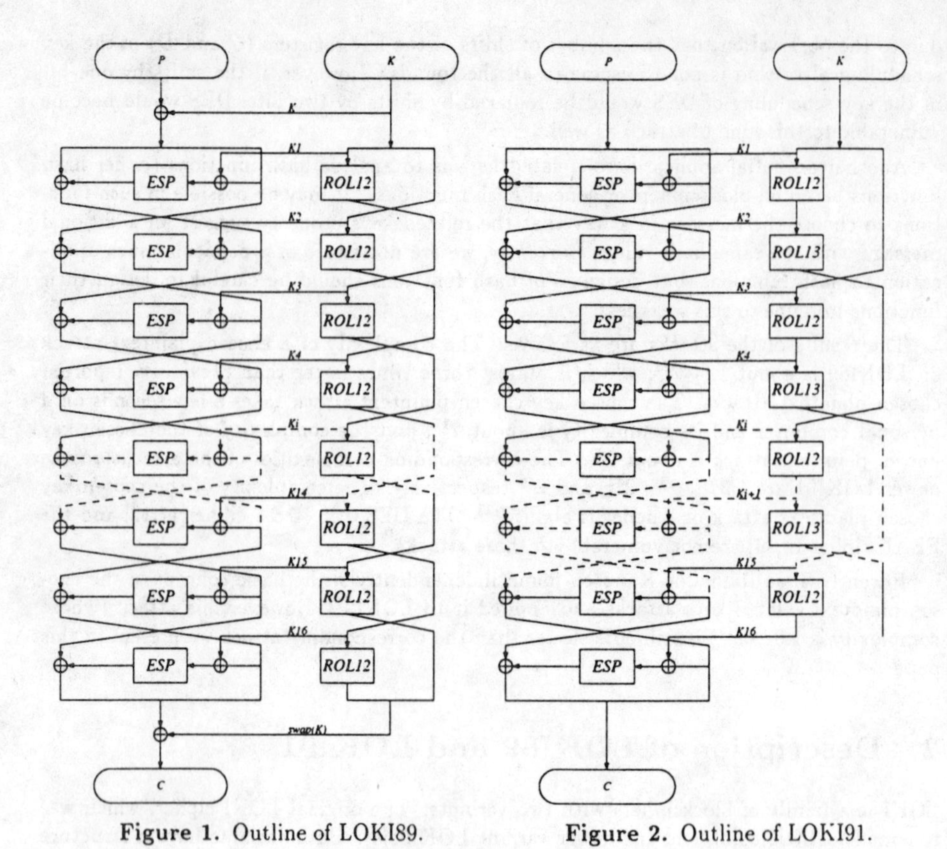

Figure 1. Outline of LOKI89. **Figure 2.** Outline of LOKI91.

3 The Chosen Key Attacks

In the chosen key attacks, two related keys with certain relationship are used and several plaintexts are encrypted under each of them. The attacker knows only the relationship between the two keys, but not the keys themselves. He receives the ciphertexts and use them to find both keys. Two kinds of chosen key attacks are studied: a chosen key known plaintext attack in which only the relation between the keys is chosen by the attacker, and a chosen key chosen plaintext attack in which the attacker chooses the relation between the keys as well as the plaintexts to be encrypted. These attacks are independent of the exact number of rounds of the attacked cryptosystem, and even if the number of rounds is enlarged (and especially if doubled), the resulting cryptosystem remains vulnerable to the same attack.

3.1 LOKI89

In LOKI89, every choice of two subkeys, one from an odd round and one from an even round, have a corresponding 64-bit key. Since all the algorithms of deriving the subkeys from the two preceding subkeys are the same, the position of the rounds in which two subkeys present does not affect the derivation of the following subkeys (nor the preceding ones). If we only fix two subkeys K2 and K3 of a key K, and define a second key K^* by choosing $K1^* = K2$ and

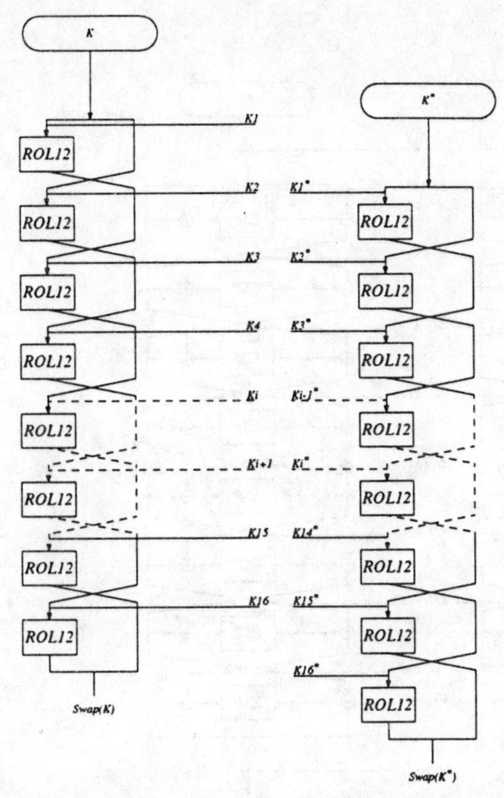

Figure 3. Relations of subkeys in the key scheduling algorithm of LOKI89.

$K2^* = K3$. then the values of the subkeys Ki^* of the key K^* are the same as of the following subkeys $K(i+1)$ of the key K. In this case, $K^* = (K2, K3) = (K_R, \mathrm{ROL12}(K_L))$. Therefore, the following property holds for any two such related keys: If the data before the second round in an encryption under the key K equals the data before the first round in an encryption under the key K^*, then the data and the inputs to the F functions are the same in both executions with a difference of one round. In this case, if the plaintext P is encrypted under the key K, then the data before the second round is $(P_R \oplus K_R, P_L \oplus K_L \oplus F(P_R \oplus K_R, K_L))$. This data equals the data before the first round in the other encryption under the key K^*, whose value is $P^* \oplus K^* = (P_L^* \oplus K_R, P_R^* \oplus \mathrm{ROL12}(K_L))$, and thus in such a pair

$$P^* = (P_R, P_L \oplus K_L \oplus \mathrm{ROL12}(K_L) \oplus F(P_R \oplus K_R, K_L)). \qquad (1)$$

We see that the right half of P equals the left half of P^* and that the relation between the other halves depends on the keys. In such a pair. there is also a similar a relation between the ciphertexts

$$C^* = (C_R \oplus K_L \oplus \mathrm{ROL12}(K_L) \oplus F(C_L \oplus K_R, K_L), C_L). \qquad (2)$$

Figures 3 and 4 describes the relations between the subkeys of the two keys and the relations between the values during the two encryptions.

A chosen key chosen plaintext attack based on this property chooses a 32-bit value P_R, 2^{16} plaintexts $P_0, \ldots, P_{65535}$ whose right halves equal P_R and whose 32-bit left halves are

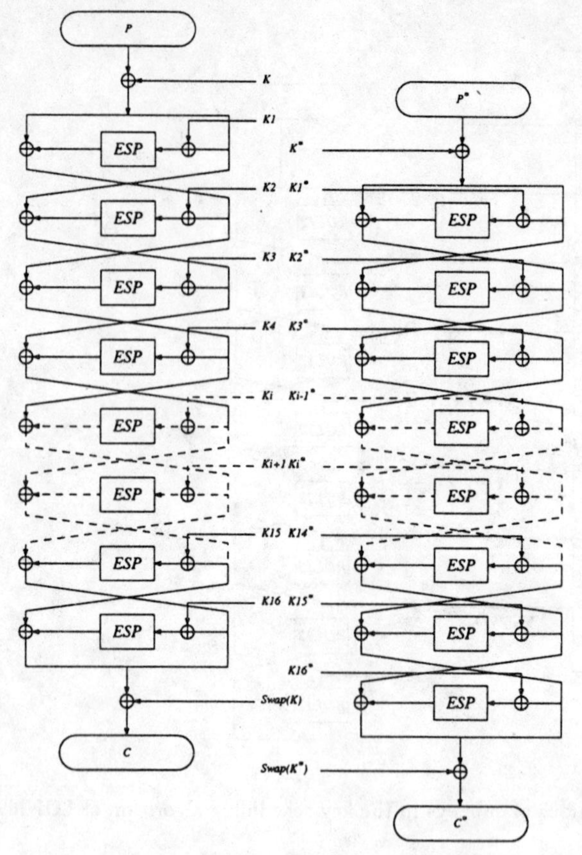

Figure 4. Relations during LOKI89 encryption.

randomly chosen, and 2^{16} plaintexts $P^*_0,\ldots,P^*_{65535}$ whose left halves equal P_R and whose 32-bit right halves are randomly chosen. Two unknown related keys are used to encrypt these plaintexts on the target machine: a key K is used to encrypt the first 2^{16} plaintexts and the key $\mathrm{K}^* = (\mathrm{K}_R, \mathrm{ROL12}(\mathrm{K}_L))$ is used to encrypt the other 2^{16} plaintexts. In every pair of plaintexts P_i and P^*_j we are guaranteed that $P^*_{jL} = P_{iR}$ and by the birthday paradox with a high probability there exist two plaintexts P_i and P^*_j such that $P^*_{jR} = P_{iL} \oplus \mathrm{K}_L \oplus \mathrm{ROL12}(\mathrm{K}_L) \oplus F(P_{iR} \oplus \mathrm{K}_R, \mathrm{K}_L)$. In such a pair the data is the same in both executions shifted by one round. It is easy to identify this pair, if it exists, by checking whether $C^*_R = C_L$. This test has a probability of 2^{-32} to pass accidentally, and thus only few pairs may pass this test.

A pair with this property relates the values of the two plaintexts and of the two ciphertexts to the key by equations 1 and 2. Thus, such a pair reveals the value of

$$F(P_R \oplus \mathrm{K}_R, \mathrm{K}_L) \oplus F(C_L \oplus \mathrm{K}_R, \mathrm{K}_L) = P^*_R \oplus P_L \oplus C^*_L \oplus C_R$$

in which the only unknown value is $\mathrm{K}_L \oplus \mathrm{K}_R$. Out of the 2^{32} possible values of $\mathrm{K}_L \oplus \mathrm{K}_R$, only few values satisfy the equation. Using differential cryptanalytic[1,2] optimization techniques, such as difference distribution tables and storing the possible pairs of each of their entries in a special preprocessed table, the identification of the value of $\mathrm{K}_L \oplus \mathrm{K}_R$ can be done in few operations. After we find $K_L \oplus K_R$, it is easy to calculate $K_L \oplus ROL12(K_L)$ by equations 1 and 2 and to derive K and K^*.

A similar chosen key known plaintext attack uses 2^{32} known plaintexts P_i encrypted under an unknown key K, and 2^{32} known plaintexts P_j^* encrypted under the related key $K^* = (K_R, ROL12(K_L))$. By the birthday paradox there is a high probability to have a pair in which the property holds. It is easy to identify this pair by the 32 common bits of the plaintexts and the 32 common bits of the ciphertexts. This pair may be used to reveal the keys in the same way as in the chosen key chosen plaintext attack.

3.2 LOKI91

The key scheduling algorithm of LOKI91 derives the subkeys of the even rounds in a different way than the subkeys of the odd rounds. In particular, the subkey of an even round i is just a rotated value of the subkey of the previous round $i-1$, and it is independent of the value of round $i+1$. Thus, we cannot shift the subkeys and the data by one round in the second execution without changing their values. However, we can shift them by two rounds instead. In this case we define the second set of subkeys to be $K1^* = K3$, $K2^* = K4$ and so on. The keys themselves are K and $K^* = (K_R, ROL25(K_L))$. Since the shift is increased to two rounds, we cannot easily identify the pair we are looking for. Luckily, the two subkeys K1 and K2 share the same 32 bits, and thus given a plaintext P, we have only to guess 32 bits of the key in order to find the data before the third round, which we define as P^*.

In the chosen key chosen plaintext attack, the attacker chooses a random plaintext P. For each one of the 2^{32} possible values of K1 and K2, he calculates the data before the third round, and defines these 2^{32} values as P_i^*. Given the ciphertexts $C = LOKI91(P, K)$ and $C_i^* = LOKI91(P_i^*, K^*)$, he searches for the plaintext P_i^* which was defined by the real values of K1 and K2 by verifying the relationships between the ciphertexts. The subkeys $K15^*$ and $K16^*$ share the bits of K1 and K2, which he has already guessed. Thus, it is easy to find the real values of K1 and K2 by calculating further from C into C_i^* by each possibility, and only the possibilities which result with the encrypted value of C_i^* might have the real value. After we find these 32 bits of the keys, it is easy to find the other 32 bits by exhaustive search.

The best choice of a chosen key known plaintext attack uses 2^{16} plaintexts P_i and 2^{48} plaintexts P_j^*. With high probability there is a pair P_i and P_j^*. for which P_j^* equals the data before the third round during the encryption of P_i. The attacker can identify the key by guessing the 32 bits of K1 and K2 and verifying which of which is possible in a way similar to the chosen key chosen plaintext attack.

4 The Chosen plaintext attacks

In this section we describe a chosen plaintext attack which reduces the complexity of exhaustive search using related keys. This attack can be combined with the attacks based on complementation properties[7] and its fastest variant is almost three times faster than the corresponding attacks based only on complementation properties. When this attack is used against 64-bit blockciphers, it requires about 2^{32}–2^{37} chosen plaintexts, whose corresponding ciphertexts are to be stored in random access memory during the analysis.

The idea is similar to the attack based on the complementation property of DES[7]. The complementation property of DES suggests that whenever a plaintext P is encrypted under a key K into a ciphertext $C = DES(P, K)$, then the complement of P is encrypted by the complement of K into the complement of C: $\bar{C} = DES(\bar{P}, \bar{K})$. The attack chooses a

Cycle Size	Number of Cycles	Number of Elements in the Cycles
1	16	16
2	120	240
4	16,320	65,280
8	33,546,240	268,369,920
Total	33,562,696	268,435,456†

† $268.435.456 = 2^{28}$

Table 1. Cycles of half-keys for LOKI89.

complementary pair of plaintexts P_1 and $P_2 = \bar{P}_1$. Given their ciphertexts $C_1 = \mathrm{DES}(P_1, \mathrm{K})$ and $C_2 = \mathrm{DES}(P_2, \mathrm{K})$ under the same key K, the attacker searches for the key K by trying all the keys K′ whose most significant bits are zero (i.e., half of the key space). For each such key, he encrypts P_1 into C' by $C' = \mathrm{DES}(P_1, \mathrm{K}')$. If $C' = C_1$, it is very likely that $\mathrm{K} = \mathrm{K}'$. In addition, the attacker can predict the ciphertext of P_1 under the key $\bar{\mathrm{K}}'$ to be $\bar{C}_2$ without an additional encryption. If $C' = \bar{C}_2$, it is very likely that $\mathrm{K} = \bar{\mathrm{K}}'$, since due to the complementation property $\bar{C}_2 = \mathrm{DES}(P_1, \bar{\mathrm{K}})$. Otherwise, neither K′ nor $\bar{\mathrm{K}}'$ can be the key K. This attack can be carried out even under a known plaintext attack[1], given about 2^{33} known plaintexts, since it is very likely that two complementary plaintexts exist within 2^{33} random plaintexts due to the birthday paradox.

LOKI89 has several complementation properties[3,1,4,10,8]. A key complementation property causes any key to have 15 equivalent keys which encrypt any plaintext to the same ciphertext. These 15 keys are the original key XORed with the 15 possible 64-bit hexadecimal numbers whose digits are identical. Encryption with these keys results with the same inputs to the F functions in all the 16 executions. Therefore, most of the keys are redundant and a known plaintext attack can be carried out with a complexity of 2^{60} rather than 2^{64}.

Another complementation property of LOKI89 is due to the observation that XORing the key with an hexadecimal value $gggggggghhhhhhhh_x$ and XORing the plaintext by $iiiiiiiiiiiiiiii_x$ where $g \in \{0_x, \ldots, F_x\}$, $h \in \{0_x, \ldots, F_x\}$ and $i = g \oplus h$ results in XORing the ciphertext by $iiiiiiiiiiiiiiii_x$. This property can be used to reduce the complexity of a chosen plaintext attack by a further factor of 16 to 2^{56}.

The attack we present in this section can predict the values of additional ciphertexts generated from additional plaintexts under related keys. Let P be any plaintext, K be any key and C be the ciphertext $C = \mathrm{LOKI89}(P, \mathrm{K})$. Let $\mathrm{K}^* = (\mathrm{K}_R, \mathrm{ROL12}(\mathrm{K}_L))$ and let P^* be the plaintext whose data before the first round of the encryption under K^* is the same as the data before the second round during the original encryption of P under K. Then, the first 15 rounds of the encryption $C^* = \mathrm{LOKI89}(P^*, \mathrm{K}^*)$ have exactly the same data and subkeys as the last 15 rounds of the original encryption of P, and the right half of C^* equals the left half of C (i.e., $C^*_R = C_L$).

For each key, there is one equivalent key whose four most significant bits are zero, and one complement key whose four most significant bits of its both halves are zero. In the following definition, the *next* operation rotates an half-key by 12 bits (as is done in the key scheduling algorithm every round) and finds the supposed equivalent value of the result.

Definition 1 The *next* operation takes a 32-bit value, rotates it 12 bits to the left (ROL12) and XORs it with an 32-bit hexadecimal number whose all digits are equal, such that the four most significant bits of the result are zero.

Table 1 shows the cycle size, number of cycles and the total number of elements in the cycles generated by the next operation. We see that almost all the cycles have the maximal size eight.

In our attack we use the observation that the shift of one round of the data and the subkeys can be done in both the backward and forward directions. Therefore. each trial key can predict ciphertexts under three related keys, and thus the attack requires to try about a third of the number of keys required by the attack based on the complementation properties. Usually it is not possible to find a subset of the keys which satisfy the related keys conditions and contain exactly a third of all the keys. The best choice for LOKI89 is to try 3/8 of the keys. We preprocess a list of half-keys $\{L_i\}$, with the properties: (1) The four most significant bits of all the values in the list are zero, and (2) The list contains exactly one value from each cycle of the next operation. This list contains about 2^{25} half-keys. The list can either be stored in memory using about $4 \cdot 2^{25} = 2^{27}$ bytes, or stored as a bitmap using $2^{28}/8 = 2^{25}$ bytes[2].

The attack requires 2^{37} chosen plaintexts whose ciphertexts are stored in a random access memory (whose size is 2^{40} bytes). The attack is as follows:

1. Choose any plaintext P_0, and calculate the 15 plaintexts P_i, $i \in \{1_x, \ldots, F_x\}$, by $P_i = P_0 \oplus iiiiiiiiiiiiiiii_x$.
2. For each plaintext P_i, choose the additional 2^{32} plaintexts $P_{i,k} = (P_{iR}, P_{iL} \oplus k)$ whose left halves are the right half of P_i and whose right halves receive all the possible values by XORing all the possible 32-bit values k to the left half of P_i.
3. For each plaintext P_i, choose the additional 2^{32} plaintexts $P^*_{i,k} = (P_{iR} \oplus k, P_{iL})$ whose right halves are the left half of P_i and whose left halves receive all the possible values by XORing all the possible 32-bit values k to the right half of P_i.
4. Given the ciphertexts $\{C_i\}$, $\{C_{i,k}\}$, $\{C^*_{i,k}\}$, try for each pair of half-keys (L_i, L_j) all the 24 keys K′ of the form $K' = (\text{ROR}m(L_i), \text{ROL}n(L_j))$, where m is a multiple of four and n is either $n = m$, $n = m + 4$, or $n = m + 8$. Figure 5 shows the choice of such 24 trial keys, which are denoted by †. All the keys are covered by the trial keys. The trial keys K′ are covered by themselves. The other keys are covered by the key relation property by trial keys created from (L_j, L_i). These keys are surrounded together with the swapped value of their trial keys. Examples of four such triples are marked in gray. The keys denoted by $*$ are covered by two trial keys. The figure should be interpreted cyclically through its edges. This is the best coverage possible for LOKI89.
5. Encrypt the plaintext P_0 under each trial key K′ into $C' = \text{LOKI89}(P_0, K')$.
6. If C' equals one of the values $C_i \oplus iiiiiiiiiiiiiiii_x$, the original key is likely to be either $K = K' \oplus 00000000iiiiiiii_x$ or any one of its 15 equivalent keys.
7. Fix k to be the output of the F function in the first round of the encryption of P_0 under the key K′. If C'_L equals one of the 16 values $C_{i,kR} \oplus iiiiiiii_x$, continue encryption of P_0 with a seventeenth round (just calculate one additional round from C' using the subkey K17′ which can be easily derived from the key K′), and if the result C'' equals $C_{i,k} \oplus iiiiiiiiiiiiiiii_x$, then the original key is likely to be $K = (K'_R, \text{ROL12}(K'_L) \oplus iiiiiiii_x)$ or any one of its 15 equivalent keys.
8. Calculate one additional round backwards from P_0 using the subkey K0′ which can be easily derived from the key K′, and fix k to be the output of the F function in this round. If the data after the fifteenth round during the encryption of P_0 under the key

[2]We have also devised an additional algorithm which chooses the trial keys on the fly and does not require a list of potential half-keys. However, due to technical details, the efficiency of the attack is reduced. Since the memory space required by the attack is much larger than the list of potential half-keys, this list does not affect the space complexity of the attack.

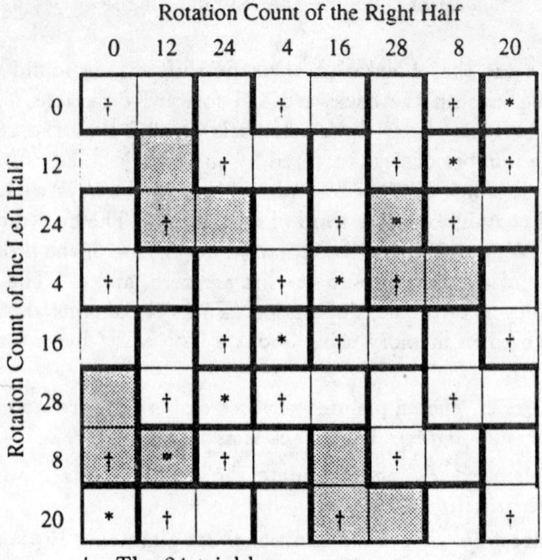

† The 24 trial keys.
∗ Keys covered by two trial keys.

Figure 5. The choice of the trial keys.

K′ equals one of the 16 values $C_{i,k} \oplus iiiiiiiiiiiiiii_x$ (for some i), the original key is likely to be K = (ROR12(K'_R), $K'_L \oplus iiiiiii_x$) or any one of its 15 equivalent keys[3].

The encryption key K must be recognized by either step 6, step 7, or step 8. Since the comparisons in steps 6, 7 and 8 are much faster than a trial encryption, the complexity of the attack is the total complexity of step 5. If the number of elements in each cycle would have been divisible by three, the complexity of the attack would decrease relatively to the attack based on complementation properties by a factor of 1/3. In the case of LOKI89 this factor is only 3/8, and the complexity of the attack is $(3/8) \cdot (2^{28})^2 = 1.5 \cdot 2^{54}$ trial encryptions[4].

The application of this attack to LOKI91 is similar. The main two differences are: (1) The plaintexts are chosen with respect to changes in the half of the key which affects the first two rounds, rather than changing the plaintext's halves directly. (2) The trial keys generated in step 4 of the attack cover only half of the keys, since there is only one complementation property. Thus, we use twice as many trial keys, namely the original trial keys and the same trial keys whose right halves are complemented. The complexity of this attack on LOKI91 is $1.375 \cdot 2^{61}$, which is about 5.8 times faster than exhaustive search.

[3]This part of the algorithm requires the calculation of one additional round for any trial key K′, which slows the attack by a factor of 17/16. In parallel hardware implementations, this behavior can be solved easily. In software, it can also be solved by using large hash tables and checking the existence of ciphertexts before the calculation of the additional round.

[4]If we choose carefully the order of trying the trial keys, we can reduce the average case complexity slightly from $1.5 \cdot 2^{53}$ by about 8%.

Round	1	2	3	4	5	6	7	8	9	10	11	12	13	14	15	16
Shifts	1	1	2	2	2	2	2	2	1	2	2	2	2	2	2	1
Modified variant	2	2	2	2	2	2	2	2	2	2	2	2	2	2	2	2

Table 2. The numbers of shifts in the key scheduling algorithm of DES, and a modified variant vulnerable to the related keys attacks.

5 Application to Lucifer

In Lucifer[16], the subkeys are derived by shifting the key by a fixed number of bits every round and selecting specific bits as the subkeys, and thus the derivation algorithm of the subkeys from the previous subkeys is the same in all the rounds. Therefore, chosen key attacks similar to the attacks on LOKI89 are applicable, but since the blocksize and the size of the key are 128 bits, the complexities become 2^{33} for the chosen key chosen plaintext attack and 2^{65} for the chosen key known plaintext attack. Lucifer has no complementation properties, and thus the complexity of the chosen plaintext attack is $1.5 \cdot 2^{126}$.

6 DES

DES[14] is not vulnerable to the related keys attacks since the number of shifts in the key scheduling algorithm is not the same in all the rounds. While usually the key registers are shifted by two bits after each round, they are shifted only by one bit after the first, the ninth and the fifteenth rounds. However, if we modify this shift pattern to shift the key registers by the same number of bits in all the rounds (either one or two or any other number including seven which was suggested in [6]), the resultant cryptosystem becomes vulnerable to the related keys attacks. Table 2 describes the numbers of shifts before each round in the key scheduling algorithm of DES and a variant with modified shift numbers which is vulnerable to the related keys attacks.

7 Summary

In this paper we described new cryptanalytic attacks which are applicable to the LOKI family of blockciphers and to Lucifer. These new attacks are based on the structure of the key scheduling algorithms. These attacks are independent of the number of the rounds of the cipher. The same attacks could be applicable to DES if only minor changes would be made to the shift pattern of its key scheduling algorithm, and thus these attacks show how so small points in the design of a cipher can contribute to its strength. The results of the related keys attacks are summarized in table 3.

References

[1] Eli Biham, Adi Shamir, *Differential Cryptanalysis of the Data Encryption Standard*, Springer-Verlag, 1993.

[2] Eli Biham, Adi Shamir, *Differential Cryptanalysis of DES-like Cryptosystems*, Journal of Cryptology, Vol. 4, No. 1, pp. 3–72, 1991.

	LOKI89	LOKI91	Lucifer	Modified DES*
Chosen Plaintext: Related Keys				
Complexity of Attack	$1.5 \cdot 2^{54}$	$1.375 \cdot 2^{61}$	$1.5 \cdot 2^{126}$	$1.43 \cdot 2^{53}$
Chosen Plaintexts Required	2^{37}	2^{34}	2^{65}	2^{34}
Chosen Key				
Chosen Plaintexts	2^{17}	2^{32}	2^{33}	2^{17}
Known Plaintexts	2^{33}	2^{48}	2^{65}	2^{33}
Previously known attacks				
Differential Cryptanalysis	–	–	–	2^{47}
Complementation Properties†	2^{56}	2^{63}	–	2^{55}
Exhaustive Search‡	2^{60}	2^{64}	2^{128}	2^{56}

* The shift pattern of the key registers in the key scheduling algorithm is modified to have the same number of shifts in all the rounds (see Table 2).

† Either two chosen plaintexts required, or 2^{33} known plaintexts. For LOKI89, 16 chosen plaintexts are required to achieve this complexity.

‡ One known plaintext required.

Table 3. Results of the related keys attacks.

[3] Eli Biham, Adi Shamir, *Differential Cryptanalysis of Snefru, Khafre, REDOC-II, LOKI and Lucifer (extended abstract)*, Lecture Notes in Computer Science, Advances in Cryptology, proceedings of CRYPTO'91, pp. 156–171, 1991.

[4] Lawrence Brown, Matthew Kwan, Josef Pieprzyk, Jennifer Seberry, *Improving Resistance to Differential Cryptanalysis and the Redesign of LOKI*, Lecture Notes in Computer Science, Advances in Cryptology, proceedings of ASIACRYPT'91, to appear.

[5] Lawrence Brown, Josef Pieprzyk, Jennifer Seberry, *LOKI - A Cryptographic Primitive for Authentication and Secrecy Applications*, Lecture Notes in Computer Science, Advances in Cryptology, proceedings of AUSCRYPT'90, pp. 229–236, 1990.

[6] Lawrence Brown, Jennifer Seberry, *Key Scheduling in DES Type Cryptosystems*, Lecture Notes in Computer Science, Advances in Cryptology, proceedings of AUSCRYPT'90, pp. 221–228, 1990.

[7] M. E. Hellman, R. Merkle, R. Schroppel, L. Washington, W. Diffie, S. Pohlig and P. Schweitzer, *Results of an Initial Attempt to Cryptanalyze the NBS Data Encryption Standard*, Stanford university, September 1976.

[8] Lars Ramkilde Knudsen, *Crypanalysis of LOKI*, Lecture Notes in Computer Science, Advances in Cryptology, proceedings of ASIACRYPT'91, to appear.

[9] Lars Ramkilde Knudsen, *Crypanalysis of LOKI91*, Lecture Notes in Computer Science, Advances in Cryptology, proceedings of AUSCRYPT'92, to appear.

[10] Matthew Kwan, Josef Pieprzyk, *A General Purpose Technique for Locating Key Scheduling Weakness in DES-Like Cryptosystems*, Lecture Notes in Computer Science, Advances in Cryptology, proceedings of ASIACRYPT'91, to appear.

[11] Xuejia Lai, James L. Massey, Sean Murphy, *Markov Ciphers and Differential Cryptanalysis*, Lecture Notes in Computer Science, Advances in Cryptology, proceedings of EUROCRYPT'91, pp. 17–38, 1991.

[12] Xuejia Lai, *On the Design and Security of Block Ciphers*, Ph.D. thesis, Swiss Federal Institue of Technology, Zurich, 1992.

[13] Shoji Miyaguchi, Akira Shiraishi, Akihiro Shimizu, *Fast Data Encryption Algorithm FEAL-8*, Review of electrical communications laboratories, Vol. 36, No. 4, pp. 433–437, 1988.

[14] National Bureau of Standards, *Data Encryption Standard*, U.S. Department of Commerce, FIPS pub. 46, January 1977.

[15] Akihiro Shimizu, Shoji Miyaguchi, *Fast Data Encryption Algorithm FEAL*, Lecture Notes in Computer Science, Advances in Cryptology, proceedings of EUROCRYPT'87, pp. 267–278, 1987.

[16] Arthur Sorkin, *Lucifer, a Cryptographic Algorithm*, Cryptologia, Vol. 8, No. 1, pp. 22–41, January 1984.

Secret-Key Reconciliation by Public Discussion

Gilles Brassard * and Louis Salvail **

Département IRO, Université de Montréal
C.P. 6128, succursale "A",Montréal (Québec), Canada H3C 3J7.
e-mail: {brassard}{salvail}@iro.umontreal.ca.

Abstract. Assuming that Alice and Bob use a secret noisy channel (modelled by a binary symmetric channel) to send a key, reconciliation is the process of correcting errors between Alice's and Bob's version of the key. This is done by public discussion, which leaks some information about the secret key to an eavesdropper. We show how to construct protocols that leak a minimum amount of information. However this construction cannot be implemented efficiently. If Alice and Bob are willing to reveal an arbitrarily small amount of additional information (beyond the minimum) then they can implement polynomial-time protocols. We also present a more efficient protocol, which leaks an amount of information acceptably close to the minimum possible for sufficiently reliable secret channels (those with probability of any symbol being transmitted incorrectly as large as 15%). This work improves on earlier reconciliation approaches [R, BBR, BBBSS].

1 Introduction

Unlike public key cryptosystems, the security of quantum cryptography relies on the properties of the channel connecting Alice and Bob. Physical imperfections in a quantum channel introduce noise into the messages passing through it. The presence of an eavesdropper wiretapping the channel disrupts communication even more. Assuming that Alice and Bob are using a quantum channel or any noisy channel to send a secret key, they would need to reconcile their keys to make them identical. Reconciliation is the process of finding and correcting discrepancies between the secret key sent by Alice and the one received by Bob. This is done by public discussion. In this paper we focus on secret noisy channels modelled by binary symmetric channels.

An eavesdropper can gain information about the secret key both by wiretapping the quantum channel and by listening to the public reconciliation. This information can be eliminated using a privacy amplification protocol [R, BBR] at the cost of reducing the secret key size proportionally to the amount of information potentially known by an eavesdropper. Since using a quantum channel

* Supported in part by NSERC's E. W. R. Steacie Memorial Fellowship and Québec's FCAR.
** Supported by an NSERC scholarship.

is expensive, we would like to minimize the information that a reconciliation protocol divulges.

A quantum public key distribution protocol is described in [BBBSS], which also discusses a way to combine together reconciliation and privacy amplification. The problem of reconciliation has been previously studied in [R, BBR, BBBSS]. Key distribution using independent channels [M] also requires reconciliation.

In section 3 we define the problem and introduce the notion of optimality; in section 4 we show how to construct optimal protocols. In section 5 we discuss efficiency; in sections 6 and 7 we present protocols that can be used in practice.

2 Preliminaries

Let $\{p(x)\}_{x\in X}$ be a probability distribution over a finite set X. The entropy of X, denoted $H(X)$, is defined as

$$H(X) = -\sum_{x\in X} p(x)\log p(x)$$

(where all logarithms are to the base 2). In particular, $H(X)$ is the expected value of the number of bits required to specify a particular event in X. It is easy to observe that

$$H(X) \leq \log|X|$$

with equality iff $p(x) = 1/|X|$ for each $x \in X$. When X is a Bernoulli trial with parameter p, we denote $H(X)$ by $h(p)$.

Given two sets X and Y and a joint probability distribution $\{p(x,y)\}_{x\in X, y\in Y}$, the conditional entropy $H(X|Y)$ is defined as

$$H(X|Y) = -\sum_{y\in Y}\sum_{x\in X} p(y)p(x|y)\log p(x|y).$$

A binary symmetric channel (BSC) permits transmission of a string of bits, each independently exposed to noise with probability p. Let A be the string sent by Alice and let B be the one received by Bob. If each bit from string A is randomly and independently chosen, it is clear that

$$H(A) = |A|.$$

The conditional entropy of A given B is

$$H(A|B) = H(A\oplus B) = nh(p)$$

where $n = |A| = |B|$. Henceforth we will denote a binary symmetric channel with parameter p by $BSC(p)$.

The quantum channel is an example of a secret binary symmetric channel even if an eavesdropper introduces noise in a non-symmetric way. Before Alice and Bob reconcile their strings, they perform a sample of the transmitted bits to estimate the error rate of the actual quantum communication. If the estimate is acceptably close to the expected error rate of the channel then they publicly and randomly permute their respective string on which reconciliation is then applied. Bob's string can now be assumed to be the result of a transmission over a *BSC*. If the estimate is not sufficiently close to the expected error rate then they rule out the communication and retry later (consult [BBBSS] for more details).

For $0 \leq \lambda \leq \frac{1}{2}$ the tail inequality is

$$\sum_{k=0}^{\lfloor \lambda n \rfloor} \binom{n}{k} \leq 2^{nh(\lambda)}.$$

For X a random variable with finite variance $V(X)$ and expected value $E(X)$ and for $a > 0$ the Chebyshev inequality is

$$prob(|X - E(X)| \geq a) \leq \frac{Var(X)}{a^2}.$$

The Hamming distance $dist(A, B)$ between A and B is the number of places in which A and B differ. The weight $w(A)$ of A is its number of nonzero positions.

3 The Problem

Suppose there is a secret channel between Alice and Bob on which Alice transmits to Bob an n-bit string A such that $H(A) = n$. We model the secret channel with a $BSC(p)$ for some p. Bob then receives an n-bit string B such that $H(A|B) = nh(p)$. Using public discussion Alice and Bob want to share an n-bit secret string S obtained from A and B. This can be done assuming an unjammable public channel without making any other security assumptions. Our goal is to find protocols minimizing the information on S that an eavesdropper with unlimited computing power can get by listening on the public channel.

A reconciliation protocol R^p is defined by Alice's and Bob's algorithms. R^p runs on strings A and B to produce string S by exchanging some information Q on the public channel. This will be denoted by $R^p = [S, Q]$ or by $R^p(A, B) = [S, Q]$ when a specific A and B are considered. If the protocol fails to produce $S \in \{0,1\}^n$ we will write $S = \perp$. The amount of *leaked information* $I_E(S \mid Q)$ is the expected amount of Shannon information that an eavesdropper E can get on S given Q.

Definition 1. A reconciliation protocol R^p is *ε–robust*, $0 \leq \varepsilon \leq 1$, if

$$(\exists N_0(\varepsilon))(\forall n \geq N_0(\varepsilon)) \sum_{\alpha,\beta \in \{0,1\}^n} prob(A = \alpha, B = \beta) prob(R^p(\alpha, \beta) = [\perp, \cdot]) \leq \varepsilon.$$

3.1 Optimality

It is easy to define the optimality property of reconciliation protocols. Theorem 2 is a direct consequence of an elementary result in information theory, namely the noiseless coding theorem.

Theorem 2. *$(\forall p \leq \frac{1}{2})(\forall$reconciliation protocol $R^p)$ if there exists $0 \leq \varepsilon < 1$ such that $R^p = [S, Q]$ is ε-robust then*

$$\lim_{n \to \infty} \frac{I_E(S \mid Q)}{nh(p)} \geq 1$$

where n is the length of the transmitted string.

A reconciliation protocol is optimal if the expression in theorem 2 is an equality.

Definition 3. A protocol R^p is *optimal* if

$$(\forall \varepsilon > 0)\,[R^p = [S, Q] \text{ is } \varepsilon\text{-robust}]$$

and

$$\lim_{n \to \infty} \frac{I_E(S \mid Q)}{nh(p)} = 1$$

where the public channel is a BSC(p).

The next section shows how to construct a family of optimal protocols.

4 Optimal Protocols

One way of constructing optimal protocols is to associate a random label of length approximately $nh(p)$ bits to each n-bit string. Alice tells Bob the label of her string A over the public channel. In order to decode string B, Bob computes a string B' of minimal Hamming distance from B, having the same label as A. Protocol 1 implements this process.

Obviously such a protocol is useless in practice since it requires Alice and Bob to choose randomly among 2^{m2^n} functions.

Protocol 1

1. Alice and Bob choose a random function among all functions from $\{0,1\}^n \to \{0,1\}^m$, where m is a parameter to be determined. The description of this function is exchanged publicly.
2. Alice sends $f(A)$ to Bob on the public channel.
3. Bob decodes his string resulting in string B' such that $dist(B, B')$ is minimal over all strings D such that $f(D) = f(A)$.

The proof of theorem 4 is similar to earlier ones showing the Shannon noisy coding theorem for BSC (see [W]).

Theorem 4. *Protocol 1 is optimal for an adequate choice of parameter m.*

Proof (sketch). Let p be the BSC parameter. Let p_e be the decoding error probability. Let $C = A \oplus B$ and E be the event associated with a decoding error. Clearly $w(C) \approx Bin(n,p)$[3]. A simple counting argument shows that

$$prob(\neg E \mid w(C) \leq r) \geq (1 - 1/2^m)^{\sum_{j=1}^{r} \binom{n}{j}}.$$

For correct decoding to occur it is sufficient that all the strings B' such that $dist(B', B) \leq dist(B, A)$ with $B' \neq A$ be distributed among those $2^m - 1$ elements of the image of f that are not equal to $f(A)$.

The decoding error probability is:

$$\begin{aligned} p_e &= prob(w(C) \leq r) prob(E \mid w(C) \leq r) \\ &\quad + prob(w(C) > r) prob(E \mid w(C) > r) \\ &\leq prob(E \mid w(C) \leq r) + prob(w(C) > r). \end{aligned}$$

Let $r = \lfloor np + n\varepsilon_n \rfloor$ and $\varepsilon_n = 1/\log n$. The Chebyshev inequality gives

$$\begin{aligned} prob(w(C) > r) &= prob(w(C) > \lfloor np + n\varepsilon_n \rfloor) \\ &\leq prob(\mid w(C) - np \mid \geq n\varepsilon_n) \\ &\leq \frac{p(1-p)}{n\varepsilon_n^2} \\ &= \frac{(\log n)^2 p(1-p)}{n}. \end{aligned}$$

If $w(C) \leq r$, the decoding probability error is bounded by:

$$\begin{aligned} prob(E \mid w(C) \leq r) &\leq 1 - prob(\neg E \mid w(C) \leq r) \\ &\leq 1 - (1 - 1/2^m)^{\sum_{j=1}^{r} \binom{n}{j}}. \end{aligned}$$

For $m = \lceil \log \delta_n + nh(p + \varepsilon_n) \rceil$ and $\delta_n = \lceil \log n \rceil$ we obtain:

$$prob(E \mid w(C) \leq r) \leq 1 - \left(\tfrac{1}{e}\right)^{2^{-m} \sum_{j=0}^{r} \binom{n}{j}}.$$

By the tail inequality the decoding probability error is bounded by:

$$prob(E \mid w(C) \leq r) \leq 1 - e^{-2^{nh(p+\varepsilon_n)-m}}.$$

When $n \to \infty$ we have $p_e = 0$, and protocol 1 is ε–robust for all $\varepsilon > 0$. It is easy to see that the resulting amount m of leaked information is asymptotically equal to $nh(p)$. □

To solve the problem that choosing a random function from a huge set is unreasonable, we choose it from a universal$_2$ class of hash functions [CW].

[3] $Bin(n,p)$ is the binomial probability distribution.

Definition 5 ([CW]). Let H be a class of functions from F to G. We say that H is universal$_2$ if for all $x, y \in F$ such that $x \neq y$, the number of functions f in H such that $f(x) = f(y)$ is less or equal than $\#H/\#G$.

Wegman and Carter [CW, WC] also describe classes for which choosing and evaluating functions can be achieved efficiently. The following theorem shows that choosing f among a universal$_2$ class ensures the protocol's optimality.

Theorem 6. *Protocol 1 is optimal for an adequate choice of parameter m if Alice and Bob choose f among an universal$_2$ class of functions.*

Proof. In order to bound the decoding error probability we must bound the probability that:

$$f(C) \neq f(x_1) \wedge f(C) \neq f(x_2) \wedge \ldots \wedge f(C) \neq f(x_l)$$

for $C \in F$, $x_i \in F$ where $x_i \neq C$ for each i. From definition 5, for any $C \in F$ and $x_1 \in F$ the following is true:

$$\#\{f \in H \mid f(C) = f(x_1)\} \leq \frac{\#H}{\#G}.$$

Therefore, the number of functions where $f(C) \neq f(x_1)$ is greater than $\#H(1 - 1/\#G)$. Among the $\#H - \#H/\#G$ remaining functions, there are at most $\#H/\#G$ functions such that $f(C) = f(x_2)$. Applying this argument over the l points in the domain of f gives $\#\{f \in H \mid f(C) \neq f(x_i) \text{ pour } 1 \leq i \leq l\} \geq \#H(1 - l/\#G)$. Let $X(r) = \{x \in \{0,1\}^n \mid w(x) \leq r \text{ et } x \neq C\}$ be the set of strings of weight r or less. Similar to the proof of theorem 4 we have:

$$\begin{aligned} prob(E \mid w(C) \leq r) &= 1 - prob(\neg E \mid w(C) \leq r) \\ &\leq \frac{\#X(r)}{\#G}. \end{aligned}$$

The proof follows by setting $F = \{0,1\}^n$, $G = \{0,1\}^m$ for r, ε_n, δ_n and m set as in theorem 4 (since $\#X(r) \leq 2^{nh(p+\varepsilon_n)}$). □

Thus we have a way to automatically generate optimal reconciliation protocols by specifying a universal$_2$ class in a short and efficient way. The problem with this approach is that there are no known efficient algorithms for Bob to compute the decoded string B'.

5 Efficiency

Finding a class of functions for which protocol 1 is optimal and such that Alice and Bob can reconcile efficiently is comparable to finding efficient decodable error correcting codes. This is due to similarities between these two problems when a non-interactive protocol such as protocol 1 is being considered. The non-interactive scheme is relevant for some applications such as quantum oblivious transfer [BBCS]. We will see that using H_3 (defined below, for more details consult [CW]) yields a decoding time complexity equivalent to that of solving the general problem of decoding linear codes.

Definition 7. An R^p reconciliation protocol is:

1. *efficient* if there is a polynomial $t(n)$ such that $\overline{T}^{R^p}(n) \leq t(n)$ for n sufficiently large, where n is the length of the strings transmitted over the secret channel;
2. *ideal* if it is both optimal and efficient.

where $\overline{T}^{R^p}(n)$ represent the expected running time of R^p given an n-bit long input string.

Theorem 8 gives an (unlikely) hypothesis that is necessary and sufficient for protocol 1 to be ideal when used along with class H_3. Recall that H_3 is the set of functions $f : \{0,1\}^n \rightarrow \{0,1\}^m$ whose i^{th} bit $f_i(x)$ of $f(x)$ is such that $f_i(x) = \bigoplus_{j=1}^{n} \lambda_{i,j} x_j$ for $\lambda_{i,j} \in \{0,1\}$. When m and n are fixed, choosing f consists of choosing $\lambda_{i,j} \in_R \{0,1\}$ for all $i \in [1..m]$ and $j \in [1..n]$.

Theorem 8. *Protocol 1 used with the universal*$_2$ *class of functions* H_3 *is ideal if and only if* **NP** $\subseteq$ **BPP**.

Proof. If C is a class of decision problems, let C^* denote the class of problems that are polynomially equivalent to some problem in C. Let X be the problem of executing step 3 of protocol 1 and let X^{H_3} be the same problem when H_3 is used. [BMT] shows that determining least-weight solution in a system of linear equations in $GF(2)$ is **NP**–hard. This problem is equivalent to X^{H_3}. Moreover we can easily show that $X \in (\sum_2^p)^*$. We want to show that $X^{H_3} \in \mathbf{BPP}^* \Leftrightarrow$ **NP** $\subseteq$ **BPP**. The left to right direction of this statement is obviously true since X^{H_3} is **NP**–hard. To prove the other direction we use a result of [Z] showing that **NP** $\subseteq$ **BPP** $\Leftrightarrow$ **PH** $\subseteq$ **BPP** combined with the fact that $X \in (\sum_2^p)^*$. □

6 Almost-Ideal Protocols

To be useful in practice a reconciliation protocol need not be optimal. Before execution of the protocol, Alice and Bob can agree on an arbitrarily small amount of additional information relative to the theoretical bound, which they are willing to reveal during execution. If the resulting protocol is also efficient then we have a protocol that might be useful in practice. For example, if the bits exchanged over the secret channel are costly, then it might be better to spend more time computing during reconciliation, thus saving some more of these costly bits.

We will formalize the latter property, and construct in section 6.2 a family of protocols that satisfy it.

6.1 Definition

An almost-ideal protocol has an error probability approaching 0 as the length of the strings increases, but its amount of leaked information is allowed to be slightly greater than the theoretical bound. Alice and Bob indicate it by choosing a parameter ζ. They must be able to set ζ such that the corresponding amount of leaked information is as close to the theoretical bound as they wish. Once the parameter ζ is set, the expected running time $\overline{T}^{R^p_\zeta}$ of protocol R^p_ζ must be bounded by a polynomial.

Definition 9. A reconciliation protocol R^p_ζ is *almost-ideal* if for all $\zeta > 0$ we have

1. $(\forall \varepsilon > 0)\left[R^p_\zeta = [S, Q] \text{ is } \varepsilon\text{-robust}\right]$
2. $\lim_{n \to \infty} \frac{I_E(S \mid Q)}{nh(p)} \leq 1 + \zeta$ and
3. $(\exists \text{ polynomial } t)(\exists N_0(\zeta))(\forall n \geq N_0(\zeta))\left[\overline{T}^{R^p_\zeta} \leq t(n)\right]$

for n the length of the strings transmitted over the BSC(p).

6.2 Protocol Shell

Protocol **Shell** uses interaction on the public channel to correct efficiently the secret strings by dividing them into blocks of fixed length. If the blocks have been corrected using a subprotocol with a decoding error probability δ_k (for blocks of length k), then **Shell** will correct them by producing an additional amount of leaked information proportional to δ_k and $\frac{1}{k}$.

Shell is constructed from a few simple interactive primitives described in the following sections.

`BINARY`. When strings A and B have an odd number of errors, then Alice and Bob can perform an interactive binary search to find an error by exchanging fewer than $\lceil \log n \rceil$ bits over the public channel in the following manner:

1. Alice sends Bob the parity of the first half of the string.
2. Bob determines whether an odd number of errors occurred in the first half or in the second by testing the parity of the first half and comparing it to the parity sent by Alice.
3. This process is repeatedly applied to the half determined in step 2. An error will be found eventually.

The reconciliation protocol described in [BBBSS] uses `BINARY` as the main primitive.

`CONFIRM`. If the strings of Alice and Bob are different, `CONFIRM` tells them with probability $\frac{1}{2}$; if they are identical, `CONFIRM` says so with probability 1.

1. Alice and Bob choose a random subset of corresponding bits from their strings.
2. Alice tells Bob the parity of her subset.
3. Bob checks that his subset has the same parity.

They can apply this process k times to convince themselves that their strings are identical. This test will fail with probability 2^{-k}.

BICONF^s. Combining BINARY and CONFIRM gives us another primitive that can correct several errors. BICONF^s runs CONFIRM s times. Each time CONFIRM shows a subset for which Alice's and Bob's string have different parities they run BINARY on this subset and thus correct an error. Let $\Delta^s(l|e)$ be the probability that BICONF^s corrects l errors if there are e errors. We have:

$$\Delta^s(l \mid e) = \begin{cases} \binom{s}{l} 2^{-s} & \text{if } l \neq e \\ \sum_{j=e}^{s} \binom{j-1}{e-1} 2^{-j} & \text{if } l = e. \end{cases} \tag{1}$$

Shell. Protocol **Shell** runs a basic protocol R^k_{bas} on blocks of size k. First Alice and Bob divide their strings into k-bit long "primary" blocks. In pass 1 they apply R^k_{bas} followed by BICONF^1 on each of these blocks. If BICONF^1 detects an error, all bits from Bob's corresponding block are set to equal Alice's bits. In pass s ($s > 1$), Alice and Bob join pairs of adjacent blocks from the preceding pass to form the blocks for the current pass. On each of these new blocks they execute BICONF^1 s times. When an error is found, the primary block containing the erroneous bit is replaced by Alice's corresponding primary block. The process is repeated until there is only one block at the current pass (at pass $s = \lceil \log \frac{n}{k} \rceil$).

6.3 An Almost Ideal Protocol

Suppose that R^k_{bas} has an error probability $\delta_k \leq 1/2$ and leaked an amount I^k_{bas} of information such that $\frac{I^k_{bas}}{kh(p)} = \tau$. Executing BICONF^1 s times in pass s (i.e. on blocks of size $2^{s-1}k$) is equivalent to the execution of BICONF^s with respect to the distribution of the number of erroneous primary blocks that will be corrected. Therefore, if there are e erroneous primary blocks in a current block during pass s, the probability of correcting l of these is $\Delta^s(l|e)$ as defined by equation 1. Let $p'_s(e)$ be the probability of having e erroneous primary blocks in a current block after completion of pass s, and let $p_s(e)$ be the same probability before execution of pass s. It follows that

$$p'_s(e) = \sum_{j=e}^{e+s} p_s(j)\Delta^s(j - e \mid j)$$

with

$$p_s(e) = \sum_{j=0}^{e} p'_{s-1}(j)p'_{s-1}(e - j).$$

Moreover

$$p'_1(0) = 1 - \frac{\delta_k}{2}, p'_1(1) = \frac{\delta_k}{2}.$$

We can prove the following theorem by induction on s and e (see [Sa] for the proof):

Theorem 10. $(\forall \delta_k \leq \frac{1}{2})(\forall e > 0)(\forall s \geq 1) \left[p'_s(e) \leq \frac{1}{2^{s+e-1}} \right]$.

The failure probability is less than $\frac{k}{n}$ and tends to 0 as n increases. Hence the condition 1 from the definition of an almost ideal protocol is met. A bound on the amount of information $I_S(n)$ leaked by **Shell** can be obtained using theorem 10:

$$I_S(n) \leq nh(p)\tau + \delta_k n + \frac{4n(\lceil \log k \rceil + 2)}{k}$$

and we have that

$$\lim_{n \to \infty} \frac{I_S(n)}{nh(p)} \leq \tau + \frac{\delta_k}{h(p)} + \frac{4(2 + \lceil \log k \rceil)}{kh(p)}.$$

If protocol 1 is used as a basic protocol then there is a k for which $\frac{\delta_k}{h(p)} + \frac{4(2+\log k)}{kh(p)} \leq \varepsilon$ for any $\varepsilon > 0$. In addition the amount I^k_{bas} of leaked information can be chosen such that $\frac{I^k_{bas}}{kh(p)} \leq \tau$ for all $\tau > 1$. It is clear that **Shell** works in polynomial time for any fixed k.

Corollary 11. *If we use protocol 1 as a basic protocol then* **Shell** *is almost-ideal.*

The size of the blocks grows too fast as the amount of leaked information approaches the theoretical bound for **Shell** to be used together with protocol 1 in an efficient implementation. However it is possible to use **Shell** with other types of basic protocol, such as a systematic error-correcting code. For example, Alice and Bob can agree on a systematic (N, k')–code [4] of distance d. Afterwards they can compute the bound on the amount of leaked information produced by **Shell** on the primary blocks of size $k = tk'(t > 0)$, with $\tau = \frac{N-k'}{k'h(p)}$ and $\delta_k = 1-(1-\varepsilon)^t$, where ε is the probability that a k'-bit-long block holds more than $\lfloor \frac{d}{2} \rfloor$ errors. However **Shell** is no longer almost-ideal with this type of basic protocol.

7 A Practical Protocol

In section 7.1 we present protocol **Cascade**, which can easily be implemented. **Cascade** leaks an amount of information close to the theoretical bound on a $BSC(p)$ when p is as big as 15%. In section 7.2 a rough analysis shows how to choose protocol parameters for which the error probability decreases exponentially fast as a function of the number of passes. We also give a table comparing the amount of information leaked empirically by **Cascade** (given these parameters) to the theoretical bound.

7.1 Cascade Description

Cascade proceeds in several passes. The number of passes is determined by Alice and Bob before execution. This choice is related to parameter p. Let $A = A_1, \ldots, A_n$ and $B = B_1, \ldots, B_n$ (with $B_i, A_i \in \{0, 1\}$) be Alice's and Bob's strings respectively.

[4] An (N, k')–code has N–bit codewords to encode $2^{k'}$ messages.

In pass 1, Alice and Bob choose k_1 and divide their string into blocks of k_1 bits. The bits whose position is in $K_v^1 = \{l \mid (v-1)k_1 < l \leq vk_1\}$ form block v in pass 1. Alice sends the parities of all her blocks to Bob. Using **BINARY** Bob corrects an error in each block whose parity differs from that of Alice's corresponding block. At this point all of Bob's blocks have an even number of errors (possibly zero). This part of the protocol is taken from [BBBSS]. However, in that paper the leaked information about the secret string is eliminated during execution by removing one bit of each subset for which the parity is known. In our protocol all the bits are kept. Saving this information from pass to pass allows us to correct more errors.

At each pass $i > 1$, Alice and Bob choose k_i and a random function $f_i : [1..n] \to [1..\lceil \frac{n}{k_i} \rceil]$. The bits whose position is in $K_j^i = \{l \mid f_i(l) = j\}$ form block j in pass i. Alice sends Bob

$$a_j = \bigoplus_{l \in K_j^i} A_l$$

for each $1 \leq j \leq \lceil \frac{n}{k_i} \rceil$. Bob computes his b_j's in the same way and compares them with the a_j's. For each $b_j \neq a_j$ Alice and Bob execute **BINARY** on the block defined by K_j^i. Bob will find $l \in K_j^i$ such that $B_l \neq A_l$ and correct it. All the blocks K_v^u for $1 \leq u < i$ such that $l \in K_v^u$ will then have an odd number of errors. Let $\mathcal{K}$ be the set of these blocks. Alice and Bob can now choose the smallest blocks in $\mathcal{K}$ and use **BINARY** to find another error. Let l' be the position of this error in strings A and B. After correcting $B_{l'}$, Bob can determine set $\mathcal{B}$ formed by the blocks containing $B_{l'}$ from each pass from 1 to pass i. He can also determine the set $\mathcal{K}'$ of blocks with an odd number of errors by computing

$$\mathcal{K}' = \mathcal{B} \nabla \mathcal{K}.[5]$$

If $\mathcal{K}' \neq \emptyset$ then Bob finds another pair of errors in the same way. This process is repeated until there are no more blocks with an odd number of errors, at which point pass i ends, and each block in passes 1 through i has an even number of errors (perhaps zero).

7.2 Using Cascade

In this section a simple analysis using only one of **Cascade**'s properties shows its usefulness in practice. This analysis yields a particular choice of block size such that the probability that a block K_v^1 has one or more errors decreases exponentially with respect to the number of passes.

The property we use is that in the passes following pass 1, correcting an error in K_v^1 implies that a second one from the same block K_v^1 will be corrected.

For parameters $k_1, \ldots, k_\omega$ chosen in a manner that depends on p, we will try to determine $\delta_i(j)$, the probability that after the pass $i \geq 1$, $2j$ errors remain in K_v^1. $\delta_1(j)$ is easily determined for $X \approx Bin(k_1, p)$:

$$\delta_1(j) = prob(X = 2j) + prob(X = 2j+1)$$

[5] $\mathcal{B} \nabla \mathcal{K} = (\mathcal{B} \cup \mathcal{K}) \setminus (\mathcal{B} \cap \mathcal{K})$

Let E_i be the expected number of errors in K_v^1 after completion of pass i. For pass 1 we have:

$$E_1 = 2\sum_{j=1}^{\lfloor \frac{k_1}{2} \rfloor} j\delta_1(j) = k_1 p - \frac{(1-(1-2p)^{k_1})}{2}.$$

If the functions f_i with $i > 1$ are randomly chosen from $\{f \mid f : [1,\ldots,n] \to [1,\ldots,\frac{n}{k_i}]$ then for $n \to \infty$, we can determine a bound on the probability γ_i of correcting at least 2 errors at pass $i > 1$ in a block K_v^1 still containing errors after completion of pass $i-1$. Since errors are corrected two by two in passes $i > 1$, we have

$$\gamma_i \geq 1 - \left(1 - (1-\frac{k_i}{n})^{\frac{nE_{i-1}}{k_1}}\right)^2$$
$$\approx 1 - \left(1 - e^{-\frac{k_i E_{i-1}}{k_1}}\right)^2$$

We can bound $\delta_i(j)$ using γ_i for $i > 1$

$$\delta_i(j) \leq \left(\sum_{l=j+1}^{\lfloor \frac{k_1}{2} \rfloor} \delta_{i-1}(l)\right) + \delta_{i-1}(j)(1-\gamma_i).$$

Suppose that k_1 is chosen such that

$$\sum_{l=j+1}^{\lfloor \frac{k_1}{2} \rfloor} \delta_1(l) \leq \tfrac{1}{4}\delta_1(j) \tag{2}$$

and let $k_i = 2k_{i-1}$ for $i > 1$. We have

$$\delta_i(j) \leq \tfrac{1}{4}\delta_{i-1}(j) + (1 - e^{-2^{i-1}E_{i-1}})^2 \delta_{i-1}(j).$$

If in addition the choice of k_1 is such that

$$E_1 \leq -\frac{\ln \frac{1}{2}}{2} \tag{3}$$

it follows that

$$\gamma_i \geq 1 - (1 - e^{-2E_1})^2 \geq \tfrac{3}{4}.$$

When $k_i = 2k_{i-1}, i > 1$ and k_1 satisfies 2 and 3, we have $\delta_i(j) \leq \frac{\delta_{i-1}(j)}{2} \leq \frac{\delta_1(j)}{2^{i-1}}$ since $E_i \leq \frac{E_{i-1}}{2}$.

We can bound the amount of information $I(\omega)$ per block of length k_1 (per block K_v^1) leaked after ω passes, with parameters k_i set as above (ω must not depend on n for the argument to apply), as follows:

$$I(\omega) \leq 2 + \frac{1-(1-2p)^{k_1}}{2}\lceil \log k_1 \rceil + 2\sum_{l=2}^{\omega}\sum_{j=1}^{\lfloor \frac{k_1}{2} \rfloor} \frac{j\delta_1(j)}{2^{l-1}}\lceil \log k_1 \rceil .$$

To completely eliminate the errors, we choose ω large enough that we can use **Shell** on blocks formed by concatenating a large number of blocks K_v^1. Thus by corollary 11 the total amount of leaked information will approach that of **Cascade**.

Table 1 gives the values of k_1(the largest one satisfying 2 and 3) for $p \in \{0.15, 0.10, 0.05, 0.01\}$ and the values of $I(4)$ are computed. In addition the average amount of leaked information $\widehat{I}(4)$ for 10 empirical tests (with $n = 10,000$) under the same conditions is reported. For each of these tests all errors were corrected after pass 4.

Table 1. Cascade benchmark

p	k_1	$\widehat{I}(4)$	$kh(p)$	$I(4)$
0.01	73	6.47	5.89	6.81
0.05	14	4.60	4.01	4.64
0.10	7	3.81	3.28	3.99
0.15	5	3.80	3.05	4.12

8 Conclusions

The reconciliation problem is a variant of the noisy coding problem. The extension of the noisy coding theorem [Sh] due to Elias [E] shows that there exist optimal linear codes. Thus, it is not surprising that there exist optimal reconciliation protocols. One must use the systematic version of an optimal linear code to obtain an optimal reconciliation protocol. While these results from information theory are non-constructive, all our results are constructive.

Theorem 8 gives an (unlikely) hypothesis for which non-interactive ideal reconciliation schemes exist. If we consider other classes of hash functions, it is possible to obtain ideal protocols based on weaker hypotheses. High performance non-interactive reconciliation protocols would be useful for efficient implementation of quantum oblivious transfer [BBCS].

From a practical point of view, **Cascade** is an efficient protocol that leaks less information than the best error-correcting-codes-based reconciliation protocols. It is an improvement on the protoocol used in [BBBSS] in a true quantum setting. It would be of interest to have a detailed analysis of **Cascade**'s performance that would tell how to choose the parameters so as to minimize the amount of leaked information while maintaining a low failure probability.

9 Acknowledgements

The authors wish to thank Pierre McKenzie, Hiroki Shizuya and Claude Crépeau for helpful discussions. Special thanks to François Bessette and Daniel Simon for their suggestions and for their great help in redaction.

References

[BBBSS] C. H. Bennett, F. Bessette, G. Brassard, L. Salvail, J. Smolin, *Experimental Quantum Cryptography*, Journal of Cryptology, Vol. 5, No. 1, 1992, pp. 3–28.

[BBR] C. H. Bennett, G. Brassard, J.-M. Robert, *Privacy Amplification by Public Discussion*, SIAM Journal on Computing, Vol. 17, No. 2, 1988, pp. 210–229.

[BBCS] C. H. Bennett, G. Brassard, C. Crépeau, M.-H. Skubiszewska, *Practical Quantum Oblivious Transfer*, In proceedings of Crypto '91, Lecture Notes in Computer Science, vol 576, Springer Verlag, Berlin, 1992, pp. 351-366.

[BMT] E. R. Berlekamp, R. J. McEliece, H. C. A. van Tilborg, *On the Inherent Intractability of Certain Coding Problems*, IEEE Transaction on Information Theory, Vol. IT-24, No. 3, 1978, pp. 384–386.

[CW] J. L. Carter, M. N. Wegman, *Universal Classes of Hash Functions*, Journal of Computer and System Sciences, Vol. 18, 1979, pp. 143–154.

[E] P. Elias, *Coding for Noisy Channels*, IRE Convention Record, 1957, pp. 46–47.

[M] U. M. Maurer, *Perfect Cryptographic Security from Partially Independent Channels*, In proceedings of 23*rd* Symposium on Theory of Computing, 1991, pp. 561–571.

[Sh] C. E. Shannon, *A Mathematical Theory of Communication (Part I)*, Bell System Technical Journal, Vol. 27, 1948, pp. 379–423.

[Sa] L. Salvail, *Le Problème de Réconciliation en Cryptographie*, Master thesis, Département d'informatique et de recherche opérationnelle, Université de Montréal, 1991.

[R] J.-M. Robert, *Detection et Correction d'Erreurs en Cryptographie*, Master thesis, Département d'informatique et de recherche opérationnelle, Université de Montréal, 1985.

[WC] M. N. Wegman, J. L. Carter, *New Hash Functions and Their Use in Authentication and Set Equality*, Journal of Computer and System Sciences, Vol. 22, 1981, pp. 265–279.

[W] D. Welsh, *Codes and Cryptography*, Oxford Science Publications, 1989.

[Z] S. Zachos, *Probabilistic Quantifiers Games*, Journal of Computer and System Sciences, Vol. 36, 1988, pp. 433–451.

Global, Unpredictable Bit Generation Without Broadcast

Donald Beaver *	Nicol So †
Penn State University	Penn State University

Abstract

We investigate the problem of generating a global, unpredictable coin in a distributed system. A fast, efficient solution is of fundamental importance to distributed protocols, especially those that rely on broadcast channels. We present two unpredictable bit generators, based on the Blum-Blum-Shub generator, that can be evaluated non-interactively; that is, each bit (or group of bits) requires each processor merely to send one message to the other processors, without requiring a broadcast or Byzantine Agreement.

The unpredictability of our generators (and the security of our protocols) are based provably on the QRA or the intractability of factoring. Remarkably, their structure seems to violate an impossibility result of [8], but our generators escape that lower bound because they achieve a slightly weaker goal: producing unpredictable bits directly, rather than producing "shares" of random bits. In doing so, they avoid the extra machinery (eg., "sharing shares") of similar results discovered independently in [8].

1 Introduction

Randomness has a variety of purposes in cryptography and computer science:

- **avoiding exhaustive search**: *eg.* finding a witness that a number is composite;
- **circumventing worst-case analysis**: *eg.* choosing a random pivot in Quicksort;
- **breaking symmetries**: *eg.* choosing a leader in a ring of processors;
- **hiding information**: *eg.* one-time pads;
- **measuring information**: *eg.* indistinguishability and Turing-like tests;
- **unpredictability**: *eg.* defeating an adversary's committed attack through unpredictable future choices.

For these and other reasons, a great deal of attention has been focused on ways to expand short (and scarce) random strings into long pseudorandom sequences.

For many applications, such as Byzantine Agreement [13, 4, 16] – the problem of agreeing on a common value in a network with unreliable nodes – randomness itself is simultaneously not enough and too much. Secret sharing, multiparty protocols, reliable decentralized databases, multicasts, and even timestamping protocols can depend very strongly on agreed-

*317 Pond Laboratory, Penn State University, University Park, PA 16802; (814) 863-0147; beaver@cs.psu.edu.

†Computer Science Dept., Pond Laboratory, Penn State University, University Park, PA 16802; so@cs.psu.edu.

upon information in a system. Thus, in addition to making efficient use of scarce random sources, we must also be worried about agreeing about the bits and protecting them from manipulation.

On the other hand, for Byzantine Agreement, the backbone of decentralized network protocols, it is not *randomness* but merely *unpredictability* that suffices to prevent an adversary's attack from having more than 50-50 chance of success. There are ingenious solutions [4, 16] requiring only two expected rounds of communication if a common but unpredictable coin is available.

In this paper, we show how to generate unpredictable bits efficiently and in one round of communication. In fact, this one round of communication is very simple: it requires each processor to send the same message to all the other processors, but it does not require the system to check whether each processor did so. We call this weak form of broadcast "*dissemination*," and we show how to achieve an unpredictable coin using one round of dissemination.

We present two solutions, one based on the QRA[1] and one based on the intractability of factoring. (The first solution is more efficient, but basing it on the weaker assumption (factoring) is not an obvious task.) Each solution uses a homomorphic technique for secret sharing in the style of Feldman [9]. (Caution is required, however; applying Feldman's VSS scheme as a "subroutine" leads to an easily predictable generator.) In a nutshell, it applies a generalized form of the Blum-Blum-Shub (BBS) pseudorandom number generator [7] to a hidden seed, and reveals the bits in a coordinated manner: no bit is revealed until some $n - t$ processors decide it is time to do so, according to their programs.

In [8], Cerecedo, Matsumoto and Imai have independently developed a method to generate secretly-shared bits using "no interaction." (Informally, a value is "secretly shared" if any collection of $n - t$ processors can determine it while any t or fewer cannot, where t is a bound on faulty processors.) Their method seems related to ours, but it requires more expensive computation to generate and verify the shares of the bits, as well as an order of magnitude increase in message size.

Furthermore, because our construction is based on the BBS generator, it is of a type that an impossibility result in their work excludes. It escapes impossibility partly because it attempts to achieve a weaker goal (obtain bits, not "shares of bits"). This weaker goal is sufficient to provide a very efficient implementation of Byzantine Agreement as well as any other protocol that relies on common, unpredictable bits to overcome an adversary.

2 Unpredictable Random Number Generators

This section provides background and technical lemmas for the main results described in §4.

We depart somewhat from the standard treatment of so-called cryptographically-strong pseudorandom number generators, in order to consider situations in which more than just the number stream itself becomes public. In particular, when a seed is revealed, it is clear that a pseudorandom bit stream no longer is indistinguishable from a uniformly random bit stream. It might be useful to permit a seed to be revealed at some later point (although, from the point of cryptographers bent on protecting communications, this seems to have no immediate use). In this case, we can no longer employ a standard treatment of pseudorandomness, but must turn to something new.

---

[1]The Quadratic Residuosity Assumption (QRA) states that no poly-size circuit family can determine whether x is a square modulo n with probability greater than $1/2 + k^{-c}$ – asymptotically, for any $c > 0$, for randomly chosen x with Jacobi symbol $+1$, and for n a randomly chosen Blum integer of size k.

The key property that we intend to capture is that the bits are unpredictable – namely, an adversary has little better than a 50-50 chance to predict the next bit, although once it is generated, the sequence may be clearly distinguishable from uniformly random bits because other (useless) information is present.

The Blum-Blum-Shub [7] generator provides a nice example. Briefly, it is based on selecting a random residue x modulo a Blum integer n, repeatedly squaring x, and outputting the least-significant bit of each successive value in reverse order ($\text{LSB}(x^{2^B}), \ldots, \text{LSB}(x^{2^1})$). The bits themselves form a sequence indistinguishable from uniformly random bits [7, 1], but if the powers of x are revealed, even at some later point when the bits can be "permissibly" compromised, the whole view is clearly identifiable as the output of a generator (most truly random sequences would have no corresponding x and n which give rise to them).

This approach to randomness is suitable for problems such as Byzantine Agreement which require coin flips that are unpredictable, but the machinery used for generating the coins can be revealed afterward without harm. That is, the bits are used for algorithmic purposes rather than privacy. They should have the properties of a random sequence (50-50 chance of 0 or 1), and they should not be predictable, but they may be distinctly identifiable *after the fact* as the output of a generator.

Definition 1 *An* **Unpredictable Random Number generator (URN)** *G is a triple $(\hat{G}, b, B)$, where $\hat{G}$ and b are poly-time algorithms and $B(k)$ is a polynomial, and such that on input k, integer i between 1 and $B(k)$, and a string x of length k, $\hat{G}$ outputs a string $\hat{G}(k,i,x)$, and $b(\hat{G}(k,i,x))$ outputs a bit. For initialization purposes, G may also output a string $\hat{G}(k,0,x)$.*

We call $b(\hat{G}(k,i,x))$ the ith bit of G and we also write it as $G(k,i,x)$.

We define a poly-strong ("cryptographically-strong") URN as follows. Let P be a predictor, namely a poly-size circuit family that takes an input i and a string v_i (representing the output of the generator at the ith step and the "revealable" information, *eg.* successive squares) and outputs a bit p_{i+1}. Let D be a distinguisher, namely a family of circuits that outputs 0 or 1.

Briefly, P attempts to predict the next bit at each stage; its guesses are compared to the answers, and the grades it gets ("right" *vs.* "wrong," as opposed to the answers themselves) should be pseudorandom.

Definition 2 *The* **prediction pattern** *of predictor P with respect to an URN G is the sequence of bits obtained as follows. Run P on $\hat{G}(k,0,x)$, $\hat{G}(k,0,x) \circ \hat{G}(k,1,x)$, $\hat{G}(k,0,x) \circ \hat{G}(k,1,x) \circ \hat{G}(k,2,x)$,* etc., *obtaining the* **predicted bits** *$a_1, a_2, \ldots, a_B$; The prediction pattern is the sequence:*

$$\langle P, G\rangle(k) = (p_1, p_2, \ldots, p_{B(k)}) = (a_1 \oplus G(k,1,x), a_2 \oplus G(k,2,x), \ldots, a_{B(k)} \oplus G(k, B(k), x)).$$

Let UNIF denote the uniform distribution on $\{0,1\}$.

Definition 3 *An URN is* **poly-strong** *if for all poly-size predictors P and distinguishers D, for all $c > 0$,*

$$\left| D(\langle P, G\rangle(k)) - D(\text{UNIF}^{B(k)}) \right| < \frac{1}{k^c},$$

where $D(Z(k))$ indicates the probability that D outputs 0 on strings sampled according to distribution $Z(k)$.

In the sequel, we often omit (k) for the sake of clarity.

An URN is stronger than a CSPRG, as is easily seen in the following:

Theorem 1 *If $(\hat{G}, b, B)$ is a poly-strong URN, then the bits output by G (ie. $G(k,1,x), \ldots, G(k,B,x)$, not the strings given by $\hat{G}$) form a cryptographically-strong pseudorandom sequence.*

Proof. If this sequence of bits fails a next-bit test, then the next-bit test can be used directly as an equally successful predictor against $(\hat{G}, b, B)$: given $\hat{G}(k,0,x), \ldots, \hat{G}(k,i,x)$, simply feed $b(\hat{G}(k,1,x)), \ldots, b(\hat{G}(k,i,x))$ to the next-bit test. □

2.1 A Poly-Strong URN Based on Blum-Blum-Shub

For purposes that will be clear in §4, we show that a generalized form of the BBS random bit generator provides an unpredictable number generator of a form suitable for distributed evaluation. Rather than repeatedly squaring, we repeatedly raise to a power that is itself a fixed power of two. Roughly speaking, we start with a slightly modified seed and follow the BBS approach, skipping over many of the bits that the BBS generator gives. We output these bits in reverse order.

Specifically, given $B(k)$, k, an integer n chosen randomly from BLUM$_k$, and poly-time computable functions $\alpha(k)$ and $\beta(k)$ with $\beta(k)$ odd, we use arithmetic over $\mathbf{Z}_n^*$ to define the URN $G_{\alpha,\beta}$ as follows (omitting k's for clarity):

$$\hat{G}_{\alpha,\beta}(k,i,x) = x^{2^{(B-i)\alpha}}$$

$$b(y) = \text{LSB}(y^{2^{\alpha-1}\beta})$$

summarized by

$$G_{\alpha,\beta}(k,i,x) = \text{LSB}(x^{2^{(B-i)\alpha+\alpha-1}\beta}).$$

The BBS generator is the special case of $\alpha(k) = \beta(k) = 1$. Like the BBS generator, $G_{\alpha,\beta}$ provides a poly-strong pseudorandom sequence of bits, under the QRA. Our unsurprising but necessary goal in this section is to show that, even knowing the bits and $x^{2^{(B-i)\alpha}}$ values used so far, a predictor cannot predict the next bit to come.

Lemma 2 *$G_{\alpha,\beta}$ is a poly-strong URN unless the QRA fails.*

Proof. We first show that a successful predictor would provide a way to compute LSB(z^β) from z^2 (compare computing LSB(z) from z^2 in the BBS case). Next, we show how to determine quadratic residuosity using the ability to compute LSB(z^β) from z^2.

First, however, assume by way of contradiction that $G_{\alpha,\beta}$ is not a crypto-strong URN. Then there exists a $c > 0$, a distinguisher D, and a predictor P such that

$$\left|\Pr[D(\langle P, G_{\alpha,\beta}\rangle(k)) = 0] - \Pr\left[D(\text{UNIF}^{B(k)}) = 0\right]\right| \geq \frac{1}{k^c}$$

for infinitely-many k. We omit the absolute value symbols without loss of generality.

Define a hybrid distribution by taking i bits of $\langle P, G_{\alpha,\beta}\rangle$, concatenating $B - i$ uniformly random bits to the end, and then running D on the result. The probability that D outputs 0 on the result is denoted by:

$$D_i = \Pr\left[D(\langle P, G_{\alpha,\beta}\rangle[1..i] \circ \text{UNIF}^{B-i}) = 0\right].$$

Now, given some y, we wish to compute LSB(z^β) where $z^2 = y$ and z is the principal square root of y. On input y, choose i from $\{1, \ldots, B\}$ at random, and run P on

$$y^{2^{(i-1)\alpha}}, y^{2^{(i-2)\alpha}}, \ldots, y.$$

P will output i guesses, $a_1, \ldots, a_i$; let $p_1, \ldots, p_{i-1}$ be the exclusive-or's with the known "correct" values (that is, $p_j = a_j \oplus \text{LSB}(y^{2^{(i-1-j)\alpha}\beta})$). Now, choose bits $p_i, \ldots, p_B$ uniformly at random. Run D on $(p_1, \ldots, p_B)$, and output $D \oplus p_i \oplus a_i$. (Intuitively, a 0 from D is more likely with distribution D_i than with D_{i-1}, which indicates that $p_i = a_i \oplus \text{LSB}(z^\beta)$.) Through a fairly standard argument, the probability that this method gives the correct answer for $\text{LSB}(z^\beta)$ is at least

$$\Pr[\text{correct}] \geq \sum_{i=1}^{B} \frac{1}{B} \cdot \left(\frac{1}{2} + D_i - D_{i-1} - o\left(\frac{1}{k^c}\right)\right)$$

In other words, for infinitely many k and for some $c_1 > 0$, we have:

$$\Pr[\text{correct}] \geq \frac{1}{2} + D_B - D_0 - o\left(\frac{1}{k^c}\right) \geq \frac{1}{2} + \frac{1}{k^{c_1}}.$$

Finally, we must show how to turn this non-negligible advantage into a method to determine residuosity. Following [7], on input x, set $z = x^2$ and let w denote the square root of z that is a quadratic residue. Thus, $x = \pm w$ (else x has Jacobi symbol -1, and clearly is not a quadratic residue). Calculate $\text{LSB}(x^\beta)$ and compare it to the guess for $\text{LSB}(w^\beta)$ obtained by running the algorithm described above on z. If the bits are the same, then output "$x \in \text{QR}_n$," else output "$x \notin \text{QR}_n$." Observing that $x = -w \Leftrightarrow x^\beta = -w^\beta$ because β is odd, and that $\text{LSB}(-w^\beta) \neq \text{LSB}(w^\beta)$, we see:

$$x = w \Leftrightarrow \text{LSB}(x^\beta) = \text{LSB}(w^\beta),$$

so we obtain a correct answer with probability exceeding $\frac{1}{2} + \frac{1}{k^{c_1}}$, infinitely often.

Although at first glance the techniques of [1] might seem adaptable to show that we could actually factor n, their argument does not apply directly to the case at hand (with $\text{LSB}(z^\beta)$ rather than $\text{LSB}(z)$), and a fix is not obvious. □

3 Unpredictable Global Bits in Distributed Systems

We turn to the problem of generating global coin flips in a distributed system. This section defines the problem and gives a primitive solution; our solution is found in §4.

Consider a network of N processors ("players"), connected by private channels but lacking a broadcast channel. For simplicity, we assume that a trusted host is available to initialize the network. Both of these assumptions can be weakened by using encryption and initial secret computation protocols, but we leave this to another level of analysis.

At the end of each of B *phases*, each processor should receive a random bit – and all processors should receive the same bit. (We omit the goal of generating more than one bit per phase, as it is a simple modification to our analysis.)

If a trusted host (incorruptible processor) were available, the direct solution would be for it to flip a coin during each phase and simply send the result to each processor. A slightly less direct solution might be for the trusted host to generate a pseudorandom sequence from an initial seed x, and then reveal these bits to the network, one by one. Or it might just send out a string during each phase, from which the current bit can be calculated. The important property is that any attempt to predict the bits will fail.

Definition 4 *An* **ideal URN protocol** *for URN G is a protocol that operates in phases. On input N (network size) and k a trusted host TH selects a k-bit seed at random, and evaluates a poly-strong URN, G. At the end of phases 0 through $B(k)$, TH sends $\hat{G}(k,i,x)$ to all players, who compute their local output, $b(\hat{G}(k,i,x))$.*

Trusted hosts are at best available only for initialization, if at all. Thus we consider networks in which no trusted host is available. If a protocol can be reduced to an ideal URN protocol, via security-preserving reductions such as "relative resilience" or related notions [2, 12], we shall say it is a *global URN protocol.* We quickly outline the formalization.

A protocol α is *as resilient as* a protocol β if there exists an interface $\mathcal{I}$ such that for any adversary $\mathcal{A}$ attacking α, there is an attack by $\mathcal{I}(\mathcal{A})$ on β yielding "equivalent" results. In particular, the vector of random variables describing $\mathcal{A}$'s view and the outputs of nonfaulty players should be indistinguishable (in some sense) in the two scenarios. We refer the reader to [2, 12] for details.

Definition 5 *A* **global bit protocol** *or* **global URN protocol** *is a protocol that is computationally as resilient as an ideal URN protocol for some URN G.*

Definition 6 *A global URN protocol is* **non-interactive** *if each phase (apart from initialization) requires each processor to send at most one message to each other processor (order-independent) per phase.*

Definition 7 *A player is said to* **disseminate** *message m in round ρ if it sends m to all other players in round ρ.*

3.1 A Simple Solution For Constant t

A very simple solution exists when at most a constant number of faults can occur. Interestingly, this solution permits random access to the bits (*ie.* they can be revealed in any order), but nevertheless violates the impossibility result of [8], apparently because the bits are not represented as "shares" in the particular form considered in that work.

For clarity, consider $N = 3$ and $t = 1$, and consider only omission-failures (*ie.* no player sends bad messages, but messages can be dropped). Let G be any pseudorandom bit generator (whether in the sense of the CSPRG's of [6] or an URN as defined above).

To initialize the system, a trusted host generates seeds s_1, s_2, s_3 randomly and gives (s_1, s_2) to player 1, (s_2, s_3) to player 2, and (s_3, s_1) to player 3.

Bit b_j is defined as $G(k, s_1, j) \oplus G(k, s_2, j) \oplus G(k, s_3, j)$.

To generate bit b_j, player 1 disseminates $(G(k, s_1, j), G(k, s_2, j))$, player 2 disseminates $(G(k, s_2, j), G(k, s_3, j))$, and player 3 disseminates $(G(k, s_3, j), G(k, s_1, j))$. Because $t = 1$, even if one pair is omitted, all three values $G(k, s_1, j), G(k, s_2, j), G(k, s_3, j)$ are present. Each player calculates b_j from the appropriate exclusive-or.

Clearly, any $t = 1$ or fewer players cannot predict b_j. It is not hard to see that this technique generalizes for any constant minority t: each $(N - t)$-subset σ is given a seed s_σ from which each member can calculate $G(k, s_\sigma, j)$. Because $N > 2t$, for each σ there is always a nonfaulty player who disseminates $G(k, s_\sigma, j)$ at the appropriate time. On the other hand, for any t-subset τ there is always a σ that excludes τ, so that curious players cannot predict the bit until the appropriate time.

Since t is constant, this protocol uses polynomial time and messages. It uses only dissemination at each round and is thus "non-interactive."

4 Main Result: Fast, Non-interactive Global Bit Generation

Our distributed bit generator is based on the URN $G_{\alpha,\beta}$ described in §2.1, with α and β defined by $2^\alpha \beta = 4(N!)^2$. Based on a secret seed x, the N-player network computes and reveals the following numbers, phase by phase:

$$x^{2^{(B-1)\alpha} \cdot 2(N!)^2}, x^{2^{(B-2)\alpha} \cdot 2(N!)^2}, \ldots, x^{2^{0 \cdot \alpha} \cdot 2(N!)^2},$$

The parities of these numbers are the desired, unpredictable bits. In fact, the parities of these numbers are the bits output by the URN $G_{\alpha,\beta}$. (The strings output by the URN are predecessors (roots) of these values, but the bit extraction function ultimately takes the LSB's of the values shown here.)

In an ideal protocol, a trusted host could broadcast these numbers (or equivalently, the strings given by $G_{\alpha,\beta}$) phase by phase. Without a trusted host, we focus on revealing and agreeing upon each number using simple, non-interactive dissemination (*ie.* in one round without broadcast channels).

For simplicity, let us assume the initialization is taken care of either by a trusted party or an initial multiparty protocol requiring broadcast. Thereafter, neither trust nor broadcast is needed.

Initialization(N, t, k, B)
// N: number of players; $t < N/2$: fault tolerance bound;
// $k = \Omega(N \log N)$: security parameter; B: number of bits.
  Let Q,M be such that $2^Q M = N!$ and M is odd.
  $W \leftarrow 2Q + 2$ // $\alpha = W, \beta = M^2$
  $n \leftarrow \text{RANDOM}(\text{BLUM}_k)$
  $g \leftarrow \text{RANDOM}(\mathbf{Z}_n^*)$ // see footnote[a]
  $(a, a_1, a_2, \ldots, a_t) \leftarrow \text{RANDOM}(\mathbf{Z}_{\phi(n)}^*)^{t+1}$
  Let $f(u) = a + a_1 u + a_2 u^2 + \cdots a_t u^t$
  for $i = 1..N$ **do**
    $y_i \leftarrow g^{f(i)N!}$ // represents $X = x^{N!} = g^{f(0)N!} = (g^a)^{N!}$
    $z_i \leftarrow y_i^{2^{BW}}$
  for $i = 1..N$ **do**
    give $(n, y_i, \langle z_1, z_2, \ldots, z_N \rangle)$ to player i.

[a]The values n and g should be chosen so that the powers of g span a polynomial fraction of $\mathbf{Z}_n^*$, but we omit such considerations for the sake of presentation.

After initialization, each party holds a "share" y_i of a secret value $X = g^{N!a} = x^{N!}$, similar to Feldman's VSS scheme [9]. Any other piece y_j can be "checked" by raising it to an appropriate power and comparing the result to z_j, thus avoiding the need to share the shares for verification purposes.

More to the point, because this scheme is homomorphic, each player i can generate a share of X raised to any desired power, simply by raising y_i to that power. Thus, without interaction, each player can generate shares of $X^{2^{(B-1)W}}, \ldots, x^{2^0}$, simply by computing $y_i^{2^{(B-1)W}}, \ldots, y_i^{2^0}$.

At each phase, each player disseminates its share of the current power of X to be revealed. Unfortunately, it is not clear how to interpolate X (or a power of it) directly from the pieces, since we do not know how to take arbitrary roots. Instead, $X^{2(N!)}$ (and its various powers) are computed. The $N!$ arises from making sure that a unique interpolation can be done over the integers without taking roots (*ie.* division in the exponents). The factor of 2 arises from squaring the revealed pieces to prevent malicious players from disseminating $-y$ instead of y (since the pieces are verified by repeated squaring, an adversary cannot successfully reveal a valid piece other than $\pm y$).

For a (correct) subset $\gamma \subseteq \{1, 2, \ldots, N\}$, define the following functions over $\mathbf{Z}$:

$$L_{i,\gamma}(u) = N! \prod_{j \in \gamma - \{i\}} \frac{u - j}{i - j}.$$

(A straightforward argument shows that each coefficient in $L_{i,\gamma}$ is not only integral but divisible by $i\binom{n}{i}$.) If $f(u)$ is a polynomial of degree t, then as long as γ contains $t+1$ or more elements, we can "interpolate" a unique value:[2]

$$N!f(0) = \sum_{i \in \gamma} f(i) L_{i,\gamma}(0).$$

> Protocol URNPROTO – code for player i
> **Player**(i, N, k, B)
> // i: id; N: number of players; k: security parameter; B: number of bits.
> Receive $(n, y_i, \langle z_1, z_2, \ldots, z_N \rangle)$ at startup.
> $\langle v_1, v_2, \ldots, v_N \rangle \leftarrow \langle z_1, z_2, \ldots, z_N \rangle$
> $\gamma = \{1, 2, \ldots, N\}$ // Correct players.
> **for** phase $j = 1..B$ **do**
> { Generate bit b_j }
> Disseminate $y_i^{2^{(B-j)W}}$
> Receive $(w_1, w_2, \ldots, w_N)$ // Apparent pieces
> // Verify pieces against known, higher squares
> $\gamma \leftarrow \{l \in \gamma \mid w_l^{2^W} \equiv v_l\}$
> $(v_1, v_2, \ldots, v_N) \leftarrow (w_1, w_2, \ldots, w_N)$
> // Interpolate j^{th} value $X^{2^{(B-j)W} 2N!}$
> // using $(w_l)^2$ values to avoid malicious negation.
> $x(j) \leftarrow \prod_{l \in \gamma} w_l^{2L_{i,\gamma}(0)}$
> $b_j \leftarrow \text{LSB}(x(j))$

Because an adversary cannot substitute false pieces, the interpolation of $x(j) = X^{2^{(B-j)W} 2N!}$ gives the LSB produced by the URN:

$$\begin{aligned} x(j) &= \prod_{l \in \gamma} w_l^{2L_{l,\gamma}(0)} = g^{\sum_{l \in \gamma} 2L_{l,\gamma}(0)(N!f(l)) 2^{(B-j)W}} = (X^{2(N!)})^{2^{(B-j)W}} \\ &= x^{2^{(B-j)W} 2(N!)^2} = x^{2^{(B-j)W+W-1} M^2}. \end{aligned}$$

4.0.1 An Optimization

Because the bits are output in reverse order, the brute-force computation of the powers of each y_i would require $\Theta(B^2)$ repeated-squarings, or B per bit. Alternatively, storing the intermediate results would use $\Theta(B)$ space – a factor whose disadvantages themselves motivate using pseudorandom number generators rather than "one-time pads."

In the full paper, we present a simple technique we call "signposting," which uses a total of $\Theta(B \log B)$ repeated-squarings, or $\log B$ repeated-squarings per bit, but $\Theta(\log B)$ space overall. This algorithmic optimization is merely a faster local computation; it does not change the protocol in any way.

4.0.2 A Second Solution

At the cost of some extra exponentiation, we can base our results on the weaker assumption that factoring is intractable. Letting $x = g^a$, observe that $G_{\alpha,\beta}$ outputs numbers of the form

$$x(j) = x^{2^{(B-j)W+W-1} M^2}.$$

[2] The value we obtain is unique regardless of γ; but it is $N!f(0)$, *not* $f(0)$.

Because of the odd factor M^2 in the exponent, we have thus far only been able to base unpredictability on the QRA. If, instead, we use numbers of the form

$$v(j) = x^{2^{(B-j)W+W-1}},$$

namely if we omit the odd factors in the exponent, then there is a straightforward modification of the arguments of §2.1 and [1] to show that predicting the LSB's is as hard as factoring. Let us call this second generator $V_{\alpha,\beta}$.

The problem is that the polynomial interpolation used in the secret sharing introduces an $N!$ factor in the exponent, apparently making the odd power unavoidable. But let us say that the number $u(j-1) = x^{2^{(B-(j-1))W}}$ is available – this next higher "power of two" already played a role in the bits revealed earlier. Then $v(j)$ can be calculated from $x(j)$ and $u(j-1)$:

$$v(j) = x(j)/u(j-1)^{\frac{M^2-1}{2}}$$

as is demonstrated by the following:

$$x(j)/u(j-1)^{\frac{M^2-1}{2}} = x^{2^{(B-j)W+W-1}M^2-(2^{(B-(j-1))W})\frac{M^2-1}{2}} = x^{2^{(B-j)W+W-1}}.$$

Thus, at additional cost, we can in fact calculate a "pure" power-of-two exponent of the original seed. The cost is one division and an exponentiation to the power $\frac{M^2-1}{2}$, where M is the odd part of $N!$. If we make $u(0) = x^{2^{BW}}$ available at the start, subsequent $u(j)$'s can be derived using the same trick as shown above.

4.1 Proof of Security

Theorem 3 *Protocol* URNPROTO *is a non-interactive global bit protocol, unless the QRA fails.*

Proof. Non-interaction holds by definition. We show a security reduction from URNPROTO to an ideal URN protocol (call it THURN) in which the trusted host broadcasts the sequence generated by the URN $G_{\alpha,\beta}$ of §2.1. For clarity, we consider only the case of a static adversary (the set of faulty players is chosen in advance). Given an adversary $\mathcal{A}$, we must show how an interface $\mathcal{I}$ maps $\mathcal{A}$'s attacks on URNPROTO to attacks on THURN. The main job of $\mathcal{I}$ is to construct initial $(y_i, \langle z_1, z_2, \ldots, z_N \rangle)$ vectors for corrupted players and to simulate the shares $y_i^{2^{(B-j)W}}$ disseminated in each phase j by nonfaulty processors. We show how $\mathcal{I}$ does this accurately, conditioned on $\mathcal{A}$ being unable to take square roots (a condition violated with negligible probability).

In short, $\mathcal{I}$ selects $\hat{a}, \hat{a}_1, \ldots, \hat{a}_t$ at random, sets $\hat{f}(u) = \hat{a} + \hat{a}_1 u + \cdots + \hat{a}_t u^t$, and for each faulty player i computes the value

$$h_i = g^{\hat{f}(i)}.$$

Note that $\mathcal{I}$ cannot compute "real" shares of a polynomial with free term a because it will not receive g^a in the ideal protocol until it's too late. Now, when player i becomes corrupted, $\mathcal{I}$ pretends i's piece is $y_i = h_i^{N!}$. This will cause no problem, as long as: (1) $\mathcal{A}$ never substitutes an undetected but incorrect piece (± 1 factors are acceptable); (2) $\mathcal{I}$ can accurately generate pieces disseminated by nonfaulty processors. Event (1) occurs with negligible probability (else factoring is easy), so we focus on (2).

At round j, then, $\mathcal{I}$ must generate fake but convincing values $y_i^{2^{(B-j)W}}$ from nonfaulty players. The $N!$ interpolation factor would normally prevent $\mathcal{I}$ from directly interpolating consistent pieces, but the initial exponentiation (representing $X = x^{N!}$ rather than x) permits $\mathcal{I}$ to get away with introducing an $N!$ factor. Interpolating from $h_1^{2^{(B-j)W}}, \ldots, h_t^{2^{(B-j)W}}$ and $x^{2^{(B-j)W}}$, $\mathcal{I}$ computes $y_i = g^{N! f(i) 2^{(B-j)W}}$ where the polynomial $f(u)$ agrees with $\hat{f}(u)$ at t

places but satisfying $f(0) = a$. This provides fake pieces from correct players at any phase (including the initial verification values, z_i).

The result is a distribution identical to that obtained when $\mathcal{A}$ attacks URNPROTO, conditioned on event (1) not occurring. When event (1) is included, the distributions are statistically indistinguishable. Intuitively, we have shown that an attack by $\mathcal{A}$ on URNPROTO achieves nothing that an attacker against the ideal trusted host protocol can't achieve – and in particular, predicting the bits is impossible. □

5 Discussion

We have presented two unpredictable bit generators requiring neither broadcast nor interaction other than simple dissemination (after an interactive initialization stage). The local computation is reasonably fast – repeated squaring, like the BBS generator, along with "interpolation" that requires some exponentiation – yet there is no need for "shares of shares" or other complicated constructs. A share is verified simply by squaring it.

Our bit generators differ from the secret bit generators discussed in [8] in that they do not provide "random access" (the constant-t generator of §3.1 is a notable exception), and they focus on unpredictability rather than the subtly different notion of pseudorandomness.

Although the unpredictability of the simpler of our generators is based on the QRA, we conjecture that it is in fact unpredictable unless factoring is easy. The methods of [1] do not seem to apply immediately, and further work is required.

We also conjecture that the number of squarings can be reduced by 50%; our protocol includes an apparently superfluous $N!$ factor to facilitate the *proof* of security.

Recently, a fast, non-interactive, global URN based on elliptic curves over rings has been developed and investigated in [3]. This solution shares the property of ours that extra "verification" procedures (such as shares of shares) are not needed. As shown in [8] for the case of doubly shared bits, we conjecture that our methods will apply to any homomorphic scheme.

References

[1] W. Alexi, B. Chor, O. Goldreich, C.P. Schnorr. "RSA and Rabin Functions: Certain Parts are as Hard as the Whole." *SIAM J. Computing,* **17**:2 (1988), 194–209.

[2] D. Beaver. "Foundations of Secure Interactive Computing." *Proc. of Crypto 1991*, 377–391.

[3] D. Beaver, H. Shan. In preparation.

[4] M. Ben-Or. "Another Advantage of Free Choice." *Proc. of* 2nd *PODC,* 1983.

[5] Blakley, "Security Proofs for Information Protection Systems." *Proceedings of the the 1980 Symposium on Security and Privacy,* IEEE Computer Society Press, NY (1981), 79–88.

[6] M. Blum, S. Micali. "How to Generate Cryptographically Strong Sequences of Pseudo-Random Bits." *SIAM J. Comput.* **13** (1984), 850–864.

[7] L. Blum, M. Blum, M. Shub. "A Simple Unpredictable Pseudo-Random Number Generator." *SIAM J. Computing,* **15**:2 (1986), 364–383.

[8] M. Cerecedo, T. Matsumoto, H. Imai. "Non-Interactive Generation of Shared Pseudorandom Sequences." To appear, Auscrypt 92.

[9] P. Feldman. "A Practical Scheme for Non-Interactive Verifiable Secret Sharing." *Proc. of the* 28th *FOCS,* IEEE, 1987, 427–437.

[10] O. Goldreich, S. Goldwasser, S. Micali. "How to Construct Random Functions." *JACM* **33**:4 (1986), 792–807.

[11] S. Goldwasser, S. Micali. "Probabilistic Encryption and How to Play Mental Poker Keeping Secret All Partial Information.

[12] S. Micali, P. Rogaway. "Secure Computation." *Proc. of Crypto 1991*, 392–404.

[13] M. Pease, R. Shostak, L. Lamport. "Reaching Agreement in the Presence of Faults." *JACM* **27**:2 (1980), 228–234.

[14] T. Pedersen. "Non-Interactive and Information-Theoretic Secure Verifiable Secret Sharing." Proceedings of CRYPTO 91, 129–140.

[15] M.O. Rabin. "Digitalized Signatures and Public-Key Functions as Intractable as Factorization." Technical Report LCS/TR-212, MIT, January, 1979.

[16] M.O. Rabin. "Randomized Byzantine Generals." *Proc. of the* 24th *FOCS,* IEEE, 1983, 403–409.

[17] A. Shamir. "How to Share a Secret." *Communications of the ACM,* **22** (1979), 612–613.

On Schnorr's Preprocessing for Digital Signature Schemes*

Peter de Rooij
PTT Research†

Abstract
Schnorr's preprocessing algorithms [6, 7] are designed to speed up the 'random' exponentiation often performed by the prover/signer in identification and signature schemes.

In this paper, an attack on these preprocessing algorithms is presented. For the proposed parameters, the attack requires about 2^{31} steps, and 700 identifications or signatures to retrieve the secret key. Here the underlying scheme may be Schnorr, Brickell-McCurley, ElGamal or DSS.

1 Introduction

At Crypto'89, a discrete log based identification protocol and a related signature scheme especially suited for use on smart cards were proposed by C. Schnorr [6]. One optional feature of both was the possibility of using a preprocessing algorithm that substantially speeds up the prover's calculations. However, use of the proposed preprocessing algorithm enables an attack that retrieves the prover's secret key [5].

An improved preprocessing algorithm, resistant to this attack, was proposed in [7]. In this paper, the attack is adapted to the new preprocessing algorithm. This new attack still is applicable to the old preprocessing, and is substantially faster than the old attack from [5].

The preprocessing algorithms can be used in any protocol where a random power of a fixed base is to be computed. The proposed attack can be adapted to any such protocol that satisfies a (rather mild) assumption. Specifically, it is shown to be applicable to Brickell-McCurley, ElGamal and DSS signatures [1, 2, 3].

2 Description of the Algorithm

2.1 Schnorr's Identification Protocol and Signature Scheme

We briefly describe Schnorr's identification protocol and the related signature scheme. For details, see [7].

The following parameters are public: a large prime p, a prime q that divides $p-1$, a primitive qth root of unity $\alpha \in \mathbb{Z}_p$ and a security parameter t. In [7], it is proposed to

*. Part of the results of this paper were presented at Eurocrypt'91 [5].
†. P.O. Box 421, 2260 AK Leidschendam, the Netherlands. E-mail: P.J.N.deRooij@research.ptt.nl

take p and q in the order of 512 bits and 140 bits respectively, and $t = 72$. Each user $\mathcal{A}$ has a secret key $s_{\mathcal{A}} \in \mathbb{Z}_q^*$ and a corresponding public key $v_{\mathcal{A}} = \alpha^{-s_{\mathcal{A}}} \bmod p$.

The identification protocol. A *prover* $\mathcal{A}$ that wants to prove her identity to *verifier* $\mathcal{B}$ first picks a random number $r^* \in \mathbb{Z}_q^*$ and sends the *initial commitment* $x^* = \alpha^{r^*} \bmod p$ to $\mathcal{B}$. Then $\mathcal{B}$ returns a random *challenge* $e \in \{0, \ldots, 2^t - 1\}$, to $\mathcal{A}$. Next, $\mathcal{A}$ sends $y = r^* + s_{\mathcal{A}} \cdot e \bmod q$ to $\mathcal{B}$. Finally, $\mathcal{B}$ checks $\mathcal{A}$'s proof of identity by verifying if $x^* = \alpha^y v_{\mathcal{A}}^e \bmod p$.

Below, the index $\mathcal{A}$ is dropped, as no confusion is possible. Furthermore, all calculations are performed modulo q, except where indicated otherwise.

The signature scheme. The identification protocol is extended to a signature scheme by replacing the challenge from the verifier by a t-bit hash value $e = h(x^*, m)$ of the initial commitment x^* and the message m to be signed. The signature consists of this hash value and of y as in the identification protocol.

In the sequel, a pair (y, e) is called a *signature*, even if it is constructed in the course of an instance of the identification protocol.

2.2 The preprocessing algorithm

The purpose of the preprocessing algorithm is the reduction of the computational effort of the prover/signer, which is determined by the random exponentiation $\alpha^{r^*} \bmod p$. The preprocessing simulates this random exponentiation by a pseudorandom one that requires a few multiplications only. This is achieved by taking, instead of a random r^*, a linear combination of several independently and randomly chosen numbers $r_i \in \mathbb{Z}_q^*$ for which $x_i = \alpha^{r_i} \bmod p$ is precomputed.

For this purpose, each user initially stores a collection of k such pairs (r_i, x_i), $0 \leq i < k$. Here k is a security parameter; the proposed value [6, 7] is $k = 8$. Then, for each signature, the pair (r^*, x^*) is chosen as a combination of the currently stored pairs (r_i, x_i), while the stored collection is 'rejuvenated' by replacing one of these pairs by a similar combination.

We denote this as follows. At time ν, $\nu \geq k$, the pair (r_ν, x_ν) replaces $(r_{\nu-k}, x_{\nu-k})$; both the pair (r^*, x^*) used in the initial commitment and the signature (y, e) receive an index ν. The preprocessing algorithm from [7] is as follows.

Initialization. Load k pairs (r_i, x_i) as above, $0 \leq i < k$; set $\nu := k$;

1. Pick a random permutation $(a_\nu(0), \ldots, a_\nu(k-1))$ of $\{0, \ldots, k-1\}$;
   $a_\nu(k) := 0$; $a_\nu(k+1) := k-1$;
   ($a_\nu(\cdot)$ determines the order of use of the stored pairs.)
2. $r_\nu^* := r_{\nu-k} + 2r_{\nu-1} \bmod q$; $\qquad r_\nu := \sum_{i=0}^{k+1} 2^i r_{a_\nu(i)+\nu-k} \bmod q$;
   $x_\nu^* := x_{\nu-k} \cdot x_{\nu-1}^2 \bmod p$; $\qquad x_\nu := \prod_{i=0}^{k+1} \left(x_{a_\nu(i)+\nu-k}\right)^{2^i} \bmod p$;
3. Keep (r_ν^*, x_ν^*) ready for the next signature;
   Keep (r_i, x_i), $\nu - k + 1 \leq i \leq \nu$ stored; (i.e., replace $(r_{\nu-k}, x_{\nu-k})$ with (r_ν, x_ν))
4. $\nu := \nu + 1$; goto 1 for the next signature.

(The preprocessing from [6] differs in the form of $a_\nu()$ only.) This preprocessing requires $2k + 2$ multiplications modulo p and a storage of k pairs (r_i, x_i) [7].

3 A New Attack

3.1 Dependencies between signatures

This section generalizes the attack from [5] to the new preprocessing. Assume that p, q, α, t and some number of successive signatures are available to the enemy. The following equations hold:

$$r_\nu^* = r_{\nu-k} + 2r_{\nu-1}, \qquad \nu = k,\, k+1, \ldots \tag{1}$$

$$r_\nu = \sum_{j=0}^{k+1} 2^j r_{a_\nu(j)+\nu-k}, \qquad \nu = k,\, k+1, \ldots \tag{2}$$

$$y_\nu = r_\nu^* + se_\nu, \qquad \nu = k,\, k+1, \ldots \tag{3}$$

From (1) and (3) it follows that $r_{j-1} = \frac{1}{2}(y_j - se_j - r_{j-k})$ for all $j \geq k$. Repeated substitution using this equality yields the *linking equation*:

$$r_{j+c(k-1)} = (-2)^{-c} r_j - \sum_{l=1}^{c} (-2)^{l-c-1} (y_{j+l(k-1)+1} - se_{j+l(k-1)+1}). \tag{4}$$

Given the (y_i, e_i)-pairs that occur in the corresponding linking equation (4), any equation of the same form as (2) can be rewritten to a linear equation in $r_0, \ldots, r_{k-2}$ and s. For details we refer to the final paper [4].

So, given n different equations of the same form as (2), say for the sequential numbers i_j, $i_0 < i_1 < \cdots < i_{n-1}$, and given the signatures (y_i, e_i) with sequential numbers $i_0 < i \leq i_{n-1} + 1$, a set of equations of the following form can be derived:

$$\left[\begin{array}{c|c} & E'_{i_0} \\ M & \vdots \\ & E'_{i_{n-1}} \end{array}\right] \cdot \left[\begin{array}{c} r_0 \\ \vdots \\ r_{k-2} \\ \hline s \end{array}\right] = \left[\begin{array}{c} Y'_{i_0} \\ \vdots \\ Y'_{i_{n-1}} \end{array}\right]. \tag{5}$$

Here the E'_* and Y'_* are known linear combinations of the e_i's respectively y_i's, and M is an $n \times (k-1)$ matrix with entries depending on the i_j and $a_{i_j}(\cdot)$ only, see [4].

In the old preprocessing it is possible that only two $a_i(l)$'s are nonzero. The corresponding row of M then has a support of weight 2. In the attack described in [5], three rows with the same support of weight 2 are collected. Then (5) reduces to a system of three equations in three unknowns, which can be solved with high probability. For details, see [5].

That attack will not work here, as it is guaranteed by the choice of the $a_i(l)$'s that all rows of M have a support of weight $k-1$. This implies that the attack will require a $k \times (k-1)$ matrix M. The work factor then increases to about $(k!)^k$ steps, as there are that many different possible M-matrices. This is infeasible for the proposed $k = 8$.

The attack presented here will collect two *identical* rows of M. Given that, the entries of M cancel out by subtraction of the two rows i_0 and i_1 of (5). This yields $s(E'_{i_0} - E'_{i_1}) = Y'_{i_0} - Y'_{i_1}$, which provides s if $E'_{i_0} \neq E'_{i_1}$, which happens with overwhelming probability. Two rows of M are identical if and only if $i_0 \equiv i_1 \pmod{k-1}$ and if the corresponding permutations $a_{i_0}(\cdot)$ and $a_{i_1}(\cdot)$ are identical, see [4].

Thus, if one possesses the signatures (y_i, e_i) for $i_0 < i \leq i_1 + 1$ and if $i_0 \equiv i_1 \pmod{k-1}$, one has a *candidate* to determine the secret key s.

Given a candidate, one can just assume that the related $a_i(\cdot)$ are equal. Then for each of the $k!$ possible values of these permutations, one obtains an estimate $\bar{s}$ for s. The correctness of an estimate is verified by checking whether $\alpha^{-\bar{s}}$ equals the public key v. This modular exponentiation determines the workload per estimate. After checking all $k!$ estimates for a candidate, one either has obtained s (this happens with probability $1/k!$), or the certainty that the permutations are unequal. This whole process is called the *checking of the candidate*. The result of this checking will be either s or "failure".

3.2 An attack

Suppose the enemy possesses a number of consecutive signatures. Each pair of signatures with the same index modulo $k-1$ provides a candidate. An attack could proceed by checking all those candidates in some order. (An example is given in [4].)

3.3 The Complexity of the Attack

The number of checks $\alpha^{-\bar{s}} \stackrel{?}{=} v$ that determines the workload is $k!$ per candidate, and must we check $k!$ candidates, as the probability of success per candidate is $1/k!$. Therefore, the expected number of steps of the attack is $(k!)^2$. For $k = 8$ this is approximately 2^{31}.

The number of required signatures is $\sqrt{\frac{1}{2}\pi(k-1)k!}$ (by the birthday paradox; the factor $k-1$ in the root stems from the fact that we have $k-1$ simultaneous birthday attacks, one for each index modulo $k-1$. For details, see [4]). For $k = 8$, this is amounts to 665 consecutive signatures.

3.4 Generalization

Other preprocessing algorithms. The attack will work for any preprocessing where Equations (1) and (3) hold, using $\sim \sqrt{\frac{1}{2}\pi(k-1)N}$ signatures and with complexity N^2 steps, if the analogue of Equation (2) allows N possibilities. For example, the preprocessing from [6] with the proposed parameters has $N = 2^{12}$, so the number of required signatures is about 212; the complexity is 2^{24}. (Compare this to the 2000 signatures and workload of 2^{38} from [5].)

Relaxing the assumptions. The requirement that one collects consecutive signatures by the same user can be dropped. This goes at the expense of increasing the number of required signatures to $k!$ sets of k consecutive ones; the workload remains the same.

Adaption to other signature schemes. The preprocessing can be applied in any situation where random powers of fixed bases are to be computed. This will occur in most discrete log based identification protocols and signature schemes. In general, Equation (3) will in then be different. However, if it is linear in r^* (i.e., its coefficient is known), it is possible to adapt the attack. This will almost always hold, as r^* will appear in an exponent in a verification equation; for ElGamal [2], DSS [3] and Brickell-McCurley [1] signatures this is the case. The workload is essentially the same in all three cases.

4 Conclusions

The idea of preprocessing the computation of a random power of a fixed base is interesting, as it reduces the effort to only a few multiplications. Especially for smart cards this is worth while. One must take care, however, that not too much information leaks: the replacement of a 'true random number' by a pseudorandom number always provides side information. The preprocessing algorithm as proposed in [7] is not sufficiently secure, as the attack described above shows.

The attack does not seem to depend very heavily on the underlying signature scheme, as it only requires a linear equation in the random exponent. A Schnorr, Brickell-McCurley, ElGamal or DSS signature provides such an equation, as it must hold for a successful verification.

Acknowledgements
The author would like to thank Johan van Tilburg, Arjen Lenstra and Jean-Paul Boly for their many useful comments and Yacov Yacobi for pointing out the issue of adapting the attack to other signature schemes.

References

[1] E. F. Brickell and K. S. McCurley, "An interactive identification scheme based on discrete logarithms and factoring", *Journal of Cryptology* 5 (1992), no. 1, pp. 29–39.

[2] T. ElGamal, "A public key cryptosystem and a signature scheme based on discrete logarithms", *IEEE Transactions on Information Theory* **IT-31** (1985), no. 4, pp. 469–472.

[3] National Institute of Technology and Standards, *Specifications for the Digital Signature Standard (DSS)*, Federal Information Processing Standards Publication XX, US. Department of Commerce, February 1 1993.

[4] P. de Rooij, "On Schnorr's preprocessing for digital signature schemes", *Journal of Cryptology*, to appear.

[5] ———, "On the security of the Schnorr scheme using preprocessing", *Advances in Cryptology – Proceedings of Eurocrypt'91* (D. W. Davies, ed.), Lecture Notes in Computer Science, vol. 547, Springer-Verlag, 1991, pp. 71–80.

[6] C. P. Schnorr, "Efficient identification and signatures for smart cards", *Advances in Cryptology – Proceedings of Crypto'89* (G. Brassard, ed.), Lecture Notes in Computer Science, vol. 435, Springer-Verlag, 1990, pp. 239–251.

[7] ———, "Efficient signature generation by smart cards", *Journal of Cryptology* **4** (1991), no. 3, pp. 161–174.

Cryptanalysis of the Chang-Wu-Chen key distribution system

Mike Burmester

RH – University of London, Egham, Surrey TW20 OEX, U.K.

Abstract. Chang-Wu-Chen presented at Auscrypt 92 a conference key distribution system based on public keys. We show that this scheme is insecure and discuss ways to fix it.

1 The CWC key distribution system

This system [3] uses a discrete logarithm setting with prime modulus p and primitive element g. Each party U_j, $j = 1, 2, \ldots, n$, has a secret key $x_j \in Z_{p-1}$ and a public key $y_j = g^{x_j} \bmod p$. A chairperson U_0 with secret key $x_0 \in Z_{p-1}$ and public key $y_0 = g^{x_0} \bmod p$ picks a random $r \in Z_{p-1}$ and computes $Y = \prod_{i=1}^{n} (y_i)^r \bmod p$. The conference key is $k \equiv Y^{-1} (\bmod p)$. Then the chairperson sends each U_j: $c_1 = g^r \bmod p$, $c_2 = (y_0)^k \bmod p$, and $Y_j \equiv Y/(y_j)^r (\bmod p)$. The parties U_j can easily compute k, since $k \equiv (Y_j \cdot (c_1)^{x_j})^{-1} (\bmod p)$. To validate k, U_j checks that $c_2 = (y_0)^k \bmod p$.

2 A cryptanalytic attack

We have

$$\prod_{i=1}^{n} Y_i \equiv \prod_{i=1}^{n} (Y/(y_i)^r) \equiv Y^{n-1} \equiv k^{1-n} \ (\bmod p) .$$

So a passive eavesdropper can easily compute $k^{n-1} \bmod p$. Since it is feasible [7, 1] to compute $(n-1)$–th roots in Z_p, the eavesdropper will succeed in finding the key k (with non-negligible probability) when $n \geq 2$.[1]

3 Authentication

In a key distribution system each party should know with whom it is exchanging the key. With the CWC system it is clear that the chairperson can substitute some of the parties $U_1, U_2, \ldots, U_n$ without the others finding this out from the key distribution system. So it is essential that the parties trust the chairperson. However the chairperson U_0 is not authenticated. Indeed the secret key x_0 of U_0 is not needed to compute either the validator c_2, or any of c_1, Y_j and the key k. So anyone can easily masquerade as U_0 (by substituting its messages).

[1] Edward Zuk from Telecom Research Laboratories, Australia, has found this attack independently. This was pointed out to me by Jennifer Seberry.

4 Fixing the system

4.1 Using a prime modulus

We get some protection from the attack in Section 2 by replacing Y and Y_j by $\tilde{Y} \equiv (y_0)^{r(n-1)} \cdot \prod_{i=1}^{n}(y_i)^r \pmod{p}$ and $\tilde{Y}_j \equiv \tilde{Y}/(y_j)^r \pmod{p}$ respectively, and by taking $\tilde{k} \equiv \tilde{Y}^{-1} \pmod{p}$ as the key.

Consider a variant of the CWC system for which the chairperson U_0 sends U_j *only* c_1 and $\tilde{Y}_j$, and *not* c_2 which, as observed earlier does not authenticate U_0 (in this case U_0 must be authenticated some other way – see Section 4.3). We shall show that cracking this variant by a passive eavesdropper (a 'ciphertext attack') is as hard as cracking the Diffie-Hellman [6] problem,

Input: g, p, $g^a \bmod p$, $g^b \bmod p$; **Output:** $g^{ab} \bmod p$.

Indeed suppose that it is easy to crack the modified key distribution system and let $g^a \bmod p$, $g^b \bmod p$ be an instance of the Diffie-Hellman problem. Set $c_1 = g^a \bmod p$, $y_0 = g^b \bmod p$ and $y_i = g^{t_i - b} \bmod p$, for $i = 1, 2, \ldots, n$, where the $t_i \in Z_{p-1}$ are chosen randomly. Then $g^r \equiv g^a \pmod{p}$ and $\tilde{Y}_j \equiv (g^a)^{T-t_j} \pmod{p}$, where $T = \sum_{i=1}^{n} t_i$. We are assuming that it is easy to compute the key,

$$\tilde{k} \equiv \tilde{Y}^{-1} \equiv (g^b)^{a(1-n)} \cdot g^{a(nb-T)} \equiv g^{ab-aT} \pmod{p},$$

so it is easy to compute $g^{ab} \equiv \tilde{k} \cdot (g^a)^T \pmod{p}$, and hence to find a solution for the Diffie-Hellman problem. The reduction in the reverse direction is straightforward: if it is easy to crack the Diffie-Hellman problem, then it is easy to compute $\tilde{k} \equiv ((y_0)^{n-1} \cdot \prod_{i=1}^{n} y_i)^{-r} \pmod{p}$, from $(y_0)^{n-1} \cdot \prod_{i=1}^{n} y_i \pmod{p}$ and $g^{-r} \bmod p$.

For this variant of the CWC system we also get some protection from known key attacks ('plaintext attacks') by active adversaries. This follows by observing that 'old-session' information: $c_1 = g^r \bmod p$, $\tilde{Y}_j \equiv ({y_0}^{n-1} \cdot \prod_{i \neq j} y_i)^r \pmod{p}$, and $\tilde{k} \equiv \left({y_0}^{n-1} \cdot \prod_{i=1}^{n} y_i\right)^r \pmod{p}$, can be simulated, and that therefore the argument used in [8] for 'non-paradoxical' systems applies. However it should be pointed out that there is a flaw [5] in the proof given in [8], and consequently the proposed variant *may* not be 'proven secure' for known key attacks (in the general case).

4.2 Using a composite modulus

To prevent the attack in Section 2 we may also use a composite modulus $m = pq$, where p, q are appropriate primes, and take g to be an element of large order, e.g. a primitive element of Z_p and Z_q.[1] Then it is not necessary to modify Y and Y_j. For a 'provably secure' protocol we may use the variant in the previous section with composite modulus (however in this case the probability distributions are not uniform and we must use randomized reductions [2] as in [8]). Again we get some protection from known key attacks.

4.3 Addressing the authentication problem

As pointed out in Section 3 the validator c_2 does not authenticate the chairperson U_0, since anybody can compute it without knowing the secret key x_0. To prevent this we may replace c_2 by $\tilde{c}_{2j} = (y_j)^{\tilde{k}x_0} \bmod p$. Clearly it is hard to compute $\tilde{c}_{2j}$ without knowing $\tilde{k}$ and either x_0 or x_j, provided the Diffie-Hellman problem is hard. Furthermore any U_j can easily validate $\tilde{k}$, since $\tilde{c}_{2j} \equiv (y_0)^{\tilde{k}x_j} \pmod p$. However this modification offers no protection against insider attacks. Indeed any U_i can compute $(y_j)^{x_0} \equiv (y_0)^{x_j} \pmod p$ from $\tilde{c}_{2j}$ (obtained by eavesdropping) and from the key $\tilde{k}$ [7, 1]. Then, at any later time, U_i can impersonate U_0, or forge any key $\tilde{k}$.

There seems to be no obvious way of solving the authentication problem without using a separate authentication system. The scheme in [4] addresses this problem and other more general issues.

Acknowledgement. The author wishes to thank Yvo Desmedt and Dieter Gollmann for many helpful discussions.

References

1. L. Adleman, K.M. Manders, and G.M. Miller. On taking roots in finite fields. *Annual Symposium on Foundations of Computer Science*, Vol. 18, 1977, pp. 175–178.
2. S. Ben-David, B. Chor, O. Goldreich, M. Luby. On the theory of Average case Complexity. *Proceedings of the twenty first annual ACM Symp. Theory of Computing, STOC*, 1977, pp. 175–178.
3. Chin-Chen Chang, Tzong-Chen Wu, and C.P. Chen. The Design of a Conference Key Distribution System. Presented at Auscrypt 92, Gold Coast, Australia, December 13–16, 1992. To appear in: *Advances in Cryptology, Lecture Notes in Computer Science*, Springer-Verlag.
4. M. Burmester, Y. Desmedt. An Efficient and Secure Conference Key Distribution System. Manuscript, April 1993.
5. Y. Desmedt, M. Burmester. Towards practical 'proven secure' authenticated key distribution. To appear in, Proceedings of the 1st ACM Conference on Communications and Computing Security, Fairfax, Virginia, November 3–5, 1993.
6. W. Diffie and M. E. Hellman. New directions in cryptography. *IEEE Trans. Inform. Theory*, IT–22(6), 1976, pp. 644–654.
7. D. Shanks. Five number theoretic algorithms. *Proc. Second Manitoba Conf. Numerical Mathematics*, 1972, pp. 51–70.
8. Y. Yacobi. A Key Distribution Paradox. In *Advances in Cryptology — Crypto '90, Proceedings (Lecture Notes in Computer Science #537)*, 1991, A.J. Menezes and S.A. Vanstone, Eds, Springer-Verlag, pp. 268–273.

An Alternate Explanation of two BAN-logic "failures"

Paul C. van Oorschot

Bell-Northern Research, P.O. Box 3511 Station C, Ottawa K1Y 4H7 Canada
paulv@bnr.ca

Abstract. Boyd and Mao ("On a Limitation of BAN Logic", in these proceedings) suggest that it is easy to use the authentication logic of Burrows, Abadi and Needham to approve protocols that are in practice unsound, and present two examples. We illustrate that the problem in the first example can be traced to a violation of pre-conditions in the BAN analysis (involving ill-founded trust in a trusted server), while in the second the idealization is simply incorrect. For the latter, a general guideline is proposed to avoid similar problems in the future.

1 Introduction

The BAN logic [3] was the first of several logics (including e.g. AT [1] and GNY [6]) designed to facilitate more rigorous analysis of cryptographic protocols than is possible by informal, ad hoc methods. It allows reasoning about beliefs held by the parties (principals) involved in the protocols. BAN analysis proceeds by a four-stage process. First the protocol in question is "idealized" — the actual or concrete protocol is expressed as a sequence S^* of formal steps ($A \rightarrow B: X$), where A and B are principals and X is a statement in the syntax of the logic. Second, the set of assumptions Q under which the protocol operates are identified and formally expressed. These typically include formalizations of assumptions such as "each principal will not disclose its private keys to other entities". In order to attain the goals established by the formal proofs, these formal assumptions must hold. Third, the goals G of the protocol are identified and formally expressed. A typical goal is the establishment of a cryptographic key shared exclusively with another specifically identified principal. Finally, a proof of the form $Q.S^*.G$ is constructed, using the inference rules of the logic, showing that given the formal assumptions Q, and upon carrying out one or more protocol steps S, the goals G are attained.

It has been suggested that the BAN logic is unable to distinguish secure and flawed versions of some protocols; two illustrative examples were given [2]. We show that in the first example, it is the failure to verify the formal assumptions one would obtain in a detailed BAN analysis that leads to this conclusion, rather than a failure of the BAN technique itself; the true source of the problem is an inappropriate assumption about the trusted server. The problem in the second example shown to be due to incorrect idealization, and a general guideline to follow during BAN idealization is offered to avoid similar problems in the future. In Section 2 we review the Otway-Rees protocol of the first example, and examine the BAN analysis of it. Section 3 discusses the second example — a simplified version of the first protocol. Section 4 concludes this note.

2 BAN analysis of the Otway-Rees protocol

The Otway-Rees protocol examined in [2] is repeated here for reference. S is a trusted server, which generates a symmetric secret key K_{AB} intended for use by A and B. K_{AS} and K_{BS} are symmetric secret keys shared a priori between S and principals A and B, respectively. N_A and N_B are nonces chosen by A and B, intended to allow detection of replayed messages. The field M is not of concern in the present discussion. The messages to be exchanged in a proper run of the protocol are given below. The cleartext identifiers A and B in messages 1 and 2 are used by S to retrieve keys K_{AS} and K_{BS}:

1. $A \rightarrow B$: $M, A, B, \{N_A, M, A, B\}_{K_{AS}}$
2. $B \rightarrow S$: $M, A, B, \{N_A, M, A, B\}_{K_{AS}}, \{N_B, M, A, B\}_{K_{BS}}$
3. $S \rightarrow B$: $\{N_A, K_{AB}\}_{K_{AS}}, \{N_B, K_{AB}\}_{K_{BS}}$
4. $B \rightarrow A$: $\{N_A, K_{AB}\}_{K_{AS}}$

A suggested attack on this protocol is as follows. An opponent C impersonates B by intercepting message 2 and substituting his own message in its place, replacing cleartext identifier "B" by "C" (but leaving both enciphered versions of both identifiers "A" and "B" as before), replacing the nonce N_B by his own nonce N, and using key K_{CS} in place of K_{BS}. The opponent is then able to recover the secret key K_{AB} upon intercepting message 3, as it will be encrypted under the key K_{CS} which C shares a priori with S. Whether this attack is successful or not depends on the actions taken by the server S:

Case 1.[1] S simply checks that the values obtained by decrypting the identifier fields (A, B) under the two different keys (K_{AS}, K_{BS}) in message 2 are equal. In this case the attack will succeed.

Case 2. S checks that the values in the cleartext identifier fields (A, B) are equal to the values obtained by decrypting the corresponding identifier fields under each of the keys (K_{AS}, K_{BS}). In this case the attack will not succeed. Clearly this is the desirable version of the protocol.

The analysis given by the BAN logic appears the same in both cases, which, according to [2], suggests a problem in the BAN idealization process. We argue that the BAN logic is indeed capable of distinguishing between these cases. Indeed, consider the following details.

Step 3 of the protocol might be idealized in BAN as

3. $S \rightarrow B$: $\{N_A, A \overset{K_{AB}}{\leftrightarrow} B, \ldots\}_{K_{AS}}, \{N_B, A \overset{K_{AB}}{\leftrightarrow} B, \ldots\}_{K_{BS}}$

Here the symbol $A \overset{K_{AB}}{\leftrightarrow} B$ asserts that K is a good cryptographic key for use by A and B. The portion of the idealization which is not of concern has been deleted ("$\ldots$"). If a detailed analysis, such as that carried out for the X.509 authentication protocol by Gaarder and Snekkenes [4], is carried out here, then one is forced to record the formal assumption

Q1. $A \mid\equiv S \Rightarrow A \overset{K_{AB}}{\leftrightarrow} B$

---

1. This is discussed in the section "A Faulty Implementation of the Otway-Rees Protocol" in [2].

which states that A *believes* that S *has jurisdiction over* (i.e. can be trusted regarding, or has control over) statements concerning shared keys between principals A and B. This is required because the formal proof of the final goal $A \models A \overset{K_{AB}}{\leftrightarrow} B$ requires use of the BAN jurisdiction rule, with $A \overset{K_{AB}}{\leftrightarrow} B$ in the place of X:

$$\text{Jurisdiction rule:} \quad \frac{A \models S \Rightarrow X, \ A \models S \models X}{A \models X}$$

This inference rule states that if A trusts S on a statement X, and if A believes that S believes X, then A should believe X. Note that assumption Q1 means that A delegates to S responsibility regarding statements about shared keys with B, and trusts S on such matters. Exploring this further with $A \overset{K_{AB}}{\leftrightarrow} B$ in place of X, this means that A trusts S to properly authenticate B; that is, A delegates authentication of B to the server. It should be clear now that in Case 1 (i.e. the flawed version) of the protocol, this trust is ill-founded, and in fact S should *not* be trusted on statements regarding a shared key with B; however in Case 2 (i.e. the secure version), S is trustworthy on this matter.

In summary, the Otway-Rees protocol requires trust in a server S, and the formal BAN analysis properly captures this requirement through assumption Q1. In the flawed version, Q1 is violated, and thus the intended goal is not reachable, i.e. the proof that the goal is reachable is invalidated. Clearly, the BAN approach does not claim to guarantee that formal assumptions always hold, but rather simply asserts that *if* the formal assumptions hold, *then* proofs regarding goals, which use such assumptions, are valid. Proof of the validity of assumptions is beyond the BAN approach itself.

While it would be helpful if the BAN logic provided automatic verification of all identified assumptions, this is an unrealistic expectation of any analysis tool. We submit that assumptions delegating trust to third parties need be carefully examined for validity in any system, and such verification does not appear easily amenable to automation. We note, however, that a more detailed BAN-like analysis of this protocol would replace assumption Q1 with a proof that step 2 in the protocol allows S to properly authenticate B. This then raises the issue that the actions carried out by S should be more clearly specified in the protocol description. This would focus attention on Case 1 vs. Case 2, and again the BAN approach would indeed distinguish the two cases.

3 BAN analysis of a simplified version of the Otway-Rees protocol

Boyd and Mao [2] also consider a simplified version of the Otway-Rees protocol, in which the nonce N_B in message 2 is no longer part of the message encrypted under K_{BS}, but rather simply sent as cleartext information. Under suitable assumptions, they then describe a possible attack, and conclude that the idealization stage of BAN logic has a fundamental difficulty.

While an attack is indeed possible, this conclusion seems unjustified. The attack outlined is through no fault of BAN-like logics or analysis; the protocol as interpreted in [2] is simply flawed, notwithstanding the fact that the BAN authors themselves apparently suggest the modification that leads to it (see [3], p.17). In all fairness, they do so in a brief concluding note, and there is some ambiguity as to exactly what protocol modifications are suggested, what type(s) of nonce(s) and nonce verification are used, and which parties are responsible for verifying nonces. Nonetheless, the issues raised are significant, and we now examine the protocol more carefully. We assume that the server does indeed carry out the check as outlined in Case 2 above. The assumption of trust in the server is then well-founded, and we must search elsewhere for the problem.

With N_B no longer encrypted in message 2, upon reception of message 3 B is still able to conclude that (i) the key K_{AB} is fresh (since it is bound with the nonce N_B); that (ii) the key is known to S; and that (iii) S intends to make the key known to one other party besides B. However, B no longer has any indication who this other party is. The problem is that this simplified protocol does not allow S to convey to B the identity of the other party the key K_{AB} has been made available to. In the original protocol, this is done implicitly, through the cryptographic binding of the nonce N_B to K_{AB} in message 3, and to the pair of identifiers (A, B) in message 2. This allows both B and S to indirectly associate K_{AB} with principal identifiers (A, B). B trusts S to make the key K_{AB} available to only those parties identified in the last two positions in the encrypted segment $\{N_B, M, A, B\}_{K_{BS}}$. However, in the simplified version, no common understanding between S and B is possible regarding the parties associated with nonce N_B (and thus with key K_{AB}), as N_B is not cryptographically bound to any identifiers in message 2 or 3. This prevents S and B from determining a common instantiation of the identifiers "A" and "B" as principals in the symbol $A \overset{K_{AB}}{\leftrightarrow} B$ in idealized message 3 above.

It is now seen that the idealization of message 3 specified above is simply incorrect for the simplified protocol. To avoid such incorrect idealizations in the future, we offer the following remarks as a guideline to be followed during BAN idealization.

In BAN idealization, a key K_{AB} in a concrete protocol is often replaced by the symbol $A \overset{K_{AB}}{\leftrightarrow} B$ in the idealized protocol. It is important to note that this latter symbol implies not only a key, but also the identities of two specific principals. A key denoted K_{AB} sent in a message, e.g. from a server to a principal B, should not be idealized as $A \overset{K_{AB}}{\leftrightarrow} B$ unless it is possible for both the message originator and intended recipient to instantiate the identifiers (A and B) either directly (by actual identifiers sent along with and cryptographically bound to K_{AB}) or indirectly (e.g. as through the nonce N_B in the original protocol, or implicitly through use of another shared secret). We suggest that if this is not possible, then the idealization is incorrect and unsound. Note also that the subscript identifier AB in a key symbol K_{AB} is typically purely notational, and for the purpose of formal analysis would best be deleted to avoid confusion between the parties *intended* to share this key, and the parties who *actually* end up sharing it.

Finally, note that the *implicit* association of N_B in message 3 to two parties as specified by a binding in message 2, could be made *explicit* if S returned, within the encrypted portions of message 3, the actual identifiers of the parties which S was making the new key available to. This idea was discussed in a preliminary draft of [2].

4 Conclusion

Regarding semantics, the BAN authors state that the assumption $S \mid\equiv A \overset{K_{AB}}{\leftrightarrow} B$ "indicates that the server initially knows a key which *is to become* [emphasis ours] a shared secret between A and B" ([3], p.16). This differs from the actual definition of $A \overset{K_{AB}}{\leftrightarrow} B$ (A and B may use the shared secret K_{AB} as a good cryptographic key), and illustrates an ambiguity between the intention of sharing a secret key, and the actual state of that key already being shared and/or secret. This a potential cause of problems in many idealizations, including [7], of protocols in which one party is trusted to choose and transport a secret key to another. That this implies that the server is responsible for authenticating these parties is apparently often overlooked in analysis. The guideline offered above attempts to address this in the case of symmetrically generated keys, and a proposal to clarify this confusion in the case of a jointly established key is included in [8].

The nature of formal analysis using BAN depends heavily on the details of the formalization of initial assumptions, and on protocol idealization. The latter appears difficult to automate or prove correct, and remains the most critical step. However, verification of the validity of formal assumptions is also essential, as the resulting conclusions are conditional upon them. Although appearing straight-forward, analysis by BAN logic does require attention to detail; however we do not believe that it is fatally deficient as suggested in [2]. The cited failures can be linked to the failure to verify the validity of formal assumptions, which to its credit, BAN analysis requires one to record explicitly; and to improper idealization. The latter can be avoided by exercising caution in the use of the symbol $A \overset{K}{\leftrightarrow} B$ in idealization, taking due care to note that this symbol has implications about both the *quality* of the key K as a shared secret, and the *identity* of the parties which supposedly share it. Finally, we note that many of the known weaknesses of BAN-like logics have previously been discussed in the literature [5], along with proposed logic improvements including those aimed at simplifying idealization and providing more detailed handling of cleartext in messages (e.g. see [6]).

5 References

[1] M. Abadi, M. Tuttle. "A semantics for a logic of authentication". *Proc. 1991 ACM Symp. on Principles of Distributed Computing*, 201-216.

[2] C. Boyd, W. Mao, "On a Limitation of BAN Logic", presented at *Eurocrypt'93*, Lofthus, Norway, 1993 May 24-26 (to appear in these proceedings).

[3] M. Burrows, M. Abadi, R. Needham, "A logic of authentication", *ACM Trans. Computer Systems* **8** (Feb. 1990), 18-36. A more detailed version is available in: M. Burrows, M. Abadi and R. Needham, "A logic of authentication", *Digital Systems Research Centre SRC Report #39* (1990 Feb. 22), 62 pages.

[4] K. Gaarder, E. Snekkenes, "Applying a formal analysis technique to CCITT X.509 strong two-way authentication protocol", *J. Cryptology* 3 (Jan. 1991), 81-98.

[5] V. Gligor, R. Kailar, S. Stubblebine, L. Gong. "Logics for cryptographic protocols — virtues and limitations". *Proc. IEEE 1991 Computer Security Foundations Workshop* (Franconia, New Hampshire).

[6] L. Gong, R. Needham, R. Yahalom. "Reasoning about belief in cryptographic protocols". *Proc. 1990 IEEE Symp. on Security and Privacy* (Oakland, CA), 234-248.

[7] D.M.Nessett. "A critique of the Burrows, Abadi and Needham logic". *Operating Systems Review* 24 (1990), 35-38.

[8] P. Van Oorschot. "Extending cryptographic logics of belief to key agreement protocols". *Proc. 1st ACM Conference on Communications and Computer Security* (Fairfax, Virginia, Nov. 3-5 1993).

The Consequences of Trust in Shared Secret Schemes[1]

Gustavus J. Simmons
P.O. Box 365
Sandia Park, New Mexico 87047

Abstract. By accepting a shared secret or shared control scheme specified by an access structure Γ, an issuing authority also implicitly accepts all of the access structures that can be realized as a result of trust relations that may exist among some of the participants in Γ. An algorithm is presented here that makes it possible to fully analyze the consequences of trust to such schemes.

1 Introduction

At Eurocrypt '90 Ingemarsson and Simmons [1] described a protocol that made it possible for a group of participants to set up a shared secret scheme without the assistance of a trusted issuing authority to generate and issue the shares. To accomplish this end, they first showed how the participants could jointly set up a unanimous consent scheme in which each of them held an equal and essential share. Each participant could then, acting as his own issuing authority, set up a private secondary shared secret scheme to share his personal share in the unanimous consent scheme with the other participants -- according to the trust he had in them to faithfully represent his interests. This two step protocol made it possible for the participants to jointly set up a shared secret scheme which accurately reflected their trust (or lack of trust) in the other participants.

The Ingemarsson-Simmons protocol has an obvious generalization: Given any shared secret or shared control scheme, Γ, what other schemes can be reached from Γ as a consequence of trust relations that may exist among some of the participants in Γ? The answer to this question is of crucial concern to an issuing authority setting up a shared secret scheme since he has no way of knowing or of controlling the trust relations that may exist among the participants to whom he issues shares. In other words, anyone can share anything they know with anyone they trust. As the present author has shown [2,3] trust defines a partial ordering and a lattice $\mathcal{L}_n$ on the monotone access structures, Γ_i, for n participants in which $\Gamma_1 > \Gamma_2$ if Γ_2 can be "reached" from Γ_1 as a consequence of some set of trust relations between participants in Γ_1. Monotonicity says that control, as represented by an access structure

[1]Work supported by the ISS '90 Foundation

Γ_1, can never be strengthened by trust, but it can be weakened. Clear as this notion is, it is unfortunately infeasible to use, even for a modest number of participants. For example, there are 7579 monotone access structures on 5 participants, i.e. 7579 irreducible and distinct ways of entrusting a secret to subsets of 5 participants. Each of these 7579 structures is a possible trust relation by which a participant in a scheme involving 6 participants might conceivably be willing to entrust his share in a shared control scheme to subsets of the other participants. Consequently, given any access structure, Γ, on 6 participants, there are potentially $(7579)^6 \approx 1.9 \times 10^{23}$ possible sets of trust relations to be considered to determine which other structures could be reached from Γ. On the other hand there are only on the order of three and a half million access structures in all on 6 participants. As a result, until now there has been no hope of exploring $\mathcal{L}_6$. On the other hand, schemes involving 6 or more participants are very likely to occur in practice.

A system designer wishing to set up a shared control scheme, Γ, needs to know which access structures are "reachable" from the scheme he has chosen as reflecting the control (and risk -- of unauthorized use) he is willing to accept. If Γ is the access structure he has chosen, there is a sublattice (in $\mathcal{L}_n$) of states that can be reached from Γ. If one or more of the access structures in this sublattice represent unacceptably weak control according to the designer's objectives (as will always be the case since Φ -- representing no control at all -- is the lowest element in the lattice itself and in all of the sublattices generated by the different access structures), he must examine the family of trust relations that carry the acceptable scheme into an unacceptable one, and decide whether in his judgement this family of trust relations are sufficiently likely to occur to make the risk of the unacceptable scheme occurring be unacceptable to him or not. If not, the original scheme represents an acceptable risk, while if so, another choice must be made for the initial scheme. Given two access structures Γ_1 and Γ_2, where $\Gamma_1 > \Gamma_2$, it is a simple write down procedure to write out the full family of trust relations that carry the first scheme into the second. While the number of terms in this family may grow as the product of the number of terms in the two schemes, the formal manipulation to do the write down operation is simple logical set manipulation of symbols. It is this utility for the analysis and design of shared control schemes that makes $\mathcal{L}_n$ so important. Unfortunately, as already been indicated, this has only been an intellectual exercise until now because of the total infeasibilty of actually calculating in whole, or even in any large part, interesting sized $\mathcal{L}_n$; i.e. for any $n > 5$.

2 The Algorithm

The object of this note is to describe an elegant, but simple, way to construct both the lattice $\mathcal{L}_n$ and the sublattices of $\mathcal{L}_n$ generated by specific access structures -- up to sizes that probably exceed our ability to meaningfully specify shared control schemes. Even more important from a design standpoint, given any specified access structure, Γ, this technique will generate and exhibit only the sublattice reachable from Γ in $\mathcal{L}_n$. This means that a designer will generate and see only those structures that he needs to consider, but he will see all of these.

Given a lattice **L** on a set S, and a subset $A \subset S$, define the closure of A , denoted by A^c, to be the set of all elements in S that dominate at least one element of A in **L**. We adopt the convention that $x > x$, so that $A \subset A^c$. The complement of A^c in S is denoted by $S \backslash A^c$. Define two sets, $\lfloor A \rfloor$ and $\lceil A \rceil$, which we will call by the names usually given these quantities in numerical analysis; the floor of A and the ceiling of A respectively. As will be apparent in a moment this nomenclature has an intuitive appeal in the setting in which it is used here. $\lfloor A \rfloor$ is the maximal set of independent elements in **L** of members of A^c, i.e. a maximal independent subset of A^c such that every element in A^c dominates at least one element in $\lfloor A \rfloor$. Another way of defining $\lfloor A \rfloor$ is that it is the set consisting of the lowest element in all maximal length chains in the sublattice on A^c in **L**. Similarly, $\lceil A \rceil$ is the minimal set of independent elements in **L** on members of $S \backslash A^c$, i.e. the set containing highest element in all maximal length chains in the sublattice on the elements of $S \backslash A^c$ in **L**. In a very natural sense $\lfloor A \rfloor$ and $\lceil A \rceil$ define the boundaries of a proper partition of **L** induced by the set A: no element in $\lceil A \rceil$ dominates any element in $\lfloor A \rfloor$ and every element in $\lfloor A \rfloor$ dominates at least one element in $\lceil A \rceil$. You can see now why I said the floor and ceiling nomenclature was so natural. A^c and $S \backslash A^c$ properly partition **L** as defined above. $\lfloor A \rfloor$ is the floor of A^c in the sense that it is the subset of A^c that lies below everything else in the sublattice on A^c. Similarly, $\lceil A \rceil$ is the ceiling of $S \backslash A^c$ in the sense that it is the subset of $S \backslash A^c$ that lies above everything else in the sublattice on $S \backslash A^c$ in **L**.

If we take $\mathbf{L}_n$ to be the lattice defined on the 2^n subsets of n elements partially ordered by the usual ordering of set inclusion, then $\mathbf{L}_n$ is isomorphic to the binary n-dimensional hypercube $\mathbf{H}_n$. The first result is that the family of all possible floor functions in $\mathbf{H}_n$ is precisely the family of all monotone access structures on n participants. The following theorem is the substance of this note.

Theorem:

In the lattice, $\mathcal{L}_n$, of monotone access structures on n participants, Γ_1 directly dominates Γ_2 if and only if

$$\Gamma_2 = (\Gamma_1 \backslash S_x) \cup x \qquad x \in \lceil \Gamma_1 \rceil \text{ and } S_x = (y \mid y \in \Gamma_1 , y>x \text{ in } \mathcal{L}_n).$$

$\lceil \Gamma_1 \rceil$ is easy to calculate. In 1991 Jackson, Martin and Simmons [4] proved that every shared secret scheme has a perfect realization by a geometrical scheme. As an essential step in their proof they defined a quantity Γ^* which was computed from Γ by interchanging the operations of $\times$ and $+$. $\lceil\Gamma\rceil$ in H_n is simply the term by term set complement (in S) of Γ^*.

Example:

$$\Gamma = AB + ACD$$

$$\Gamma^* = A + BC + BD$$

$$\lceil\Gamma\rceil = AC + AD + BCD.$$

Using the theorem, we can now write down with no difficulty the access structures directly dominated by Γ in $\mathcal{L}_4$: AB + AC, AB + AD and AB + ACD + BCD .The reader may ask, Why doesn't AB +CD appear in this list since the other two pairs of elements from the term ACD do occur paired with the term AD in schemes directly dominated by Γ (besides the fact that the theorem doesn't generate this structure)? The answer is that the structure AB + ACD + BCD which is directly dominated by Γ in turn directly dominates AB + CD as can easily be verified by applying the theorem to AB + ACD + BCD. The beauty of the theorem is that it is constructive -- using the constructive characterization for $\lceil\Gamma\rceil$ given above. Nice as it is to have a practical means of directly constructing $\mathcal{L}_n$ (one merely starts by applying the theorem to the unanimous consent scheme on n participants, so that no recursion on lower order lattices is needed or involved) the practical merit of the result is that we can start with an arbitrary access structure, Γ, and by applying the theorem iteratively construct precisely the sublattice of $\mathcal{L}_n$ generated by Γ, and then by using simple write down rules mechanically express all of the families of trust relations that can weaken Γ to reach these schemes.

Conclusion

The algorithm described here makes it easy to directly calculate either the lattice $\mathcal{L}_n$ of monotone access structures on n participants or the sublattice of $\mathcal{L}_n$ generated by a particular access structure, Γ, for values of n that probably exceed our ability to meaningfully specify shared control scheme in practical situations.

References

1. I. Ingemarsson, G. J. Simmons: A Protocol to Set Up Shared Secret Schemes Without the Assistance of a Mutually Trusted Party, Lecture Notes in Computer Science 473; Advances in Cryptology: Proceedings of Eurocrypt '90, I. Damgard, Ed. Aarhus, Denmark, May 21-24 ,1990, Springer-Verlag, Berlin, 1991, pp. 266-282

2. G. J. Simmons: Geometric Shared Secret and/or Shared Control Schemes, Lecture Notes in Computer Science 537; Advances in Cryptology: Proceedings of Crypto '90, S. A. Vanstone, Ed. Santa Barbara, CA, August 11-15, 1990, Springer-Verlag, Berlin 1991, pp. 216-241

3. G. J. Simmons: An Introduction to the Mathematics of Trust in Security Protocols, Proceedings of the IEEE Computer Security Foundations Workshop VI, June 15-17 1993, Franconia NH, IEEE Computer Society 1993, pp. 121-127

4 G. J. Simmons, W. A. Jackson and K. Martin: The Geometry of Shared Secret Schemes, Bulletin of the Institute of Combinatorics and its Applications (ICA), Vol. 1, No. 1, 1991 pp. 71-88

Markov Ciphers and Alternating Groups

G. Hornauer, W. Stephan,
R. Wernsdorf

SIT Gesellschaft für Systeme der Informationstechnik mbH
15537 Grünheide (Mark), Germany
Charlottenstraße 7

Abstract. This paper includes some relations between differential cryptanalysis and group theory. The main result is the following:
If the one-round functions of an r-round iterated cipher generate the alternating or the symmetric group, then for all corresponding Markov ciphers the chains of differences are irreducible and aperiodic.
As an application it will be shown that if the hypothesis of stochastic equivalence holds for any of these corresponding Markov ciphers, then the DES and the IDEA(32) are secure against a differential cryptanalysis attack after sufficiently many rounds for these Markov ciphers.
The section about IDEA(32) includes the result that the one-round functions of this algorithm generate the alternating group.

The theoretic foundations in group theory and Markov chains are described for instance in [Wie 64] and [Fel 58].

1 Properties of Markov Chains and Markov Ciphers

Let us recall some definitions and properties of Markov chains. The definitions follow the notations of [LMM 91]. In this section we will briefly review parts of this paper.

A sequence of discrete random variables $v_0, v_1, \ldots, v_r$ is a Markov chain if for $0 \leq i < r$ (where $r = \infty$ is allowed):

$$P(v_{i+1} = \beta_{i+1} | \; v_i = \beta_i, v_{i-1} = \beta_{i-1}, \ldots, v_0 = \beta_0) = P(v_{i+1} = \beta_{i+1} | \; v_i = \beta_i).$$

A Markov chain is called homogeneous if $P(v_{i+1} = \beta | \; v_i = \alpha)$ is independent of i for all pairs (α, β).

Let $\Pi = \|p_{ij}\|$ denote the transition probability matrix of a finite homogeneous Markov chain with M states and p_{ij} the transition probabilities.

A finite Markov chain with the transition matrix Π is irreducible if for any (i, j) there is an r such that the (i, j) entry in the r-th transition matrix Π^r, $p_{ij}^{(r)} > 0$.

The chain is aperiodic if $\gcd\left(r_i = \min_r \{p_{ii}^{(r)} > 0\}\, ;\ 1 \le i \le M \right) = 1.$

Theorem 1 completes the probability theoretic background.

THEOREM 1 [Fel 58]:

If a finite, homogeneous, irreducible aperiodic Markov chain has a doubly stochastic transition matrix $\Pi = \|p_{ij}\|$, then in the limit all states become equally probable, i.e. $p_j^\infty = \frac{1}{M}$ (j=1, ... , M). (Doubly stochastic means, every row sum and every column sum of Π is 1.)

The encryption of a pair of plaintexts by an r-round iterated cipher is shown in the following scheme [LMM 91]:

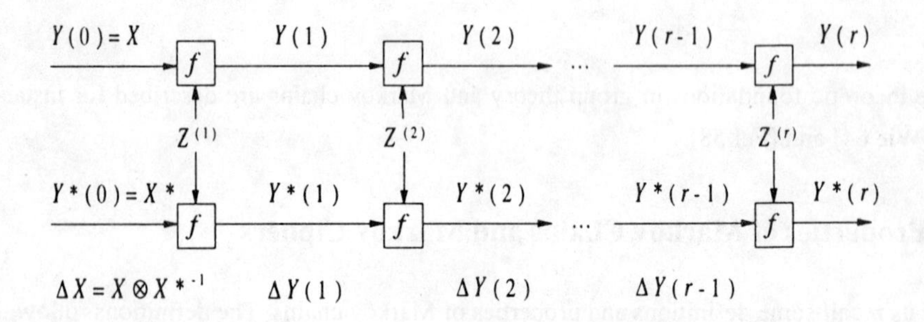

Figure 1

The *one-round function* $Y = f_Z(X)$ generates for every round subkey Z a one to one correspondence between X and Y.

Let the *"difference"* ΔX between two plaintexts (or two ciphertexts) X and X^* be defined as $\Delta X = X \otimes X^{*-1}$ where $\otimes$ denotes a specified group operation on the set of plaintexts (= set of ciphertexts) and X^{*-1} denotes the inverse of the element X^* in the group.

DEFINITION:

If there is a difference ΔX such that for all choices of α $(\alpha \neq e)$ and β $(\beta \neq e)$ $P(\Delta Y = \beta \mid \Delta X = \alpha, X = \gamma)$ is independent of γ when the subkey Z is uniformly random, then an iterated cipher with round function $Y = f_Z(X)$ is a *Markov cipher* in relation to the difference ΔX.

If there exists more than one difference ΔX which generates a Markov cipher, then all of these Markov ciphers are called *corresponding to the one-round function* f_Z.

THEOREM 2 [LMM 91]:

If an r-round iterated cipher is a Markov cipher and the r-round subkeys are independent and uniformly random, then the sequence of differences $\Delta X = \Delta Y(0), \Delta Y(1), \ldots, \Delta Y(r)$ is a homogeneous Markov chain.

THEOREM 3 [Lai 92]:

The transition matrix Π of a Markov cipher is doubly stochastic.

THEOREM 4 [Lai 92]:

For a Markov cipher the chain of differences is irreducible, if for every plaintext pair (X, X^*), $X \neq X^*$, and every ciphertext pair (Y, Y^*), $Y \neq Y^*$, there is an integer r_0 and a choice of subkeys for the first r_0 rounds such that, under the first r_0 rounds of the cipher with the chosen subkeys, X is encrypted to Y and X^* is encrypted to Y^*.

The cryptographic background needs the *Hypothesis of Stochastic Equivalence:*

For virtually all high probability (r-1)- round differentials (α, β),

$$P\left(\Delta Y(r-1) = \beta \middle| \Delta X = \alpha\right)$$
$$\approx P\left(\Delta Y(r-1) = \beta \middle| \Delta X = \alpha, Z^{(1)} = z_1, \ldots, Z^{(r-1)} = z_{r-1}\right)$$

holds for a substantial fraction of the subkey values $(z_1, \ldots, z_{r-1})$.

In the following it is shown that if the one-round functions $Y = f_Z(X)$ fulfil special algebraic properties, then the conditions of Theorem 1 come true for the chain of differences of a Markov cipher.

2 Group Theoretic Conditions for the One-Round Function

In the following we consider an arbitrary r-round iterated cipher on a finite set X, which is a Markov cipher in relation to a given difference. We derive sufficient conditions for the corresponding Markov chain of differences to be irreducible and aperiodic. These conditions are independent of the given difference.

For some of the following notations we refer to standard books on group theory like [Wie 64] and [Rob 82].

Let $G := \langle \{ f_Z \mid Z \in Z \} \rangle$ (Z - set of the round subkeys)

be the permutation group on X generated by the one-round functions f_Z and

$\forall\, t \in \mathbb{N}$: $H_t := \left\langle \left\{ f_{Z_1} \circ f_{Z_2} \circ \ldots \circ f_{Z_t} \mid (Z_1, Z_2, \ldots, Z_t) \in Z^t \right\} \right\rangle$

be the permutation group generated by the t-round functions.

LEMMA 1:

For every $t \in \mathbb{N}$ either $G = H_t$ or the group H_t is a proper normal subgroup of G.

LEMMA 2:

Let the Markov chain of differences be irreducible and periodic. Then there exists a $t \in \mathbb{N} \setminus \{1\}$ such that H_t is not doubly transitive.

From these two Lemmas we obtain

THEOREM 5:

Let $|X| \geq 3$ and every normal subgroup of G (except the identity group $\langle \{Id\} \rangle$) be doubly transitive. Then for all corresponding Markov ciphers the Markov chains of differences are irreducible and aperiodic.

Some special cases of G are considered in the following corollaries.

COROLLARY 1:

a) If G is a doubly transitive and simple group, then for all corresponding Markov ciphers the Markov chains of differences are irreducible and aperiodic.

b) If G is 4-transitive and $|X| \geq 5$, then for all corresponding Markov ciphers the Markov chains of differences are irreducible and aperiodic.

In the next two sections we will apply:

COROLLARY 2:

If G is the symmetric or the alternating group on X and $|X| \geq 5$, then for all corresponding Markov ciphers the Markov chains of differences are irreducible and aperiodic.

3 Application to the DES

The following scheme shows the one-round function f_Z of the DES [NBS 77]:

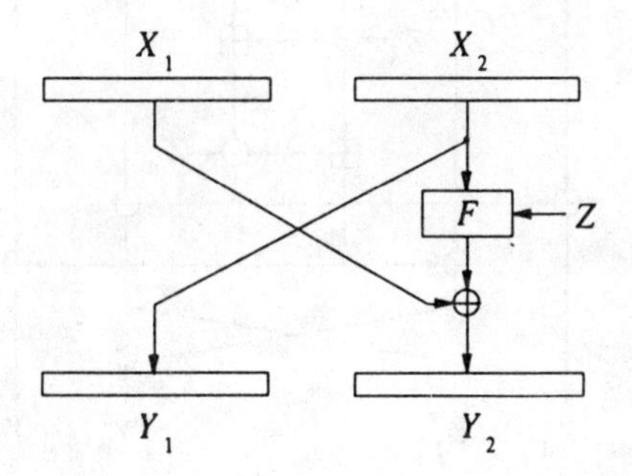

Figure 2

In [Wer 93] it is shown that the set $\{f_Z | Z \in Z\}$ generates the alternating group, and therefore the following theorem is an immediate consequence of Corollary 2.

THEOREM 6:

If f_Z is the one-round function of the DES, then for all corresponding Markov ciphers the chain of differences is irreducible and aperiodic, i. e. after sufficiently many rounds all differences will be roughly equally probable.

CONCLUSION 1:

If the hypothesis of stochastic equivalence (see Section 1) holds for a part of the corresponding Markov ciphers, then the DES is secure against a differential cryptanalysis attack after sufficiently many rounds for all of these Markov ciphers.

The results hold for all r-round iterated ciphers, if the one-round functions are DES-like functions which generate the alternating group (see [EG 83]).

4 Application to the IDEA

In [Lai 92] the block cipher algorithm IDEA is defined (in [LMM 91] it is called "IPES"). A difference between plaintexts is given such that the IDEA is a Markov cipher. The considered one-round functions f_Z are demonstrated in the following figure:

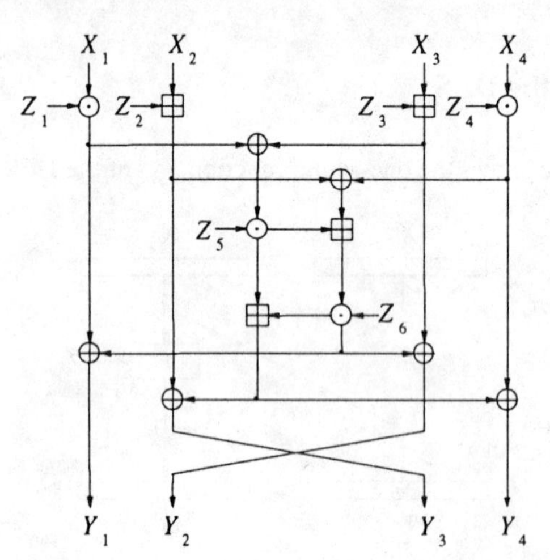

Figure 3

In order to treat several analytical approaches, the author of [Lai 92] introduced so called "mini" IDEA-ciphers IDEA(8), IDEA(16) and IDEA(32). These ciphers differ from IDEA = IDEA(64) only in their block lengths, which are 8, 16 and 32 respectively. Taking the analogous differences they are Markov ciphers too.

For the ciphers IDEA(8) and IDEA(16) it is proved that the Markov chains based on the given differences are irreducible and aperiodic [Lai 92]. That gives evidence to conjecture the same properties for IDEA(32) and IDEA(64). In [Lai 92] this remained an open question.

In the following, we will answer this question for IDEA(32).

The corresponding group G of the one-round functions (see Section 2) has the following properties:

LEMMA 3:

a) G is transitive.

b) G contains only even permutations.

For the arbitrary chosen permutation $f_Z \in G$ with the key parameters

$(Z_1, Z_2, \ldots, Z_6) = (218, 133, 79, 184, 97, 113)$

we computed a part of the cycle representation. The computations took nearly 30 hours on a SUN workstation.

Starting from the zero vector we found a cycle with a length of

$3\,639\,977\,669 = 211 \cdot 17\,251\,079$.

On the basis of this cycle length, of Theorem 13.10 in [Wie 64] and of the results of Lemma 3 we proved Theorem 7.

THEOREM 7:

For IDEA(32) we have $G = A_{2^{32}}$. ($A_{2^{32}}$ = Alternating group on $X := \{0,1\}^{32}$.)

The result of Theorem 7 provides the possibility to apply Corollary 2.

THEOREM 8:

If f_Z is the one-round function of the IDEA(32), then for all corresponding Markov ciphers the chain of differences is irreducible and aperiodic, i. e. after sufficiently many rounds all differences will be roughly equally probable.

CONCLUSION 2:

If the hypothesis of stochastic equivalence (see Section 1) holds for a part of the corresponding Markov ciphers, then IDEA(32) is secure against a differential cryptanalysis attack after sufficiently many rounds for all of these Markov ciphers.

5 Comments

- The result of Theorem 7 implies that several other imaginable cryptanalytic "shortcuts" of IDEA(32) can be excluded or restricted (see [Wer 93]). Though the results of Section 4 do not refer to the full IDEA(64), they may serve as arguments for the cryptographic strength of this algorithm.
- For an r-round iterated cipher for some special keys the one-round functions $(f_{Z_1},\ldots,f_{Z_r})$ may generate only a subgroup $G' \subset G$ that is not doubly transitive. This provides that the hypothesis of stochastic equivalence does not always hold (see also [Lai 92]). Hence the DES and the IDEA(32) may not be secure for all differences which fulfil Theorem 6 and Theorem 8 respectively.
- The behaviour of differences which do not generate Markov chains is unknown. Its analysis is an open, interesting problem.

References

[EG 83] Even, S.; Goldreich, O.
DES-like Functions Can Generate the Alternating Group
IEEE Transactions on Information Theory, IT-29, Nr. 6, 1983, 863 - 865

[Fel 58] Feller, W.
An Introduction to Probability Theory and Its Applications
Volume I, Second Edition 1958
John Wiley & Sons, Inc., New York

[Lai 92] Lai, X.
On the Design and Security of Block Ciphers
ETH Series in Information Processing, v.1 (Dissertation)
Hartung-Gorre Verlag, Konstanz, 1992

[LMM 91] Lai, X.; Massey, J. L.; Murphy, S.
Markov Ciphers and Differential Cryptanalysis
Proc. EUROCRYPT '91, LNCS 547, 1991, 17 - 38

[NBS 77] Data Encryption Standard (DES)
US NBS, FIPS PUB 46, 1977, Washington

[Rob 82] Robinson, D. J. S.
A Course in the Theory of Groups
Graduate Texts in Mathematics, Springer, 1982, New York

[Wer 93] Wernsdorf, R.
The One-Round Functions of the DES Generate the Alternating Group
Proc. EUROCRYPT '92, LNCS 658, 1993, 99 - 112

[Wie 64] Wielandt, H.
Finite Permutation Groups
Academic Press, 1964, New York and London

On Key Distribution and Authentication in Mobile Radio Networks

Choonsik Park, Kaoru Kurosawa, Tatsuaki Okamoto† and Shigeo Tsujii

Department of Electrical and Electronic Engineering,
Faculty of Engineering, Tokyo Institute of Technology
2-12-1 O-okayama, Meguro-ku, Tokyo, 152 Japan
Email: parkcs@ss.titech.ac.jp

† NTT Network Information Systems Laboratories,
Nippon Telegraph and Telephone Corporation
1-2356, Take, Yokosuka-shi, Kanagawa-ken, 238-03 Japan

Abstract. Mobile communication networks need public key cryptosystems that offer both low computation cost and user authentication. Tatebayashi et al. showed such a key distribution protocol for such networks at CRYPTO'89 based on low exponent RSA. This paper shows that their protocol is not secure. We also present a new secure and efficient key distribution protocol.

1 Introduction

Security in digital mobile communication systems has two major characteristics that must be be achieved, low computation cost and user authentication. Theoretically, A(lice) and B(ob) who have never met can share a cryptographic key by using a public key cryptosystem. The disadvantage of current public key cryptosystems is that encryption and decryption take too long. This disadvantage is serious in mobile communication networks because each user has very small computational power. The user authentication problem is also important to avoid charges of fraudulent usage. Another property of mobile communication networks, which is a good news for us, is that each user communicates with each other through a network center. Our goal is to design a public key cryptosystem for such networks that realizes both low computation cost and user authentication.

Until recently, however, only slight attention has been paid to this problem. The key distribution protocol shown by Tatebayashi et al.[1] at the 1989 CRYPTO conference is the only product known to the authors.

This paper first shows that their protocol is not secure. In the protocol of [1], A and B send initial information to the network center by using a low exponent RSA. However, we show that B can find the secret of A, needed for user authentication, easily by using this low exponent property. We then present a new key distribution protocol which realizes both low computation cost and user authentication by introducing a simple nonlinear function.

Our technical contribution is as follows. Hastad [3] showed that we can solve the following simultaneous equations,

$$(\alpha_i X + \beta_i)^3 = c_i \pmod{N_i}, (i = 1, \cdots, 7),$$

for X in polynomial time if $\gcd(N_i, N_j) = 1$ for $i \neq j$ and if the number of equations is seven. Our analysis of [1] shows that we can obtain the following simultaneous equations,

$$(\alpha_i X + \beta_i)^3 = c_i \pmod{N}.$$

Since N is common for each i, Hastad's attack cannot be applied. We show a method for this problem which finds X in polynomial time if the number of equations is three. We propose a key distribution protocol which is secure for both attacks.

$\circ$ denotes concatenation. $|X|$ denotes the bit length of X.

2 Review of Tatebayashi et al.'s scheme

The key distribution scheme of [1] was developed in the following process. First, network center C generates an RSA cryptosystem $e = 3$. Let

$$E(M) = M^3 \bmod N(= pq),$$

where N is the public key of C and p, q are the secret keys of C. Suppose that A and B want to share a key K. X and Y are opponents.

2.1 KDP1

Their basic protocol KDP1 is as follows.
[KDP1]

1. A chooses a random number r_1 and computes $Z_a = E(r_1)$.
   A sends Z_a to C.
2. B chooses a key K randomly and computes $Z_b = E(K)$.
   B sends Z_b to C.
3. C decrypts Z_a and Z_b. It computes $u = r_1 + K \bmod N$.
   C send u to A.
4. A computes the key K as $K = u - r_1 \bmod N$.

2.2 Simmons attack

Simmons showed one attack on KDP1.
[Simmons' attack]

1. X and Y monitor Z_b.
2. X chooses a random number R and computes $Z_x = E(R)$.
   X sends Z_x to C.
3. Y sends Z_b to C.
4. C then sends $u' = R + K \bmod N$ to X.
5. X can compute from u' and R the key K of A and B.

2.3 User authentication

We should also consider the problem of user authentication because it is important to avoid charges of fraudulent usage. To eliminate this problem, KDP1 can be modified as follows. Let f be a pseudorandom function which is the secret of C. Let ID_a and ID_b be the identity of A and B, respectively. In the preprocessing stage, C computes

$$S_a = f(ID_a), \quad S_b = f(ID_b)$$

C sends S_a to A secretly. Similarly, C sends S_b to B secretly.
[KDP1*]

1. A chooses a random number r_1 and computes $Z_a = E(S_a \circ r_1)$.
   A sends Z_a and ID_a to C.
2. B chooses a key K randomly and computes $Z_b = E(S_b \circ K)$.
   B sends Z_b and ID_b to C.
3. C decrypts Z_a and Z_b. It checks that $S_a = f(ID_a)$ and $S_b = f(ID_b)$.
   It the check passes, C computes $u = r_1 + K$ and send u to A.
4. A computes the key K as $K = u - r_1$.

It is easy to see that Simmons' attack can also be applied to KDP1*.

2.4 KDP2

Finally, Tatebayashi et al. developed the following protocol KDP2 which uses timestamps to avoid Simmons' attack.
[KDP2]

1. A chooses a random number r_1 and computes $Z_a = E(T_a \circ S_a \circ r_1)$, where T_a is A's timestamp information.
   A sends Z_a and his identity ID_a to the center C.
2. C decrypts the ciphertext and verifies the identity of A and the timestamp information. C then calls B.
3. B chooses a key K randomly and computes $Z_b = E(T_b \circ S_b \circ K)$, where T_b is B's timestamp information.
   B sends Z_b and his identity ID_b to the center C.
4. C checks T_b and S_b. C then computes $u = r_1 + K$ and sends u to A.
5. A computes K as $K = u - r_1$.

3 Attack on KDP2

This section shows that KDP2[1] is not secure. Actually, we show that B can find A's secret information S_a if B executes KDP2 with A three times.

After repeating KDP2 three times, B obtains

$$Z_{ai} = E(T_{ai} \circ S_a \circ r_{1i}), \quad u_i = r_{1i} + K_i, \quad (i = 1, 2, 3)$$

by listening to the conversation between A and the center C. B then gets

- r_{1i} from u_i and K_i.
- T_{ai} because it is the time at which A sends Z_{ai} to C.

The unknown constant is only S_a. Suppose that

$$|r_{1i}| = l, \quad |S_a| + |r_{1i}| = m,$$

Z_{ai} is written as follows.

$$(T_{a1} \times 2^m + S_a \times 2^l + r_{11})^3 = Z_{a1} \pmod N$$
$$(T_{a2} \times 2^m + S_a \times 2^l + r_{12})^3 = Z_{a2} \pmod N$$
$$(T_{a3} \times 2^m + S_a \times 2^l + r_{13})^3 = Z_{a3} \pmod N$$

Let $X_a = S_a \times 2^l$ and $Y_i = T_{ai} \times 2^m + r_{1i}$. We then obtain

$$X_a^3 + 3X_a^2 \times Y_1 + 3 \times (Y_1)^2 \times X_a = Z_{a1} - (Y_1)^3 \pmod N,$$
$$X_a^3 + 3X_a^2 \times Y_2 + 3 \times (Y_2)^2 \times X_a = Z_{a2} - (Y_2)^3 \pmod N,$$
$$X_a^3 + 3X_a^2 \times Y_3 + 3 \times (Y_3)^2 \times X_a = Z_{a3} - (Y_3)^3 \pmod N.$$

The only unknown constant is X_a. We can view the above equations as linear simultaneous equations on X_a^3, X_a^2 and X_a. B can easily solve these equations in polynomial time. B can compute S_a from the solution X_a.

This attack works even if T_a, S_a and r_1 are interleaved. It also works for $E(M) = M^e \bmod N$ if e is small.

Similarly, A can find the secret information S_b of B because KDP2 is symmetric for A and B.

4 Proposed scheme

The reason why our attack succeeds is that B can obtain r_{1i} (A can obtain K_i) from the equation

$$u_i = r_{1i} + K_i. \tag{1}$$

We can prevent our attack if we introduce a nonlinear function h and modify eq.(1) as follows.

$$u_i = h(r_{1i}) + h(K_i).$$

For example, the following simple h is enough.

$$h(x_1 \circ x_2) = x_1 + x_2 \bmod 2^{l/2}$$

where $|x_1| = |x_2| = l/2$ and $\circ$ denotes concatenation.

It is information theoretically impossible to determine x_1 and x_2 from $x_1 + x_2$. Based on this observation, we propose a key distribution protocol as follows.
(E is the public key of the center. See the beginning of section 2.)

1. A chooses random numbers (r_1, r_2) such that $|r_1| = |r_2| = l/2$ and computes $Z_a = E(T_a \circ S_a \circ r_1 \circ r_2)$, where T_a is A's timestamp information.
   A sends Z_a and his identity ID_a to the center C.

2. C decrypts the ciphertext and verifies the identity of A and the timestamp information. C then calls B.
3. B chooses (K_1, K_2) randomly and computes $Z_b = E(T_b \circ S_b \circ K_1 \circ K_2)$, where T_b is B's timestamp information.
   B sends Z_b and his identity ID_b to the center C.
4. C checks T_b and S_b. C then computes $u = r_1 + r_2 + K_1 + K_2 \bmod 2^{l/2}$ and send u to A.
5. The session key is given by $K = K_1 + K_2 \bmod 2^{l/2}$. A computes K as $K = u - r_1 - r_2 \bmod 2^{l/2}$.

We can use the Rabin cryptosystem instead of RSA because the plaintext has a special data structure.

Security

Suppose that the proposed protocol is executed I times. Let the ith parameters be Z_{ai},T_{ai},r_{1i},r_{2i},Z_{bi},T_{bi},K_{1i} and K_{2i}. B knows that

$$Z_{ai} = E(T_{ai} \circ S_a \circ r_{1i} \circ r_{2i}).$$

He knows the values of Z_{ai},T_{ai} and $r_{1i} + r_{2i} \pmod{2^{i/2}}$ by monitoring A's communication. However, he cannot know $r_{1i} \circ r_{2i}$. ($\circ$ denotes concatenation.) Therefore, what B can have is the following type of equations,

$$(\alpha_i X + Y_i + \beta_i)^3 = c_i \pmod{N}, (i = 1, \cdots, I),$$

where $X = S_a$,$Y_i = r_{1i} \circ r_{2i}$, $\alpha_i = 2^l$, $\beta_i = T_{ai} \times 2^{(l+|S_a|)}$,$c_i = Z_{ai}$. Here, X and Y_i are unknown variables. α_i,β_i and c_i are known values.

Then, the number of equations is I and that of unknown variables is $I + 1$. Hence, B cannot find $X(= S_a)$. Similarly, A cannot find S_b.

References

1. Tatebayashi, M., Matsuzaki, N., Newman, Jr., D.B.: Key Distribution Protocol for Digital Mobile Communication Systems. Advances in Cryptology, Proceedings of Crypto'89 (1989) 324–334
2. Moore, J. H.: Protocol Failures in Cryptosystems. Proc. of IEEE, Vol.76, No.5 (1988) 594–602
3. Hastad,J.: On using RSA with Low exponent in a public key network. Advances in Cryptology, Proceedings of Crypto'85 (1985)403–408
4. Simmons,G.J.: A 'weak' privacy protocol using the RSA cryptoalgorithm. Cryptologia, Vol.7 (1983) 180–182
5. Beller,M.J. Chang,L.F., Yacobi,Y. : Privacy and Authentication on a Portable Communication System. IEEE GLOBECOM '91 conference (1991) 1922 – 1927

List of Authors

Printing: Weihert-Druck GmbH, Darmstadt
Binding: Buchbinderei Schäffer, Grünstadt

Lecture Notes in Computer Science

For information about Vols. 1–690
please contact your bookseller or Springer-Verlag

Vol. 691: M. Ajmone Marsan (Ed.), Application and Theory of Petri Nets 1993. Proceedings, 1993. IX, 591 pages. 1993.

Vol. 692: D. Abel, B.C. Ooi (Eds.), Advances in Spatial Databases. Proceedings, 1993. XIII, 529 pages. 1993.

Vol. 693: P. E. Lauer (Ed.), Functional Programming, Concurrency, Simulation and Automated Reasoning. Proceedings, 1991/1992. XI, 398 pages. 1993.

Vol. 694: A. Bode, M. Reeve, G. Wolf (Eds.), PARLE '93. Parallel Architectures and Languages Europe. Proceedings, 1993. XVII, 770 pages. 1993.

Vol. 695: E. P. Klement, W. Slany (Eds.), Fuzzy Logic in Artificial Intelligence. Proceedings, 1993. VIII, 192 pages. 1993. (Subseries LNAI).

Vol. 696: M. Worboys, A. F. Grundy (Eds.), Advances in Databases. Proceedings, 1993. X, 276 pages. 1993.

Vol. 697: C. Courcoubetis (Ed.), Computer Aided Verification. Proceedings, 1993. IX, 504 pages. 1993.

Vol. 698: A. Voronkov (Ed.), Logic Programming and Automated Reasoning. Proceedings, 1993. XIII, 386 pages. 1993. (Subseries LNAI).

Vol. 699: G. W. Mineau, B. Moulin, J. F. Sowa (Eds.), Conceptual Graphs for Knowledge Representation. Proceedings, 1993. IX, 451 pages. 1993. (Subseries LNAI).

Vol. 700: A. Lingas, R. Karlsson, S. Carlsson (Eds.), Automata, Languages and Programming. Proceedings, 1993. XII, 697 pages. 1993.

Vol. 701: P. Atzeni (Ed.), LOGIDATA+: Deductive Databases with Complex Objects. VIII, 273 pages. 1993.

Vol. 702: E. Börger, G. Jäger, H. Kleine Büning, S. Martini, M. M. Richter (Eds.), Computer Science Logic. Proceedings, 1992. VIII, 439 pages. 1993.

Vol. 703: M. de Berg, Ray Shooting, Depth Orders and Hidden Surface Removal. X, 201 pages. 1993.

Vol. 704: F. N. Paulisch, The Design of an Extendible Graph Editor. XV, 184 pages. 1993.

Vol. 705: H. Grünbacher, R. W. Hartenstein (Eds.), Field-Programmable Gate Arrays. Proceedings, 1992. VIII, 218 pages. 1993.

Vol. 706: H. D. Rombach, V. R. Basili, R. W. Selby (Eds.), Experimental Software Engineering Issues. Proceedings, 1992. XVIII, 261 pages. 1993.

Vol. 707: O. M. Nierstrasz (Ed.), ECOOP '93 – Object-Oriented Programming. Proceedings, 1993. XI, 531 pages. 1993.

Vol. 708: C. Laugier (Ed.), Geometric Reasoning for Perception and Action. Proceedings, 1991. VIII, 281 pages. 1993.

Vol. 709: F. Dehne, J.-R. Sack, N. Santoro, S. Whitesides (Eds.), Algorithms and Data Structures. Proceedings, 1993. XII, 634 pages. 1993.

Vol. 710: Z. Ésik (Ed.), Fundamentals of Computation Theory. Proceedings, 1993. IX, 471 pages. 1993.

Vol. 711: A. M. Borzyszkowski, S. Sokołowski (Eds.), Mathematical Foundations of Computer Science 1993. Proceedings, 1993. XIII, 782 pages. 1993.

Vol. 712: P. V. Rangan (Ed.), Network and Operating System Support for Digital Audio and Video. Proceedings, 1992. X, 416 pages. 1993.

Vol. 713: G. Gottlob, A. Leitsch, D. Mundici (Eds.), Computational Logic and Proof Theory. Proceedings, 1993. XI, 348 pages. 1993.

Vol. 714: M. Bruynooghe, J. Penjam (Eds.), Programming Language Implementation and Logic Programming. Proceedings, 1993. XI, 421 pages. 1993.

Vol. 715: E. Best (Ed.), CONCUR'93. Proceedings, 1993. IX, 541 pages. 1993.

Vol. 716: A. U. Frank, I. Campari (Eds.), Spatial Information Theory. Proceedings, 1993. XI, 478 pages. 1993.

Vol. 717: I. Sommerville, M. Paul (Eds.), Software Engineering – ESEC '93. Proceedings, 1993. XII, 516 pages. 1993.

Vol. 718: J. Seberry, Y. Zheng (Eds.), Advances in Cryptology – AUSCRYPT '92. Proceedings, 1992. XIII, 543 pages. 1993.

Vol. 719: D. Chetverikov, W.G. Kropatsch (Eds.), Computer Analysis of Images and Patterns. Proceedings, 1993. XVI, 857 pages. 1993.

Vol. 720: V.Mařík, J. Lažanský, R.R. Wagner (Eds.), Database and Expert Systems Applications. Proceedings, 1993. XV, 768 pages. 1993.

Vol. 721: J. Fitch (Ed.), Design and Implementation of Symbolic Computation Systems. Proceedings, 1992. VIII, 215 pages. 1993.

Vol. 722: A. Miola (Ed.), Design and Implementation of Symbolic Computation Systems. Proceedings, 1993. XII, 384 pages. 1993.

Vol. 723: N. Aussenac, G. Boy, B. Gaines, M. Linster, J.-G. Ganascia, Y. Kodratoff (Eds.), Knowledge Acquisition for Knowledge-Based Systems. Proceedings, 1993. XIII, 446 pages. 1993. (Subseries LNAI).

Vol. 724: P. Cousot, M. Falaschi, G. Filè, A. Rauzy (Eds.), Static Analysis. Proceedings, 1993. IX, 283 pages. 1993.

Vol. 725: A. Schiper (Ed.), Distributed Algorithms. Proceedings, 1993. VIII, 325 pages. 1993.

Vol. 726: T. Lengauer (Ed.), Algorithms – ESA '93. Proceedings, 1993. IX, 419 pages. 1993

Vol. 727: M. Filgueiras, L. Damas (Eds.), Progress in Artificial Intelligence. Proceedings, 1993. X, 362 pages. 1993. (Subseries LNAI).

Vol. 728: P. Torasso (Ed.), Advances in Artificial Intelligence. Proceedings, 1993. XI, 336 pages. 1993. (Subseries LNAI).

Vol. 729: L. Donatiello, R. Nelson (Eds.), Performance Evaluation of Computer and Communication Systems. Proceedings, 1993. VIII, 675 pages. 1993.

Vol. 730: D. B. Lomet (Ed.), Foundations of Data Organization and Algorithms. Proceedings, 1993. XII, 412 pages. 1993.

Vol. 731: A. Schill (Ed.), DCE – The OSF Distributed Computing Environment. Proceedings, 1993. VIII, 285 pages. 1993.

Vol. 732: A. Bode, M. Dal Cin (Eds.), Parallel Computer Architectures. IX, 311 pages. 1993.

Vol. 733: Th. Grechenig, M. Tscheligi (Eds.), Human Computer Interaction. Proceedings, 1993. XIV, 450 pages. 1993.

Vol. 734: J. Volkert (Ed.), Parallel Computation. Proceedings, 1993. VIII, 248 pages. 1993.

Vol. 735: D. Bjørner, M. Broy, I. V. Pottosin (Eds.), Formal Methods in Programming and Their Applications. Proceedings, 1993. IX, 434 pages. 1993.

Vol. 736: R. L. Grossman, A. Nerode, A. P. Ravn, H. Rischel (Eds.), Hybrid Systems. VIII, 474 pages. 1993.

Vol. 737: J. Calmet, J. A. Campbell (Eds.), Artificial Intelligence and Symbolic Mathematical Computing. Proceedings, 1992. VIII, 305 pages. 1993.

Vol. 738: M. Weber, M. Simons, Ch. Lafontaine, The Generic Development Language Deva. XI, 246 pages. 1993.

Vol. 739: H. Imai, R. L. Rivest, T. Matsumoto (Eds.), Advances in Cryptology – ASIACRYPT '91. X, 499 pages. 1993.

Vol. 740: E. F. Brickell (Ed.), Advances in Cryptology – CRYPTO '92. Proceedings, 1992. X, 593 pages. 1993.

Vol. 741: B. Preneel, R. Govaerts, J. Vandewalle (Eds.), Computer Security and Industrial Cryptography. Proceedings, 1991. VIII, 275 pages. 1993.

Vol. 742: S. Nishio, A. Yonezawa (Eds.), Object Technologies for Advanced Software. Proceedings, 1993. X, 543 pages. 1993.

Vol. 743: S. Doshita, K. Furukawa, K. P. Jantke, T. Nishida (Eds.), Algorithmic Learning Theory. Proceedings, 1992. X, 260 pages. 1993. (Subseries LNAI)

Vol. 744: K. P. Jantke, T. Yokomori, S. Kobayashi, E. Tomita (Eds.), Algorithmic Learning Theory. Proceedings, 1993. XI, 423 pages. 1993. (Subseries LNAI)

Vol. 745: V. Roberto (Ed.), Intelligent Perceptual Systems. VIII, 378 pages. 1993. (Subseries LNAI)

Vol. 746: A. S. Tanguiane, Artificial Perception and Music Recognition. XV, 210 pages. 1993. (Subseries LNAI).

Vol. 747: M. Clarke, R. Kruse, S. Moral (Eds.), Symbolic and Quantitative Approaches to Reasoning and Uncertainty. Proceedings, 1993. X, 390 pages. 1993.

Vol. 748: R. H. Halstead Jr., T. Ito (Eds.), Parallel Symbolic Computing: Languages, Systems, and Applications. Proceedings, 1992. X, 419 pages. 1993.

Vol. 749: P. A. Fritzson (Ed.), Automated and Algorithmic Debugging. Proceedings, 1993. VIII, 369 pages. 1993.

Vol. 750: J. L. Díaz-Herrera (Ed.), Software Engineering Education. Proceedings, 1994. XII, 601 pages. 1994.

Vol. 751: B. Jähne, Spatio-Temporal Image Processing. XII, 208 pages. 1993.

Vol. 752: T. W. Finin, C. K. Nicholas, Y. Yesha (Eds.), Information and Knowledge Management. Proceedings, 1992. VII, 142 pages. 1993.

Vol. 753: L. J. Bass, J. Gornostaev, C. Unger (Eds.), Human-Computer Interaction. Proceedings, 1993. X, 388 pages. 1993.

Vol. 754: H. D. Pfeiffer, T. E. Nagle (Eds.), Conceptual Structures: Theory and Implementation. Proceedings, 1992. IX, 327 pages. 1993. (Subseries LNAI).

Vol. 755: B. Möller, H. Partsch, S. Schuman (Eds.), Formal Program Development. Proceedings. VII, 371 pages. 1993.

Vol. 756: J. Pieprzyk, B. Sadeghiyan, Design of Hashing Algorithms. XV, 194 pages. 1993.

Vol. 757: U. Banerjee, D. Gelernter, A. Nicolau, D. Padua (Eds.), Languages and Compilers for Parallel Computing. Proceedings, 1992. X, 576 pages. 1993.

Vol. 758: M. Teillaud, Towards Dynamic Randomized Algorithms in Computational Geometry. IX, 157 pages. 1993.

Vol. 759: N. R. Adam, B. K. Bhargava (Eds.), Advanced Database Systems. XV, 451 pages. 1993.

Vol. 760: S. Ceri, K. Tanaka, S. Tsur (Eds.), Deductive and Object-Oriented Databases. Proceedings, 1993. XII, 488 pages. 1993.

Vol. 761: R. K. Shyamasundar (Ed.), Foundations of Software Technology and Theoretical Computer Science. Proceedings, 1993. XIV, 456 pages. 1993.

Vol. 762: K. W. Ng, P. Raghavan, N. V. Balasubramanian, F. Y. L. Chin (Eds.), Algorithms and Computation. Proceedings, 1993. XIII, 542 pages. 1993.

Vol. 763: F. Pichler, R. Moreno Díaz (Eds.), Computer Aided Systems Theory – EUROCAST '93. Proceedings, 1993. IX, 451 pages. 1994.

Vol. 764: G. Wagner, Vivid Logic. XII, 148 pages. 1994. (Subseries LNAI).

Vol. 765: T. Helleseth (Ed.), Advances in Cryptology – EUROCRYPT '93. Proceedings, 1993. X, 467 pages. 1994.

Vol. 766: P. R. Van Loocke, The Dynamics of Concepts. XI, 340 pages. 1994. (Subseries LNAI).

Vol. 767: M. Gogolla, An Extended Entity-Relationship Model. X, 136 pages. 1994.

Vol. 768: U. Banerjee, D. Gelernter, A. Nicolau, D. Padua (Eds.), Languages and Compilers for Parallel Computing. Proceedings, 1993. XI, 655 pages. 1994.